DESIGN AND ANALYSIS OF ALGORITHMS

BISWAJIT R BHOWMIK

DESIGN AND ANALYSIS OF ALGORITHMS

BISWAJIT R BHOWMIK

Research Scholar
Department of Computer Science & Engineering
Indian Institute of Technology Guwahati
Assam, India

S.K. KATARIA & SONS®

Publisher of Engineering Books
4885/109, Prakash Mahal, Dr. Subhash Bhargav Lane,
Daryaganj, New Delhi-110002 (India)
Phone: +91-11-43551243, Mobile: +91-9871775858, 7678425209
e-mail: katariabook@gmail.com, katariabooks@yahoo.com
Website: www.skkatariaandsons.com
(e-books, subscriptions and bulk buying also available on our website)

Published by:

S.K. Kataria & Sons®

4885/109, Prakash Mahal, Dr. Subhash Bhargav Lane, Daryaganj,
New Delhi-110002 (India)
Phone: +91-11-43551243, Mobile: +91-9871775858, 7678425209
e-mail: katariabook@gmail.com, katariabooks@yahoo.com
Website: www.skkatariaandsons.com

First Edition: 2011, Second Edition: 2012; Reprint: 2026

Laser Type Setting by:

SARA ASSIGNMENTS, Patparganj, Delhi-110092.

This Book Is Dedicated To

My Mother

I Love You Maa With All My Heart...

Contents

Preface to the Second Edition

This new edition is a major revision of the book that brings it fully up to date, expands the coverage, and includes numerous improvements to the original material. The readers have benefitted from using the first edition as text book. In writing this preface to the second edition it is assumed that the readers have read the original preface. In the first edition of this book I have received numerous email messages and letters from readers commenting on the book and suggesting how it could be improved. Therefore, the changes reflect my own experiences in using the book, as well as suggestions received from readers. With the aid of all this information I have completely revised the book. The most obvious changes in this second edition are the addition of six new chapters.

When dealing with an interdisciplinary area it is tempting to cover the topics as clearly as possible. I explained my philosophy in the preface to the first edition, but allow me a few more words here. I also have what I regard as a higher aim, to convey to the readers an enjoyment of the subject *Algorithms*. I try to put up an elegant and logically tight framework and put a light touch on a heavy subject.

The book has been divided into Twenty chapters. The First Edition included Fourteen chapters. This edition includes six new chapters on Advanced Algorithms along with all the chapters of previous edition. The changes to the book can be summarized as follows.

Chapter 15 is anticipated for advanced manipulation of integers and is named as *Large Integers Multiplication.* We introduce here the divide-and-conquer based approach for finding efficient way to multiplying large integers.

Chapter 16 covers broad areas of *File Compression.* The chapter includes specially Huffman Coding. It covers both fixed and variable length encoding, decoding. Also various applications and simple algorithms have been discussed.

Chapter 17 discusses an important application area of Depth First Search approach. In this chapter we will discuss *Biconnectivity.* Here we are interested in finding the *articulation point.*

Chapter 18 aims designing efficient algorithms that show how to do decomposition a graph into *strongly connected components* using two depth-first searches. After decomposition, the algorithm is run separately on each strongly connected component. The solutions are then combined according to the structure of connections between components.

Chapter 19 is planned for a *Topological Sort* of a directed acyclic graph that can be viewed alternatively as an ordering of its vertices along a horizontal line so that all directed edges go from left to right.

Chapter 20 has been discussed with the extension of *Single Source Shortest Path Problem* using another algorithm design technique – Greedy Approach. We focus on it with the well famous Dijkstra's algorithm.

In addition to the inclusion of the above six chapters, other major changes in this edition are the followings. Every chapter covers more simple questions and answers. Also additional solved examples are included in this edition. Every chapter ends with semester questions of several years of different universities. The more attractive of this second edition is that it includes all the questions of GATE related to this subject Design and Analysis of Algorithms.

I hope the readers will enjoy reading this new edition, and benefit from the additional material.

June, 2012 **BISWAJIT R BHOWMIK**

Preface to the First Edition

Electronic computers first started its journey during late 1950s. Resources of computers at that time were very limited and they in spite of resulted additional impetus for devising efficient computer algorithms as the computer engineers, scientists, researchers started to pay attention towards the betterment of program performances. The field of computer algorithms are being flourished since then.

This book provides a comprehensive opening to the study of computer algorithms in most acceptable form. Most of the algorithms have been discussed in considerable depth keeping in mind the readers at all levels. All the algorithm design techniques have been explained simply and shown their utility in the solution to many problems. I have tried to present each technique with its context where it can be applied, special features that make it unique apart, comparison with other techniques where possible, and illustration by applying to several problems.

Although the main intention of the book is to be familiar with algorithm design techniques, it also places interest on another major component - the analysis of algorithms. All the algorithms have been analysed in detail. Since efficiency is the desired algorithm design criterion, I carefully analyse the running times of all the algorithms. Being my focus is on practical applications of the design techniques I illustrated each technique by straightforward and use pseudocode similar to the syntax of any structured programming languages such that anyone can write a program of the correspondence.

The book is based according to the syllabi of the all major technical universities in India. The book in this First Edition has been divided into Fourteen chapters. Each chapter covers those algorithm design techniques, their applications that have common objectives. Each chapter has been organised as:

Chapter 1 sets the stage for the rest of the book providing the background of subsequent chapters. In this chapter I have tried to show how to write an algorithm, its features, representations, model of computations, etc.

Chapter 2 is devoted to the study of Asymptotic Notations which are extremely fundamental to algorithmic complexities analysis. This chapter is basically dedicated to the studies on basic mathematical backgrounds required for the analysis of algorithms. Algorithms of some fundamental problems have been encountered in this chapter. These problems include searching, and sorting.

Chapter 3 is dedicated to recursive design techniques and recurrences which are indispensable tool in the field of computer science: recursion. Also I have attempted to cover this chapter in details with the most common summations and recurrence relations usually appeared across when analyzing algorithms.

Chapter 4 covers intuitive and natural design technique Divide-and-Conquer which is more essential in solving sorting problems. More emphasis has been given to the solutions as they are basic in the analysis of a large class of algorithms referred as divide-and-conquer algorithms.

Chapter 5 is concerned with the Dynamic Programming – the optimization technique come up with efficient algorithms for optimization problems. Some examples of the use of dynamic programming are considered in this chapter. Recursion is first used in this design technique to model the solution of the problem. The recursive model thus obtained is then converted into an efficient iterative algorithm. Dynamic programming is applied to solve some of the following problems: matrix chain multiplication, the shortest path problems, etc.

Chapter 6 is intended for Greedy Approach – another optimization technique and results better over dynamic programming in some specific optimization problems. In this chapter, we study in detail some algorithms of well-known problems in computer science that give optimal solutions.

Chapter 7 proposes Backtracking that is required sometimes when intermediate state for finding solutions of some problems becomes retreated. This is basically a solution space search technique that works for the problems whose solution space is large. Examples solved in this chapter include graph colouring problem, the 8-queens problem, etc.

Chapter 8 mainly deals with Branch-and-Bound that solves optimization problems by making a bound on the solution. Like backtracking it is also a solution space search technique. An acceptable solution to the problem is gained by exhaustively searching through a large but finite number of possibilities. In this chapter, a well-known branch-and-bound algorithm for 15-puzzle problem is presented.

Chapter 9 projects on Lower Bound Theory which emphasizes on theoretical studies of finding a lower bound of various problems. In this chapter, two models of computations are employed for the purpose. First, the decision tree model that establishes lower bounds for comparison-based problems like searching and sorting. Second, the more powerful algebraic decision tree model sets lower bounds on some other problems.

Chapter 10 aims designing efficient algorithms for two operations – union and find on Disjoint Sets – a building block in the design of many efficient algorithms, especially graph algorithms.

Chapter 11 is planned for Graph Traversals that are handy especially in graph and geometric problems. We will see traversing the graphs, visiting its vertices, etc are problem specific and the solutions are made by making local decisions.

Chapter 12 is anticipated for manipulation of polynomials with the introduction of Fourier transforms. Also Integer multiplication schemes have been comprised in this chapter.

Chapter 13 covers up a class of problems called NP-completeness problems and in search of possible proving their intractability. In this chapter, we study numerous problems in this class drawn from many problem domains. These problems share the property that if any one problem in the class is solvable in polynomial time, then all other problems in the class are solvable in polynomial time. Informally I have covered this chapter that is no specific computation model is assumed. This chapter is somewhat advanced one.

Chapter 14 pays attention to Approximation Algorithms for some NP-hard problems instead of finding their optimal solution. In this chapter we see the ratio of the optimal solution to the approximate solution is bounded by a small, reasonable constant. For the knapsack problem, we show a polynomial approximation scheme.

The book is intended as a text primarily for use in undergraduate or graduate courses in the field of Design and Analysis of Algorithms. The material is best suitable for self-study. Every chapter ends with exercises and each exercise contains varieties of problems — short questions, self-check outs, and substantially suitable as assigned homework.

July, 2011 **BISWAJIT R BHOWMIK**

Acknowledgments

This book is written for the students those who are facing problems and fear the subject algorithms. I thought at the beginning that this project would simply be a matter of converting my existing notes into a refined text. However, this was in practice not the real truth.

I first acknowledge to my beloved students who have requested several times and inspired me to do something for them.

With a sense of deepest gratitude I do express immense indebtedness to Sovana who always guided me in proper way, and grew interests in life. I cannot forget her continuous encouragements that keep me energetic. Without her devotion and effort it would become very difficult to complete the manuscript in time.

Especially I would like to be evidence for my honor to Dr. R. N. Bhowmik, Associate Professor, Department of Physics, University of Pondicherry, Mr. Pradip K. Maity (my favorite Pradipda) who helped and supported me by all the ways to become what I am at present.

With inmost appreciation I would like to ample my sincere thankfulness to Dr. G. Barua, Professor & Director, Dr. Hemangee Kapoor, Associate Professor, Dr. K. Sreenivasan, Visiting Professor, Dr. P. Bhaduri, Associate Professor & Head, Dr. S. Nandi, Professor, Dr. P. Mitra, Professor, Dr. S. V. Rao, Associate Professor, Dr. S. Ranbir Singh, Assistant Professor, Dr. V. Vijaya Saradhi, Assistant Professor, and other faculty members of the Department of Computer Science & Engineering, Indian Institute of Technology Guwahati for their precious ideas, and opinions in the direction of success of this project.

I am also thankful to Soumen Bag, Research Scholar, Department of Computer Science & Engineering, Indian Institute of Technology, Kharagpur, Niladri Sett, Chandan Kalita, Pradeep K. Biswal, Mousum Handique, and all the other Research Scholars, and Sanjib Das, Parikshit Saikia, Ganesh Khade, Nayanmani Sharma, and all the other M.Tech students, Department of Computer Science & Engineering, Indian Institute of Technology, Guwahati for their cooperation, suggestion, and friendliness.

With profound glee I extend my thanks to my colleagues and friends — Aparajita Dutta, Debleena Mitra, Tanushree Chakraborty, Sayani Mondal, Mitali Ghata, Sarada Mondal, Gayatri Parvat, Deepa V. Naik, Salini Shukla, Dr. Sudha Singh, Anindita Das, Smita Roy, Tulika Bhattacharjee, Neepa Biswas, Anandaprova Ghoshal, Parna

Chatterjee, Riya Roychwdhury, Arnab Sarkar, Syed Zahir Hasan, Mrinmoy Sen, Palash Paik, Anupam Ghosh, Sk Abdul Rahim, Koushik Roy Choudhury, Abhisek Pal, Biplab Palye, Umacharan Karamakar, Pramit Majumdar, Amartya Mukherjee, Suprobhat Mondal, Santu Kundu, Pardipta Banerjee, Soumya Kanti Bhattacharya, Rathin Pailan, Kalyan Chatterjee, Susovan Patra, Sudipta Mondal, D. Dasgupta, Dinkar, and Dr. R. N. Mukherjee who have lots of contributions to prepare the text. They carefully read the manuscript, subjected it to their scrutiny, reviewed, suggested, and helped me with the rewrite of some essential chapters. I am really thankful to all of them for this edition.

A special word of thanks to our adored Head, Dr. S. Dasgupta for his important tips failing which this endeavor would not be completed.

The preparation of this book has been possible due to continuous support of our beloved Director, Dr. A. C. Ganguli.

Sincere thanks are extended to Parag Kumar Guha Thakurta, Asst. Professor, Bibhas Sen, Asst. Professor, Sanjib Sadhu, Asst. Professor, and Dr. Goutam Sanyal, Professor, Department of Computer Science & Engineering, Nationsal Institute of Technology, Durgapur for their new innovative ideas, criticisms, and valuable comments to strengthen the book.

I would also like to thank Nikhil Bandhu Pal, Pratap Sharma, Kamal Sen, Rajesh Mondal, Jyotirmoy Gupta, Bapi Chatterjee, and other technical and non-technical staffs for online, hardware, software supports and their heartily co-operations.

I do convey my heartily gratitude to Dr. Subrata Trivedi, Asst. Professor, Dept. of Biotechnology of our institute for his amiable behavior, enthusiastic facilitates, valuable suggestions and remarks towards improvement of the quality of this book.

In particular, I am obliged to Pramod Kumar Hota, Librarian, Jishnu Mondal, Asst. Librarian, and all the staff members of BCET Central Library for their co-operation.

I must articulate my warm thanks to Sanjeev Kataria and the entire team of S. K. Kataria & Sons for being so professional, friendly and their continuous efforts throughout the process.

Before I finish, I would like to share that when one writes a book then he realizes the true power of MS-Word from grammar checks to replace all. Simply, without this software, this book would not be written. Thanks to the Philanthropist Bill Gates, and his Microsoft Corporation!

June, 2012 **BISWAJIT R BHOWMIK**

Suggestions

I have attempted to clean up almost every error, mistake, ambiguity in this second edition, but as what happens with human beings—a few obscure troubles may remain. I would appreciate hearing from all the teachers and students those who are studying the subject about any textual errors or omissions that you can identify. Any suggestions regarding improvements I would be glad to hear from you. Please send correspondence to Research Scholar, Department of Computer Science & Engineering, Indian Institute of Technology, Guwahati, Assam, India, 781 039. Readers can also e-mail me directly to either of following e-mail ids: biswajitbhowmik@gmail.com or b.bhowmik@iitg.ernet.in. Readers can also post their comments, suggestions regarding improvement of the book on my website: www.biswajitrbhowmik.co.nr and may visit for detailed information of the book.

Syllabus

Design and Analysis of Algorithms (WBUT, West Bengal)
CS-503 and IT-803C

1. ***Models of Computation:*** RAM, TM, etc. Time and Space Complexity
2. **Asymptotic Notations:** Big-O, Omega, Theta Etc.; Finding Time Complexity of Well Known Algorithms Like- Heap Sort, Search Algorithm Etc.
3. ***Algorithm Design Techniques:*** Recursion-Definition, Use, Limitations, Examples: Hanoi Problem, Tail Recursion.
4. ***Divide and Conquer:*** Basic Method, Use, Examples: Merge Sort, Quick Sort, Binary Search.
5. ***Dynamic Programming:*** Basic Method, Use, Examples: Matrix-Chain Multiplication, All Pair Shortest Paths, Single-Source Shortest Path and Travelling Salesman Problem.
6. ***Branch And Bound:*** Basic Method, Use, Examples: The 15-Puzzle Problem
7. ***Backtracking:*** Basic Method, Use, Examples: Eight Queens Problem, Graph Coloring Problem and Hamiltonian Problem.
8. ***Greedy Method:*** Basic Method, Use, Examples: Knapsack Problem, Job Sequencing with Deadlines, Minimum Spanning Tree -Prim's and Kruskal's Algorithms.
9. ***Lower Bound Theory:*** Bounds on Sorting and Sorting Techniques Using Partial and Total Orders.
10. ***Disjoint Set Manipulation:*** Set Manipulation Algorithm Like UNION-FIND, Union By Rank, Path Compression.
11. ***Properties of Graphs And Its Traversal Algorithms:*** BFS and DFS.
12. ***Matrix Manipulation Algorithms:*** Different Types of Algorithms And Solution of Simultaneous Equations, DFT & FFT Algorithm, and Integer Multiplication Schemes.
13. ***Notion of NP-Completeness:*** P Class, NP-Hard Class, NP-Complete Class, Circuit Satisfiability Problem, Clique Decision Problem.
14. ***Approximation Algorithms:*** Necessity of Approximation Scheme, Performance Guarantee, and Polynomial Time Approximation Schemes: 0/1 Knapsack Problem.

Design and Analysis of Algorithms (BPUT, Orissa)
BCSE-3301

Module I

Introduction to design and analysis of algorithms, Growth of Functions (Asymptotic notations, standard notations and common functions), Recurrences, solution of recurrences by substitution, recursion tree and Master methods, worst case analysis of Mergesort, Quicksort and Binary search, Design & Analysis of Divide and conquer algorithms.

Heaps, Building a heap, The heapsort algorithm, Priority Queue, Lower bounds for sorting.

Module II

Dynamic programming algorithms (Matrix-chain multiplication, Elements of dynamic programming, Longest common subsequence).

Greedy Algorithms – (Activity- selection Problem, Elements of Greedy strategy, Fractional Knapsack problem, Huffman codes).

Module III

Data structure for disjoint sets: Disjoint set operations, Linked list representation, Disjoint set forests. Graph Algorithms: Breadth first and depth-first search, Minimum Spanning Trees, Kruskal and Prim's algorithms, single – source shortest paths (Bellman-ford and Dijkstra's algorithms).

Module – IV

Fast Fourier Transform, string matching (Rabin-Karp algorithm), NP – Completeness (Polynomial time, Polynomial time verification NP – Completeness and reducibility, NP-Camplete problems (without Proofs), Approximation algorithms (Travelling Salesman Problem).

Design and Analysis of Algorithms (GBTU & MTU, Uttar Pradesh) TCS-503

Unit I

Introduction: Algorithms, analysis of algorithms, Growth of Functions, Master's Theorem, Designing of Algorithms. Sorting and order Statistics: Heapsort, Quicksort, Sorting in Linear time, Medians and Order Statistics.

Unit II

Advanced Data Structure: Red-Black Trees, Augmenting Data Structure. B-Trees, Binomial Heaps, Fibonacci Heaps, Data Structure for Disjoint Sets.

Unit III

Advanced Design and Analysis Techniques: Dynamic Programming, Greedy Algorithms, Amortized Analysis, Back Tracking.

Unit IV

Graph Algorithms: Elementary Graphs Algorithms, Minimum Spanning Trees, Single-source Shortest Paths, All-Pairs Shortest Paths, Maximum Flow, Travelling Salesman Problem.

Unit V

Selected Topics: Randomized Algorithms, String Matching, NP Completeness, Approximation Algorithms.

Design and Analysis of Algorithms (JNTU, Andhra Pradesh)

Unit I

Introduction: Algorithm, Psuedocode for expressing algorithms, Performance Analysis–Space complexity, Time complexity, Asymptotic Notation–Big oh notation, Omega notation, Theta notation and Little oh notation, Probabilistic analysis, Amortized analysis.

Unit II

Disjoint Sets: Disjoint set operations, union and find algorithms, spanning trees, connected components and biconnected components.

Unit III

Divide and conquer: General method, applications–Binary search, Quicksort, Mergesort, Strassen's matrix multiplication.

Unit IV

Greedy method: General method, applications-Job sequencing with dead lines, 0/1 Knapsack problem, Minimum cost spanning trees, Single source shortest path problem.

Unit V

Dynamic Programming: General method, applications-Matrix chain multiplication, Optimal binary search trees, 0/1 Knapsack problem, All pairs shortest path problem,Travelling sales person problem, Reliability design.

Unit VI

Backtracking: General method, applications-n-queen problem, sum of subsets problem, graph coloring, Hamiltonian cycles.

Unit VII

Branch and Bound: General method, applications Travelling sales person problem, 0/1 Knapsack problem- LC Branch and Bound solution, FIFO Branch and Bound solution.

Unit VIII

NP-Hard and NP-Complete problems: Basic concepts, non deterministic algorithms, NP-Hard and NP-Complete classes, Cook's theorem.

Design and Analysis of Algorithms (PTU, Punjab)
CS-307

Models of computation. Algorithm analysis, order arithmetic, time and space complexities and average and worst case analysis, lower bounds.

Algorithm design techniques: divide and conquer, search and traversals. Dynamic programming. Backtracking. Branch and bound.

Sorting and searching algorithms, combinatorial algorithms, string processing algorithms. Algebraic algorithms, set algorithms. Hard problems and approximation algorithms.

Problem classes P, NP, NP-hard and NP-complete, deterministic and non-deterministic polynomial time algorithms., Approximation algorithms for some NP-complete problems.

Design and Analysis of Algorithms (RTU, Rajasthan)
6CS3 & 6IT2

(*i*) **BACKGROUND:** Review of Algorithm Complexity and Order Notations and Sorting Methods.DIVIDE AND CONQUER METHOD: Binary Search, Mergesort, Quicksort and Strassen's matrix multiplication algorithms.GREEDY METHOD: Knapsack Problem, Job Sequencing, Optimal Merge Patterns and Minimal Spanning Trees.

(*ii*) **DYNAMIC PROGRAMMING:** Matrix Chain Multiplication. Longest Common Subsequence and 0/1 Knapsack Problem. BRANCH AND BOUND: Travelling Salesman Problem and Lower Bound Theory. Backtracking Algorithms and queens problem.

(*iii*) **PATTERN MATCHING ALGORITHMS:** Naïve and Rabin Karp string matching algorithms, KMP Matcher and Boyer Moore Algorithms.ASSIGNMENT PROBLEMS: Formulation of Assignment and Quadratic Assignment Problem.

(*iv*) **RANDOMIZED ALGORITHMS.** Las Vegas algorithms, Monte Carlo algorithms, randomized algorithm for Min-Cut, randomized algorithm for 2-SAT. Problem definition of Multicommodity flow, Flow shop scheduling and Network capacity assignment problems.

(*v*) **PROBLEM CLASSES NP, NP-HARD AND NP-COMPLETE:** Definitions of P, NP-Hard and NP-Complete Problems. Decision Problems. Cook's Theorem. Proving NP-Complete Problems - Satisfiability problem and Vertex Cover Problem. Approximation Algorithms for Vertex Cover and Set Cover Problem.

Design and Analysis of Algorithms (VTU, Belgaum)
06CS 43

PART-A

(*i*) **Introduction:** What is an Algorithm?, Fundamentals of Algorithmic Problem Solving, Important Problem Types, Fundamental Data Structures.

(*ii*) **Fundamentals of the Analysis of Algorithm Efficiency:** Analysis Framework. Asymptotic Notations and Basic Efficiency Classes, Mathematical Analysis of Nonrecursive and Recursive Algorithms, Example – Fibonacci Numbers.

(*iii*) **Brute Force: Selection Sort and Bubble Sort, Sequential Search and Brute-Force String Matching, Exhaustive Search. Divide and Conquer:** Mergesort, Quicksort, Binary Search. Binary tree traversals and related properties, Multiplication of large integers and Stressen's Matrix Multiplication.

(*iv*) **Decrease and Conquer:** Insertion Sort, Depth First Search, Breadth First Search, Topological Sorting, Algorithms for Generating Combinatorial Objects

PART-B

(*v*) **Transform and Conquer:** Presorting, Balanced Search Trees, Heaps and Heapsort, Problem Reduction. Space and Time Tradeoffs: Sorting by Counting, Input Enhancement in String Matching. Hashing.

(*vi*) **Dynamic Programming:** Computing a Binomial Coefficient, Warshall's and Floyd's Algorithms, The Knapsack Problem and Memory Functions.

(*vii*) **Greedy Technique:** Prim's Algorithm, Kruskal's Algorithm, Dijkstra's Algorithm, Huffman Trees.Limitations of Algorithm Power: Lower-Bound Arguments, Decision Trees. P, NP and NP-Complete Problems.

(*viii*) **Coping with the Limitations of Algorithm Power:** Backtracking, Branch-and-Bound, Approximation Algorithms for NP-Hard Problems.

Analysis and Design of Algorithms (MDU, Rohtak)
Code: CSE – 305E

Unit I

Brief Review of Graphs, Sets and disjoint sets, union, sorting and searching algorithms and their analysis in terms of space and time complexity.

Unit II

Divide and Conquer: General method, binary search, mergesort, quicksort, selection sort, Strassen's matrix multiplication algorithms and analysis of algorithms for these problems.

Unit III

Greedy Method: General method, Knapsack problem, job sequencing with deadlines, minimum spanning trees, single source paths and analysis of these problems.

Unit IV

Dynamic Programming: General method, optimal binary search trees, O/I Knapsack, the travelling salesperson problem.

Unit V

Back Tracking: General method, 8 queen's problem, graph colouring, Hamiltonian cycles, analysis of these problems.

Unit VI

Branch and Bound: Method, 0/1 Knapsack and traveling salesperson problem, efficiency considerations. Techniques for algebraic problems, some lower bounds on parallel computations.

Unit VII

NP Hard and NP Complete Problems: Basic concepts, Cook's theorem, NP hard graph and NP scheduling problems some simplified NP hard problems.

Design and Analysis of Algorithms (BATU, Raigad)

Code: CE – 3105

Unit I

Introduction: Review to data structures, Trees-AVL trees and RB tree, algorithm specification, Performance analysis.

Complexity of algorithm: Time complexity, asymptotic notations, Big O notation, Q notation, W notation, Recurrences

Unit II

Divide and conquer: General method, Mergesort, Quicksort, Selection sort, Strassen's matrix multiplication.

Sorting in linear time: Counting sort, Radixsort, Bucketsort .Analysis of these algorithms.

Unit III

The greedy method: The general method, Knapsack problem, Huffman Coding algorithm, Job sequencing with deadlines, Minimum cost spanning tree, optimal storage on tapes, Single source shortest path.

Advanced data structures: B-trees, Binomial heap and Fibonacci heap.

Unit IV

Dynamic programming: The general method, Multistage graph, optimal binary search trees, matrix chain multiplication, 0/1 Knapsack, longest common sequence, assembly line scheduling

Unit V

Direct address tables: Algorithm for inserting an element into direct address tables, deleting an element from direct address tables, searching an element .Complexities of these algorithms .

Hash tables: Comparison with direct address tables, binary search trees. Algorithm for insertion and searching, hash functions, collision resolution techniques – Chaining and open addressing.

Unit VI

Graphs: Representation of graphs in memory , traversal algorithms – depth first search and breadth first search, topological sort, spanning trees, minimum spanning tree – Kruskal's algorithm, Prim's algorithm, Analysis of these algorithms, Dijkstra's algorithm, Bellman – Ford algorithm, Floyd – Warshall algorithm,

Backtracking: 8 queen's problem, algorithm to solve n – queens problem, travelling salesman problem.

Design and Analysis of Algorithms (GTU, Ahmedabad)

Code: CS – 150703

Basics of Algorithms and Mathematics: What is an algorithm?, Mathematics for Algorithmic Sets, Functions and Relations, Vectors and Matrices, Linear Inequalities and Linear Equations.

Analysis of Algorithm: The efficient algorithm, Average and worst case analysis, Elementary operation, Asymptotic Notation, Analyzing control statement, Amortized analysis, Sorting Algorithm, Binary Tree Search.

Divide and Conquer Algorithm: Introduction, Multiplying large Integers Problem, Problem Solving using divide and conquer algorithm - Binary Search, Sorting (Mergesort, Quicksort), Matrix Multiplication, Exponential.

Greedy Algorithm: General Characteristics of greedy algorithms, Problem solving using Greedy Algorithm - Activity selection problem, Elements of Greedy Strategy, Minimum Spanning trees (Kruskal's algorithm, Prim's algorithm), Graphs: Shortest paths, The Knapsack Problem, Job Scheduling Problem.

Dynamic Programming: Introduction, The Principle of Optimality, Problem Solving using Dynamic Programming – Calculating the Binomial Coefficient, Making Change Problem, Assembly Line-Scheduling, Knapsack problem, Shortest path, Matrix chain multiplication, Longest Common Subsequence.

Exploring Graphs: An introduction using graphs and games, Traversing Trees – Preconditioning, Depth First Search - Undirected Graph, Directed Graph, Breath First Search, Backtracking – The Knapsack Problem, The Eight queens problem, General Template.

String Matching: Introduction, The naive string matching algorithm, The Rabin-Karp algorithm, String Matching with finite automata.

Introduction to NP-Completeness: The class P and NP, Polynomial reduction, NP- Completeness Problem, NP-Hard Problems.

Design and Analysis of Algorithms (KU, Kurukshetra)
Code – CSE-301 & IT-352

Unit I

Introduction: Review of elementary data structures, analyzing algorithms, asymptotic notation, recurrence relations, hash tables, Binary search trees. Sorting and Order Statistics Heapsort, Priority queues, Quicksort, Sorting in linear time, medians and order statistics, dynamic order statistics.

Unit II

Advanced Design and analysis Techniques: Dynamic programming – Elements, Matrix-chain multiplication, longest common subsequence, optimal polygon triangulation, Greedy algorithms- Elements, activity- selection problem, Huffman codes, task scheduling problem. Advanced data Structures Operations in B-Trees, Binomial heaps, Fibonacci heaps data structures for disjoint sets, strings.

Unit III

Graph Algorithms: Review of graph algorithms, topological sort, strongly connected components, minimum spanning trees- Kruskal's and Prim's, Single source shortest paths, relaxation, Dijkstra's algorithm, Bellman- Ford algorithm, Single source shortest paths for directed acyclic graphs, difference constraints and shortest paths, All pairs shortest paths- shortest paths and matrix multiplication, Floyd-Warshall algorithm, Johnsons algorithm.

Unit IV

Flow and Sorting Networks: Flow networks, Ford- Fulkerson method, Maximum bipartite matching, Sorting Networks, Comparison network, the zero- one principle, Bitonic sorting network, merging network.

Back tracking: Overview, 8-queen problem, and Knapsack problem.

Brach and bound: LC searching Bounding, FIFO branch and bound, LC branch and bound application: 0/1 Knapsack problem, Travelling Salesman Problem.

Computational Complexity: Complexity measures, Polynomial Vs non-polynomial time complexity; NP-hard and NP-complete classes, examples.

List of Figures

CHAPTER 6

CHAPTER 7

CHAPTER 8

CHAPTER 9

CHAPTER 10

CHAPTER 11

CHAPTER 13

CHAPTER 15

CHAPTER 16

CHAPTER 17

CHAPTER 18

CHAPTER 19

CHAPTER 20

Chapter 1 Introduction to Algorithms

"Computer science is the study of algorithms"

— Donald E. Knuth

1.1 INTRODUCTION

In the field of computer science, the design and analysis of algorithms have immeasurable contribution as every area in computer science depends heavily on the design of efficient algorithms. Compilers, operating systems, etc. as examples are nothing but direct implementations of special purpose algorithms.

When developing software we, as a software engineer, go through a number of intermediate phases like feasibility study, requirement analysis and specification, design, coding, testing, and maintenance. These are to be followed one by one in the sequence as they are mentioned but not haphazardly.

Again in the morning after leaving bed we go for brushing and to do that first we take toothpaste on the brush, then take it into mouth, then rub it left and right on the teeth, sometime later we wash our mouth with water. In this case, obviously a sequence for the purpose of brushing mouth is maintained.

In all the above cases it is seen that a sequence of steps to complete a task is gone after. This sequence is nothing but what we call an algorithm.

1.2 WHY DO WE STUDY ALGORITHM ?

"That fondness for science, ... that affability and condescension which God shows to the learned, that promptitude with which he protects and supports them in the elucidation of obscurities and in the removal of difficulties, has encouraged me to compose a short work on calculating by al-jabr and al-muqabala , confining it to what is easiest and most useful in arithmetic." - Al-Khwarizmi

A proper preplanning to complete a large number of tasks in less time and space is required with the increase in performance of a computer. This brings us to have a deep knowledge about the algorithms. Reasons are the following.

1. To write better and efficient code.
2. To solve real-world problems algorithmically.
3. To develop thinking power about solution of a problem more abstractly and mathematically.
4. To make competent computer engineers, scientists, etc to take the challenges.
5. To simulate a model intellectually.
6. To be able solve problems that could not otherwise be addressed For example, network connectivity.

1.3 ALGORITHMS: DEFINITIONS

The word *Algorithm* came from the name of the Persian mathematician Abu Ja'far Muhammad ibn Musa Al-Khwarizmi [Born: about 780 A.D. in Baghdad (now in Iraq). Died: about 850 A.D.] who wrote a book on Hindu-Arabic numerals and was among the first to use zero as a place holder in positional base notation.

An algorithm can be defined by any one of the followings as there no specific way exists to define it. However it should have some common properties.

1. An algorithm is a set of rules for proper calculation either by hand or on a machine.
2. An algorithm is a finite sequence of steps of a procedure to achieve a required result.
3. An algorithm is a sequence of computational steps that take some values as input, process that input, and transform these input into some values as output.
4. An algorithm is a sequence of operations performed on data to organize a process efficiently.
5. An algorithm is an executable abstraction of a program for a machine.
6. An algorithm is a tool for solving a well defined computational procedure in terms of input / output relationship.

1.4 ALGORITHMS : FEATURES

An algorithm must include the following criteria.

1. *Input:* An algorithm must have either zero or more finite number of physical quantities considered as input. These are generally supplied externally or internally.
2. *Output:* At least one quantity as output value must be produced by every algorithm.
3. *Definiteness:* Every instruction/step in an algorithm must be clear and unambiguous.
4. *Finiteness:* The algorithm must complete execution in finite steps, *i.e.* it must terminate after a finite number of instructions/steps have been executed.
5. *Feasibility:* An algorithm must be feasible, *i.e.* at least possible to implement.
6. *Effectiveness:* Every instruction/step must be basic so that it can be carried out by paper and pencil by a person very easily.
7. *Serializability:* Each step in the algorithm must follow each other, *i.e.* there should have a unique defined preceding and succeeding step.
8. *Correctness:* If for every input instance an algorithm halts with a correct output, etc.

1.5 ALGORITHMS : PHASES

During designing an algorithm it should be always careful to include all possible important and major activities at every phase of the algorithms. Generally, four phases are gone through to implement it.

(*i*) Design Techniques.
(*ii*) Validation.
(*iii*) Analysis.
(*iv*) Testing.

1.5.1 Algorithms: Design Techniques

Choosing an exact method for designing a suitable algorithm of a problem requires thorough understanding, identification, classification of the problem, and more experiences in the field. For a given problem there may exist more than one algorithm. It is preferable to use the most appropriate one. A major objective is thus to study various algorithm designing techniques which are very useful. Once one becomes a master on these strategies, it will be very easy to devise appropriate algorithm for a new problem identifying its type.

The followings are the most popular algorithm design approaches.

- Incremental Approach.
- Recursive Approach.
- Divide-and-Conquer Approach.
- Greedy Approach.
- Dynamic Approach.
- Backtracking Approach.
- Branch and Bound Approach.
- Brute Force Approach, etc.

1.5.2 Algorithms: Validation

Once an algorithm of a problem has been designed using an appropriate approach, it is most necessary to be valid one. Validation of a newly designed algorithm is confirmed having correct answers just produced for some possible input data set independent of the programming languages whenever will be used for implementation. If it is so, then refer this algorithm as valid algorithm.

1.5.3 Algorithms: Analysis

A specific problem might have several existing algorithms. An intelligent will choose cleverly the better one. We never say this algorithm is the best one because there may still exist much better one than these algorithms. But how do we select an algorithm over the other? In such cases, need of comparisons in terms of space and time is the possible tool.

Definition: Analysis of algorithm also called performance analysis refers to the task of measuring how much storage space and determining how much computing time an algorithm requires.

In other words, the theoretical study of measuring performance of computer-program and resource usage is nothing but the analysis of algorithms.

1.5.3.1 *Space Complexity*

Amount of computer memory required during the program execution, as a function of the input size is called space complexity of an algorithm.

Space complexity is measured by using polynomial amounts of memory. These are not the space occupied by the program in the computer secondary memory (hard disk, pen drive, etc.) rather the space in main memory required during execution of the program. These spaces can be reused and are independent of determinism and nondeterminism of the machine.

Space complexities however may depend on various factors such as memory, communication bandwidth, computer hardware, software, etc.

1.5.3.2 *Time Complexity*

Amount of time required to complete a task by an algorithm or to execute a program is called its time complexity.

Unlike space complexity, time complexities are machine dependent. It means that they depend on both determinism and nondeterminism of machines. Similar to space complexities, time complexities also depend on the factors such as memory, communication bandwidth, computer hardware, software, etc.

Most of the cases, it has been seen that the same algorithm comprises different time complexity values. These are aroused depending on the input size n, the number of steps executed, etc.

1.5.3.3 *Kinds of Analysis*

Time complexity is much more significant than space complexity. More traces therefore would be drawn on time complexity instead of space complexity. Whenever we mean a complexity from now onwards it actually means to time complexity unless specifically mentioned. Three variations of time complexity (also for space complexity) are plausible. These are:

(*i*) Best Case.
(*ii*) Worst Case.
(*iii*) Average Case.

These three cases are defined w.r.t. time complexity briefly.

- **Best Case:** The minimum amount of time taken by an algorithm on input size n is called the best case time complexity of the algorithm. Best case is possible by executing the minimum number of steps. It is denoted by the symbol Big Omega (Ω). For example, best case time complexity of quick sort algorithm is $\Omega(n\log n)$.
- **Worst Case:** The maximum amount of time taken by an algorithm on input size n is called the worst case time complexity of the algorithm. The algorithm in this case executes the maximum number of steps. It is denoted by the symbol Big Oh (O). For example, worst case time complexity of quick sort algorithm is $O(n^2)$.
- **Average Case:** The average amount of time taken by an algorithm on input size n is called the average case time complexity of the algorithm. The algorithm consists of the average number of steps. It is denoted by the symbol Big Theta (Θ). For example, average case time complexity of quick sort algorithm is $\Theta(n\log n)$. Average case time complexity indicates expected running time of algorithms on n inputs.

Some experts often neglect this average case because sometimes it needs assumptions on statistical distribution of inputs. That means its value varies from best case to worst case including them. Relationship among these three cases of the complexities is shown Figure 1.1 below.

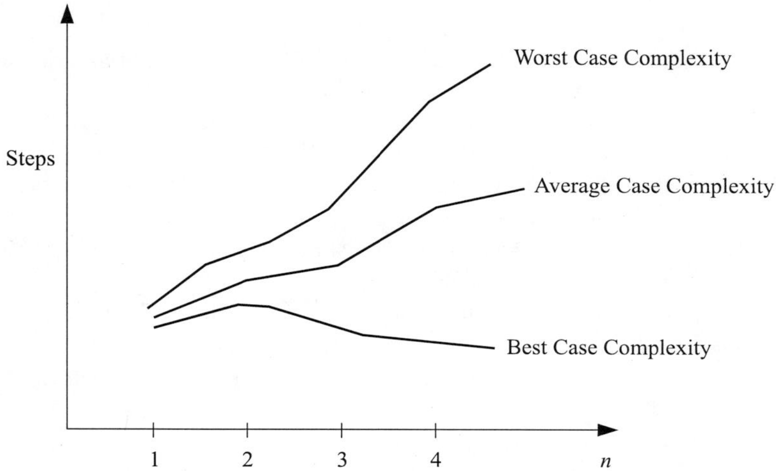

Figure 1.1 Best, worst, and average-case complexity.

1.5.4 Algorithms: Testing

This phase involves testing of a program in two sub-phases.

- **Debugging:** Determining whether a desired result is produced or not during execution of a program on sample data values. If incorrect output results, correction of the program is made.
- **Performance Measurement:** The process of executing a correct program on input data set and measuring its complexities is termed as performance measurement. It is also called *profiling*.

1.6 ALGORITHMS: REPRESENTATION

An algorithm can be described in many ways. One obvious way is natural languages like english. However, this representation is not suitable one because:

(*i*) With the increase in input size, the problem becomes very descriptive.

(*ii*) The execution sequence can't be viewed and maintained.

Another way of representing an algorithm is a graphical representation which we know as flowchart. This is again more suitable and works well only if the problem is small and simple one.

Another way is some real programming language. However it is not suitable. Therefore, we should follow a way of representation for the algorithms which is easily understandable, problem oriented, and programming languages independent. This handy representation tool is pseudocode.

1.7 PSEUDOCODE

A pseudocode can be defined as compact and informal high-level description of a computer programming algorithm. Simply, the representation of an algorithm that uses the conventions of some structural or high level programming languages and is intended for human understanding rather than machine understanding is the pseudocode. In a line, English-like statements used for low-level program design is pseudocode.

Some essential features that can be kept in mind are:

1. Pseudocode typically omits some part such as variable declarations, system-specific code and subroutines, etc. of the algorithm.
2. The purpose of using pseudocode is to make easier understanding of the problems for human beings rather than conventional programming language code.
3. Pseudocode should be a compact and problem independent description.
4. Pseudocode must include the key part of an algorithm.
5. Pseudocode is a sketched outline in planning computer program development before writing the actual code.
6. There is no standard way for syntaxes of a pseudocode.
7. A pseudocode is not actually an executable program.
8. Pseudocode may sometimes resemble skeleton programs which can be compiled without errors.
9. This may include some dummy code.
10. Graphical alternative to a pseudocode can be thought of as flowcharts.

Example 1. Write a pseudocode to find sum of two numbers.

Pseudocode: Initialize Sum=0

Input a, b.

Sum = $a + b$

Output Sum

1.8 PSEUDOCODE NOTATIONS

Some useful notations for pseudocode are used to describe an algorithm in brief in this section purposely. The possible notations used here are mainly for pseudocode representations.

1. **Title:** Every algorithm should have a title. And an algorithm is identified by its title. A title is a meaningful word that should mean to related problem type. For example, to sort a list of elements using Quicksort approach, the possible titles are as - Algorithm: Quick_Sort(), Algorithm: QuickSort(), Algorithm: Quicksort(), etc. Title must be started with a capital alphabet. Here we consider Algorithm: Quick_Sort() for explanation.
2. **Parameters:** A list of parameters may be passed to the title of the algorithm just like passing parameters in functions of some high level programming languages. Consider the above title then it would look like Algorithm: Quick_Sort(L, n).
3. **Description:** Every algorithm must be described with a meaningful title, a parameter list, and any assumptions if considered.

   For example, Algorithm: Quick_Sort (L, n)

   Description: Where L = a list of elements to be sorted.

   n = number of elements in L.
4. **Body:** Body contains essential codes. These are enclosed with { } where {, and } are opening and closing braces used for beginning and ending of the body respectively.
5. **Variables:** Two types of variables viz. local and global variables are used. All the variables used inside the body of the algorithm are generally local variables.

6. **Comments:** Comments are used to describe statements, steps. The notations // for single line statements and /* ...*/ for multiple statements are respectively used for comment lines. For example, if $a > b$ its comment line will be: // compares a & b.
7. **Steps:** Every algorithm should have a finite number of steps. Each step performs a particular task and is labelled with a numeric value. For example,

   Step 1: if $a > b$

   Output a is bigger than b.
8. **Assignment:** Mainly two symbols = or ← are used to assign values to variables. For example, $i = 6$ or $i \leftarrow 6$.
9. **Operators:** Operators are mainly Assignment(= , ←), Boolean(true, false), Conditional(>, <, ≤, ≥, ==, ≠), Logical (&&, ||, !), Arithmetic(+, –, *, /, %, ^), Exchange(↔).
10. **Input:** Input is taken using "read" or "input" word. For example, read x, input x that means value of x is reading or taking.
11. **Output:** Output or result is displayed by using "display" or "output" or "write" or "print" word. For example, write x, display y, output z.

1.9 ALGORITHM VS. PROGRAM

Sometimes we used algorithm, sometimes called program alternately. But basic comparisons can be made between an algorithm and a program. Basic differences are:

(*i*) Program is written in programming language whereas algorithm is in pseudo-code.

(*ii*) Algorithms describe solutions where as programs implement them in a programming language.

(*iii*) Unlike a program, an algorithm is independent of programming language or machine.

(*iv*) Study of algorithm can be classified in four distinct areas namely how to devise, validate, analyze, and test algorithms.

(*v*) For a given problem, design phase produces algorithm and implementation phase produces program of that algorithm.

Figure 1.2 describes the above facts comparatively more easily.

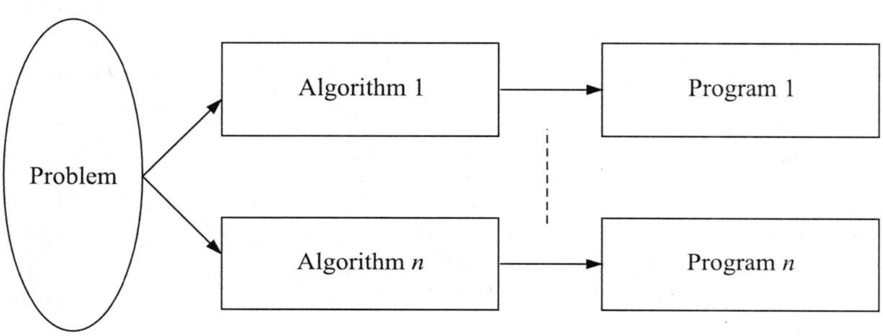

Figure 1.2 Relation between Algorithm and Program.

1.10 MODELS OF COMPUTATION

Analyzing an algorithm means predicting the resources which results in terms of space and time complexity. Studying these complexities needs a general computing device for the classification of the problems. Several models of computation are in our hand for these resources and their cost. Some of these are:

(*i*) Finite Automaton,
(*ii*) Turing Machine,
(*iii*) RAM tmodel,
(*iv*) PRAM model,
(*v*) External memory model,
(*vi*) Boolean Logic Circuit,
(*vii*) Clocked Circuit,
(*viii*) Cellular Automaton,
(*ix*) Biological Computing model, etc.

Here only two important computational models: Turing Machine and RAM are illustrated.

1.10.1 Turing Machine Model

The English mathematician A. Turing in 1936 before the invention of programmable computers, historically introduced the first "infinite" model of computation. To honour him this model was named as the Turing machine. The model can be thought of as a finite automaton (control unit) equipped with an infinite one-dimensional tape (array of cells) as storage (memory).

Each cell of the tape contains one symbol from some finite alphabet S. The Turing machine reads and rewrites the scanned symbol in the cell by tape head. The head moves either one cell to the left, one cell to the right, or it remains on the current cell at each scanning. The actions of the Turing machine are specified by its finite state control. The control unit makes all local changes to the scanned memory cell. A Turing machine at any time remains in some state.

One of the great importances of this model is that it can accomplish all computations. In fact, Turing machines are the simplest universal machines. At the same time, they differ from real-life computers in several important aspects. So they are less appropriate for the description of concrete algorithms.

For further study readers are advised to refer any books on automata theory.

1.10.2 Random Access Machine Model

The most important weakness of the Turing machine in comparison to real computers is that its memory is not accessible immediately meaning that in order to read a distant memory cell, all intermediate cells must also be read which consumes extra time. This problem can be overcome by the introduction of another model of computation - Random Access Machine (RAM) model.

The RAM can reach an arbitrary memory cell in a single step. It can be considered as a simplified model for real world computers along with the abstraction that it has the capability to store arbitrarily large integers in each of its memory cells, *i.e.* unbounded memory.

To read a memory register immediately, we must address it. This address is again stored in another register. As the number of memory registers is not bounded, the address can't be bounded. Therefore, an arbitrarily large number are to allow in the register which is containing these addresses. The content of this register must itself change during the running of the program which indicates indirect addressing. This further increases the power of the RAM.

It is seen that both the Turing machine and the RAM are equivalent from many points of view like the same functions are computable on Turing machines and on the RAM. Unfortunately, these properties of the RAM machine bring to costly.

The RAM can be programmed in some specified programming language. As the RAM is closest to real program writing, it is more practical to use it for the description of algorithms.

1.11 MORE EXAMPLES

Example 2. Write a suitable algorithm for n!

Algorithm: Fact (n)

Where n = any positive integer.

Step 1: Start

Step 2: Initialize fact=1.

Step 3: Read n.

Step 4: If $n = 0$
(*a*) Go to step 6.

Step 5: For I = 1 to n
(*a*) fact = fact * I

Step 6: Write fact

Step 7: Stop.

Example 3. Write a pseudocode for $\boldsymbol{S = P(1+R)}$

Algorithm: Interest (P, r)

Where P = principal amount.
R = rate of interest.

Step 1: Start

Step 2: Initialize S=0.0

Step 3: Input P, R

Step 4: Assign $S = P*(1+R)$

Step 5: Output S

Step 6: Stop

Example 4. Write an algorithm to multiply two matrices. Assume rule of matrix multiplication.

Algorithm: Matrix_Multiplication (A, B, C)

Where A = a mxp matrix.
B = a pxn matrix
C = a result matrix = $A \times B$. Hence, it is of $m \times n$.

Step 1: (a) Repeat for $i = 1$ to m

- Repeat for $j = 1$ to n
  - (i) $C[i, j] = 0$
  - (ii) Repeat for $k = 1$ to p
    - $C[i, j] = C[i, j] + A[i, k]*B[k, j]$

Step 2: Return C

Step 3: Exit

1.12 SOME COMMON QUESTIONS

Question 1. What is a good algorithm?

Answer: An algorithm is said to be a good algorithm if it must be

(i) Correct - it should compute the answer it is supposed to.

(ii) Feasible - each step should be executed in a reasonable time;

(iii) Efficient - the algorithm should use as low complexities as possible within the limits of an acceptable development time.

Question 2. Why do we study algorithms?

Answer: The simple answer is that analysis of algorithms is fundamental to computer science and it is very practical too.

Question 3. What are the characteristics of an algorithm?

Answer: Basic characteristics of an algorithm include

(*a*) Simplicity

(*b*) Generality

(*c*) Definiteness

(*d*) Finiteness

(*e*) Effectiveness

Question 4. What are three basic parts of algorithm?

Answer: Basic parts of an algorithm include

(*a*) Pre Conditions

(*b*) Body of Algorithm

(*c*) Post Conditions

Question 5. What are the criteria used to identify performance of an algorithm?

Answer: The criteria are

(*a*) Order of Growth

(*b*) Time Complexity

(*c*) Space Complexity

Question 6. What are three components of Space Complexity?

Answer: Space Complexity indicates how much memory an algorithm needs to execution. This complexity has three components. They are instruction space, data space and environment space.

Question 7. Find the time complexity for summation of N numbers.

Answer: The time complexity for summation of N numbers is O(N).

Question 8. What is the Space Complexity for the following algorithm?

```
Void N()
{
    Int P, A, B, C;
    P = A + B – C ;
    Printf("%d", P);
}
```

Answer: Space Complexity = Space needed for P, A, B, C = 8.

Exercises

1. Write algorithms for the followings to compute.
   (*a*) The sum of all numbers from 1 to *n*.
   (*b*) 2^n, *n* is a non-negative integer.
   (*c*) Any number to the power of a non-negative integer.
   (*d*) The *n*th Fibonacci number.
   (*e*) LCM of two given integers.
2. Write an algorithm to find the maximum and minimum from a sequence of *n* integers stored in a list *L*[1.....*n*].
3. Modify the algorithm of Exercise 1.2 to find an element in *L* which is neither maximum nor the minimum.
4. Given a set of integers. Give an efficient algorithm that determines whether two of them are equal.
5. The following pattern of numbers is called Pascal's triangle.

   1
   1 1
   1 2 1
   1 3 3 1
   1 4 6 4 1
   ....

   Write a procedure for this triangle.
6. Define algorithm. How does it differ from flowchart?
7. How can we design algorithms?
8. Why do we study algorithms?
9. Write an algorithm for the polynomial $f(x) = 2^k$, $0 \leq k \leq n$ for some x.
10. What are the factors that affect complexity of algorithms?
11. What do you mean by model of computation? Give some examples.

12. Write short notes on:
    (*a*) RAM
    (*b*) Turing Machine.
    (*c*) Pseudocode.
13. Make difference between algorithm and program.
14. Define complexities of algorithms.
15. Compare three cases of complexities. Which one is more significant?
16. Space complexity or time complexity – which one is most important than the other? Why?
17. Write a suitable procedure that evaluates the following polynomial.

$$F(x) = a_n x^n + a_{n-1} i x^{n-1} + ... + a_1 x + a_0$$

University Questions

1. What is an algorithm? *[PTU'08, DOEACC'09]*
2. What do you mean by complexity? Explain its types. *[UPTU'10, PTU'08, 09, '11, JNTU'10]*
3. Explain various characteristics of algorithm. *[RTU'11, DOEACC'09]*
4. Write short notes on Turing machines. *[WBUT'09]*
5. What do you understand by algorithm evaluation? *[PTU'07]*
6. What is asymptotic time complexity? *[PTU'07]*
7. Enumerate five important characteristics of an algorithm. *[IGNOU'09]*
8. Write the sequential search and binary search algorithm on the data stored in an array. Compare these two methods. *[UPTU'07]*
9. Write an algorithm that computes the number of elements and sum of elements in a linear linked list. *[BPUT'05]*
10. Explain how to validate and analyze the algorithms. *[PTU'08, 09]*
11. Write an algorithm to determine a given string is a palindrome. Note that a string is a palindrome if it can be read forward and backward with the same meaning. Capitalization and spacing are ignored. For example, *anna* and *godog* are palindromes. *[IETE ALCCS'11]*
12. Explain the algorithm of a non-deterministic finite automation. *[PTU'07, 09]*
13. Write algorithm for insertion of an element into a binary tree represented in an array. *[RTU'11]*
14. What are the various steps in the design of an algorithm? *[PTU'08]*
15. Explain the relationship between Turing Machine and RAM models. *[PTU'07]*
16. What are various techniques used for designing and improving performance of algorithms? *[RTU'08]*
17. Suppose a linked list consists of some numeric values. Design an algorithm to find maximum value in the list. *[UPTU'07]*
18. Write a procedure to merge two singly linked lists whose elements are sorted in ascending order to produce a single singly linked list sorted in ascending order. *[UPTU'06]*
19. Write an algorithm for deleting duplicate numbers from a linear array. *[UPTU'07]*

20. With an example explain how you will measure the efficiency of an algorithm? ***[PTU'09]***
21. How the deletion can be performed from a binary search tree? Write and explain the algorithm. ***[UPTU'07, JNTU'08, DOEACC'07]***
22. Write an algorithm for RIGHT-ROTATE in the construction of AVL trees. Explain with one example. ***[DOEACC'07]***
23. Write an algorithm for inserting a new element into a queue. ***[PTU'05]***
24. What is re-entrant program? ***[PTU'07, '09]***
25. A programming language does not accept parenthesis in the expressions. However, it has a facility of outputting fully parenthesized expressions of the input expressions. The programmer can then check the correctness of the input expression. Develop the algorithm for outputting full parenthesized expressions from an expression without parenthesis. Please ensure that the precedence rules are not violated. Show the correctness of your algorithm taking 2 examples. ***[RTU'11]***
26. Write an algorithm for selection sort. Explain how the selection sort is more efficient than bubble sort. ***[UPTU'07]***
27. Write short notes on Time Space Trade-off. ***[UPTU'10, PTU'05]***
28. Consider a Binary tree with names attached to the vertices. Write an algorithm to print the names in (*i*) Pre-order, (*ii*) Inorder, (*iii*) Post order. ***[PTU'07]***
29. Give an Example of an algorithm which is infinite in nature. ***[PTU'07, '08, '09]***
30. Write an algorithm to compute the frequency of occurrence of ASCII characters in a given string. No space will be used for the characters not present in the input string and the list shall contain the frequencies in the order of ASCII value of characters in the input stream. ***[RTU'11]***
31. What do you mean by time complexity and space complexity of an algorithm? ***[BPUT'05, '07, DOEAC'08]***
32. Briefly discuss the Algorithm Analysis Framework? What are different ways of measuring the running time of an algorithm? ***[BPUT'08]***
33. Write an algorithm for finding solution to Tower's of Hanoi problem. Explain the working of the algorithm for 4 disks. ***[IETE ALCCS'11]***
34. Write an efficient algorithm to determine the second largest element of n numbers. Determine the number of comparisons needed to find it. ***[DOEACC'07]***
35. Give an algorithm for finding an unknown number x in a sequence of supplied numbers. ***[DOEACC'09]***
36. How can we find out K^{th} smallest number from an array of n numbers in expected linear time? ***[DOEACC'06,'09]***
37. Write an algorithm that reverses a given string of length n. ***[IETE ALCCS'11]***

❑❑❑

Chapter 2 Asymptotic Notations

2.1 INTRODUCTION

There are several methods available for proper simplification and asymptotic analysis of algorithms. Certain notations for measuring efficiency and performance are used in a standard way. These notations are called asymptotic notations since they show the complexities behaviour or nature of algorithms for any input instances. Remember that it does not provide any numerical complexity value.

Definition: The notations used to describe asymptotic behaviour of complexities of an algorithm are called asymptotic notations. Asymptotic notations are defined in terms of functions, whose domains belong to natural number *N*.

Several asymptotic notations are available in practice. Our interest is basically on the three cases of running time complexities of the algorithms. Only Big Oh (O), Big Omega (Ω), and Big Theta (Θ) for worst case, best case, and average case time complexities are the respective notations. These are very popular as well as efficient.

Consider two functions f and g defined on N to R^+. Then these three notations can be defined one by one as follows.

2.2 BIG OH (O) NOTATION

The Big *Oh* (*O*) notation expresses the asymptotic upper bound on the running time of an algorithm. By design, this Big *Oh* (*O*) notation is used to represent worst case time complexity of any algorithm. Remember that it may not be indicative of the actual running time of an algorithm.

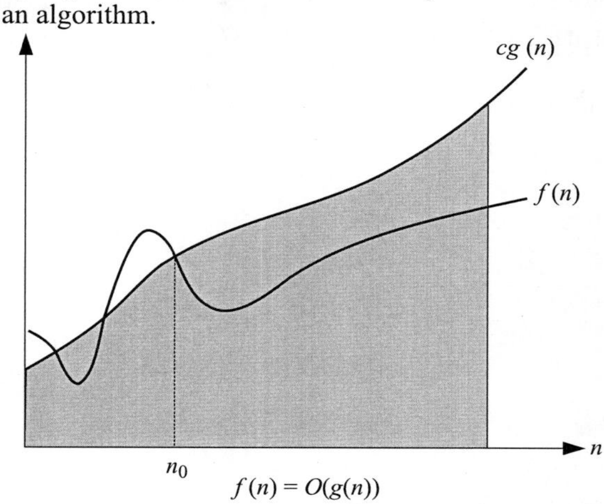

Figure 2.1 Illustration of Big Oh (O) notation.

Definition: The function $f(n) = O(g(n))$ read as "f of n is Big Oh of g of n" is said to exist iff $\exists$ a non-negative constant $c > 0$ and initial value n_0 such that $f(n) \leq c*g(n)$, $\forall\ n \geq n_0$.

Consequently, if $\lim_{n\to\infty} \frac{f(n)}{g(n)}$ exists and it is not equal to ∞, then $f(n) = O(g(n))$.

Whenever $f(n) \in O(g(n))$, $f(n)$ is at least c times of the function $g(n)$. It means $f(n)$ grows no faster than c times $g(n)$. The Figure 2.1 describes the above definition.

2.2.1 Big Oh (O): Features

A list of features describing Big Oh (O) notation may be listed as below.

- Big Oh notation allows us to compare a function with a set of functions to describe the performance of a particular algorithm. These functions are again bounded by some other function.
- Constant factors are not so important during analyzing the performance of programs on large input sets. So, the notation omits information about constant factors. This allows us to bound the function $f(n)$ by $g(n)$, saying that $f(n)$ grows at most as fast as $g(n)$. Thus, $g(n)$ provides an asymptotic upper bound.
- Typically a simple polynomial functions such as n, n^2, etc. are referenced for comparison because the constant factor c does not matter when using Big Oh notation.
- The notation $O(g(n))$ represents set of all functions for which the above condition is satisfied.
- This notation does not say anything about how much this bound is good.
- Big-O notations express an ordering property among functions, etc.

2.2.2 Big Oh (O): Some Examples

Some common asymptotic functions given and their Big Oh value are derived as below.

Example 1. Constant Function: $f(n) = 15$

Solution: Given $f(n) = 15 \leq 16$

$$= 16*1$$

Here, $c = 16$, $n_0 = 0$, and $g(n) = 1$. So, $f(n) = O(1)$

Example 2. Linear Function: $f(n)= 5n + 5$

Solution: Given, $f(n) = 5n + 5 \leq 6n$ for $n \geq 5$

Here, $c = 6$, $n_0 = 5$, and $g(n) = n$. So $f(n) = O(n)$

Example 3. Quadratic Function: $f(n)= 7n^2 + 5n$

Solution: Given, $f(n) = 7n^2 + 5n$

Now $5n \leq n^2$ implies $n \geq 5$

So, $f(n) \leq 8n^2$

Here $c = 8$, $n_0 = 5$, and $g(n) = n^2$. So, $f(n) = O(n^2)$.

Example 4. Cubic Function: $f(n) = 2n^3 + n^2 + 2n$

Solution: Given, $f(n) = 2n^3 + n^2 + 2n$

$$\leq 2n^3 + n^2 + n^2 \quad \text{for } n^2 \geq 2n \text{ and } n \geq 2$$
$$\leq 2n^3 + 2n^2$$
$$\leq 2n^3 + n^3 \quad \text{for } n^3 \geq 2n^2 \text{ and } n \geq 2$$
$$\leq 3n^3$$

Here, $c = 3$, $n_0 = 2$, and $g(n) = n^3$.

So, $f(n) = O(n^3)$.

2.2.3 Big Oh (O): Some Rules

When we are willing to consider the growth rate of a function for asymptotically upper bound using Big Oh notation following assumptions must be kept in our mind during computation.

- First, ignore all the lower order terms and the coefficients of the highest order term
- No need to specify the base of logarithm if it occurs.
- Changing the base from one constant to another changes the value of the logarithm by only a constant factor.
- If $T_1(n) = O(f(n))$ and $T_2(n) = O(g(n))$, then
  - (*i*) $T_1(n) + T_2(n) = max(O(f(n)), O(g(n)))$
  - (*ii*) $T_1(n) * T_2(n) = O(f(n) * g(n))$
  - (*iii*) $O(T_1(n)) = T_1(n)$
  - (*iv*) $O(f(n).g(n)) = f(n)O(g(n))$
  - (*v*) $O(f(n) + O(f(n))) = O(f(n))$
  - (*vi*) $C^*O(f(n)) = O(f(n))$ for some constant C.

These rules can also be applicable for other notations.

2.2.4 Big Oh (O): Some Common Plots

Time complexity value of a function $f(n)$ is referred by $T(n)$ meaning that if $f(n) = O(n)$ then $T(n) = O(n)$. Some common Big Oh Plots are shown below.

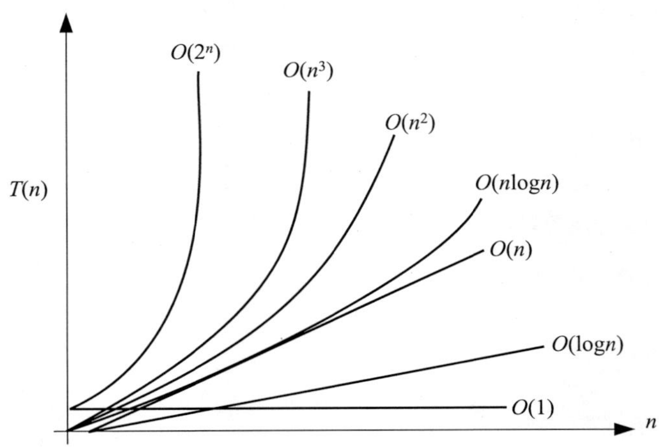

Figure 2.2 Some common Big Oh Plots.

2.2.5 Big Oh (O): More Examples

Some more examples to derive worst case complexity from a given function have been discussed.

Example 5. Prove that $T(n) = 3n^3 + 2n^2$ is $O(n^3)$

Solution: Given, $T(n) = 3n^3 + 2n^2 < cn^3$

Let $c = 5$, $n_o = 1$, then for all $n > 1$

$$3n^3 + 2n^2 < cn^3$$

$$2n^2 < 3n^3$$

$$n^2 < n^3$$

$$1 < n$$

Thus, $T(n) = 3n^3 + 2n^2 < 5n^3$ for $c = 5$ and $n > 1$

Therefore, $T(n) = O(n^3)$

Example 6. Solve $T(n) = 6n^3 + 5n^2 + 4n + 2$ in terms of big Oh.

Solution: Given, $T(n) = 6n^3 + 5n^2 + 4n + 2$

$$= 6n^3 + 5n^2 + 5n \qquad (\therefore\ 4n + 2 \le 5n)$$

$$= 6n^3 + 5n^2 + 5n^2 \qquad (\therefore\ 5n \le 5n^2)$$

$$= 6n^3 + 10n^2$$

$$= 6n^3 + 10n^3 \qquad (\therefore\ 10n^2 \le 10n^3)$$

$$= 16n^3$$

$$\le 17n^3$$

Here, $c = 17$, $n_0 = 1$

$\therefore\ T(n) = O(n^3)$ for $c = 17$, $n \ge 1$.

Example 7. Prove that $f(n) = n! = O(n^n)$

Solution: We know,

$$n! = n*(n-1)*(n-2)* \ldots 2*1.$$

$$= (n^2 - n)(n-2)(n-3) \ldots \qquad \text{// multiplying first two terms}$$

$$= (n^3 - 3n^2 - 2n)(n-3) \quad \ldots(2.1) \qquad \text{// multiplying first three terms}$$

$$\ldots \ldots \ldots \ldots \ldots \ldots$$

$$\le cn^n$$

$$= O(n^n).$$

2.3 BIG OMEGA (Ω) NOTATION

Big Omega (Ω) notation is exactly opposite to the Big Oh (O) notation and provides an asymptotic lower bound for the function $f(n)$ for n input instances. To improve into the best case, one instance of each input size is solved efficiently. Accordingly the algorithm is modified to test whether the input is special instance, and then output the correct answer. The best-case complexity of the algorithm is denoted by this notation. It also may not be the indicative of the actual running time of an algorithm.

Definition: The function $f(n) = \Omega(g(n))$ read as "f of n is Big Omega of g of n" is said to exist iff $\exists$ a non-negative constant $c > 0$ and initial value n_0 such that $f(n) \geq c*g(n)$, $\forall\ n \geq n_0$.

Consequently, if $\lim_{n\to\infty} \frac{f(n)}{g(n)}$ exists and it is not equal to 0, then $f(n) = \Omega(g(n))$.

Now $\Omega(g(n))$ is a set of function $f(n)$ from the non-negative integers to non-negative real numbers thus $f(n)$ is a member of $\Omega(g(n))$ *i.e.* $f(n) \in \Omega(g(n))$. For any value of $n \geq n_0$, $f(n)$ is at most c times of the function $g(n)$ *i.e.* $f(n)$ grows at least as fast as c times of $g(n)$. It is clear from the definitions that $f(n) = \Omega(g(n))$ iff $g(n) = O(f(n))$.

The Figure 2.3 given below illustrates the behaviour between $f(n)$ and $g(n)$ in terms of Big Omega (Ω) notation for the input instances n.

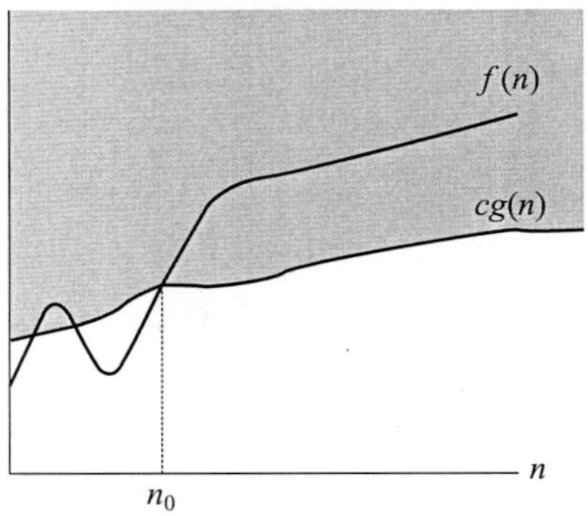

$f(n) = \Omega(g(n))$

Figure 2.3 Illustration of Big Omega (Ω) notation.

2.3.1 Big Omega (Ω): Some Examples

Some common asymptotic functions and their Big Omega value are given below.

Ecample 8. Constant Function: $f(n) = 15$

Solution: Given, $f(n) = 15 \geq 14 = 14*1$

Here, $c = 14$, $n_0 = 0$, and $g(n) = 1$. So $f(n) = \Omega(1)$

Example 9. Linear Function: $f(n) = 5n + 5$

Solution: Given, $f(n) = 5n + 5 \geq 5n$ for $n \geq 1$

Here, $c = 5$, $n_0 = 1$, and $g(n) = n$.

So, $f(n) = \Omega(n)$

Example 10. Quadratic Function: $f(n) = 7n^2 + 5n$

Solution: Given, $f(n) = 7n^2 + 5n \geq 7n^2$ for $n \geq 5$

Here, $c = 7$, $n_0 = 5$, and $g(n) = n^2$.

So, $f(n) = \Omega(n^2)$.

Example 11. Cubic Function: $f(n) = 2n^3 + n^2 + 2n$

Solution: Given, $f(n) = 2n^3 + n^2 + 2n \geq 2n^3$ for $n \geq 1$

Here, $c = 2$, $n_0 = 1$, and $g(n) = n^3$.

So, $f(n) = \Omega(n^3)$.

2.4 BIG THETA (Θ) NOTATION

Both the upper and lower bound for the function $f(n)$ for input size n are in general denoted by Big Theta (Θ) Notation. This is also called asymptotic tight bound for the functions. Because, for input size n the asymptotic tight bound is considered as the average number of operations on the overall problem instances. Thus, the average-case time complexity of an algorithm is denoted by this notation. This running time can be bounded below by $c_1 g(n)$ and above by $c_2 g(n)$ including them, where $0 < c_1 \le c_2$. This notation naturally expresses the exact order/bound on running time of an algorithm. Formally it is defined as below.

Definition: The function $f(n) = \Theta(g(n))$ read as "f of n is Big Theta of g of n" is said to exist iff ∃ two non-negative constants $c_1, c_2 > 0$ and initial value n_0 such that $c_1 * g(n) \le f(n) \le c_2 * g(n)$, $\forall\ n \ge n_0$.

Consequently, if $\lim_{n \to \infty} \dfrac{f(n)}{g(n)}$ exists and it is equal to some constant $c > 0$, then $f(n) = \Theta(g(n))$.

An important consequence can be drawn from the above definition that $f(n) = \Theta(g(n))$ iff $f(n) = O(g(n))$ and $f(n) = \Omega(g(n))$.

However, the average-case is difficult to compute. Even its meaning often arises a subject to debate. Because what does the term "average" exactly mean? Here after it is better to avoid all these complexities and readers should show their interests to obtain a very useful result by just considering the worst case.

2.4.1 Big Theta (Θ): Features

Some of the common features for big Theta (Θ) are listed below.

1. $f(n) = \Theta\ g(n))$ iff $f(n) = O(g(n))$ and $f(n) = \Omega(g(n))$.
2. The growth rate of $f(n)$ equals with the growth rate of $g(n)$.
3. This notation bounds a function to within constant factors.
4. Big Theta (Θ) means the bound is the tightest possible, etc.

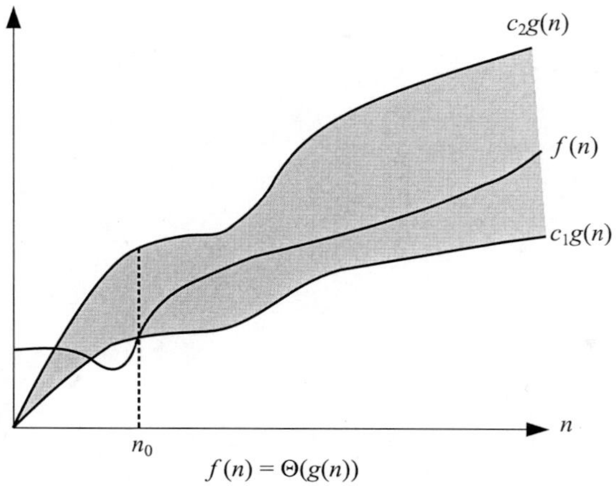

Figure 2.4 Illustration of Big Theta (Θ) notation.

2.4.2 Big Theta (Θ): Some Examples

Some common asymptotic functions and their Big Theta value are given below.

Example 12. Constant Function: $f(n) = 15$

Solution: Given, $f(n) = 15$

$\Rightarrow \quad 14*1 \le f(n) \le 16*1$

Here, $c_1 = 14, c_2 = 16, n_0 = 0$

$\Rightarrow \quad f(n) = \Theta(1).$

Example 13. Linear Function: $f(n) = 5n + 5$

Solution: Given, $f(n) = 5n + 5$

$\Rightarrow \quad 5n \le f(n) \le 6n$ for $n \ge 5$

Here, $c_1 = 5, c_2 = 6, n_0 = 5$

$\Rightarrow \quad f(n) = \Theta(n).$

Example 14. Quadratic Function: $f(n) = 7n^2 + 5n$

Solution: Given, $f(n) = 7n^2 + 5n$

$\Rightarrow \quad 6n^2 \le f(n) \le 8n^2$ for $n \ge 5$

Here, $c_1 = 6, c_2 = 8, n_0 = 5$

$\Rightarrow \quad f(n) = \Theta(n^2).$

Example 15. Cubic Function: $f(n) = 2n^3 + n^2 + 2n$

Solution: Given, $f(n) = 2n^3 + n^2 + 2n$

$\Rightarrow \quad 2n^3 \le f(n) \le 3n^3$ for $n \ge 2$

Here, $c_1 = 2, c_2 = 3, n_0 = 2$

$\Rightarrow \quad f(n) = \Theta(n^3).$

2.4.3 Big Theta (Θ) : More Examples

Example 16. Show that $(n + a)^b = \Theta(n)^b, b > 0$.

Solution: To prove the above expression, we find constants $c_1, c_2, n_0 > 0$ such that $0 \le c_1 n^b \le (n + a)^b \le c_2 n^b$ for all $n \ge n_0$.

Now, when $|a| \le n$, $\quad n + a \le n + |a|$

$$\le 2n$$

When $|a| \ge n$, $\quad n + a \ge n - |a|$

$$\ge \frac{n}{2}$$

Thus, $0 \le \frac{n}{2} \le n + a \le 2n$

$\Rightarrow \quad 0 \le (n/2)^b \le (n + a)^b \le (2n)^b \quad$ [since $b > 0$]

$\Rightarrow \quad 0 \le (1/2)^b (n)^b \le (n + a)^b \le (2)^b (n)^b$

$\therefore \quad c_1 = (1/2)^b, c_2 = 2^b$ and $n_0 = 2|a|$

$\Rightarrow \quad (n + a)^b = \Theta(n)^b$ (shown).

Example 17. Show that $f(n) = n^2 \log n$ is asymptotically bound on $\Theta(n^2)$.

Solution: Given, $f(n) = n^2 \log n$

Suppose, two constants $c_1, c_2 > 0$ and initial value on *n, i.e.* n_0 such that

$$c_1 n^2 \leq f(n) \leq c_2 n^2 \;\; \forall\, n \geq n_0$$

$$\Rightarrow \qquad c_1 n^2 \leq n^2 \log n \leq c_2 n^2$$

$$\Rightarrow \qquad c_1 \leq \log n \leq c_2$$

Now, with increase of n, $\log n$ increases. Therefore, $\log n \leq c_2 \Rightarrow n \leq 2^{c_2}$ cannot be satisfied large enough n since c_2 is constant. Up to some values of n, this inequality is correct.

Thus, $f(n) = \Theta(n^2)$.

2.5 PERFORMANCE EVALUATION

Solving a problem in large scale like developing software for any system usually a judgement of an algorithm is accomplished based on the following criteria.

(*i*) Does it fulfil our needs?
(*ii*) Is the code reusable?
(*iii*) Is the code readable?
(*iv*) Are the methodologies used logical?
(*v*) Does it work correctly as per requirements?
(*vi*) Is there any document available describing how the algorithm is implemented and works, etc?

Naturally we expect the answer "yes". Besides these criteria, most essential criteria that have more direct relationship to performance are:

(*i*) Ease of programming.
(*ii*) Time efficiency.
(*iii*) Space efficiency.
(*iv*) No ambiguity.
(*v*) Simplicity, etc.

Performance is however considered in terms of space and time complexity only. During measuring complexities of a program, following behaviours are studied:

(*i*) If the number of I/O increases.
(*ii*) If the size of I/O increases.
(*iii*) Which characteristics of the data affect complexities?

The next subsections show how this performance is evaluated.

2.5.1 Evaluation of Space Complexity

Space complexity is defined as the amount of memory an algorithm needs to run to completion. It excludes the space allocated to hold the input because size of input is basically to distinguish between algorithms that use "less than" linear space throughout their computation.

All definitions of order of growth and asymptotic bounds relevant to time complexity are legacy to space complexity. Writing algorithms into each memory cell requires at least a constant amount of time. So, space complexity cannot exceed the time complexity of the algorithm. If $T(n)$ and $S(n)$ denote the time and space complexities of an algorithm respectively, then $S(n) = O(T(n))$. This is usually measured in bytes, KB or MB and is denoted by $S(n)$ for an input of size n.

2.5.1.1 *Why Space Complexity is Necessary?*

Space complexity of algorithms is generally computed due to the following reasons.

- Programs may be run on a multi-user system in a fixed allocation scheme.
- Insufficient memory on our computer.
- Multiple solutions may have different space requirements.
- Program may handle the space complexity for an upper bound on the data.
- Program space = Instruction space + data space + stack space occupied in memory.

2.5.1.2 *Factors Affecting Space Complexity*

The space needed by an algorithm depends on the following components.

- **Fixed Part:** These are the independent characteristics of the problem.
  (*i*) Input size
  (*ii*) Output size
  (*iii*) Space for variables, etc.
- **Variable Part:** During execution of the program sometimes some additional space is required by some components. This amount of space is varied and is dependent on the problem characteristics being solved. This variable extra memory may be:
  (*i*) Space needed by reference variable.
  (*ii*) Stack space, etc.

  Thus, Space = fixed part + variable part

  Our concentration should be specially on estimating this variable part during analyzing space complexity.

Example 18. Estimate the space complexity S(P) of the problem P, algorithm of which is given below.

Algorithm: Sum (L, n)

Where L = list of elements.

n = number of elements in L.

Step 1: Start.

Step 2: Initialize sum=0

Step 3: Repeat for $i = 0$ to $n - 1$

sum = sum + $L[i]$

Step 4: Write sum

Step 5: End

From the above algorithm it is seen that space complexity S(Sum) depends on

(*i*) Size n, requires one word.
(*ii*) Counter i, requires one word.
(*iii*) Result sum, requires one word.
(*iv*) List L, requires n word.

Hence $S(\text{Sum}) \geq n + 3 = \Theta(n)$

Example 19. In linear searching, searched result is stored in only one memory cell. Adding local variables, e.g. for looping, it does not change. It is concluded that the space complexity needed is $\Theta(1)$. This is true for Binary Searching, Insertion Sorting, Selection Sorting, etc.

Example 20. Merging two sorted arrays, we need exact n amount of auxiliary storage if the input size is n. Consequently, the space complexity is $\Theta(n)$.

Example 21. Let in binary searching, middle element A[mid] of an array $A[1 \ldots n]$ is stored using an auxiliary array B after comparing it with a key $\boldsymbol{x}$ in each iteration. B can be sorted later. The number of comparisons i in this case is at most $(\log n + 1)$. Thus, space complexity, i.e. the size of B is $O(\log n)$.

Example 22. An algorithm that outputs all permutations of a given n characters needs $\Theta(n)$ space.

2.5.2 Evaluation of Time Complexity

Time complexity is the amount of time a program needs to run. This is actually sum of both compile and execution time. We may however neglect compile time of the program because a compiled program may be executed several times without every time compilation that is recompilation. And the execution time is usually measured by the number of times a statement executed. $T(n)$ notation represents time complexity of a problem of size n .

2.5.2.1 *Why Do We Care About Time Complexity?*

We do not simply run the program to measure run time but we should care about the followings too. Because:

- Some computers require upper limits for program execution times.
- Some programs require a real-time response.
- Among many solutions, the quickest one is preferable.
- Change of I/O data affect running time.

2.5.2.2 *Factors Affecting Time Complexity*

The time needed by an algorithm depends on the following components.

- **Fixed Part:** These are the independent characteristics of the problem.
  - (*i*) Input size.
  - (*ii*) Output size.
  - (*iii*) Compile time.
  - (*iv*) Computer speed, etc.
- **Variable Part:** During execution of the program, similar to space complexity, time required by some parameters is varied. These are dependent on the problem characteristics being solved. These are also unknown during the program is concerned but are reasonable to estimate during execution. These parameters are:
  - (*i*) Speed of different machines.
  - (*ii*) Characteristics of the compiler.
  - (*iii*) Number of steps executed.
  - (*iv*) Types of statement.
  - (*v*) Number of operations.
  - (*vi*) Types of operations such as addition, load, store, etc.

2.5.2.3 *Determination of Time Complexity*

The running time of a program for some particular input size n depends on the primitive operations or the steps executed. Following conventions for determination of time complexity are assumed.

(*i*) Every statement is machine independent.
(*ii*) A constant time generally one unit for a statement is assigned.
(*iii*) A statement sometimes may require more than one unit time.
(*iv*) Comment lines will not require any time unit.
(*v*) After an operation a unit time complexity is increased.

$T(n)$ of a program can be determined by the following approaches.

- Step counts – count the number of steps
- Operation counts – count the number of operations
- Asymptotic complexity.

A. Steps Count Method

The step-count method counts the time spent in all parts of the program. This can be done in two ways.

(*i*) Counter method
(*ii*) Tabular method

- ***Counter Method***

A global variable namely "COUNT" at the beginning of a program is introduced and set it initially as zero. Every time a statement is executed COUNT is incremented by one. This is an easy method that tells us how a running time changes with the change of this COUNT variable. If it is increased by 5 times then $T(n)$ is increased by 5 times and so on. Therefore the run time grows linearly with input size rather instance characteristics.

- ***Tabular Method***

The second easily understandable method to determine the steps executed of an algorithm is tabular Method. In this technique we will build a table that contains five attributes namely

(*i*) Steps.
(*ii*) Algorithm.
(*iii*) Steps per Execution represented as s/e.
(*iv*) Number of times a statement executed called frequency.
(*v*) Total number of steps executed for each statement. It is obtained by multiplying s/e and frequency. We say this as steps count.

By adding steps count in property (*v*) for all the statements the total step count of the algorithm is achieved. When we are familiar with this methodology, we need not further to consider properties (*ii*) and (*iii*), directly we may consider property (*iv*).

Example 23. Find out time complexity $T(n)$ for the following algorithm.

Algorithm: Sum(L, n)

Where L: list of elements.
n: number of elements in L.

Step 1: Start.

Step 2: Initialize $S = 0$

Step 3: Repeat for $i = 0$ to n

$S = S + L[i]$

Step 4: Write S

Step 5: End

Now, how time complexity is computed using above methods is given below.

Using Counter Method

Initially COUNT = 0

Step 1: Start

COUNT = 0 // since start does not perform any operation.

Step 2: Initialize $S = 0$

COUNT = COUNT + 1 // incremented due to initialization

Step 3: Repeat for i=0 to n

(*i*) COUNT = COUNT + 1 // counts n+1 times

(*ii*) $S = S + L[i]$

(*iii*) COUNT = COUNT + 1 // counts n+1 times

COUNT = COUNT + 1 // for last execution of loop

Step 4: Write S

COUNT = COUNT + 1 // incremented by 1

Step 5: End

COUNT = COUNT // no increment

Thus in

Step 1: COUNT = 0

Step 2: COUNT = 1

Step 3: COUNT = $1 + n + 1 + n + 1 + 1 = 2n + 4$

Step 4: COUNT = $2n + 4 + 1 = 2n + 5$

Step 5: COUNT = $2n + 5$

Hence $T(n) = O(n)$

Example 24. Consider the following algorithm Count_A() that returns count by step 2.

Algorithm: Count_A(n)

Where: $n = 2^k$, for some positive integer k.

Output: count — number of times Step *4*

Step 1: count = 0

Step 2: While $n > 1$

(*i*) Repeat for $i = 1$ to n

(*a*) count = count + 1

(*ii*) $n = n/2$

Step 3: Return count

Step 4: End

The step 2 executes $k + 1$ times, where $k = \log n$. The for loop is executed as n times, $n/2$ times, $n/4$ times, ..., 1.

$$\text{Thus } T(n) = \sum_{i=0}^{n} \frac{n}{2^i} = n \sum_{i=0}^{n} \frac{1}{2^i} = \Theta(n)$$

Example 25. Consider the following algorithm Count_B() that returns count value.

Algorithm: Count_$B(n)$

Where: $n = a$ positive integer.

Step 1: count = 0

Step 2: Repeat for $i = 1$ to n

(*i*) $m = n/i$

(*ii*) Repeat for $j = 1$ to m

(*a*) count = count + 1

Step 3: Return count

Step 4: End

The inner for loop is iterating as n, $n/2$, $n/3$ ... 1.

$$\text{Thus } T(n) = \sum_{i=1}^{n} \frac{n}{i} = n \sum_{i=1}^{n} \frac{1}{i} = \Theta(n \log n) \quad [\therefore \sum_{ni=1}^{n} \frac{1}{i} = O(\log n) \text{ and } \sum_{i=1}^{n} \frac{1}{i} == \Omega(\log n)]$$

Tabular Method

Steps	*Algorithm*	*s/e*	*Frequency*	*Step count*
1	Start	0	0	0
2	$S = 0$	1	1	1
3	Repeat for $i = 0$ to n $S = S + L[i]$	1 1	$n + 2$ $n + 1$	$n + 2$ $n + 1$
4	Write S	1	1	1
5	End	0	0	0
			Total	**$2n + 5$**

Therefore, $T(n) = O(n)$.

B. Operation Count Method

The operations may be of floating, real, complex, etc. types. Depending on types and number of operations time complexity of algorithms varies. E.g., the efficiency of a numerical method depends on floating point operations. Every operation like addition, subtraction, multiplication, division, or a square root is counted as a single operation respectively. When the data are complex then the operations become complex operations. Consider them as real operations as the data are real.

The execution time of an algorithm in fact depends not only on the operation counts but also much more on what we know as computers. For this reason, this operation count method is only used for measuring efficiency, though it may not be fair and accurate in practice. Also, we will only count the highest orders in terms of the sizes of matrices, i.e. the order of operations. Three ways to analyze time complexity in this method are:

- Worst case count that results maximum count
- Best case count that results minimum count
- Average count that lies between them.

Generally the operation-count method omits accounting for the time spent on all but the chosen operation. These three methods are described considering following examples.

Worst Case Operation Count

Consider the pseudocode

```
For ( i = 1; i < n; i++)
For (j = i – 1; j > = 0)
  a[j + 1] = a[j];
```

In outer loop i is compared n times. But inner loop iterates $n - 1$ times. Number of operations (comparisons) depends on i.

$$\text{Thus, total number of comparisons} = 1+2+3+\ldots+(n-1)$$
$$= (n-1)n/2$$
$$\Rightarrow T(n) = O(n^2).$$

Best Case Operation Count

Consider the pseudocode

```
For (i = n; i > = 0; i--)
   If(key = a[i])
     Write "key found"
```

Let $a = [1, 3, 2, 4]$ and key = 1 then number of comparisons= 1

Hence $T(n) = O(1)$

Average Case Operation Count

Consider the pseudocode

```
For (i = n; i >= 0; i--)
   If (key = a[i])
     Write success
```

Let $a = [1, 2, 3, 4]$ and key = any value ≤ 4 then number of comparisons may be 1, 2, 3, or 4.

In general for $a = [1, 2, 3, 4\ldots, n]$ and t = any value $\leq n$, then total number of comparisons will also be $\leq n$. Hence $T(n) = O(n)$.

C. Asymptotic Complexity

Asymptotic analysis means a line that tends to converge to a curve which may or may not touch the curve. This line is defined within a bound. Different growth rate functions for varying input size are generally considered. Among them clearly the better one is chosen. In three ways it can be done. Already we have discussed these three techniques namely Big Oh, Big Omega, and Big Theta notations.

➢ Some Rules for Asymptotic Complexity

Suppose two programs have complexities $c_1n^3 + c_2n^2 + c_3n$ and c_4n^2 respectively

- The program with complexity c_4n^2 will be faster than the first one for sufficiently large values of n due to low power.
- For small values of n, either program could be faster and depends on the values of constants c_1, c_2, c_3, and c_4c_1.

➢ A drawback of asymptotic analysis

Even though an algorithm with lower growth rate is assumed as faster than an algorithm with higher growth rate when input size is very large, but it cannot be demanded that the algorithm with lower growth rate is always better.

➢ Some Common Growth Rate Functions

Here is a list of some common growth rate functions given below.

- Constant (1): growth is independent of the problem size n.
- Logarithmic (log n): growth increases slowly compared to the problem size, *e.g.*, binary search
- Linear (n): directly proportional to the size of the problem.
- $n \log n$: some Divide-and-Conquer approaches *e.g.,* Merge sorting
- Quadratic (n^2): typical in nested loops
- Cubic (n^3): three nested loops
- Exponential (2^n): growth is extremely rapid and possibly impractical.

Growth rates of these common functions are measured in nanoseconds.

➢ Some Common Asymptotic Functions Values

log n	n	$n \log n$	n^2	n^3	2^n
0	1	0	1	1	2
1	2	2	4	8	4
2	4	8	16	64	16
3	8	24	64	512	256
4	16	64	256	4096	65,536
5	32	160	1024	32,768	4,294,967,296

Figure 2.5 Values of common asymptotic functions.

➢ **Plots of Some Common Asymptotic Functions**

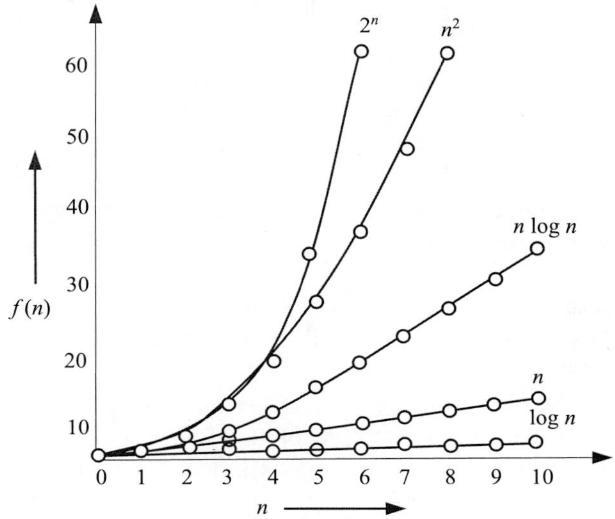

Figure 2.6 Plots of some common asymptotic functions.

➢ ***Points to Remember***

Many problems come across time-space tradeoff. More space allocation makes the algorithm the faster and vice versa. Obviously this is true within limits. Because in most of the algorithms we see that increasing the amount of space does not accelerate noticeably the run times. Oppositely, it is almost always the case that decreasing the amount of space required by an algorithm results degradation in its speed.

2.6 COMPLEXITY OF WELL KNOWN ALGORITHMS

Now we are at the point to analyze some well known and easy algorithms. Using the above methods of finding time complexity we will derive time complexities of the algorithms and show these values are the values as we have known till now without any mathematical computation. In this section mainly following are discussed.

(*i*) heap sort

(*ii*) searching

- linear
- binary

2.6.1 Heap

Heaps are the almost complete or nearly complete or complete binary tree that satisfies the following two conditions/properties.

(*i*) All the leaves are at the same level or the difference between the levels is maximum one.

(*ii*) All the nodes at the lowest level are as far to the left as possible.

These properties are called structural property of that kind of tree.

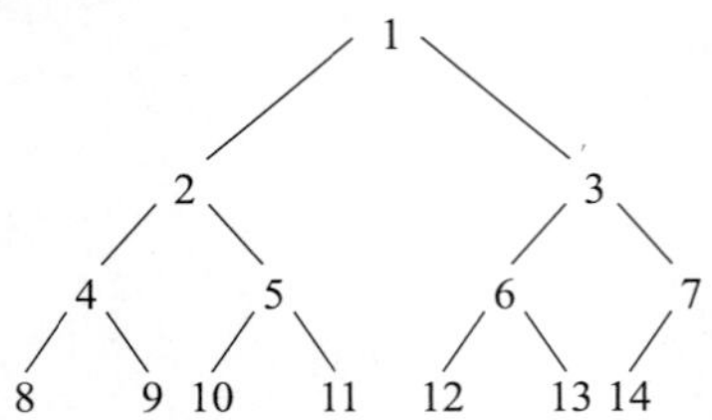

Figure 2.7 Almost complete. binary tree.

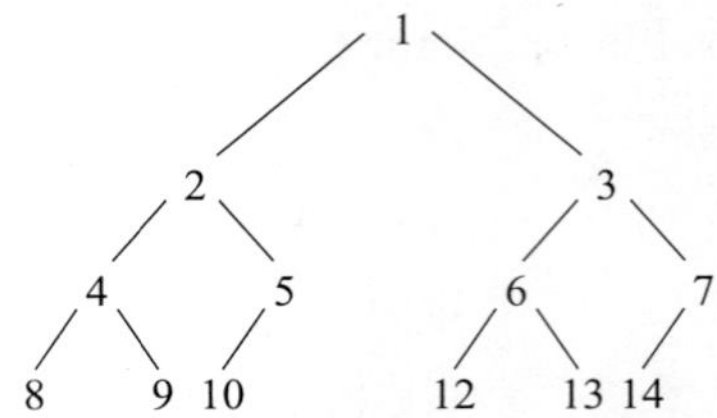

Figure 2.8 Not almost complete binary tree.

Definition: A heap tree or simply heap can be defined as an almost complete binary tree in which the tree is either empty or satisfies the following property.

The value of a node in the tree is either

(*i*) Greater than or equal to the value of its children.

(*ii*) Smaller than or equal to the value of its children.

That means $N_i \geq N_j$ or $N_i \leq N_j$ for $2 \leq j \leq n$, and $i = j/2$, *i.e.* for node x, either parent $(x) \geq x$ or parent$(x) \leq x$ respectively. These properties are called order or heap properties of heap trees. Figure 2.9 is a heap tree whereas Figure 2.10 is not because it does not satisfy structural property of almost complete binary tree.

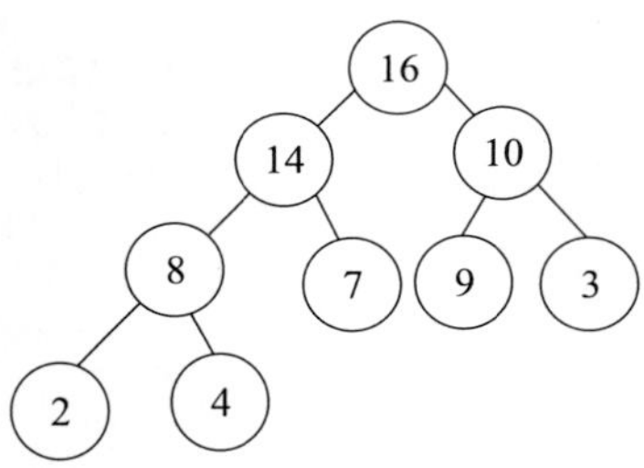

Figure 2.9 A heap tree.

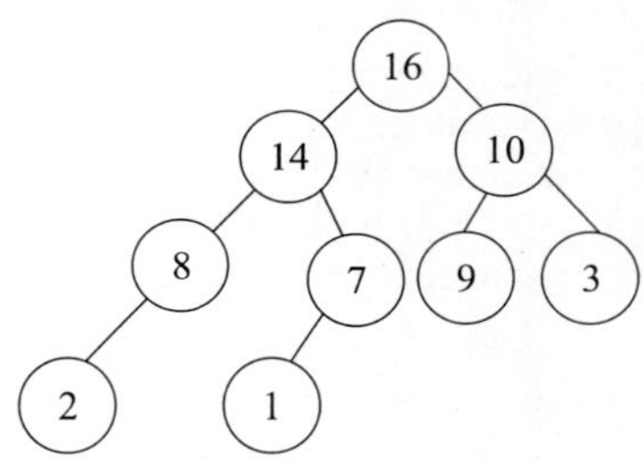

Figure 2.10 A binary tree but not a heap tree.

2.6.1.1 *Heap : Types*

Based on the heap properties it is seen that there are two types of heap tree possible.

(*i*) Max-heap

(*ii*) Min-heap

- ***Max-heap:*** A heap tree is said to be maximum heap tree or simply max-heap if the value of a parent node is greater or equal than that of its children. For a node x, parent$(x) \geq x$. Hence, the first property yields max-heap. Figure 2.11 for example is a max-heap tree.
- ***Min Heap:*** A heap tree is said to be minimum heap tree or simply min-heap if the value of a parent node is less or equal than that of its children. For a node x, parent$(x) \leq x$. Hence, the second heap property yields min-heap. Figure 2.12 for example is a min-heap tree.

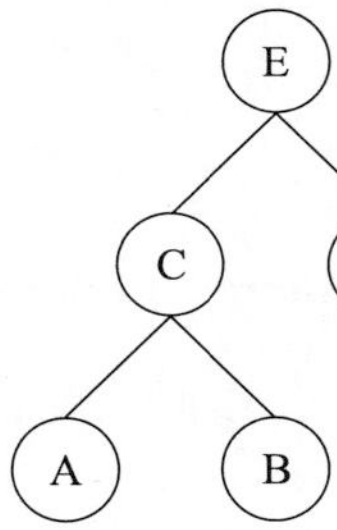

Figure 2.11 Max-heap.

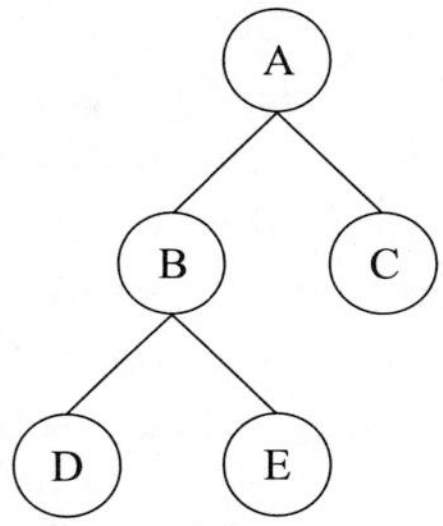

Figure 2.12 Min-heap.

2.6.1.2 *Heap : Representation*

A heap can be represented by both array and linked representation. Here only array representation is shown.

- Store all the nodes of the heap into an array L with the following conditions.
  - Root of the heap should be first element of L, *i.e.* $L[1]$ = Root
  - Left child of a node should be at $L[2i]$ position if it is at $L[i]$.
  - Right child of a node should be at $L[2i + 1]$ position if it is at $L[i]$.
  - Parent of $L[i]$ should be at $L[\lfloor i/2 \rfloor]$
  - Size of Heap $\leq |L|$
- The elements in the sub array $L[(\lfloor n/2 \rfloor+1) .. n]$ are leaves where $n = |L|$.
- The root is the maximum element of the heap for max-heap and is the minimum element of the heap for min-heap.

Consider Figure 2.13. The corresponding array representation will be as in Figure 2.14.

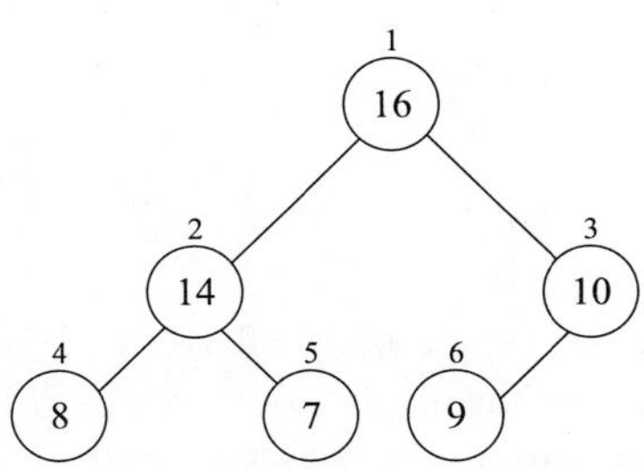

Figure 2.13 A max-heap.

index	1	2	3	4	5	6
node	16	14	10	8	7	9

Figure 2.14 Array representation of max-heap.

A heap is therefore a binary tree that is filled in order.

2.6.1.3 *Building Heap*

Creating a heap depends on the type that we want to construct. The construction is started with an element taken from a list of elements given. This element at the moment behaves as root of the tree and it is a heap one. Lets the heap we need to build is max-heap. The next element is inserted to the previous heap maintaining property of the almost complete binary tree. The inserted node is now compared with its parent node. An exchange is necessary whenever max-heap property is maintained. Next element from the list is then inserted. The process is continued. Maintaining the heap property (max or min) under consideration is termed as heapify (max-heapify or min-heapify). Thus, after insertion of an element at the bottom of the heap, we compare it with its parent, grandparent, great grandparent, and so on until heapify.

Algorithm: Build_Max-Heap(L, n)

Where L = list of elements given.
n = size of L.

Step 1: Repeat for $k = 2$ to n through step 4. // build heap

Step 2: Initilize
(*i*) $i = k$
(*ii*) temp = $L[k]$
(*iii*) $j = i/2$

Step 3: Repeat while $i > 1$ and temp > $L[j]$ //Find new parent
(*i*) $L[i] = L[j]$
(*ii*) $i = j$
(*iii*) $j = i/2$
(*iv*) If $j < 1$
• $j = 1$

Step 4: Initialize
$L[i]$ = temp

Step 5: Exit.

Analysis: The above algorithm can be analyzed as- step 1 iterates $n - 1$ times, *i.e.* $O(n)$. again comparison goes through the height of the tree $O(h)$ just built in every insertion. Now h varies through log n. Thus the running time of the Built_Max-Heap() algorithm is $O(n)$. We can express the total cost as

$$\sum_{h=0}^{\lfloor \log n \rfloor} \left\lceil \frac{n}{2^{h+1}} \right\rceil O(h) = O\left(n \sum_{h=0}^{\lfloor \log n \rfloor} \frac{h}{2^h} \right) \qquad \left[\because \sum_{h=0}^{\infty} \frac{h}{2^h} = \frac{1}{1-\frac{1}{2}} = 2 \right]$$

$$= O(n)$$

Hence, we can build a heap from an unordered list in linear time.

Example 26. Construct max-heap using the following list $L = < 3, 5, 4, 7, 6, 8, 9 >$.

Solution: Initially the tree is empty. Now the max-heap can be constructed step-by-step as given below.

(*a*) Insert first element 3 into empty tree. Now it is the new parent.

(*b*) Add next element 5 as left child to 3. Compare new child with its parent 3. Now 5 > 3, so exchange 3 and 5.

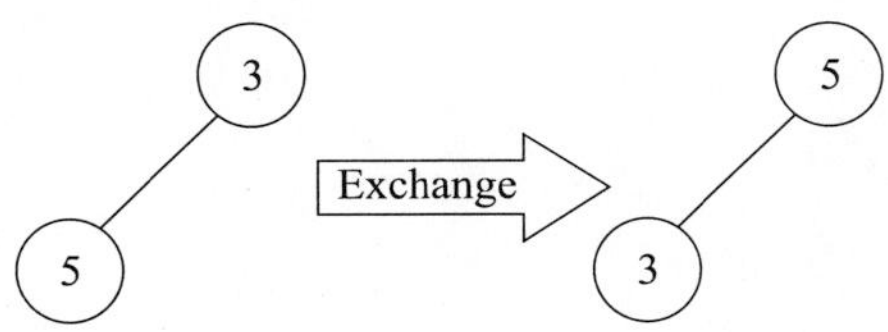

(*c*) Add next element 4 as right child to 5. No exchange is required.

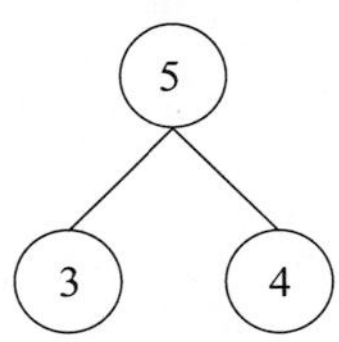

(*d*) Add 7 as left child of 3. Exchange it with 3 followed by 5 so that 7 becomes the root.

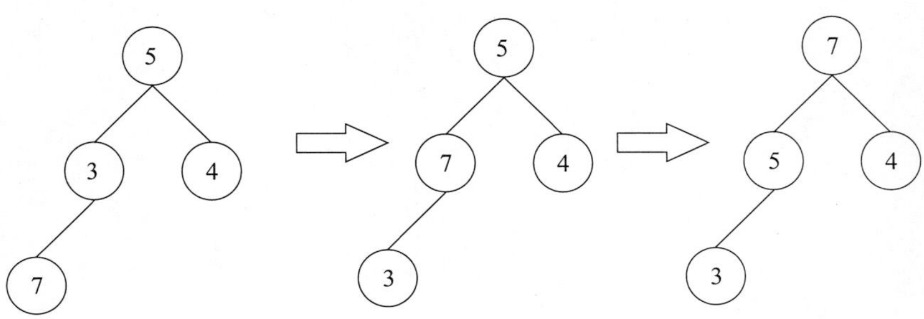

(*e*) Add 6 as right child to 5. Exchange with 5.

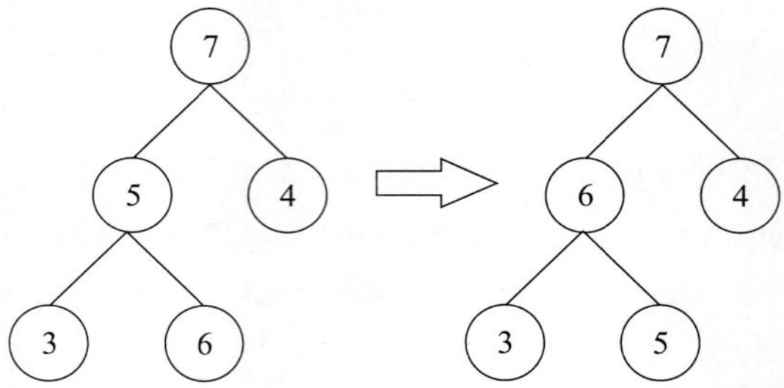

(*f*) Add 8 as left child to 4. Compare and exchange with 4 followed by 7. Now 8 becomes the root.

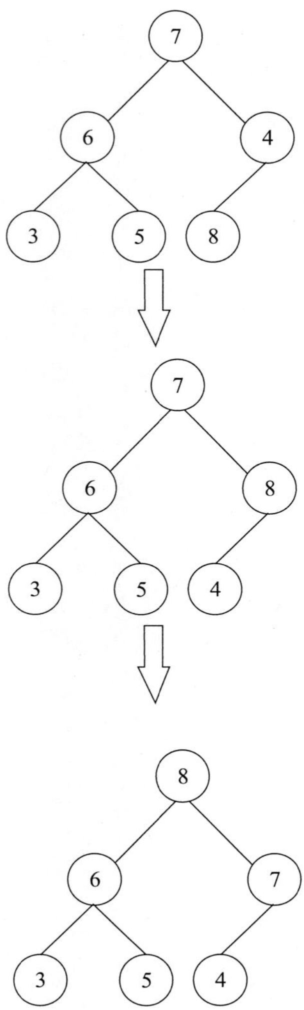

(*g*) Add 9 as right child to 7. Compare and exchange it with 7 and 8 so that it becomes the root.

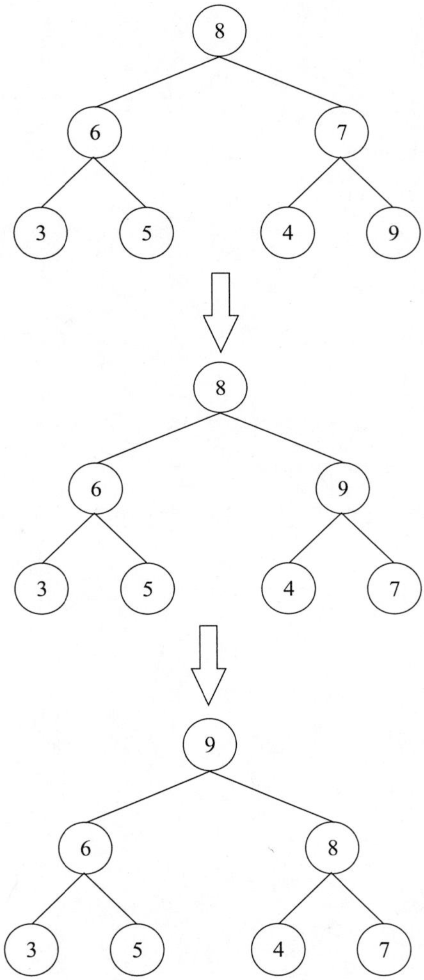

This is our Max-Heap tree.

2.6.1.4 *Heap Sort*

Heap sort starts with creating a heap (max or min) that depends on the nature of sorting, *i.e.* ascending or descending order. If a list of elements are to sort in ascending order naturally it is advisable to construct max-heap. Sorting starts by exchanging the root with last leaf at last level because the maximum element is the root. Then the tree is heapified except this new last node to be max-heap once more. Now the new root is exchanged with the second last node. Reheapify the current tree if necessary. Process is continued until all elements remain in their proper places.

Algorithm: Heap_Sort(L)

Where L = list of n elements.

Start 1: Call Build_Max-Heap(L, n)

Step 2: Repeat for $k = n$ to 2 through step 7

Step 3: Initialization and Exchange

a. temp = $L[1]$

b. $L[1] = L[k]$

c. $L[k]$ = temp

d. Root = $L[1]$

e. $i = 1$

f. $j = 2$

Step 4: Find index of new element

a. If $(j + 1) < k$

- If $L[j+1] > L[j]$
  - $j = j + 1$

Step 5: Repeat while $j \leq k-1$ and $L[j]$ > Root through step 7// reheapify

Step 6: Initialization

a. $L[i] = L[j]$

b. $i = j$

c. $j = 2*i$

Step 7: Find index of new largest child

a. If $(j + 1) < k$

- If $L[j+1] > L[j]$
  - $j = j + 1$
- Else If $j > n$
  - $j = n$
  - $L[i]$ = Root

Step 8: Exit

Analysis: The running time of the sorting algorithm is basically based on the construction of heap and reheapifying after exchange operation during sorting. Naturally, complexity depends on these operations. Now already it is seen that there are $n - 1$ iterations and every time it works on the tree of maximum height log n. Now construction of heap takes time $O(n)$ and reheapifying takes time $O(\log n)$. Therefore the time complexity of the heap sort algorithm is $O(n \log n)$.

Example 27. Apply heap sorting on the following list: < 3, 5, 4, 7, 6, 8, 9 >.

Solution: Corresponding max-heap of the list is given below.

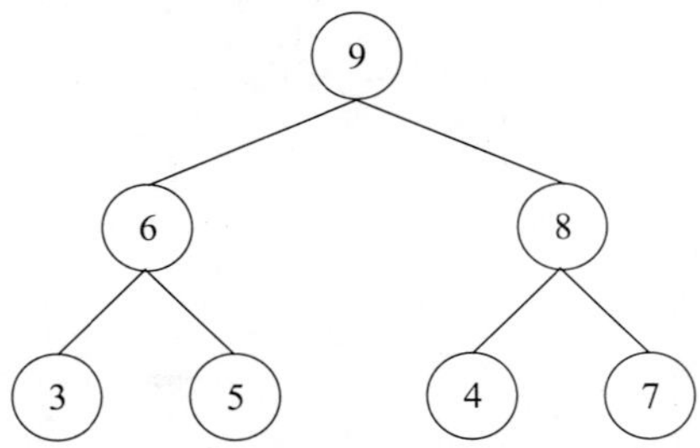

Now the sorting algorithm is applied on the above tree step-by-step as.

(*a*) Exchange root node 9 with last node 7.

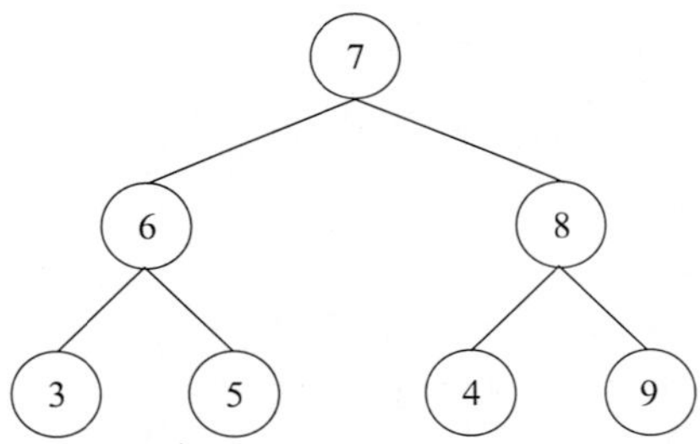

(*b*) Reheapify this new tree except the last node 9. After heapifying the tree will be as.

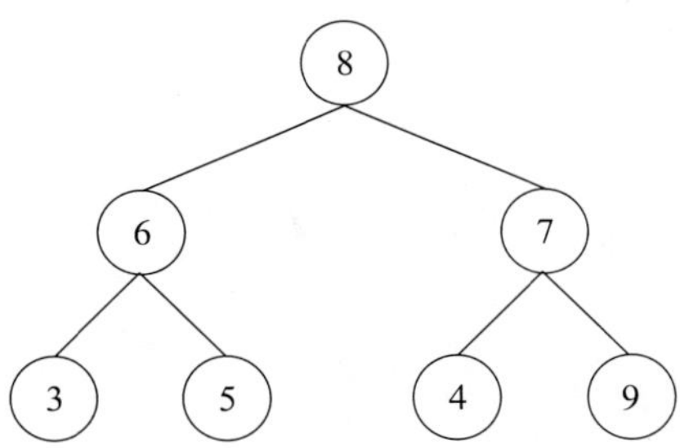

(*c*) Exchange root node 8 with last node 4.

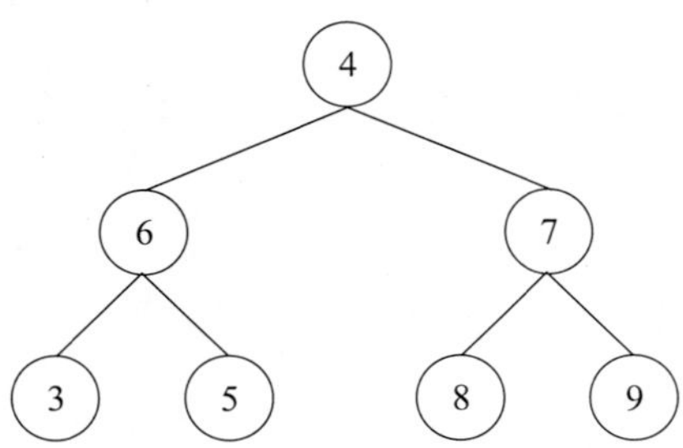

(*d*) Reheapify this new tree except the last node 8.

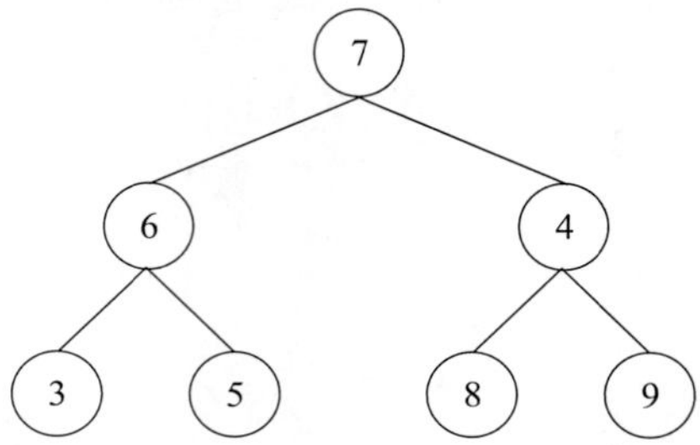

(*e*) Exchange root node 7 with last node 5.

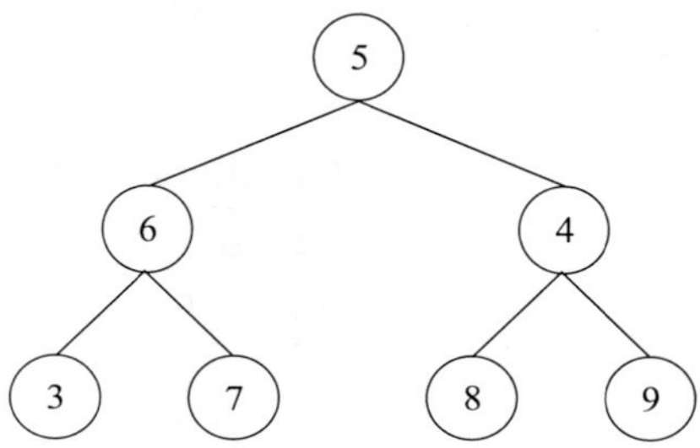

(*f*) Reheapify this new tree except the last node 7.

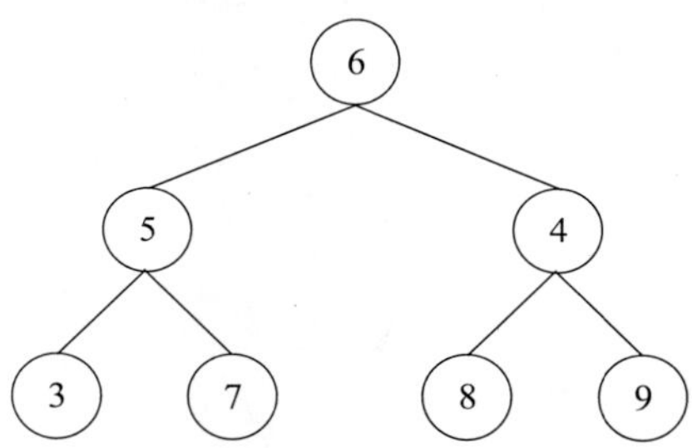

(*g*) Exchange root node 6 with last node 3.

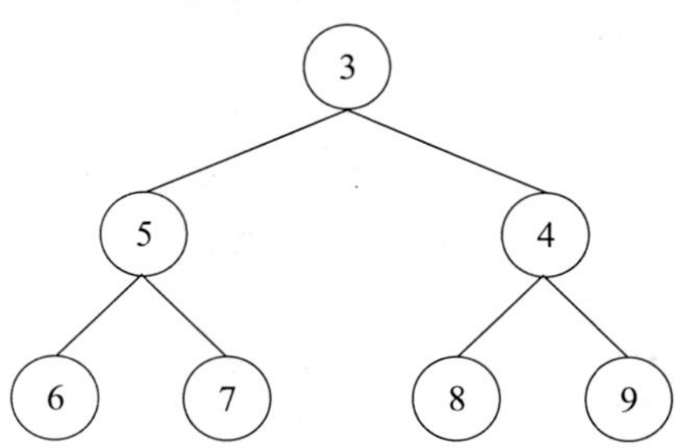

(*h*) Reheapify this new tree except the last node 6.

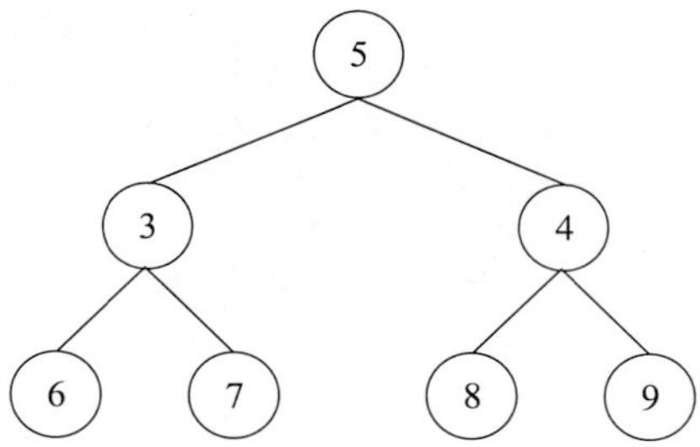

(*i*) Exchange root node 5 with last node 4.

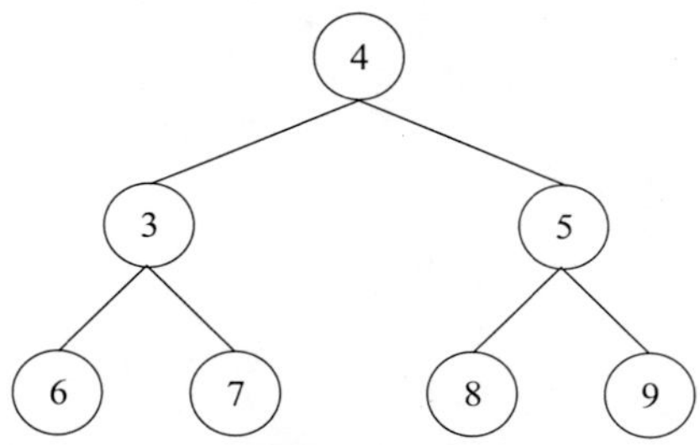

(*j*) Reheapify this new tree except the last node 5. Here it is not necessary.

(*k*) Exchange root node 4 with last node 3.

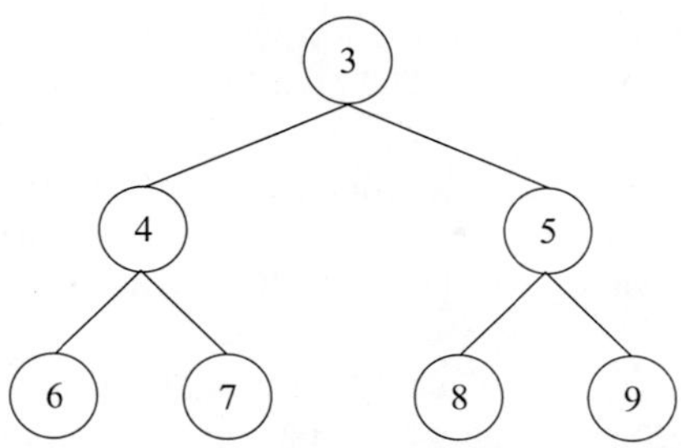

(*l*) Only one node, the root 3 remains. Thus the list obtained is in ascending order.

2.6.2 Searching

Searching is a technique that helps to find an element from a list of elements. The list may be in sorted or unsorted order. During finding an element called key element in the list, the key on the other hand may be present or may not be present in the list. Different searching techniques that are in our hand are listed below.

(*i*) Linear searching.

(*ii*) Binary searching.

(*iii*) Fibonacci searching.

In the subsequent section, our most familiar linear and binary searching techniques have been discussed.

2.6.2.1 *Linear Searching*

The technique is mainly applied to find out key element from an unsorted list of elements. Searching starts from the beginning by comparing every element of the list with the key. If matches found then further searching is stopped. It may be proceeding to the end of the list to verify if there exists more than once the same key element.

Algorithm: Linear_Search (L, n, d)

Where L: list of elements.
n: number of elements in L
d: key element.

Step 1: Start

Step 2: (*a*) Repeat for $i = 1$ to n // take input L & d
- Read $L[i]$.

(*b*) Read d.

Step 3: (*a*) Repeat for $i = 1$ to n

(*b*) If $d = L[i]$

(*i*) write successful

(*ii*) Exit

Step 4: End

Analysis: Clearly if step 2 may be ignored. The time complexity for linear searching is analyzed below.

- Best Case: If the key is present always in the first position of the list, we need only one comparison. Then it returns best case time complexity and it is $\Omega(1)$.
- Worst Case: If the key is present always in the last position of the list, then the list is traversed till its end. Number of comparisons in this case is maximum and is equal to list size n. Thus, it returns worst case time complexity and it is $O(n)$.
- Average Case: In case of average case analysis, the key may be present at any position in the list, *i.e.* between first and last position. Obviously it is equally likely to occur. Therefore, average time complexity in terms average number of comparisons is calculated as:

$$\begin{aligned} T(n) &= 1/n.1 + 1/n.2 + 1/n.3 + \ldots\ldots\ldots\ldots + 1/n.n \\ &= (1 + 2 + 3 + \ldots\ldots\ldots\ldots + n).1/n \\ &= n(n+1)/2.1/n \\ &= (n+1)/2 \qquad [\ 1/n = \text{prob. of occurrence of an element in } L] \\ &= \Theta(n) \end{aligned}$$

2.6.2.2 *Binary Searching*

Binary search algorithm sometimes called binary chop is a searching technique for searching an element called key from a sorted list in much faster time over the linear searching.

The searching starts comparing key with middle element of the sorted list. If this middle element is key element then searching is successful. If not, the sorted list is divided into two sub lists of equal or almost equal size. If key is less than the middle element of the sorted list then first sublist (all elements ≤ key) is considered as the next sorted list to be searched. Otherwise, second sublist (all elements ≥ key) is considered as the next sorted list to be searched. The process is continued until the key value is found or the searching interval is empty.

Algorithm: Binary_Search(L, min, max, d)

Where L: list of elements.

min: lowest index of $L = 1$.

max: highest index of L = number of elements in $L = n$

d: key element.

Step1: (a) Repeat while min ≤ max

- mid = (min + max) / 2 //integer division

(b) If $d = L$[mid]

- Write "searching successful".
- Return mid.

(c) Else If $d < L$[mid])

- max = mid – 1 // change upper bound

(c) Else

- min = mid + 1 // change lower bound

Step 2: Return Null

Step 3: Exit

Analysis: In each of iterations the key is compared with middle element of the list L. This operation requires maximum Θ(1) time. If this middle element is not the key element, then the algorithm iterates with either lower half or upper half of L. Number of iterations, *i.e.* maximum number of comparisons will be equal to minimum height of the binary tree if we construct with this L. Now if there are n elements in L then minimum height will be log n. Total time for log n comparisons is log n times of Θ(1), *i.e.* Θ(log n).

Hence, best case time complexity = Θ(1) since in first iteration key is found. Worst case time complexity = Θ(log n) as maximum iterations are performed. Average case time complexity is Θ(log n) as key may be present at any height of the tree, *i.e.* it requires average number of iterations.

Example 28. Consider the following sorted list L of 14 elements. Suppose key = 69. Search the using binary searching.

1	2	3	4	5	6	7	8	9	10	11	12	13	14
12	18	22	31	34	40	46	59	67	69	72	82	84	98

Searching the key = 69 proceeds in the following way.

Min = 1
Max = 14
Mid = (1 + 14)/2 = 7
L[Mid] = L[7] = 46 ≠ key, L is subdivided at Mid position.

Since key > L[Mid], searching is continued on second sub list as given below.

8	9	10	11	12	13	14
59	67	69	72	82	84	98

Now, Min = Mid + 1 = 7 + 1 = 8
Max = 14
Mid = (8 + 14) / 2 = 11
L[Mid] = L[11] = 72 ≠ key, above list is subdivided at this Mid position.

Since key < L[Mid], searching is continued on first sub list as given below.

8	9	10
59	67	69

Now, Min = 8
Max = Mid – 1 = 11 – 1 =10
Mid = (8 + 10) / 2 = 9
L[Mid] = L[9] = 67 ≠ key, above list is subdivided at this Mid position.

Since key > L[Mid], searching is continued on second sub list as given below.

10
69

Now, Min = Mid + 1 = 9 + 1 = 10
Max = 10
Mid = (10 + 10)/2 = 10
L[Mid] = L[10] = 69 = key.

Since key = L[Mid] = 69, searching is thus completed.

2.7 SOME COMMON QUESTIONS

Question 1. What are the minimum and maximum numbers of elements in a heap of height h?
Answer: A heap is a complete or almost complete binary tree, so the minimum and maximum number of elements in a heap of height h is 2^h and $2^{h+1} – 1$ respectively.

Question 2. Is the sequence < 23, 17, 14, 6, 13, 10, 1, 5, 7, 12 > a max-heap?
Answer: An almost complete binary tree of the sequence is constructed below.

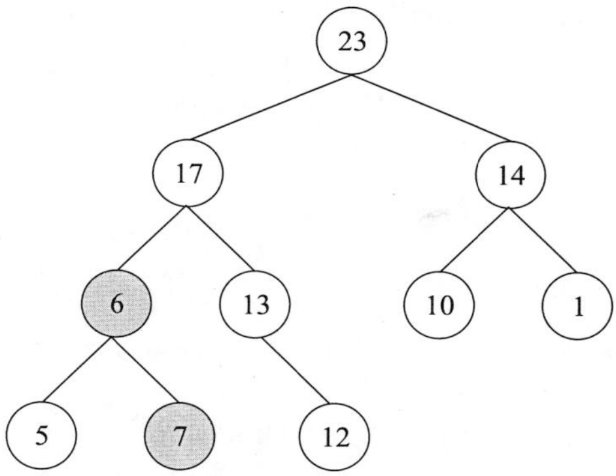

The tree above is not a max-heap, because 7 is a child of 6, but 7 > 6.

Question 3. How time complexity is determined?

Answer: Total time T of an algorithm has two parts. $T = C + N \times R$ where C is compile time, R is run time and N is a constant that indicates number add, sub, mul etc. operations.

Question 4. Make a difference between Compile and Running Time.

Answer: Difference can be made as below.

Compile Time	***Running Time***
1. Compile time refers to the amount of time required to compile a program. 2. The operations performed at compile time usually include syntax analysis, various kinds of semantic analysis and code generation. 3. Compile time occurs before link time and runtime.	1. Running Time refers to the time required for execution of a program. 2. Type checking, storage allocation, and even code generation and code optimization may be done at compile-time or upon a runtime, depending on the language and compiler. 3. Running time occurs after compile time.

Question 5. What are the various types of asymptotic notations used for complexity of an algorithm?

Answer: Generally five asymptotic notations are used.

(*a*) Big Oh (O) notation
(*b*) Big Omega (Ω) notation
(*c*) Big Theta (Θ) notation
(*d*) Little Oh (o) notation
(*e*) Little Omega (ω) notation

Question 6. What are the difference between Best Case and Worst Case Complexities?

Answer: The best case complexity of an algorithm is its efficiency for the best case for input of size N, An algorithm runs fastest among all possible inputs of that size.

The worst case complexity of an algorithm is its efficiency for the worst case input of size N, An algorithm runs longest among all possible inputs of that size.

Question 7. What are the common Running Time values for Big Oh (O) notation of input N?

Answer: The common running time for Big Oh notation with input size N

(*a*) O(N)

(*b*) O(log N)

(*c*). O(N log N)

(*d*). $O(N^C)$, C is constant > 1.

Question 8. What are the Best, Worst and Average Case complexity of Linear Search for *n* elements?

Answer: Best Case – O(1), Average Case – O(N), Worst Case – O(N)

Question 9. Define Linear Search.

Answer: Linear search is a method for finding particular element one by one and in sequence in a list until the desired element is found. Searching always starts from the beginning of the list.

Question 10. What is the height of a heap of *n* elements?

Answer: .Suppose the height of the heap is *h*. then by definition,

$$2^h \le n \le 2^{h+1} - 1 < 2^{h+1}.$$

$$\Rightarrow \quad h \le \log n \le h + 1$$

$$\Rightarrow \quad h = \lfloor \log n \rfloor$$

Question 11. How many elements are in a heap?

Answer: Since a heap is an almost complete binary tree, it has at most $2^h - 1$ elements and at least $2^h - 1 + 1 = 2^h$ elements.

Exercises

1. Show that the following functions are correct:

   (*a*) $f(n) \ge 5n^2 - 6n = \Theta(n^2)$

   (*b*) $f(n) = n! = O(n^n)$

   (*c*) $f(n) = 2n^2 2^n + n \log n = \Theta(n^2 2^n)$

2. Rewrite the following functions in increasing order of magnitude.

   (*a*) n^2

   (*b*) $n \log n$

   (*c*) $n-1$

   (*d*) $n/\log n$

   (*e*) $2^{\log n}$

   (*f*) 2^n **[Ans.** $n-1 < n/\log n < 2^{\log n} < n \log n < n^2 < 2^n$]

3. Express the following functions in big Oh (O) notation.

   (*a*) $f(n) = n^3/1000 - 100n^2 - 100n + 3$

   (*b*) $f(n) = \log n + 100$

(c) $f(n) = 10\log^2 n + 100$

(d) $f(n) = \sum_{i=0}^{n} i$

4. Show that the following inequalities are incorrect:

(a) $f(n) \geq 5n^2 - 6n = O(n^2)$

(b) $f(n) = n! = \Theta(n^n)$

(c) $f(n) \geq 2n^2 2^n + n \log n = \Omega(n^2 2^n)$

(d) $f(n) = 10n^2 + 9 = O(n)$

(e) $f(n) \geq n^2 \log n = \Theta(n^2)$

5. Prove that: if $f(n) = O(F(n))$ and $g(n) = O(G(n))$, then $f(n)/g(n) = O(F(n)/G(n))$.

6. Prove the following statement: for all functions $f(n)$ and $g(n)$, either $f(n) = O(g(n))$ or $g(n) = O(f(n))$.

7. If $f(n) = O(g(n))$ then does it imply that $2^{f(n)} = O(2^{g(n)})$?

8. Which one of the O and 0 notations is more appropriate to express the time complexity of the algorithm?

9. Show that $(\log(\log n))!$ is polynomial bounded while $(\log n)!$ is not.

10. Which are true:

(a) $n^2 \in O(n^3)$

(b) $n^2 \in \Omega(n^3)$

(c) $2^n \in \Theta(2^{n+1})$

11. Define the followings:

(a) Ω – Notation.

(b) Θ – Notation.

(c) O – Notation.

12. How is time complexity of an algorithm measured? Discuss different methods.

13. Prove that: $2^{n+1} = O(2^n)$, but $2^{2n} \neq O(2^n)$.

14. Make the difference between $O(1)$ and $\Theta(1)$.

15. Express the following functions in terms of Θ-notation.

(a) $f(n) = 2n + 31og^{100}n$.

(b) $f(n) = 7n^3 + 1000n \log n + 3n$.

(c) $f(n) = 3n^{1.5} + \sqrt[3]{n} \log n$

(d) $f(n) = 2^n + 100^n + n!$.

(e) $f(n) = 18n^3 + \log n^8$.

(f) $f(n) = (n^3 + n)/(n + 5)$.

16. Express the following functions in terms of the O-notation.

(a) $f(n) = 18n^3 + \log n^8$.

(b) $f(n) = (n^3 + n)/(n + 5)$.

(c) $f(n) = \log^2 n + \sqrt{n} + \log \log n$.

(d) $f(n) = n!/2^n + n^{n/2}$.

(e) $f(n) = 7n^3 + 1000n \log n + 3n$.

(f) $f(n) = 3n^{1.5} + \sqrt[3]{n} \log n$

(g) $f(n) = 2^n + 100^n + n!$.

17. Express the following functions in terms of the Ω-notation.

(*a*) $f(n) = 18n^3 + \log n^8$.

(*b*) $f(n) = (n^3 + n)/(n + 5)$.

(*c*) $f(n) = \log^2 n + \sqrt{n} + \log \log n$.

(*d*) $f(n) = n!/2^n + n^{n/2}$.

(*e*) $f(n) = 7n^3 + 1000\, n \log n + 3n$.

(*f*) $f(n) = 3n^{1.5} + \log n$

(*g*) $f(n) = 2^n + 100^n + n!$.

18. Suppose $f(n) = \{1/n + 1/n^2 + 1/n^3 + \ldots\}$. Express $f(n)$ in Θ-notation.
(Hint: Find a recursive definition of $f(n)$).

19. Show that: $n^{100} = O(2^n)$, but $2^n \neq O(n^{100})$.

20. Prove that: $2^n \notin \Theta(3^n)$.

21. Prove your answer.

(*a*) Is $n! = \Theta(n^n)$?

(*b*) Is $2^{n^2} = \Theta(2^{n^3})$?

(*c*) Is $\log n! \in O(n), \Theta(n)$, or $\Omega(n)$?

22. Given a set of integers. Give an efficient algorithm that determines whether two of them are equal.

23. Write a suitable procedure that evaluates the following polynomial.
$F(x) = a_n x^n + a_{n-1} i x^{n-1} + \ldots + a_1 x + a_0$, in $\Omega(n^2)$ and $O(n)$ time.

24. Let $L[1..90]$ = 11,12, ..., 100. How many comparisons are performed by Binary Search Algorithm when searching for the following values

(*a*) 33.

(*b*) 7.

(*c*) 70.

(*d*) 77.

25. Let $A[1..2000]$ = 1,2, ...,2000. How many comparisons are performed by Binary Search Algorithm when searching for the following values

(*a*) –3.

(*b*) 1.

(*c*) 1000.

(*d*) 3000.

26. Draw the decision tree for the binary search algorithm with an input of

(*a*) 12 elements.

(*b*) 15 elements.

(*c*) 20 elements.

(*d*) 35 elements.

27. What is heap? Construct a heap with the following list of elements
< 12, 3, 5, 2, 15, 20, 45, 25, 50, 13, 23, 33, 40 >.

28. Discuss heap sorting algorithm with an example.

29. Analyze heap sorting algorithm.

30. Illustrate the operation of Max-Heapify on the array
A = < 7, 17, 3, 16, 15, 11, 13, 10, 1, 5, 7, 12, 4, 14, 8, 9, 10, 18.

31. Show that the worst case time of Max-Heapify on a heap of size n is $\Omega(\log n)$.

32. Show that the worst case time of Heapsort on a heap of size n is $\Omega(n \log n)$.

33. Give an algorithm to construct a priority queue using heap with an example.

34. What are the merits and demerits of implementing a priority queue using heap?

35. Which of the following arrays are heaps?

(*a*) 8 6 4 5 3 7 2.

(*b*) 9 4 8 3 2 5 7 8.

(*c*) 1 2 3 4 5 6 7 8 9.

(*d*) 9 7 5 6 3 4 7 2 1 6 5 3.

36. Where do the following elements reside in a heap?

(*a*) Largest.
(*b*) Second largest.
(*c*) Third largest.
(*d*) Minimum.
(*e*) Second minimum.

37. How can you test whether a list of n elements is a heap? Give an efficient algorithm. What is the time complexity of your algorithm?

38. In a heap both operations – insertion and deletion have the same time complexity $O(\log n)$. Then which operation is more costly? Justify your answer.

39. Let H be the heap shown in Figure 2.15. What will the heap result when

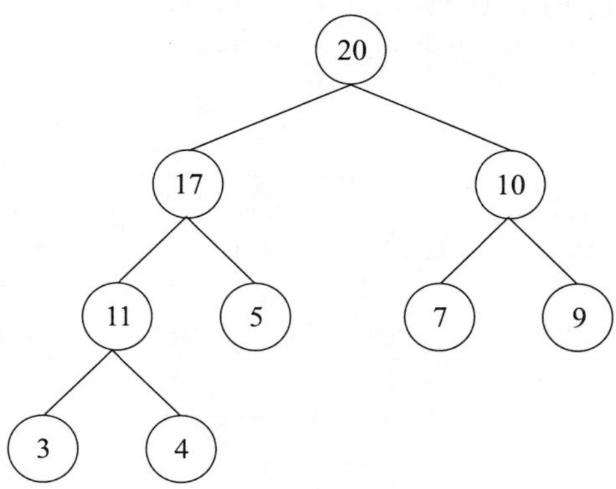

Figure 2.15 A max-heap tree.

(*a*) Deleting the element with key 17.
(*b*) Inserting an element with key 15.
(*c*) Inserting an element with key 19.
(*d*) Deleting the element with key 20.

40. Show the heap in array representation that results from deleting the maximum key in the heap shown in Figure 2.15 of exercise 39.

41. How fast is it possible to find the minimum key in a max-heap of n elements?

42. Illustrate the steps to make a heap on the array: 3 7 2 1 9 8 6 4 11 13 5.

43. Show the steps of transforming the following array into a heap: 4 3 11 2 5 1 7 6 8 3 1 6.

44. How does the Heap sorting algorithm behave when the input array is already sorted in
    (*a*) Increasing order.
    (*b*) Decreasing order.
45. Construct a binary search tree with heap property.
46. Merge two heaps of the same size into a single heap with designing an algorithm. What is the time complexity of your algorithm?
47. Compute minimum and maximum number of element comparisons performed when running Heap sorting algorithm.

University Questions

1. Write short notes on Asymptotic notation.
   [WBUT'08, UPTU'06, '09, '10, RTU'11, MDU'08, '09, JNTU'08]
2. What is the need of analysis of algorithms? Discuss various notations in brief.
   [RTU'08, RTU'11, JNTU'08, IGNOU'09]
3. Differentiate between profiling and debugging. ***[JNTU'08]***
4. Describe the performance analysis in detail. ***[JNTU'08]***
5. Explain why input size is an important consideration for algorithm complexity.
   [BPUT'05]
6. Explain the importance of asymptotic analysis for running time of an algorithm.
   [IGNOU'09]
7. An algorithm runs a given input of size n. When n is 4096 the run time is 512 milliseconds. Also when n is 16384, the run time is 2048 milliseconds. What is the complexity? What is the big-O-notation?
   [DOEACC'06, IETE ALCCS'11]
8. The input is an $N \times N$ matrix of numbers which is already in memory. Each individual row is increasing from left to right. Each individual column is increasing from top to bottom. Given an $O(N)$ worst-case algorithm that decides if a number x is in the matrix.
   [IITRG'11]
9. In algorithmic complexity show that the lower order terms and constant do not matter and are ignored. ***[IETE ALCCS'11]***
10 Differentiate between big-O, Ω, and Θ-notations used to specify algorithmic complexity.
   [IETE ALCSS'11, DOEACC'08]
11. Prove that if

$$f(n) = a_m n^m + a_{m-1} n^{m-1} + \ldots + a_1 n + a_0$$

   Then $f(n) = O(n^m)$ ***[WBUT'07, 09]***
12. Prove that $n! = O(n^n)$ ***[WBUT'08]***
13. Consider a polynomial in n of the form. Then, $f(n) = \Omega(n^m)$.

$$f(n) = \sum_{i=0}^{m} a_i n^i = a_m n^m + a_{m-1} n^{m-1} + \ldots + a_2 n^2 + a_1 n + a_0$$

   Where $a_m > 0$. ***[JNTU'08, DOEACC'07]***

14. Show that $f_1(n) + f_2(n) = O(\max(g_1(n), g_2(n))$ where $f_1(n) = O(g_1(n))$ and $f_2(n) = O(g_2(n))$. *[JNTU'08]*

15. Suppose it is known that the running time of one algorithm is $O(N \log N)$ and that the running time of another algorithm is $O(N^3)$. What does this say about the relative performance of algorithms? *[BPUT'05]*

16. What is the purpose of Big Oh notation? *[BPUT'05, '08]*

    Reorder the following complexity from smallest to largest: Justify your answer.

    $$2^n, n!, n^{10}, 2^4, n \log_2(n)$$

    [IETE ALCCS'11]

17. Arrange the following function into increasing order: $n \log n$, $(\log n)^3$, 2^n, $\log n$.*[BPUT'05]*

18. Define Big Omega Notation (A) and Little Omega Notation (ω). *[PTU'08]*

19. Calculate the big-O notation of the functions. *[IETE ALCCS'11]*

    $$5n^2 + n^{3/2} \text{ and } 3n^4 + n \log_2(n)$$

20. What is Ω notation? Compare order of growth of $n!$ and $2n$. *[BPUT'05]*

21. Find an asymptotic upper bound on the summation: *[DOEACC'08]*

    $$\sum_{k=0}^{\lfloor \log n \rfloor} \left\lceil \frac{n}{2^k} \right\rceil$$

22. How can you count number of ones in a binary string? Show that the counting algorithms for the above problem of a binary string of length n have time complexities varied from $O(n)$ to $O(1)$. *[DOEACC'09]*

23. What do you mean by average and worst case time complexity? *[WBUT'09, BPUT'08]*

24. Write an algorithm to find minimum and maximum elements simultaneously from a given list of elements. You are also required to discuss its running time. *[UPTU'10]*

25. Explain why instance characteristic is an important consideration for algorithm complexity? *[BPUT'08]*

26. Show that $(1 + n)^5 = O(n^5)$. *[BPUT'07]*

27. Find out worst-case efficiency, best-case efficiency and average-case efficiency of the sequential search algorithm. Express them using asymptotic notations. *[IETE ALCCS'11]*

28. Explain the term the order of complexity. *[PTU'09]*

29. Give an algorithm for finding an unknown number x in a sequence of supplied numbers. Find Θ, Ω, and O of the algorithm. *[DOEACC'09]*

30. Suppose you are given a sorted list of n elements followed by $f(n)$ randomly ordered elements. How large can $f(n)$ be for the entire list to be sorted in $O(n)$ time. *[IITG'11]*

31. What are the minimum number comparisons required to sort 5 elements. *[IITG'11]*

32. Justify whether the statements below are correct. *[RTU'11]*

    (*i*) $3n + 2 = O(n)$

    (*ii*) $3n + 2 = \Omega(n)$

    (*iii*) $3n + 2 = \theta(n)$

    (*iv*) $10n^2 + 2n + 2 = O(n^2)$

    (*v*) $6 \times 2^n + n^2 = \Omega(2^n)$

    (*vi*) $6 \times 2^n + n^2 \neq \Omega(2^n)$

33. Determine the frequency counts for all statements in the following program segments. Also compute the time complexity for both the segments. ***[RTU'11]***

(*i*)
```
for (i = 1; i <= n; i ++)
for (j = 1; j <= i; j ++)
for (k = 1; k <= j; k ++)
x = x + 1;
```

(*ii*)
```
i = 1;
while (i <= n)
{
    x = x + 1;
    i = i + 1;
}
```

34. Calculate the asymptotic complexity of the code segments shown below (using big-O notation) with respect to the problem size *n*: ***[IETE ALCCS'11]***

(*i*)
```
for (i = 0; i < n; i = i ×2) {
for (j = 1; j < n; j++) {
...
}
}
```

(*ii*)
```
for (i = 0; i < n – 2; i++) {
for (j = 0; j < n; j = j × 3) {
for (k = 1; k < 5000; k = k × 5) {
...
}
}
for (j = 0; j < 1000 × n; j = j + 1) {
...
}
for (j = 0; j < n/5; j = j + 5) {
}
}
```

35. Which function grows at the faster rate e^n or 2^n? Justify your answer. ***[BPUT'08]***

36. Give the worst case time and the best case time for the following algorithm. ***[IETE ALCCS'11]***

```
"ABC (n)":
 ABC (n)
 { for (i = 2 to n/2)
    { if (n% i == 0)
        break;
    }
 }
```

37. Suppose you insert n numbers into an initially empty priority queue implemented as a heap. What is the worst- case running tune? *[BPUT'08]*
38. What is the minimum number of keys that must be moved during a "remove the largest" operation in a max-heap? Use a heap of size 15 to explain your answer. *[WBUT'07]*
39. Differentiate between performance analysis and performance measurement of an algorithm. *[BPUT'07]*
40 Write down both iterative and recursive version of Binary search algorithm. What is the time-complexity in both cases? *[UPTU'06]*
41. Write an efficient algorithm to determine the second largest element of n numbers. Determine the number of comparisons needed to find it. *[RTU'10]*
42. Critically comment on $\Theta(g(n)) \subseteq O(g(n))$. *[WBUT'09]*
43. Show that for any real constants a and b, where $b > 0$, $(n + a)b = \Theta(n)b$. *[UPTU'05, DOEACC'07]*
44. Give an algorithm to count the number of leaf nodes in a binary tree t. What is its computing time? *[UPTU'06]*
45. Find the best case and worst case time complexity of deletion in a binary tree. *[UPTU'06]*
46. For $f(x) = 2x^3 + 3x2 + 1$, show that *[IGNOU'07]*

$$f(x) = O(x^n) \text{ for } n \geq 4$$
$$f(\text{x}) \neq O(x^n) \text{ for } n \leq 3$$

47. What will be the complexity of the linear search algorithm for both the worst case and average-case? *[PTU'09]*
48. Define the heap. How a priority queue be implemented using the heap? *[UPTU'07]*
49. Illustrate the operation of heap-sort on the following array : (5,13, 2, 25,7, 17, 20, 8, 4). *[UPTU'09]*
50. What is Heap property? Write an algorithm to make a Heap containing elements. Then, show that how can you insert an element into a Heap. Then, write the algorithm of Heap sort and find the running time of this algorithm. Write an algorithm to find the existence of an element into a Heap. *[WBUT'10, BPUT'05, '08]*
51. Explain the following Heap Sort. *[UPTU'10, PTU'08, MDU'09, IGNOU'07]*
52. Write the pseudocode for heapify algorithm. *[UPTU'05]*
53. Define max-heap and min-heap? How will you represent a max-heap as an array? Write an algorithm to insert an element to a max-heap. *[IETE ALCCS'11]*
54. Explain the process of Heap sort. Write a general algorithm to construct a min-heap. Construct min-heap from the following list {21, 6, 56, 61, 44, 7, 9, 76, 75, 32, 34, 4, 49, 33}. *[BPUT'05, 07, '08, DOEACC'08]*
55. Construct a max-heap from the following list. Sketch the heap after deleting 76. What is the time complexity of the whole process? {21, 6, 56, 61, 44, 7, 9, 76, 75, 32, 34 4, 4} *[BPUT'06]*
56. Write an algorithm Heap_Delete (A, I), which deletes the item in node I from heap A. *[UPTU'05]*
57. Write a program in C for binary search. What is the worst case time complexity of binary search? *[UPTU'08, BPUT'08]*
58. Write Heapsort algorithm. Analyze the running time of your algorithm. *[UPTU'08, RTU'11, BPTU'05]*

59. Insert 38 in the heap in the following figure. Also explain the method of getting the new heap after insertion. ***[IETE ALCCS'11]***

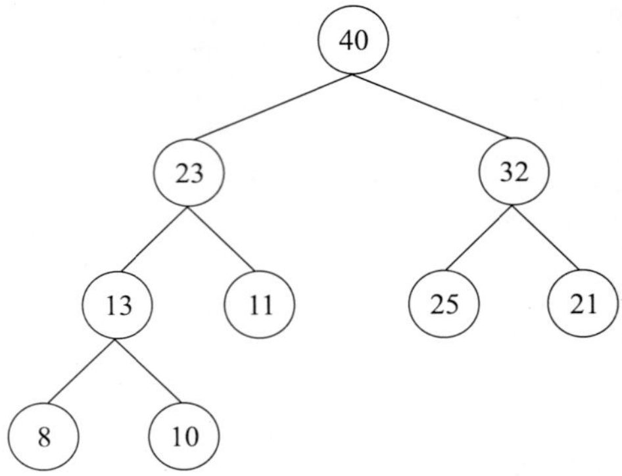

60. Sort the following list using Heapsort technique : <5, 9, 20, 8, 4, 100, 11, 3, 30>. ***[PTU'09]***

61. Write the algorithm for Heap-Sort. Also show functioning of your algorithm on the following array : <15, 10, 11, 22, 55, 44, 9>. ***[IGNOU'07]***

62. Construct a Max-Heap with the following elements {A, L, G, O, R, I, T, H, M). ***[BPUT'08]***

63. Suppose that the string < ALGORITHMS > is to be sorted by the heap-sort: Show how the above data would be arranged in the array after heap construction phase. ***[IETE ALCCS'11]***

64. What is the difference between the min-heap property and the binary search tree property? Can the min-heap property be used to print out the keys of a binary tree of n vertices in sorted order in $O(n)$ time? Justify. ***[DOEACC'06, '09]***

65. Show the steps to create the heap that results when the following keys are inserted into an initially empty heap. Keys: 41, 28, 33, 15, 25, 7, 12. ***[IETE ALCCS'11]***

66. Give an algorithm to find all nodes less than some value X in a heap. Your algorithm should run in $O(k)$ time where k is the number of output nodes. ***[IITG'11]***

❑❑❑

3

Chapter

Recursion vs Recurrence

3.1 INTRODUCTION

Recursion is a fundamental programming tool that can provide an elegant solution to certain kinds of problems. In fact, *recursion* is one of the central ideas of computer science. The term recursion is used more generally to describe a process that is repeating objects in a self similar way. For instance, when the surfaces of two mirrors are almost placed parallel to each other the nested images will occur and these are a form of infinite recursion. When recursion solves a problem the solution depends on solutions to smaller instances (partial solutions) of the same problem.

It is told that "The power of recursion evidently lies in the possibility of defining an infinite set of objects by a finite statement. In the same manner, an infinite number of computations can be described by a finite recursive program, even if this program contains no explicit repetitions."

Figure 3.1 shows a famous Sierpinski triangle that describes a recursion of triangles to form a lattice.

Figure 3.1 A Sierpinski recursion triangle to form a geometric lattice.

Definition: Recursion is an algorithmic technique where an object or entity defines or calls with some part of a task in terms of itself in order to accomplish the task. Here an object may be a function, a procedure, or an algorithm. All these terms are used synonymously.

Every recursive solution involves two major parts.

1. ***Base Case:*** In this case the problem is simple enough to be solved directly. Some values are assigned that are treated as initial conditions, or constants. This is also called thumb rule or base value.
2. ***Recursive Case(s):*** A recursive case has three components:
   (*i*) *Divide* – It involves in dividing the problem into similar, simpler, smaller parts to solve easily.
   (*ii*) *Call* – Next call the function recursively on each part to have partial solutions.
   (*iii*) *Combine* – Finally combine the partial solutions of the parts as a desired solution for the problem.

   Depending on the problem, any of these may be trivial or complex.

Example 1. The following is a recursive definition of a person's ancestors:

(*i*) *Base case:* One's parents are one's ancestors.
(*ii*) *Recursion step:* The parents of one's ancestors are also one's ancestors.

Example 2. The Fibonacci sequence is a classic example of recursion:

(*i*) Base Case:

$$Fib(0) = 0$$
$$Fib(1) = 1$$

(*ii*) Recursive Case: $Fib(n) = (Fib(n\text{-}1) + Fib(n\text{-}2))\ \forall$ integers $n > 1$.

Example 3. The Factorial ( ) function is recursively expressed as:

(*i*) Base Case: Factorial (0) = Factorial (1) = 1.
(*ii*) Recursive Case: Factorial (n) = n * Factorial (n – 1), $\forall$ integers $n > 1$.

3.2 RECURSION: WORKING PRINCIPLE

When a function is called, the computer must "remember" the place from where it was called, *i.e.* the *return address*, so that it can return results to that location once execution of the call is completed. Typically, this information is saved on the linear data structure stack which is a simple list of return locations in order of the times that the call locations were described. Sometimes, at the last a function, after completing all other operations, simply calls a function, possibly itself, and return its result.

3.3 RECURSION: FEATURES

Recursion is a way of thinking about a new way to solve problems. Some common features of a recursion are drawn as below.

1. Most of high-level computer programming languages support recursion.
2. Recursive functions are useful for some mathematical functions.
3. It may encounter certain dynamic data structures such as linked lists or binary trees. These can be easily created and accessed recursively.
4. A recursive definition states a concept clearly.
5. The code of a recursive method is structured one. It is done to handle both the base and recursive cases.
6. Each recursive call sets up a new execution environment.

7. When a method completes it execution, control returns at the point of calling to the method that invoked it which may be an earlier invocation of it.

3.4 RECURSION: CLASSIFICATION

The recursive approaches are classified based on:

1. Direct or indirect recursion—whether the function calls itself directly or indirectly.
2. Tail-recursion — whether there are pending operations at each recursive call.
3. Linear or tree-recursive — the way in which the recursion grows, *i.e.* whether pending operations are also recursive or not.
4. Collective recursion—where nothing is returned.
5. Finite or infinite recursion—how many times it will repeat finitely or infinitely.

3.4.1 Direct Recursion

Recursion can be *direct* when an entity refers to itself directly. Similarly an algorithm is *directly* recursive if it contains an explicit call to itself. In computer science a directly recursive data type contains pointers to instances of the data type.

Example 4. Following algorithm shows direct recursion.

Algorithm: Direct_Rec(x)

Where x: an integer.

Step 1: Start.
Step 2: If $x \leq 0$
Return 1
Step 3: Return Direct_Rec($x - 1$)
Step 4: End

The recursive call will occur for positive values of x.

3.4.2 Indirect Recursion

Recursion can be *indirect* when it refers to other entities which again refer to it. Similarly a function could invoke another function, which invokes original function. In this case, a function invokes another function which can invoke another function and so on. The invocation is continued until eventually the original function is invoked again. This is called *indirect recursion*. It is often more difficult to trace and debug. For example, all *mutually recursive* routines are an example of indirect recursion.

Example 5. The following pair of algorithm is indirectly recursive.

Algorithm: Indirect_Rec(x)

Where x: an integer ≥ 0.

Step 1: Start.
Step 2: If $x \leq 0$
Return 1
Step 3: Return Rec(x)
Step 4: End

Sub-Algorithm: Rec(y)

Where y: an integer.

Step 1: Start.
Step 2: Return Indirect_Rec($y - 1$)
Step 3: End

The function Indirect_Rec() is *indirectly* recursive as it contains a call to another function Rec() which ultimately calls Indirect_Rec(). Since they call each other, they are also known as *mutually recursive* functions.

3.4.3 Tail Recursion

A recursion is said to be tail recursive if there are no pending operations to be performed on return from a recursive call. A tail recursion also called tail-end recursion is a special case of recursion. The last operation of the function is recognized as "tail call". This is also a recursive call. Tail recursive functions "return the value of the last recursive call as the value of the function." It is very desirable because the amount of information which must be stored during the computation is independent of the number of recursive calls.

Example 6. The "infamous" factorial function Fact( ) can be written in a non-tail recursive manner as:

Algorithm: Fact(x)

Where x an integer >= 0.

Step 1: Start.
Step 2: If $x = 0$
  Return 1
Step 3: Return n*Fact($x - 1$)
Step 4: End

Notice that there is a "pending operation," namely multiplication, to be performed when return from each recursive call. Since there is a pending operation, the function is not tail recursive. So, information about each pending operation must be stored. And the amount of this information depends on the number of calls.

The above factorial function can be written in a tail-recursive way as:

Algorithm: Fact(x)

Where x: an integer ≥ 0.

Step 1: Start.
Step 2: Return Tail_Rec(x, 1)
Step 3: End

Sub-Algorithm: Tail_Rec(x, result)

Where x: an integer ≥ 0.
  result: factorial value of x

Step 1: Start.
Step 2: If $x = 1$
  Return result
Step 3: Return Tail_Rec($x - 1$, x*result)
Step 4: End

The secondary, sub or auxiliary function Tail_Rec() is used to keep the syntax of Fact(x) the same as before. Here the recursive function is actually Tail_Rec(), not Fact(). Note that Tail_Rec( ) has no pending operations on return from recursive calls. The value computed by the recursive call is simply returned with no modification. The amount of information which must be stored is constant. Here both x and result are independent of the number of recursive calls.

3.4.4 Conversion: Recursive Functions to Tail Recursive

A non-tail recursive function can often be converted to a tail-recursive function by using one or more "auxiliary or secondary" parameters. These parameters are used for the result. The idea is to fit these parameters in such a way that the recursive call has no longer a pending operation. The technique uses another "auxiliary" function. This simply keeps the syntax. And it hides the fact that auxiliary parameters are needed.

Example 7. A Fibonacci function can be implemented by using two auxiliary parameters for accumulating results in tail-recursive form.

Algorithm: Fibo_Tail(x, next, result)

Where x, next, result are integers ≥ 0.

Step 1: Start.
Step 2: If $x = 0$
    Return result
Step 3: Return Fibo_Tail(x–1, next + result, next)
Step 4: End

To compute Fibo(x), we call Fibo_Tail(x, 1, 0). This linearly recursive algorithm would require $O(n)$ time. The computation can also be done by tree recursive Fibo() function discussed later and requires two auxiliary parameters to compute results.

3.4.5 Linear Recursion

The simplest form of recursion is called *linear recursion*. It occurs only where an action has a simple repetitive structure consisting of some basic steps followed by the action again. For example, the `fact()` function is linearly recursive. The pending operation is multiplication by a scalar. Also, it does not involve another call.

3.4.6 Tree Recursion

Another common pattern of computing recursive functions is *tree recursion*. This recursion is also called non-linear recursion.

A recursive function is said to be *tree* recursive or *non-linearly* recursive when the pending operation does involve another recursive call to the function.

Example 8. The Fibonacci function `Fibo()` is a classic example of tree recursion.

The sequence of Fibonacci numbers in which each number is the sum of the preceding two number: 0,1,1,2,3,5,8,13,21, ... can be defined by the rule:

$$\text{Fibo}(n) = \begin{cases} 0 & \text{if } n = 0, \\ 1 & \text{if } n = 1, \\ \text{Fibo}(n-1) + \text{Fibo}(n-2), & \text{otherwise} \end{cases}$$

For example, the first five Fibonacci numbers are

Fibo(0) = 0
Fibo(1) = 1
Fibo(2) = Fibo(1) + Fibo(0) = 1
Fibo(3) = Fibo(2) + Fibo(1) = 2
Fibo(4) = Fibo(3) + Fibo(2) = 3

This leads to the following algorithm and recursion tree:

Algorithm: Fibo(x)

Where x: an integer ≥ 0.

Step 1: Start.
Step 2: If $x = 0$
Return 0
Step 3: If $x = 1$
Return 1
Step 4: Return Fibo($x - 1$) + Fibo($x - 2$)

Step 5: End

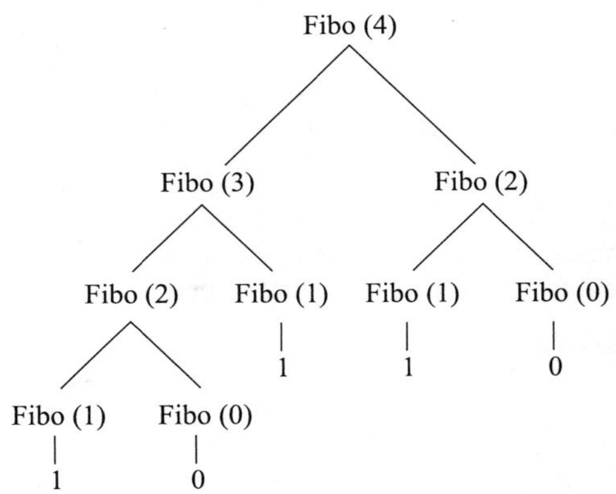

Figure 3.2 Recursion tree for first five fibonacci number.

Notice that the pending operation for the recursive call is another call to Fibo. Therefore Fibo is tree-recursive. The tree recursive Fibo() method has complexity value $O(2^n)$. In other words as x increases the problem size roughly doubles.

3.4.7 Collective Recursion

Collective Recursion is a special form of tail recursion. In this case, results are produced only during the recursive calls and nothing is returned. This recursion is not optimized one. The recursion may be optimized by using the current stack rather than creating a new stack to execute the call. Also the same can be done by de-allocating the entire recursion stack at once rather than a little at each return.

Example 9. The following algorithm prints the values greater than limit in a list.

Algorithm: Overflow(L, limit)

Where L: list of elements.
limit: maximum size of L.

Step 1: Start.
Step2: If L = Null
Return
Step 3: If limit < Head(L)
Output Head(L).

Step 4: Call Overflow(Tail(L), limit)
Step 5: End

At the return, the compiler can de-allocate all the memory for the recursion stack at once.

3.5 RECURSION: APPLICATIONS

A recursion has lots of applications in the following areas.

(*i*) Language.
(*ii*) Mathematics.
(*iii*) Computer Science.

3.5.1 Recursion in Language

In two ways we use recursion in languages.

- **Recursion in plain English:** Mainly recursion and its related terms such as recursive procedure, running the procedure, difference between them through English like sentences are described. E.g., Recursion is the process where a procedure goes through when one of the steps of the procedure involves re-running the procedure.
- **Recursive Humor:** This is a parody on references in dictionaries. e.g., suppose we want to find the term recursion in a book. Possible result may be: *recursion 86, 139, 141, 182, 202, 269.*

3.5.2 Recursion in Mathematics

Recursion in practice is defined mathematically. Some of the basic application areas of recursion in mathematics are:

- Recursively defined sets – it is given by the natural numbers.

Example 10. The canonical example of a recursively defined set is given by the natural numbers:

(*i*) 1 is in $\mathbb{N}$
(*ii*) if n is in $\mathbb{N}$, then $n + 1$ is in $\mathbb{N}$

- Functional Recursion: A function may be fully or partly defined in terms of itself. We call this function a recursive function.

Example 11. A familiar example is the Fibonacci number sequence:

$F(n) = F(n - 1) + F(n - 2)$ for all non negative integers n. For such a definition to be recursive it must lead to base values which are non-recursively defined, in this case $F(0) = 0$ and $F(1) = 1$.

Example 12. Another famous recursive function is the Ackermann function which, unlike the Fibonacci sequence, cannot be expressed without recursion.

The Ackermann function is defined recursively for non-negative integers m and n as follows:

$$A(m, n) = \begin{cases} n + 1 & \text{if } m = 0 \\ A(m - 1, 1) & \text{if } m > 0 \text{ and } n = 0 \\ A(m - 1, A(m, n - 1)) & \text{if } m > 0 \text{ and } n > 0 \end{cases}$$

Its value grows rapidly, even for small inputs. For example, $A(4, 2)$ is an integer of 19,729 decimal digits.

- Recursive Proofs—This is a powerful generalization of mathematical induction which is widely used to derive proofs in mathematical logic and computer science.
- Recursive Optimization—Dynamic programming and Greedy approach are the approaches to optimization which restate a multi-period or multi-step optimization problem in recursive form.

3.5.3 Recursion in Computer Science

Recursion in computer programming is exemplified when a function is defined in terms of simpler, often smaller versions of itself. The solution to the problem is then devised by the combining the solutions obtained from the simpler versions of the problem. Some common examples are:

(*i*) Divide-and-Conquer—Quicksort, Mergesort, etc.
(*ii*) Dynamic programming—TSP, etc.
(*iii*) Parser generators.
(*iv*) Recurrences.

3.6 RECURSION: ADVANTAGES AND DISADVANTAGES

Use of recursion in algorithms may boast the following advantages and disadvantages.

- ***Advantages***

1. The main advantage is usually simplicity.
2. Using recursion unnecessary calling of functions can be avoided.
3. Recursion solves problems in easy way over iterative solutions. Ex: Tower of Hanoi.
4. Code size is reduced.
5. Recursion replaces complex nesting code complexity when is used to divide the problem into same problem of subtypes.
6. Recursive solution is always logical.
7. Code is shorter and is closer to a mathematical definition. This helps to understand easily. Ex: Factorial.

- ***Disadvantages***

1. The main disadvantage is often that the algorithm may require large amounts of memory if the number of recursive call is very large.
2. Too many recursive functions may create confusion in the code.
3. Recursive solution is often very difficult to trace, *i.e.* debug and understand.
4. Less efficient as compared to the non-recursive approach as the overhead involved in entering, re-entering.
5. There must be an "if" statement somewhere in the code to force the function to return without the recursive call being executed otherwise the function will never return.

3.7 RECURSION: LIMITATIONS

Although a recursion technique has a wide area of application, still it is not free from limitations.

1. It consumes more space.
2. The computer may run out of memory if the recursion calls are not checked.
3. It is less efficient in terms of speed and execution time.
4. Sometimes recursion does not propose any concrete compensation over non-recursive procedures.
5. If proper precautions are not taken, recursion may be non-terminating iterations.
6. No need to invoke recursion when a problem can be through iteration.
7. As a software tool it should be used both carefully and selectively.

Remember: Whatever could be done through recursion could be done through iterative way but reverse is not true.

3.8 RECURRENCES

When a recursive definition is considered as an equation to be solved we call it *recurrence relation*. But why does it necessary? Answer is very simple. The running time of a recursive algorithm can be evaluated by a recurrence. And recurrences are the major tool for analysis of recursive algorithms.

Definition: A recurrence is an equation or inequality that describes a function in terms of its smaller values on small inputs. We generally use the terms F_n, T_n, X_n, $T(n)$, etc. to denote recurrence notation for input n.

The basis of the recursive definition is called *initial conditions* of the recurrence.

Example 13. The recursive definition of the Fibonacci sequence 1, 1, 2, 3, 5, 8, 13 ... can be represented by the recurrence as: $F_n = F_{n-1} + F_{n-2}$ where the initial conditions are $F_0 = 0$, $F_1 = 1$.

Example 14. The recurrence relation of n! can be written as:

$$T(n) = \begin{cases} 1, & n = 0 \text{ or } 1 \\ n & T(n-1), \text{ otherwise} \end{cases}$$

3.8.1 Recurrences: Applications

Knowledge of recurrences may be thought as its applications in different areas. Such as:

1. *Biology:* Some of the best-known difference equations have their origins in the attempt to model population dynamics. For example, the Fibonacci numbers were once used as a model for the growth of a rabbit population.
2. *Digital Signal Processing:* In this case, recurrence relations can model feedback in a system, where outputs sometimes become inputs. They thus arise in infinite impulse response (IIR) digital filters.
3. *Mathematics:* Recurrence relations are a fundamental mathematical tool. They can be used to represent and solve mathematical sequences efficiently. However, they cannot be easily represented non-recursively.

3.8.2 Recurrences: Solution Methodology

A solution to a recurrence relation is a sequence that verifies the recurrence. Solving recurrences means the asymptotic evaluation of their efficiency, *i.e.* finding the asymptotic bounds (big-O, big-Ω, and big-Θ) for a function represented by the recurrence say $T(n)$. The recurrences are solved using some mathematical tools. To solve a recurrence relation we use any one or combinations of the following methods.

1. Substitution Method.
2. Iteration Method.
3. Recursion Tree Method.
4. Changing variable Method.
5. Characteristics Equation Method.
6. Power or n^{α} Method.
7. Master or Cookbook Method.

3.8.2.1 *Substitution Method*

This is known as a "good guess method". For a given recurrence, its asymptotic bound using substitution method is evaluated in two steps.

- The first step is to guess a probable solution.
- The second step is to prove the correctness of the guessed solution using mathematical induction.

Example 15. Find a solution for the recurrence given below.

$$T(n) = T(n) = \begin{cases} 1, & n = 1 \\ 2T\left(\frac{n}{2}\right) + n, & n \geq 2 \end{cases}$$

Solution: Guess: $T(n) = n \log n + n$

Induction:

Basis: for $n = 1$, $T(n) = 1$; $g(n) = 1 \cdot \log 1 + 1 = 1 \rightarrow T(n) \leq c^*g(n) \Rightarrow c = 1$

Inductive step:

Hypothesis: $T(k) = k \log k + k \ \forall \ k < n$

Let us use this hypothesis:

$$\begin{aligned} T(n) &= 2T(n/2) + n \\ &= 2\left(\frac{\frac{n}{2}\log n}{2} + \frac{n}{2}\right) + n \\ &= n \log \frac{n}{2} + n + n \\ &= n(\log n - \log 2) + n + n \\ &= n \log n - n + n + n \\ &= n \log n + n \end{aligned}$$

Hence, our assumption was correct and $T(n) = n \log n + n = O(n \log n)$

Example 16. Consider $T(n) = 2T(n - 1) + 1$, $n \geq 1$ and $T(0) = 0$. Obtain a solution using substitution method.

Solution: Computing first few values, we have

n	1	2	3	4	5	6
$T(n)$	1	3	7	15	31	63

Guess: $T(n) = 2^n - 1 \; \forall \; n \geq 0$.

Induction:

For, $n = 0, T(0) = 2^0 - 1 = 0$

$n = 1, T(1) = 2^1 - 1 = 1$

$n = 2, T(2) = 2^2 - 1 = 3$

Similarly, $\forall \; k \geq 0, T(k) = 2^k - 1$

Now, by induction hypothesis,

$T(k) + 1 = 2T(k) + 1 \Rightarrow T(k+1) = 2(2^k - 1) + 1 = 2^{k+1} - 2 + 1 = 2^{k+1} - 1$

Therefore, our guess is correct. And $T(n) = 2^n - 1 = O(2^n)$.

Drawbacks: Unfortunately there is no general way to make a correct guess for solutions to recurrences using substitution method. Making a good guess:

- Needs experience and creativity.
- Uses recursion trees to generate good guesses.
- If a recurrence is similar to one already we are familiar, then guessing a similar solution is only then reasonable.
- Proves loose upper and lower bounds and then reduce the range of uncertainty.

3.8.2.2 *Iteration Method*

The iteration method is the reiteration of the recurrence. This method sometimes is also called an expansion or unfolding method. The basic ideas behind this method are:

1. Expand the recurrence formula iterating it until it can be expressed as a summation of terms dependent only on n and the initial conditions, *i.e.* boundary.
2. Use standard techniques for bounding summations to get the asymptotic cost.

Example 17. Find the solution for the recurrence using iteration method.

$$T(n) = 3T(n/4) + n, \; \forall \; n \geq 0$$

Solution: Given, $T(n) = 3T(n/4) + n$

$= n + 3T(n/4)$

$= n + 3((n/4) + 3T(n/16))$ [*expanding* $T(n/4)$]

$= n + 3((n/4) + 3((n/16) + 3T(n/64)))$ [*expanding* $T(n/16)$]

$= n + 3(n/4) + 9(n/16) + 27T(n/64)$

Hence, $T(n) \leq n + 3n/4 + 9n/16 + 27n/64 + \ldots + 3^{\log_4 n}\,\Theta(1)$

$$\leq n \sum_{i=0}^{\alpha} \left(\frac{3}{4}\right) + \Theta(3^{\log_4 n})$$

$= 4n + O(n)$

$= O(n)$ [$3^{\log_4 n} = n^{\log_4 3}$ *and* $\log_4 3 < 1 \Rightarrow \Theta(n^{\log_4 3}) = O(n)$]

Example 18: Find the solution for the recurrence using iteration method.

$$T(n) = \begin{cases} 1, & n = 1 \\ 4T\left(\dfrac{n}{2}\right) + n, & n \geq 2 \end{cases}$$

Solution: Given, $T(n) = n + 4T(n/2)$

$= n + 4[n/2 + 4T(n/4)]$ [*expanding* $T(n)$]

$= n + 2n + 4^2T(n/2^2)$

$= n + 2n + 4^2[\, n/4 + 4T(n/8)]$

$= n + 2n + 4n + 4^3T(n/2^3)$

....

$= n + 2n + 2^2n + 2^3n + \ldots\ldots.. + 2^in$

$= n \sum_{i=0}^{\log n - 1} 2^i$

$= n\, [(\, 2^{\log n - 1 + 1} - 1)/(2 - 1)] + 1$

$= n[2^{\log n} - 1] + 1$

$= n^2 - n + 1$

$= O(n^2)$.

Example 19. Find the solution for the recurrence using iteration method.

$$T(n) = n + 2T(n/2), \forall\ n \geq 0$$

Solution: *Given,* $T(n) = n + 2T(n/2)$

$= n + 2(n/2 + 2T(n/4))$

$= n + n + 4T(n/4)$

$= n + n + 4(n/4 + 2T(n/8))$

$= n + n + n + 8T(n/8)$

....

$= in + 2^iT(n/2^i)$

Now let us consider $n/2^i = 1 \rightarrow 2^i = n$

$\rightarrow i = \log n$

Hence, $T(n) = n \log n + nT(1)$

$= \Theta(n \log n)$

Example 20. Find the solution for the recurrence using iteration method.

$$T(n) = T(n/2) + c$$

Solution: Given, $T(n) = T(n/2) + c$

$= c + T(n/2)$

$= c + c + T(n/4)$

$= 2c + T(n/2^2)$

$$= c + c + c + T(n/8)$$

$$= 3c + T(n/2^3)$$

$$\ldots\ldots\ldots\ldots\ldots$$

$$= kc + T(n/2^k)$$

Assume, $n/2^k = 1 \rightarrow 2^k = n \rightarrow k = \log n$

Hence, $T(n) = \log n .c + T(1)$

$$= c \log n + T(1)$$

$$= \Theta(\log n)$$

3.8.2.3 *Recursion Tree Method*

A recursion tree method is variation of iteration method represented graphically. It is used to present a problem as a composition of subproblems. It is more appropriate to present any divide-and-conquer algorithm. Each node in the tree represents the cost of a single sub problem. Usually each level of the tree corresponds to one step of the expansion of the recursion. The complexity of the problem is characterized in terms of its cost. The steps involved towards complexity analysis are:

- *Find per level cost:* Find and sum the costs of all nodes at each level of the tree to obtain per level costs.
- *Find total cost:* Sum all the per-level costs to determine the total cost of the recursion tree. This cost is in some mathematical form.
- *Solve total cost:* Finally, generate a guess by solving the mathematical form. This guess is the probable time complexity.

Procedure

Determination of a "Good Asymptotic Bound" of a problem using recursion tree involves the following steps.

1. Draw the recursion tree based on the recurrence relation of the problem.
2. From the tree determine:
   - No. of levels in the tree.
   - Cost per level.
   - No. of nodes in the last level.
   - Cost of the last level.
3. Write down the summation using $\sum$ notation—this summation sums up the cost of all the levels in the recursion tree.
4. Simplify the summation expression just derived with your "guess" in terms of Big-O, Big-Θ, or Big-Ω depending on which type of asymptotic bound is being sought.
5. Finally use Substitution Method or Master Method to prove that the "guess" is correct.

Advantages

Some of advantages for using recursion tree method in algorithm analysis are:

(*i*) It is as simple as iteration method.

(*ii*) It replicates the time complexity in terms of costs of a recursive execution of an algorithm.

(*iii*) It is suitable for good guesses of the substitution method.

(*iv*) It promotes intuition.

Still, the recursion-tree method may be unreliable.

Example 21. Using recursion tree method, solve the following recurrence.

$$T(n) = T(n/3) + T(2n/3) + n$$

Solution: The recursion tree of the recurrence $T(n)$ is

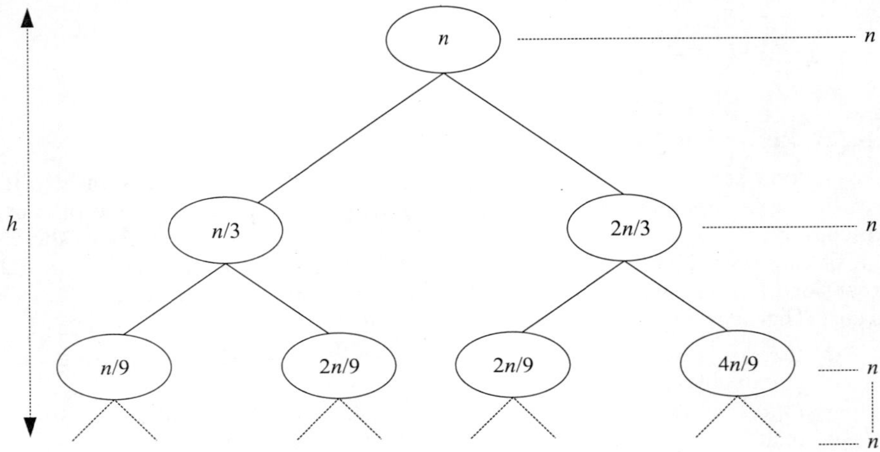

Figure 3.3 Recursion tree of *T*(*n*).

Hence $T(n) = n + n + n + \ldots\ldots\ldots\ldots\ldots + \text{upto } h \text{ terms}$

$= hn$

Now, left child is decreased by $n/3k$ and right child is decreased in $\left(\frac{2}{3}\right)^h \cdot n$ for some integer k, $h \geq 0$. Maximum height of the tree will be due to this right child.

Now, $\left(\frac{2}{3}\right)^h \cdot n = 1 \rightarrow h = \log_{3/2} n$

Therefore, $T(n) = n \log_{3/2} n$

$= n (\log n / \log_{3/2})$

$= \log_{2/3} . n \log n$

$= \Theta(n \log n)$

Example 22. Using recursion tree method, prove that the following recurrence $T(n) = 2T(n/2) + n^2$ is asymptotically bounded.

Solution: The recursion tree of the recurrence $T(n)$ is

Hence, $T(n) = n^2 + n^2/2 + n^2/4 + \ldots + \text{upto } h \text{ terms}$

$= n^2 (1 + 1/2 + 1/4 + \ldots + \text{upto } h \text{ terms})$ // it is in G.P series

$$= n^2 \cdot \frac{1}{1 - \frac{1}{2}} = 2n^2 = O(n^2)$$

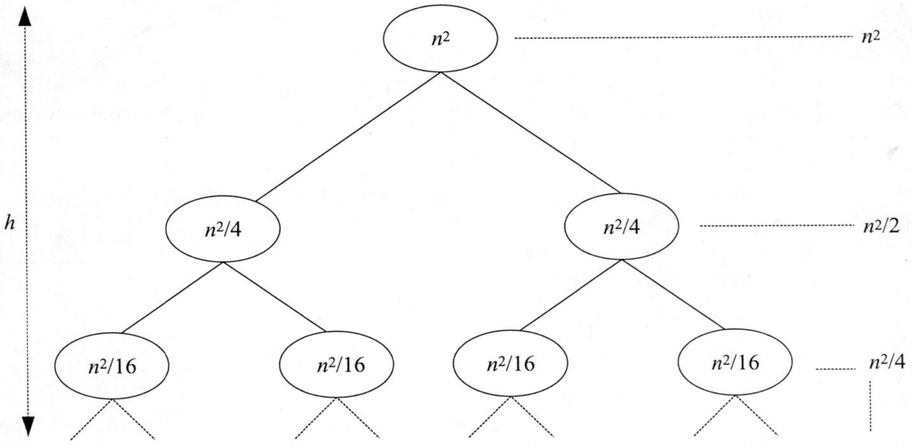

Figure 3.4 Recursion tree of *T*(*n*).

Example 23. Using recursion tree method, solve the following recurrence

$$T(n) = T(n/4) + T(n/2) + n^2 \text{ and } T(1) = \Theta(1)$$

Solution: The recursion tree for $T(n)$ is drawn below.

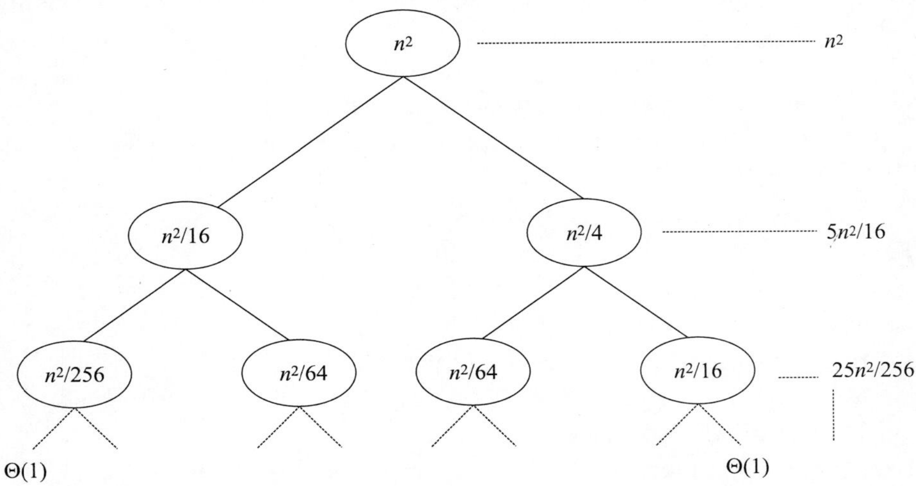

Figure 3.5 Recursion tree of *T*(*n*).

Hence,
$$\begin{aligned} T(n) &= n^2 + 5n^2/16 + 25n^2/256 + \dots \\ &= n^2(1 + 5/16 + (5/16)^2 + \dots) \\ &= n^2\left(\frac{1}{1-\frac{5}{16}}\right) \qquad \left[\, using\ 1 + x + x^2 + \dots = \frac{1}{1-x},\ |x| < 1 \right] \\ &= \frac{16}{11}n^2 \\ &= \Theta(n^2) \end{aligned}$$

3.8.2.4 *Master Theorem Method*

In the analysis of algorithms, "The Master Theorem" provides a "cookbook" method like a cook book that contains lots of recipes for preparing different foods. Similarly the master theorem is applied to solve most of the recurrences of the following form:

$$T(n) = aT(n/b) + f(n)$$

where $a \geq 1$ and $b > 1$ are constants and $f(n)$ is an asymptotically positive function. Here n/b is either $\lfloor n/b \rfloor$ or $\lceil n/b \rceil$.

In the application to the analysis of a recursive algorithm, the constants and function have the following significance:

(*i*) n is the size of the problem.

(*ii*) a is the number of subproblems in the recursion.

(*iii*) n/b is the size of each sub problem.

(*iv*) $f(n)$ is the cost of the work done outside the recursive calls, which includes the cost of dividing the problem and the cost of merging the solutions to the sub problems.

The number of levels in the recursion tree of $T(n)$ is n^E where $E = \log_b a$.

Master Theorem: Suppose $T(n)$ be defined on the positive integers set by the recurrence $T(n) = aT(n/b) + f(n)$, Where $a \geq 1$ and $b > 1$ are constants and $f(n)$ is an asymptotically positive function. Consider, $E = \log_b a$, then $T(n)$ can be asymptotically bounded as:

1. If $f(n) = O(n^{E-e})$ for some constant $e > 0$, then $T(n) = \Theta(n^E)$.
2. If $f(n) = \Theta(n^E \log^k n)$ with $k \geq 0$, then $T(n) = \Theta(f(n).\log^{k+1} n)$.
3. If $f(n) = \Omega(n^{E+e})$ with $e > 0$, and $af(n/b) \leq cf(n)$ for some constant $0 < c < 1$ then $T(n) = \Theta(f(n))$.

The above three principles collectively is called master theorem.

Example 24. Solve the recurrence using master theorem.

$T(n) = T(3n/4) + 1$ and $T(1) = \Theta(1)$

Solution: Given $T(n) = T(3n/4) + 1$

Here $a = 1, b = 4/3$, $E = \log_b a = \log_{4/3} 1 = 0 \Rightarrow n^E = n^0 = 1$

Now $f(n) = 1 = n^0 = n^E \Rightarrow f(n) = \Theta(n^E)$, case 2 is applied and $k = 0$.

Therefore, $T(n) = \Theta(f(n).\log^{k+1} n) = \Theta(\log n)$

Example 25. Using master theorem solve the recurrence $\boldsymbol{T(n) = 4T(n/2) + n}$

Solution: From master method, we have

$$a = 4, b = 2, \; E = \log_b a = \log_2 4 = 2$$
$$n^E = n^2$$

Now, $f(n) = n = n^{2-1} = O(n^{2-1}) \rightarrow$ case 1 is applicable.

Therefore, $T(n) = \Theta(n^2)$

Example 26. Find out the solution for the recurrence $T(n) = 4T(n/2) + n^3$

Solution: From the given recurrence using master theorem, we have

$$a = 4, b = 2, E = \log_b a = \log_2 4 = 2$$
$$n^E = n^2$$

Now, $f(n) = n^3 = n^{2+1} = \Omega(n^{E+e})$ where $e = 1$.

Again, $af(n/b) = 4(n/2)^3 = 4n^3/8 = \frac{1}{2}n^3 \leq cf(n)$ where $c = \frac{1}{2} < 1$.
By case 3, $T(n) = \Theta(f(n)) = \Theta(n^3)$.

Example 27. Solve the following recurrence using master theorem:

$$T(n) = 5T(n/4) + n.$$

Solution: Here, $a = 5, b = 4, E = \log_b a = \log_4 5 = 1.609$

Now, $f(n) = n = n^{1.609-0.609} = n^{E-e} = O(n^{E-e})$ where $e = 0.609$

Case 1 is applicable. Thus, $T(n) = \Theta(n^{\log_4 5}) = \Theta(n^{1.609})$

Example 28. Solve the following recurrence using master theorem:

$$T(n) = 5T(n/4) + n^2$$

Solution: Here, $a = 5, b = 4, E = \log_b a = \log_4 5 = 1.609$

Now, $f(n) = n^2 = n^{1.609+e} = n^{E+e} = O(n^{E-e})$ where $e = 2 - 1.609 = 0.391$

Therefore, case (3) seems to apply.

Now, $af(n/b) = 5(n/4)^2 = 5n^2/16 \leq cf(n)$ where $c = \frac{5}{16} < 1$.

Thus, $T(n) = \Theta(n^2)$

3.9 RECURRENCES: MORE EXAMPLES

Example 29. Solve $T(n) = T(n - a) + T(a) + cn$, where $a \geq 1$ and $c > 0$ are constants.

Solution: Recursion tree method is likely to be better approach. The recursion tree for $T(n)$ is constructed below.

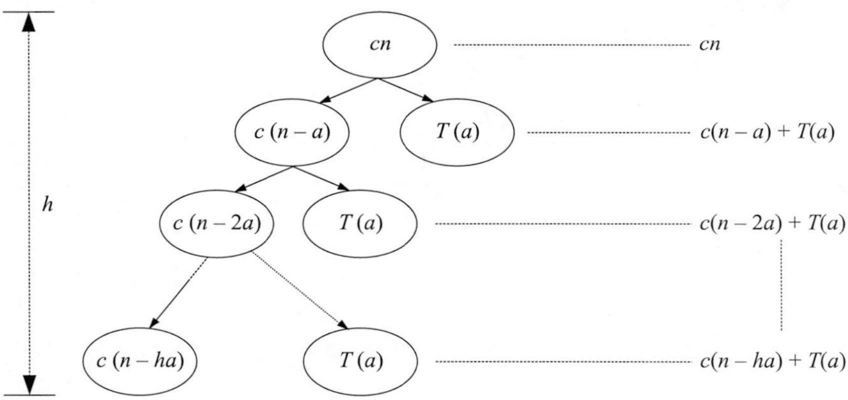

Figure 3.6 Recursion tree for *T*(*n*).

Now, assume that $T(0) = 0$ then $c(n - ha) = 0 \rightarrow h = n/a$

Thus, $T(n) = cn + c(n-a) + T(a) + c(n-a) + T(a) + \ldots + c(n-ha) + T(a)$

$= hcn - ca(1 + 2 + \ldots h) + hT(a)$

$\leq cn^2$ [solving and putting $h = n/a$]

$= O(n^2)$.

Example 30. Use a recursion tree to solve the recurrence,

$$T(n) = T(\alpha n) + T((1-\alpha)n) + cn,$$

Where α is a constant in the range $0 < \alpha < 1$ and $c > 0$ is also a constant.

Solution: the recursion tree of $T(n)$ is Figure 3.7

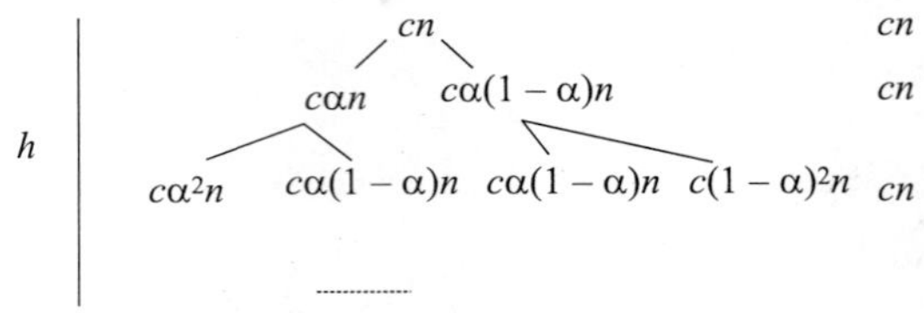

Figure 3.7 Recursion tree for T(n)

Now, $h = \log_{1-\alpha} n$ or $h = \log_{\alpha} n$

The tree will be expanded until

$$T(0) = 0, \text{ i.e. } \alpha = 0 \text{ or } 1 - \alpha = 0 \text{ but } 0 < \alpha < 1 \qquad \therefore \alpha \neq 0, 1$$

As α is uncertain, there are three situations:

(*a*) $0 < \alpha < ½$
(*b*) $\alpha = ½$
(*c*) $½ < \alpha < 1$

Since, the value of α will influences h, the height of the recursion tree, for simplicity we just consider $\alpha = 1/2$. Although the final answer is the same in all the cases.

Therefore, $T(n) = cn + cn + \ldots$ upto h terms.

$$= hcn$$

$$= \frac{\log n}{\log 2} cn \qquad \left[h = \log_{\alpha} n = \log_{1/2} n = \frac{\log n}{\log 2}\right]$$

$$= \Theta(n \log n).$$

Example 31. Consider the following recurrence relation.

(*i*) $T(a) = \Theta(1)$

(*ii*) $T(n) = T(n-a) + T(a) + n$

Find $T(n)$ asymptotically using the iteration method.

Solution: Expanding $T(n)$ using iteration method, we have

$$T(n) = T(n-a) + T(a) + n$$

$$= [T(n-2a) + T(a) + (n-a)] + T(a) + n$$

$$= [T(n-3a) + T(a) + (n-2a)] + 2T(a) + 2n - a$$

$$= T(n-3a) + 3T(a) + 3n - 2a - a$$

$$= [T(n-4a) + 4T(a) + (n-3a)] + 3T(a) + 3n - 2a - a$$

$$= T(n-4a) + 4T(a) + 4n - 3a - 2a - a$$

$$= T(n - ia) + iT(a) + in - \sum_{k=1}^{i-1} ka$$

... ...

$$= T(n - ia) + iT(a) + in - ai(i - 1)/2$$

The iterations will stop when $T(0) = 0 \rightarrow n - ia = 0 \rightarrow i = n/a$

So, $T(n) = T(0) + (n/a)T(a) + n^2/a - a(n/a - 1)(n/a)/2$
$= \Theta(1) + (n/a)\Theta(1) + n^2/a - n^2/2a + n/2$
$= \Theta(1) + O(n) + n^2/2a$
$= O(n^2)$

Example 32. Solve the following recurrence relation.

$$T(n) = \begin{cases} 1, & n = 1 \\ T(n-1) + \frac{1}{n}. & \text{otherwise} \end{cases}$$

Solution: Using iteration method, we have

$T(n) = T(n - 1) + 1/n$
$= T(n - 2) + 1/(n-1) + 1/n$
$= T(n - 3) + 1/(n - 2) + 1/(n-1) + 1/n$
... ...
$= T(n - (n - 1)) + 1/(n - (n - 1) + 1) + ... + 1/(n-2) + 1/(n-1) + 1/n$
$= T(1) + ½ + ... + 1/(n-2) + 1/(n-1) + 1/n$

$$= 1 + \sum_{k=2}^{n} \frac{1}{k}$$

$$= \log n \qquad \left[\text{using } 1 + \sum_{k=2}^{n} \frac{1}{k} = \int_1^n \frac{1}{x} dx\right]$$

$$= \Theta(\log n)$$

Example 33. Suppose, $T(n) = 3T(n/2) + n \log n$. Use the Master–Method to find $T(n)$.

Solution: From the recurrence $T(n)$, we have

$a = 3, b = 2 \rightarrow E = \log_b a = \log_2 3 = 1.5$
Now, $f(n) = n \log n \leq cn^{1.5}$
$\Rightarrow n \log n \leq cn^{1.5}$ [case 1 is applied]
$\Rightarrow \log n < cn^{0.5}$
$\Rightarrow c = 1, n_0 = 16$
Hence, $T(n) = \Theta(n^{\log 2^3}) = \Theta(n^{1.5})$

3.10 SOME COMMON QUESTIONS

Question 1. What are the types of recurrence relation?

Answer: Three types recurrence are possible.

(*a*) Geometric sequence

(*b*) Arithmetic sequence

(*c*) Linear sequence

Question 2. How could we analyze recursive algorithm?

Answer: A general way to analyze recursive algorithm.

(*a*) Decide a parameter responsible for Input Size.

(*b*) Identify the algorithms Basic Operations.

(*c*) Check how many times the basic operation is executed only on the size of an input. Compute for best, average and worst cases.

(*d*) Derive a recurrence relation, with an appropriate initial condition, for the number of times basic operation is executed.

Question 3. What are the types of recurrence relations?

Answer: Two kind of recurrence exist.

(*a*) Homogeneous recurrence relation.

(*b*) Non-homogeneous recurrence relation.

Question 4. What is Growth Order?

Answer: An order of growth of an algorithm is a set of functions whose asymptotic growth behaviour is considered equivalent to the given algorithm. E.g., $2n$ is $O(n)$.

Question 5. What do you mean the order of an algorithm?

Answer: A standard notation used to represent the computing time of the algorithm is termed as the order of an algorithm and is denoted by O.

Exercises

1. Write a function to compute GCD based on the following relations

   (*i*) $GCD(2m, 2n) = 2 * GCD(m, n)$

   (*ii*) $GCD(2m, 2n+1) = GCD(m, 2n+1)$

   (*iii*) $GCD(2m+1, 2n+1) = GCD(n-m, 2m+1)$ *if* $m < n$

   (*iv*) $GCD(m, m) = m$

2. A function is defined by the rule that:

$$f(n) = \begin{cases} n, \text{ if } n < 3 \\ f(n-1) + 2f(n-2) + 3f(n-3), \text{ if } n \geq 3 \end{cases}$$

   Write a procedure that computes *f* by means of a recursive process. Also write a procedure that computes *f* by an iterative process.

3. The following pattern of numbers is called *Pascal's triangle*

1

1 1

1 2 1

1 3 3 1

1 4 6 4 1

... ...

Write a procedure that computes elements of Pascal's triangle by means of recursion.

4. The term *ancestor* is recursive: A parent is an ancestor and an ancestor of a parent is an ancestor. Give some other recursive terms in everyday use.
5. Write a tail recursive routine.
6. What do you mean by recursion? How does it differ with iteration?
7. What are the drawbacks of using recursive procedure?
8. Classify recursions.
9. Mention some features of recursive algorithms.
10. Solve the following recurrences using substitution method.
    - (*i*) $T(n) = 2T(n/2) + O(n)$ **[Ans.** $O(n \lg n)$]
    - (*ii*) $T(n) = 2T(\lfloor n/2 \rfloor) + n$ **[Ans.** $O(n \lg n)$]
    - (*iii*) $T(n) = 2T(\lfloor n/2 \rfloor) + 20 + n$ **[Ans.** $O(n \lg n)$]
11. Solve the following recurrences using iteration method.
    - (*i*) $T(n) = T(n/2) + n$ **[Ans.** $O(n)$]
    - (*ii*) $T(n) = 3T(n/4) + n$ **[Ans.** $O(n)$]
    - (*iii*) $T(n) = 2T(n - 1) + 1$ **[Ans.** $O(2^n)$]
    - (*iv*) $T(n) = \begin{cases} \Theta(1), & n = 1 \\ 2T\left(\frac{n}{2}\right) + \Theta(n). & n > 1 \end{cases}$ **[Ans.** $\Theta(n \log n)$]
    - (*v*) $T(n) = \begin{cases} \Theta(1), & n = 1 \\ T(n-1) + \Theta(n), & n > 1 \end{cases}$ **[Ans.** $O(n^2)$]
    - (*vi*) $T(n) = 2T(n/2) + \Theta(n)$ **[Ans.** $\Theta(n \log n)$]
12. Fibonacci series is defined as follows:

    $f(0) = 0$

    $f(1) = 1$

    $f(n) = f(n-1) + f(n-2)$

    Find both iterative and recursive algorithms to compute Fibonacci(n) for a number n in Fibonacci series. Analyze the running–time of each algorithm. **Ans.** [$O(n)$, $O(2^n)$]
13. Find a close–form expression for the sum of $2i$, i=0 to n. Justify your expression by induction.

14. Sort the following functions into increasing order of magnitude.

(*i*) n^2
(*ii*) $n \log n$
(*iii*) $n-1$
(*iv*) $2^{\log n}$
(*v*) $n/\log n$
(*vi*) 2^n
(*vii*) n^3

15. Solve the following recurrences using master theorem.

(*i*) $T(n) = 3T(n/2) + n^2$ [**Ans.** $\Theta(n^2)$, (Case 3)]
(*ii*) $T(n) = T(n/2) + 2^n$ [**Ans.** $\Theta(2^n)$, (Case 3)]
(*iii*) $T(n) = 2^nT(n/2) + n^n$ [**Ans.** Does not apply, (a is not constant)]
(*iv*) $T(n) = 2T(n/2) + n \log n$ [**Ans.** $\Theta(n \log^2 n)$, (Case 2)]
(*v*) $T(n) = 2T(n/2) + n/\log n$ [**Ans.** Does not apply due to non-Polynomial difference between $f(n)$ and $n^{\log_b a}$)]
(*vi*) $T(n) = 2T(n/4) + n^{0.52}$ [**Ans.** $\Theta(n^{0.52})$, (Case 3)]
(*vii*) $T(n) = 0.5T(n/2) + 1/n$ [Does not apply as $a < 1$]
(*viii*) $T(n) = 16T(n/4) + n!$ [**Ans.** $\Theta(n!)$, (*Case* 3)]
(*ix*) $T(n) = \sqrt{2}\, T(n/2) + \log n$ [**Ans.** $\Theta(\sqrt{n})$, (*Case* 1)]
(*x*) $T(n) = 3T(n/3) + \sqrt{n}$ [**Ans.** $\Theta(n)$, (*Case* 1)]
(*xi*) $T(n) = 3T(n/4)+ n \log n$ [**Ans.** $\Theta(n \log n)$, (*Case* 3)]
(*xii*) $T(n) = 2T(n/2)+ n/2$ [**Ans.** $\Theta(n \log n)$, (*Case* 2)]
(*xiii*) $T(n) = 7T(n/3)+ n^2 \log n$ [**Ans.** $\Theta(n^2 \log n)$, (*Case* 3)]
(*xiv*) $T(n) = 4T(n/2)+ n/ \log n$ [**Ans.** $\Theta(n^2)$, (*Case* 1)]
(*xv*) $T(n) = 64T(n/8) - n^2 \log n$ [**Ans.** Does not apply since $f(n)$ is not positive]
(*xvi*) $T(n) = 4T(n/2)+ \log n$ [**Ans.** $\Theta(n^2)$, (Case 1)]

16. Suppose $T(n) = 8T(n/2) + n^2$ describes the running time of an algorithm A. Another competing algorithm A' has running time $T'(n) = aT'(n/4) + n^2$. Find largest integer value of a such that A' is asymptotically faster than A?

University Questions

1. Write short notes on Recursion tree. ***[WBUT'09]***
2. Write short notes on Removal of Recursion. ***[UPTU'10]***
3. What are recursion's drawbacks? ***[PTU'08]***
4. Define recurrence relation. ***[PTU'08]***
5. What is the Tail Recursion? Give an example. ***[WBUT'07, UPTU'08]***
6. Write short notes on Principles of recursion with example. ***[UPTU'09]***
7. Estimate the asymptotic time complexity by solving the following recurrence relations.

   (*a*) $T(n) = 2T(n-1) + n$, given $T(1) = 1$
   (*b*) $T(n) = 2T(n/2) + n \log n$, given $T(1) = 1$
   (*c*) $T(n) = \sqrt{n}\, T(\sqrt{n}) + n$, given $T(2) = 1$ ***[IITG'11]***

8. Solve the following recurrence relation using generating function: *[WBUT'07]*

   $a_n = 6a_{n-1} - 11a_{n-2} + 6a_{n-3}$ for $n >= 3$ with initial condition $a_0 = 1$, $a_1 = -1$ and $a_2 = 1$.

9. Write a recursive procedure to find the sum of first n natural numbers. *[IGNOU'09]*

10. Solve the following recurrence relation. *[BPUT'05]*

    $T(n) = n$; if $n = 0$ or $n = 1$

    $T(n) = 5T(n-1) - 6T(n-2)$; Otherwise

11. Consider the following recursive function. *[IITG'11]*

```
int C(int n, int k)
  {
    if (k==0 || k==n)
      Return(1);
    else
      Return(C(n – 1, k - 1) + C(n – 1, k))
  }
```

    Estimate the number of recursive invocations as a function of n and k.

12. Let a and b denote positive integers. Suppose a function Q is defined recursively as follows:

$$Q(a,b) = \begin{cases} 0 & \text{if } a < b \\ Q(a-b, b) + 1 & \text{if } b \le a \end{cases}$$

    Find the value of $Q(14, 3)$. What does this function do? *[IETE ALCCS'11]*

13. Show that the following equations are correct? *[WBUT'07]*

    (*a*) $n! = O(n^n)$

    (*b*) $33n^3 + 4n^2 = O(n^2)$

14. How can you convert the following recursion algorithm to tail recursion? *[WBUT'10]*

```
Algorithm: Fib (n)
  if n ≤ 1 then
    Return (n)
  else
    Return Fib (n – 1) + Fib(n – 2)
  end if
```

15. State master's theorem and find the time complexity for the following recurrence:

    $T(n) = 2T(\sqrt{n}) + \log n$

    $T(n) = 1$, if $n = 2$ *[IITG'11, WBUT'09, RTU'09, DOEACC'07]*

16. Solve the following recurrence using Master method: *[UPTU'10]*

    $T(n) = 4T(n/3) + n^2$.

17. Write short notes on Tower of Hanoi problem. *[UPTU'06, '09]*

18. State Tower of Hanoi problem with n rings. Write the recursive relation for its complexity. Prove that Tower of Hanoi problem cannot be solved fewer than $2^n - 1$ movements of rings. ***[DOEACC'06]***

19. Write a recursive function to print the reverse of a string passed to it as an argument. ***[IETE ALCCS'11]***

20. Write a non-recursive algorithm of In-order traversal of a tree and also analyze its time complexity. ***[IITG'07, '09, '11, JNTU'08]***

21. Solve the following recurrence exactly. Express your answer using the Θ-notation. ***[UPTU'06]***

$$t_n = \begin{cases} n & \text{if } n = 0, 1, 2 \text{ or } 3 \\ t_{n-1} + t_{n-3} - t_{n-4} & \text{otherwise} \end{cases}$$

22. What is Ackermann's function give example? ***[PTU'08]***

23. Suppose we have a recurrence relation. ***[WBUT'10, RTU'07]***

$$T(n) = aT\left(\frac{n}{b}\right) + f(n)$$

Show that the followings:

(*a*) If $af\left(\frac{n}{b}\right) = kf(n)$ for some constant $k < 1$. then $T(n) = O(f(n))$.

(*b*) If $af\left(\frac{n}{b}\right) = kf(n)$ for some constant $k > 1$. then $T(n) = O(n \log_b{}^a)$.

(*c*) If $af\left(\frac{n}{b}\right) = kf(n)$ for some constant $k = 1$. then $T(n) = O(f(n)) \log_b{}^n$.

24. Find the Big Omega (Ω), Big Theta (Θ) and Big Oh (*O*) Notations for the following functions. ***[RTU'07]***

(*a*) $f(n) = 6n + 200$

(*b*) $f(n) = 3n^2 + 2$

25. Define the recursion. Write a recursive and non-recursive program to calculate the factorial of the given number. ***[UPTU'08, 10, PTU'08, '09]***

26. Find whether the solution of the following recurrence relations can be obtained using Master theorem or not: ***[DOEACC'08]***

(*a*) $T(n) = 3\,T(n/4) + n \lg n$

(*b*) $T(n) = 2\,T(n/2) + n \lg n$.

27. Write a recursive algorithm that determines the height of a binary tree. ***[IETE ALCCS'11]***

28. Write a recursive function, which computes M^n where both M and n are integers. ***[IETE ALCCS'11]***

29. Solve the recurrence relation by iteration : $T(n) = T(n - 1) + n^4$. ***[UPTU'05]***

30. Can the Master Method be applied to solve the following recurrence? Why?

$$T(n) = 4T(n/2) + n^2 \log n$$ ***[UPTU'05, BPUT'07]***

31. Prove that the following are incorrect bounds: ***[RTU'07]***

(*a*) $10n^2 + 7 \neq 0(n^3)$

(*b*) $3n^2 + 5 \neq 0(n)$

32. Solve the recurrence equation given below: ***[IGNOU'07]***

$f(n) = f(n-1) + f(n-2)$

$f(0) = 0$

$f(1) = 1$

33. Solve the recurrence equation given below. ***[BPUT'07]***

$f(\text{n}) = f(n-2) + n\ ,\ n > 1$

$f(0) = c$

$f(1) = d$

34. Solve the following recurrence relation: ***[BPUT'08]***

$T(n) = 4T(n-1) - 3T(n-2) + 4n\, 3^n$

$T(0) = 1,\ T(1) = 12$

35. Write the non-recursive algorithm for finding the fibonacci sequence and derive its time complexity. ***[JNTU'08]***

36. Solve the following Recurrence using Recursive Tree Method:

$T(n) = T(n/3) + T(2n/3) + n$ ***[DOEACC'08]***

37. Show that $f(n) = 4n^2 - 64n + 288 = O(n^2)$. ***[JNTU'08]***

38. Find the complexity of the following recurrence relation. ***[UPTU'06]***

$T(n) = 9T(n/3) + n$

$T(1) = 1,\ T(n) = 4T(n/3) + n^2$ for $x \geq 2$

39. Solve the recurrence. ***[IGNOU'09]***

$T(n) = 2T(n/2) + n \geq 2$

$\quad = 1 \qquad n \geq 2$

40. What is a recurrence relation? Solve the following relation by recurrence tree method:

$T(n) = 2T(n/2) + n^2$ ***[MDU'09]***

41. Write a recursive algorithm to compute the sum of squares from given number m to n i.e. Compute $m^2 + (m+1)^2 \ldots + n^2$. ***[IETE ALCCS'11]***

❑❑❑

Chapter 4 Divide-and-Conquer Paradigm

4.1 INTRODUCTION

Most of the algorithms are recursive in nature. For solving a large problem, they call themselves repeatedly one or more times. All these algorithms follow divide-and-conquer approach to solve them. It is a powerful algorithm design technique for solving a variety of problems. Divide-and-Conquer (D&C) is based on what is said multi-branched recursion paradigm. In different cases, D&C is considered as the basis of their efficient algorithms such as sorting (Quicksort, Mergesort), searching (binary), syntactic analysis (top-down parsers), etc.

4.2 METHODOLOGY

D&C concept follows top-down technique for designing algorithms. Initially, the problem is divided into smaller sub-problems ensuring that the solutions of the subproblems are easier to find. Then organising the partial solutions, the solution of the original problem is obtained.

More formally, Divide-and-Conquer paradigm in its simplest form consists of following three major phases:

1. **Divide:** Break the whole problem into similar, smaller several sub-problems. If the subproblems are relatively still large the Divide-and-Conquer strategy is reapplied.
2. **Conquer:** Solve the sub-problem recursively with successively, separately and independently. Subproblems are conquered only if they are small enough to be solved otherwise step1 is repeated.
3. **Combine:** Finally, these partial solutions of sub-problems are combined to create a solution to the original problem.

4.3 ADVANTAGES

Several advantages are provided by this Divide-and-Conquer algorithm design technique. Some of them are listed below.

1. **Solving Difficult Problems:** Difficult problems such as the classic Tower of Hanoi problem can be solved effectively and conceptually.
2. **Improve Algorithm Efficiency:** The approach often helps in efficient algorithms design. In most of the cases like sorting, searching it leads to an improvement in the asymptotic cost of the solution in $O(n \log n)$ time.
3. **Parallelism:** It can be adapted for execution in multi-processor machines, especially shared-memory systems.

4. **Memory Access:** It uses cache memory without accessing the slower main memory once a sub-problem is small enough.
5. **Accuracy Control:** Sometimes it yields more accurate results through rounding off them in some arithmetic operations with floating point numbers than an equivalent iterative method.

4.4 USES

The main objectives of using D&C approach are basically:

(*i*) It is an efficient approach.

(*ii*) To represent running time in recurrence form.

(*iii*) To design algorithms with lower time complexity for large input n.

4.5 APPLICATIONS

Using D&C approach we generally try to solve a large problem recursively, efficiently. The problem areas mainly include:

1. Sorting: Mergesort, Quicksort.
2. Searching: Binary search.
3. Multiplication: Strassen's matrix multiplication, integer multiplication.
4. Compiler: Syntactic analysis in top-down parser.
5. Computing FFT.
6. Finding convex hull.
7. Finding maximum and minimum from a set of elements.

4.6 QUICKSORT

The basic version of Quicksort algorithm was invented by C. A. R. Hoare in 1960. However formally it was introduced in 1962. Its working principle is based on the principle of Divide-and-Conquers approach.

4.6.1 Quicksort: Advantages

Quicksort is an efficient sorting algorithm of programmers' choice in many places because of the following reasons:

- Ease of implementation.
- Serves as good "general purpose" sorting technique.
- Consumes relatively fewer resources during execution.
- In-place sorting *i.e.*, no need of auxiliary storage for the elements to be sorted.
- May use a small auxiliary stack.
- Average case complexity is $\Theta(n \log n)$ to sort n items.
- The algorithm has been subjected to a thorough mathematical analysis.
- A very precise statement can be made about performance issues.

4.6.2 Quicksort: Disadvantages

Major drawbacks of this popular sorting technique are listed below:

- Without recursion, its implementation becomes extremely complicated.
- The worst-case running time is $O(n^2) >> \Theta(n \log n)$ for large n.
- Simple mistake in the implementation make it fragile.

4.6.3 Quicksort: Description

Quicksort algorithm sorts a list of elements $L[p ..... r]$ by employing a D&C strategy. It partitions L and performs the following steps.

1. Choose randomly an element q, called a *pivot*, from the list.
2. Divide L at q into two non-empty sub-arrays $L[p ... q]$ and $L[q+1 ... r]$ such that every key in $L[p ... q] \leq$ every key in $L[q+1 .. r]$. This is called **partition** operation.
3. Recursively sort the sub-arrays $L[p .. q]$ and $L[q+1 .. r]$ using Quicksort.
4. The entire array is thus sorted. No need to combine them.

The base case may be either the list of size zero or contains single element, which is always sorted.

4.6.3.1 *Quicksort : Algorithm*

Algorithm: Quick_Sort(L, first, last)

Where L : list of elements.
n : no. of elements in L.
first : first position of $L = 1$.
last : last position of $L = n$.

Step 1: Initialise pivot
(*i*) low = first
(*ii*) high = last
(*iii*) pivot = L[(low + high)/2] // you may take any element in L

Step 2: Repeat through step 5 while low ≤ high

Step 3: Repeat while pivot > L[low]
low = low + 1.

Step 4: Repeat while pivot < L[high]
high = high – 1.

Step 5: If low ≤ high // Comparisons and Exchange
(*i*) L[low] ⟷ L[high]
(*ii*) low = low + 1
(*iii*) high = high – 1

Step 6: Recursively call sub-lists
(*i*) if first < high
(*a*) call Quick_Sort(L, first, high)
(*ii*) if low < last
(*a*) call Quick_Sort(L, low, last)

Step 7: Repeat for $i = 1$ to n // display sorted list
(*a*) Output $L[i]$

Step 8: Exit

4.6.3.2 *Quicksort: Analysis*

The running time of quicksort depends on whether the partitioning is balanced or unbalanced.

- ***Best Case:*** In this case the list is divided into two equal or almost equal sublists. The partitioning is shown in Figure 4.1. The number of levels in the corresponding recursion tree will be $h = \log n$. And at each level the partitioning is linear in size of $\Theta(n)$. The recurrence for this case is given below.

$$T(n) \leq 2T(n/2) + \Theta(n)$$
$$= \Theta(n \log n) \qquad \text{[using master theorem]}$$

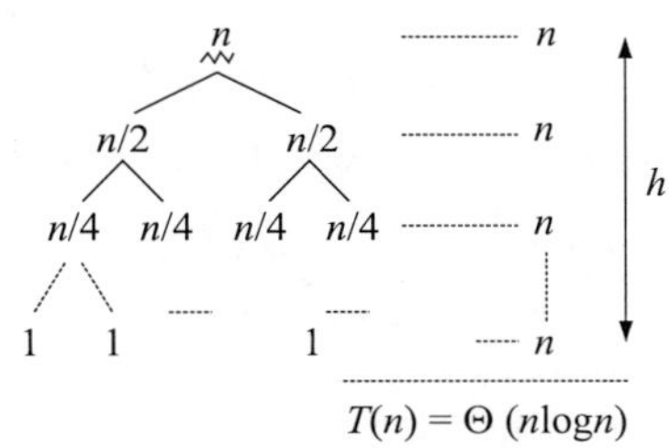

Figure 4.1 Recursion tree for best case analysis of quicksort.

- ***Average Case:*** Pivot is chosen randomly. Therefore, some of the partitions are balanced and some are not at every level. Consider the list is partitioned in 9:1 ratio. Then the recurrence is so obtained:

$$T(n) = T(9n/10) + T(n/10) + n = O(n \log n)$$

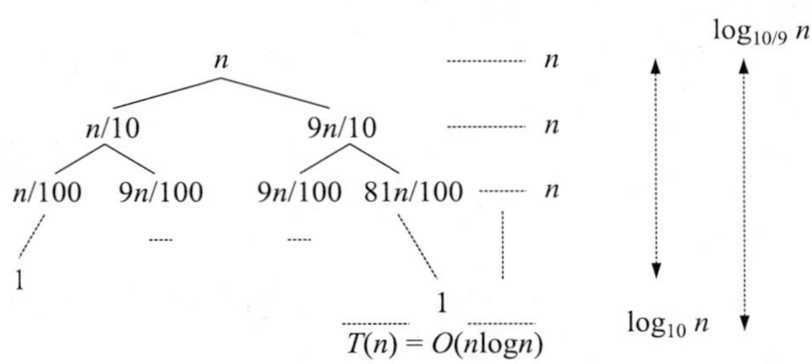

Figure 4.2 Recursion tree for average case analysis of quicksort.

- ***Worst Case:*** This case occurs when the partition is fully unbalanced. It is possible when one sub-list contains no element and another contains all elements at each level. This generates $n - 1$ levels in recursion tree. The corresponding recurrence is:

$$T(n) = T(n-1) + \Theta(n) = \sum_{k=1}^{n} k = O(n^2)$$

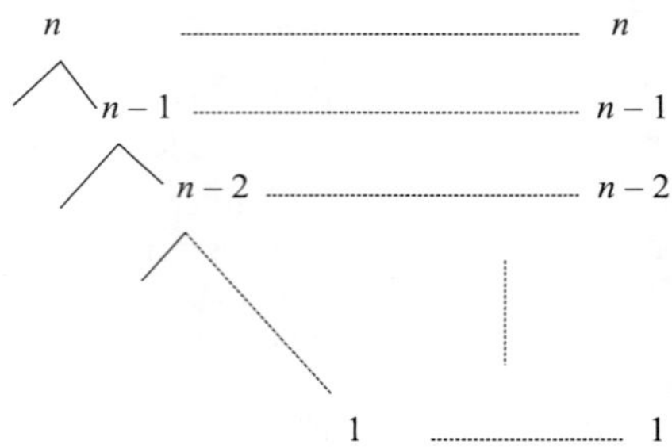

$$T(n) = n + n - 1 + ...+ 2 + 1$$
$$= O(n^2)$$

Figure 4.3 Recursion tree for worst case analysis of quicksort.

Example 1. Sort the following list of elements using quick sorting.

$$L = < 1, 2, 4, 3, 7, 6, 5, 9 >$$

Solution: Suppose the middle element is our pivot element. Then based on the quick-sort algorithm, above list has been sorted in a sequence given below.

First Partition:

Lower index first = 1.

Maximum index last = 8

Now low = first = 1, high = last = 8

Pivot $= L\left[\dfrac{\text{low} + \text{high}}{2}\right] = L\left[\dfrac{1+8}{2}\right] = \text{L}[4] = 3$

Now, continue while low ≤ high as: // Step 2

A. Pivot > L[low] → 3 > L[1] → 3 > 1 ← true // Step 3
low = low + 1 = 1 + 1 = 2.
Similarly,

B. Pivot > L[low] → 3 > L[2] → 3 > 2 ← true
low = low + 1 = 2 + 1 = 3.

C. Pivot > L[low] → 3 > L[3] → 3 > 4 ← false. So we iterate step 4.

D. Pivot < L[high] → 3 < L[8] → 3 < 9 ← true
high = high – 1 = 8 – 1= 7

Similarly,

E. Pivot < L[high] → 3 < L[7] → 3 < 5 ← true
high = high – 1 = 7 – 1= 6

F. Pivot < L[high] → 3 < L[6] → 3 < 6 ← true
high = high – 1 = 6 – 1= 5

G. Pivot < L[high] → 3 < L[5] → 3 < 7 ← true
high = high – 1 = 5 – 1= 4

H. Pivot < L[high] → 3 < L[4] → 3 < 3 ← false.
So, step 5 is considered.

I. Exchange L[low] and L[high]

$L[\text{low}] = L[\text{high}] = L[4] = 3$

$L[\text{high}] = L[\text{low}] = L[3] = 4$

low = low + 1= 3 + 1 = 4

high = high – 1= 4 – 1= 3.

Now, L = <1, 2, 3, 4, 7, 6, 5, 9 >. Thus L is partitioned into two sublists as <1, 2, 3> and < 4, 7, 6, 5, 9 >.

Second Partition:

Applying above procedure to these sublists,

A. The first sub-list L[1..3] is already sorted. No need to call Quick_Sort(L, first, high).

B. Call Quick_Sort(L, low, last).

Lower index first = low = 4.

Maximum index last = high = 8

$$\text{Pivot} = L\left\lceil \frac{\text{low} + \text{high}}{2} \right\rceil = L\left\lceil \frac{4+8}{2} \right\rceil = L[6] = 6$$

Now low ≤ high → 4 ≤ 8 ←true

a. Pivot > L[low] → 6 > L[4] → 6 > 4 ← true
low = low + 1= 4 + 1= 5.

b. Pivot > L[low] → 6 > L[5] → 6 > 7 ← false

c. Pivot < L[high] → 6 < L[8] → 6 < 9 ← true
high = high – 1= 8 – 1 = 7

d. Pivot < L[high] → 6 < L[7] → 6 < 5 ← false

e. low ≤ high → 5≤5 → true

f. Exchange

$L[\text{low}] = L[\text{high}] = 5$

$L[\text{high}] = L[\text{low}] = 7$

low = low + 1= 5 + 1 = 6

high = high – 1= 7 – 1= 6.

Now the sublist < 4, 7, 6, 5, 9 > is subdivided as < 4, 5 > and < 6, 7, 9 >

These are sorted lists. Thus sorting completed and the list is <1, 2, 3, 4, 5, 6, 7, 9 >.

Example 2. Sort the following list of elements using quick sorting.

$$L = < 2, 5, 3, 7, 6, 4 >$$

Solution: Suppose the 2nd element is our pivot element.

Now, first = low = 1

last = high = 6

Pivot $P = L[2] = 5$

Scanning L from left to right till $P >$ the elements in L

$$P > L[\text{low}] \Rightarrow 5 > 2 \Rightarrow \text{low} = \text{low} + 1 = 1 + 1 = 2$$

$$\text{Similarly, } 5 > 5 \Rightarrow \text{low} = 3$$

$5 > 3 \Rightarrow$ low = 4

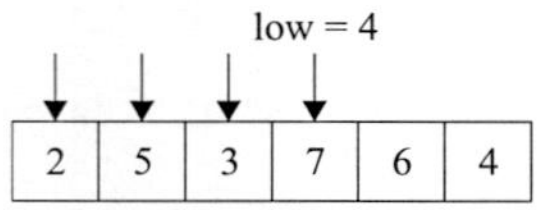

Scanning from right to left till $P <$ the elements in L.

$P < L[\text{high}] \Rightarrow 5 < 4 \Rightarrow$ false. So no change in high value *i.e.,* high = 6.

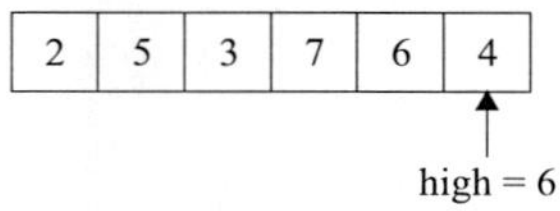

Since low ≤ high, exchange between L[low] and L[high]

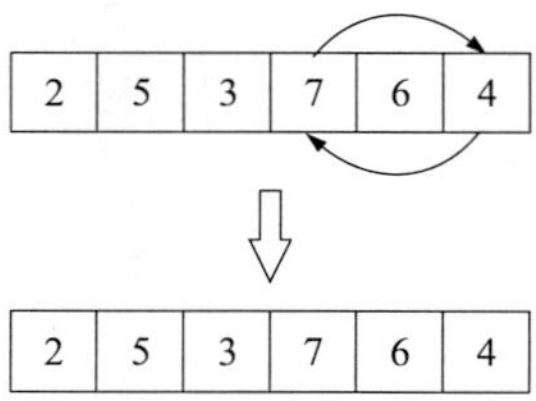

low = low + 1 = 4 + 1 = 5

high = high – 1 = 6 – 1 = 5

low ≠ high ⇒ need not to partition L. Simply L is divided as:

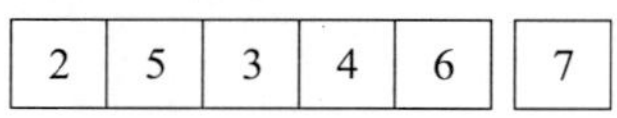

Now, first = low = 1

last = high = 5

pivot $P = L[2] = 5$

Proceeding in the same way the sorted L can be obtained as below.

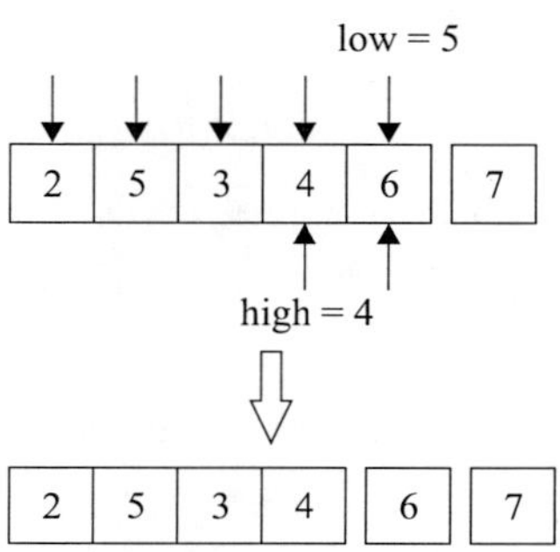

first = low = 1
last = high = 4

first = low = 1

last = high = 4

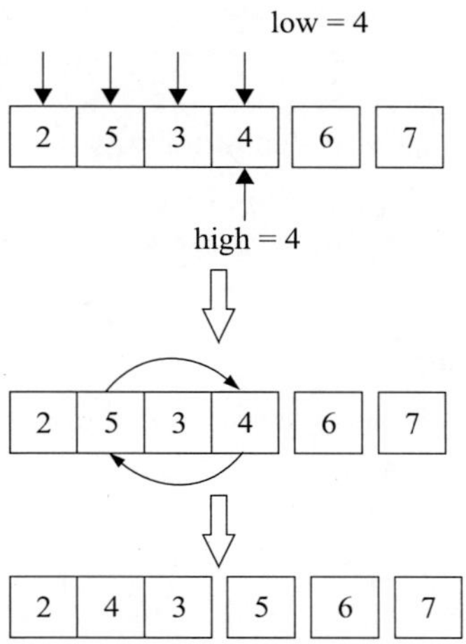

first = low = 1

last = high = 3

Pivot $P = L[2] = 4$

Now, first = low = 1

last = high = 2

Pivot $P = L[2] = 3$

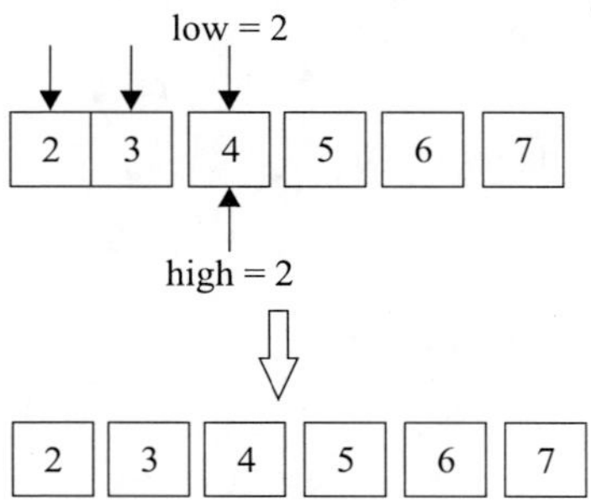

Thus, the list L is sorted.

4.7 MERGESORT

Mergesort is an $O(n \log n)$ time comparison-based sorting algorithm that is based on D&C approach. In most implementations it is stable, meaning that it preserves the input order of equal elements in the sorted output. The algorithm was invented by John von Neumann in 1945. Merge sorting works solving a problem in a top-down manner.

4.7.1 Mergesort: Description

Working principle of Mergesort can conceptually be described as follows:

1. Halves the unsorted list L into two sub-lists A and B.
2. Sort each sub-list recursively by re-applying Mergesort.
3. Merge the two sub-lists back into one sorted list. The process is continued until size of the sorted list is same as that of initially taken list.

The conquer step (step 3) of Merge sorting consists of merging two sorted sublists A and B into a sorted L. Merging these A and B, each with $n/2$ elements takes $O(n)$ time.

4.7.1.1 *Mergesort : Algorithm*

The Mergesort algorithm can be described in the following steps:

Algorithm: Mergesort (L, first, last)

Where L : list of elements.

first : lower bound of L.

last : upper bound of L.

Step 1: Initialise

(*i*) d = last – first +1

Step 2: Check L size d

(*i*) if $d \leq 1$

(*a*) Goto step 4

Step 3: If first < last

(*i*) mid = (first + last) / 2

(*ii*) Call Mergesort(L, first, mid)

(*iii*) Call Mergesort(L, mid + 1, last)

(*iv*) Call Merge(L, first, mid, last)

Step 4: Exit

Sub-Algorithm: Merge (L, first, mid, last)

Step 1: Initialise

(*i*) i = first

(*ii*) j = mid + 1

(*iii*) $k = 1$

Step 2: Repeat while i ≤ mid and j < last // starts merging

(*i*) If $L[i] < L[j]$

(*a*) temp[k] = L[i]

(*b*) $i = i + 1$

Else

(a) temp[k] = L[j]

(b) $j = j + 1$

(ii) $k = k + 1$

Step 3: Repeat while $i \leq$ mid // put rest elements of first sub-list in temporary list

(i) temp[k] = L[i]

(ii) $i = i + 1$

(iii) $k = k + 1$

Step 4: Repeat while $j <$ last // put rest elements of second sub-list in temporary list

(i) temp[k] = L[j]

(ii) $j = j + 1$

(iii) $k = k + 1$

Step 5: Repeat for i = first to last // replace all elements of temp to L

(i) L[i] = temp[i]

Step 6: Repeat for $i = 1$ to $|L|$ // display sorted list

(i) Output L[i]

Step 7: Exit.

4.7.1.2 *Mergesort: Analysis*

Merge sorting incorporates two main ideas to improve its runtime:

(a) Use of small list that takes fewer steps to sort.
(b) Fewer steps are required to construct a merged list.

Now the following cases may arise during analysing Mergesort algorithm.

(i) If input size n is small enough say $n \leq c$ where c = const. Then $T(n) = \Theta(1)$
(ii) Dividing L into two sublists requires $\Theta(1)$ time.
(iii) Solving two equal or almost equal sub-lists recursively requires $2T(n/2)$ time.
(iv) Merging two sorted lists requires $\Theta(n)$ time.

Thus, the recurrence for Mergesort is given by:

$$T(n) = \begin{cases} \Theta(1) & \text{for } n = 1 \\ 2T\left(\dfrac{n}{2}\right) + \Theta(n) + \Theta(1) & \text{otherwise} \end{cases}$$

or,

$$T(\text{n}) = \begin{cases} c, & \text{for } n = 1 \\ 2T\left(\dfrac{n}{2}\right) + cn, & \text{otherwise} \end{cases}$$

$$= O(n \log n) \qquad \text{[Using Master theorem]}$$

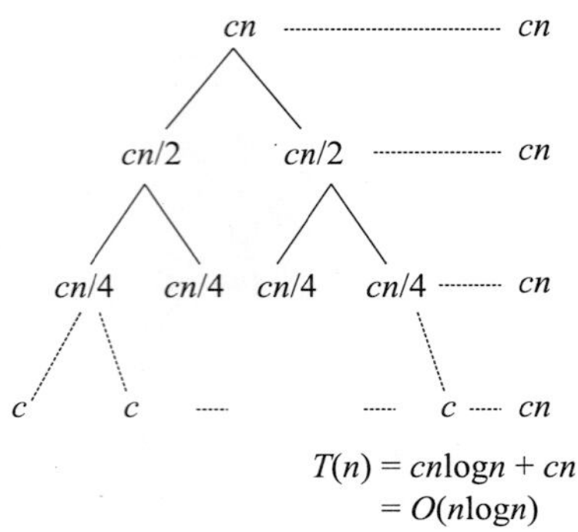

Figure 4.4 Recursion tree for mergesort.

Example 3. Sort the following list of elements 38, 27, 43, 3, 9, 82, 10. Using Mergesort algorithm in ascending order.

Solution: An execution of mergesort is depicted by a binary tree.

(*i*) Each node except the last represents a recursive call of mergesort and stores unsorted sequence before the execution and its partition.
(*ii*) Sorted list is the leaf node obtained at the end of the execution.
(*iii*) The root is the initial call of the procedure.
(*iv*) The nodes before merging are sub-sequences of size 0 or 1.

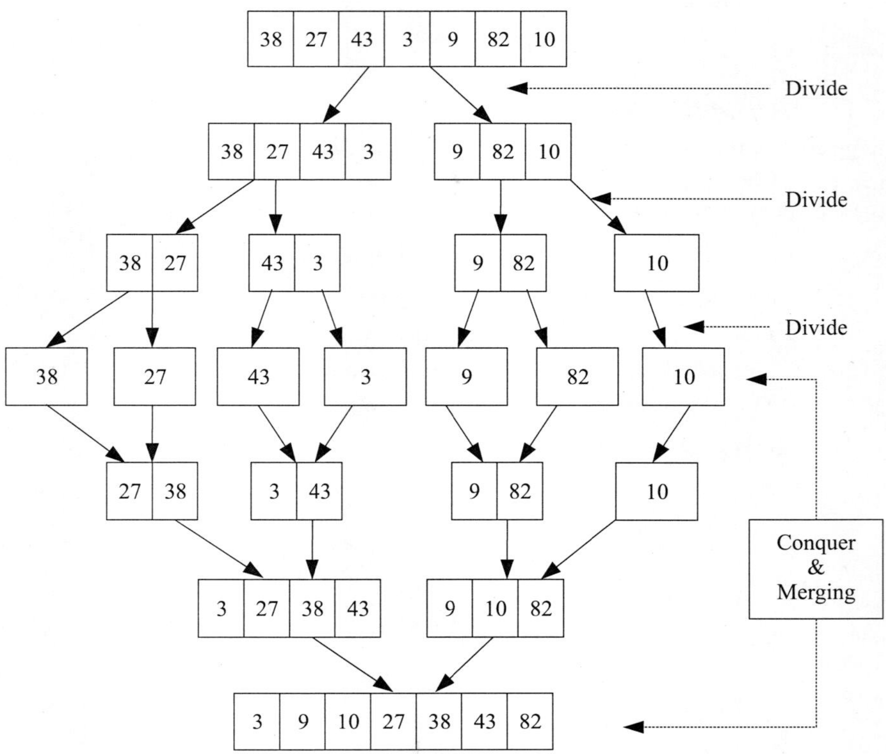

Figure 4.5 Pictorial representation of mergesort algorithm.

Example 4. Sort the following lists each having single element as 5, 2, 4, 6, 1, 3, 2, 6 respectively using Mergesort algorithm in ascending order.

Solution: To sort a lists using Merge sorting

(*i*) Leaves contain the elements (list) which are already sorted by them.
(*ii*) Merge two sorted lists into a single list.
(*iii*) Continue step (*ii*) to have a final sorted list at root.

The process is shown by the following binary tree.

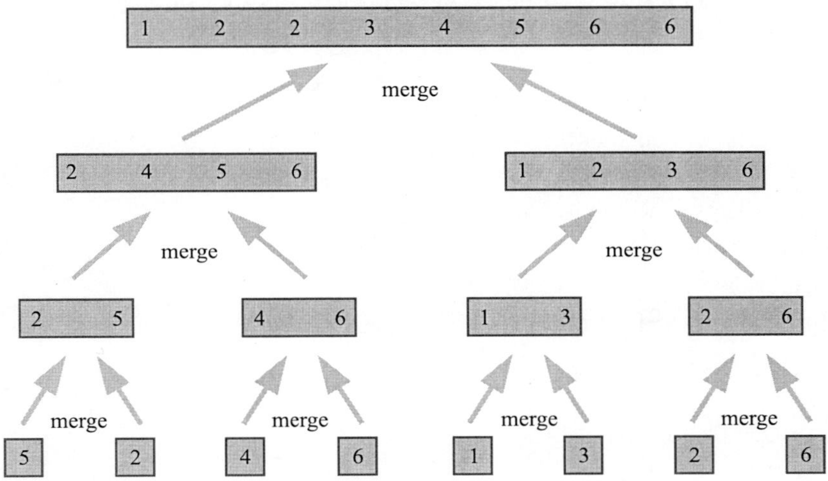

Figure 4.6 Pictorial representation of mergesort algorithm.

4.7.2 Mergesort: Drawbacks

The main drawbacks of Merge sorting in contrast to quicksort are:

(*i*) It requires an extra auxiliary array of size n. Since it requires $2n$ spaces for sorting. That's why it is not in-place sorting.
(*ii*) On average it is about twice as slow.
(*iii*) Merge sorting always executes $2n \log_2 n$ steps.

4.8 BINARY SEARCH

Binary search is one of the fundamental algorithms in computer science. It is an extremely well-known illustration of divide-and-conquer paradigm. Above these it is an efficient searching methodology as compared to linear or sequential searching on a sorted list.

4.8.1 Binary Search: Methodology

Consider a sorted list L of n elements. Then apply the following steps to search the key element d in L.

(*i*) Find the middle element of L *i.e.*, $L[\text{mid}] = L[n/2]$
(*ii*) Compare d with $L[\text{mid}]$

(*a*) If $d = L[\text{mid}]$
- Search is successful.

(*b*) $d < L[\text{mid}]$
- Divide L into two halves and search d in first half.

(*c*) $d > L[\text{mid}]$
- divide L into two halves and search d in second half.

(*iii*) Repeat above steps until d is found or search is exhausted.

4.8.2 Binary Search: Algorithm

Binary search techniques searches data element in minimum possible comparisons. So the efficient algorithm will be as:

Algorithm: Binary_Search (*L, lb, ub, d*)

Where L : sorted list of n elements.

lb : lower bound of L.

ub : upper bound of L.

d : key element to be searched.

Step 1: If L has one element

(*i*) $k = 0$

(*ii*) if $lb = ub$ and $d = L[lb]$

(*a*) $k = 1$

(*iii*) Exit.

Step 2: Divide L into halves

(*i*) $\text{mid} = (lb + ub)/2$

(*ii*) if $d = L[\text{mid}]$

(*a*) $k = 1$

(*b*) Exit

(*iii*) If $d < L[\text{mid}]$

(*a*) Call Binary_Search(*L*, *lb*, mid – 1, *d*)

(*iv*) If $d > L[\text{mid}]$

(*a*) Binary_Search(*L*, mid + 1, *ub*, *d*)

Step 3: If $k = 1$ // Status of d

(*i*) Output d is present and searching successful.

Else

(*i*) Output d is absent and searching unsuccessful.

Step 4: Exit

4.8.3 Binary Search: Analysis

In a sorted list L, the binary search eliminates half parts of the search pool L in every comparison. Initially all items are represented by the roots of the recursion trees as shown Figure 4.7, Figure 4.8, and Figure 4.9. Rather than beginning the search at the start of the list, a binary search begins in the middle of the sorted list.

Best Case: If that middle element is at the root of the tree *i.e.*, the target key is available in single step execution. Thus, the time complexity is $\Omega(1)$.

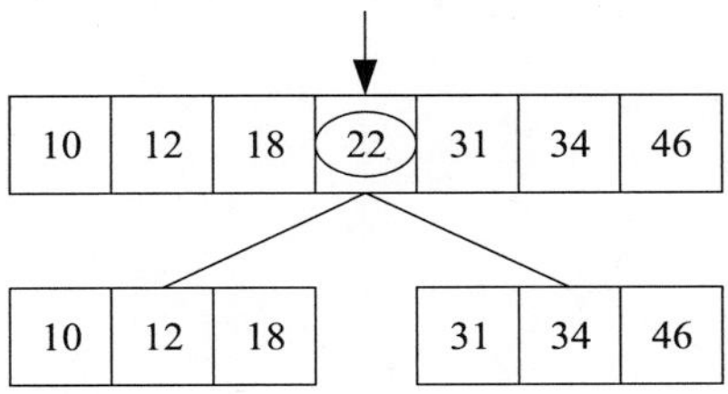

Figure 4.7 Recursion tree for best case binary search.

Average Case: If that middle element L[mid] is not the key, the searching continues. Since the list is sorted, we will eliminate half of the list depending on:

- If the key > L[mid], our search continues with upper indices ($n/2 + 1$ to n).
- If the key < L[mid], our search continues with lower indices (0 to $n/2 - 1$).

Suppose, the key = 12 and it is found in any inner node of the tree, then recurrence for average case is given by:

$$T(n) = T(n/2) + 1 = \Theta(\log n)$$

Maximum height of the tree $h = \log n - 2$. Since root and leaf levels are out of consideration.

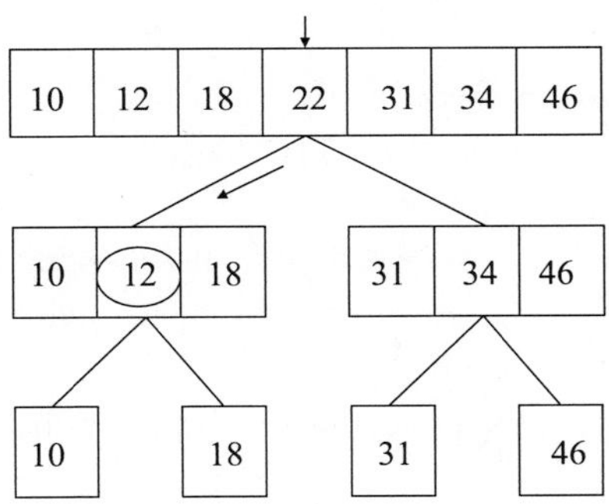

Figure 4.8 Recursion tree for average case binary search.

Worst Case: In this case the target key element is either present at the leaf node or absent in the list. We continue searching as in average case till the key is found or searching completed.

Suppose, the key = 31 and it is found in any leaf node of the tree, then recurrence for average case is given by:

$$T(n) = T(n/2) + 1 = O(\log n)$$

In this case maximum height of the tree $h = \log n$.

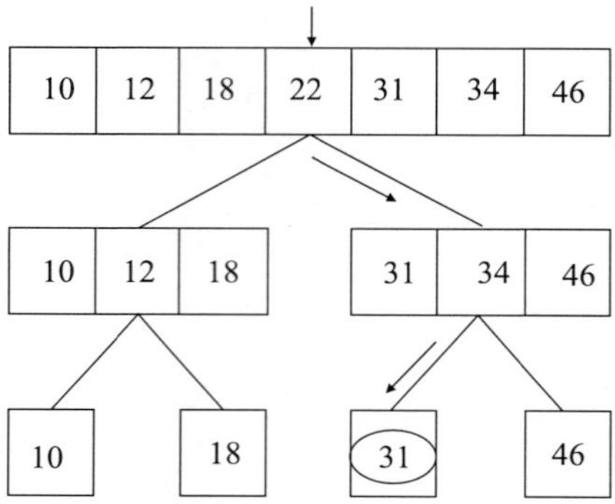

Figure 4.9 Recursion tree for worst case binary search.

4.8.4 Binary Search: Points to Remember

(*i*) Each Comparison halves the search pool or list.

(*ii*) We resume our search repeating the process, beginning again with the middle element of the remaining either of the sub-lists.

(*iii*) If still not the key found, we do eliminate half of the search list.

(*iv*) This process (steps (*ii*) and (*iii*)) is continued until the key is found.

(*v*) A binary search is left with half the data after the first comparison, one quarter of the data after the second comparison, one-eighth after the third and so on.

(*vi*) It has a complexity of order $O(\log n)$.

4.8.5 Binary vs. Linear Searching

Binary searching is more preferable over the linear searching. This is because of the following reasons.

(*i*) The binary search is clearly more efficient than the linear search.

(*ii*) Primarily binary search requires that the list of elements should be sorted.

(*iii*) It is quite easy to implement.

(*iv*) If efficiency is not a big issue, linear search is probably the best choice, for its simplicity.

4.9 SOME COMMON QUESTIONS

Question 1. Give some examples of divide and conquer method.

Answer: Some of the examples of D&C method are binary search, quicksort, merge-sort, Euclidean algorithm, Cooley-Tukey Fast Fourier transform (FFT) algorithm etc.

Question 2. What is the time complexity of binary search?

Answer: Best Case – $O(1)$, Average Case – $O(\log n)$, Worst Case – $O(\log n)$.

Question 3. Consider the binary search algorithm splits the input into three sets of sizes approximately one-third. Write down the recurrence for this ternary search algorithm followed by the asymptotic complexity of this algorithm.

Answer: The recurrence for the number of comparisons made by this ternary search is:

$$T(n) = T(n/3) + 2 = \Theta(\log n) \qquad \text{[using master theorem]}$$

Question 4. Consider the binary search algorithm that splits the input into two sets of sizes approximately one-third and two-thirds. Write down the recurrence for this search algorithm and the asymptotic complexity of this algorithm.

Answer: The searching creates two partitions, one of roughly $n/3$ elements and the other of $2n/3$ elements. Here the worst case arises when the recursive call is on the larger partition. Thus the recurrence corresponding to this worst case number of comparisons is given by:

$$T(n) = T(2n/3) + 1 = \Theta(\log n) \qquad \text{[using master theorem]}$$

Question 5. Consider the binary search algorithm that splits the input into k-almost partitions. Write down the recurrence for this search algorithm.

Answer: the same results for general k-ary search as n approaches infinity. The recurrence will be as below.

$$T(n) = T(n/k) + k - 1, \qquad \text{for best case}$$

$$T(n) = T\left(\frac{(k-1)}{k}n\right) + 1, \qquad \text{for worst case}$$

Exercises

1. Devise an iterative min-max algorithm that finds both the minimum and maximum elements from a set of n elements using only $\left(\frac{3n}{2} - 2\right)$ comparisons, where n is power of 2.
2. Write merits and demerits of quicksort.
3. Analyse complexity behaviour of binary search algorithm.
4. Why do we use Divide-and-Conquer algorithmic paradigm?
5. Write a divide-and-conquer version of Linear_Search algorithm which starts with dividing the input elements into approximately two halves. Find space complexity required by the algorithm.
6. Design a divide-and-conquer algorithm to find the sum of all elements in an array of n integers.
7. Modify Binary_Search algorithm so that it searches for two keys k_1 and k_2 present at two distinct positions x_1 and x_2 in an array A containing n elements respectively.
8. Design a searching algorithm that divides a sorted array into one-third and two-thirds instead of two halves. Analyse your algorithm.
9. Modify Binary_Search algorithm so that it divides the sorted array into three equal parts instead of two. In each iteration, the algorithm must test the element k to be searched for against two entries in the array. Analyse the time complexity of the algorithm.
10. Let $A[1.....n]$ be an array of n integers and x an integer in it. Derive a divide-and-conquer algorithm to find the number of times (the frequency) x appears in A. What is the time complexity of your algorithm?

11. Use Mergesort algorithm to sort the following arrays.

(*a*) 12 25 17 19 51 32 45 18 22 37 15 80 29 45 21

(*b*) 32 15 12 13 16 19 20 25 30 31 17 33 24 20

12. Prove that $\Theta(n)$ is space complexity of Mergesort Algorithm.

13. Using mathematical induction prove the correctness of Mergesort Algorithm. Assume that algorithm works correctly.

14. Consider that Mergesort algorithm first divides the input array A[low....high] of n elements into four parts A_1, A_2, A_3 and A_4 instead of two. It then sorts each sub-array recursively, and finally merges these four sorted parts to obtain the original array in sorted order. For simplicity assume that n is a power of 4.

(*a*) Write the modified algorithm.

(*b*) Analyse its running time.

15. What will be the running time of the modified algorithm in Exercise 4.14 if the input array A is divided into k parts instead of 4? Here, k is a *fixed* positive integer greater than 1.

16. Use Quicksort Algorithm to sort the lists.

(*a*) 24 33 24 45 12 24 12 15 18 16 20 21 13

(*b*) 3 4 5 6 7 8 10 12 13 15 17 19 20

(*c*) 3 32 7 18 45 11 63 12 19 16 25 52 14

17. Show that the space complexity of Quicksort Algorithm varies between $\Theta(\log n)$ and $\Theta(n)$. What is its average space complexity?

18. What will be the behaviour of Quicksort Algorithm if the input is already sorted in descending order? You may assume the input elements are all distinct.

19. Explain Quicksort Algorithm behaviour on the input array $A[1.....n]$ containing n identical elements.

20. Modify Quicksort Algorithm in such a way that it can solve the selection problem. Analyse the time complexity of the new algorithm.

21. Give an iterative version of Quicksort Algorithm.

22. Which of the following sorting algorithms are stable?

(*a*) Heapsort

(*b*) Mergesort

(*c*) Quicksort.

23. An *adaptive* sorting algorithm is one whose running time depends on both the number of elements n and their order. Which of the following sorting algorithms are adaptive then?

(*a*) Heapsort

(*b*) Quicksort

(*c*) Radixsort

(*d*) Mergesort

(*e*) Bubble sort

(*f*) Insertion sort

(*g*) Selection sort

24. Determine whether two given binary trees T_1 and T_2 are identical by designing a divide-and-conquer algorithm. Analyse its complexities.

25. Design a divide-and-conquer algorithm that computes the height of a binary tree.

26. Write an algorithm to find the second largest element in an array of n numbers. Your algorithm must follow divide-and-conquer principle. Derive the time complexity of your algorithm.

University Questions

1. What is the divide-and-conquer method? *[RTU'09, MDU'09, JNTU'08]*
2. Explain the basic concept of a divide-and-conquer algorithm. *[WBUT'07, PTU'08]*
3. Describe divide-and-conquer strategy to find optimal solution to the problem. To which type of problems it is applied? *[MDU'08, IETE ALCCS'11]*
4. Derive the asymptotic time complexity of a non-recursive, binary search algorithm using divide-and-conquer approach. What is the recurrence relation for the algorithm? Find the time complexity of the algorithm. *[BPUT'07, '08, IGNOU'09]*
5. Design a divide-and-conquer type algorithm to compute the maximum (MAX) and the minimum (MIN) of a set of n numbers. Estimate the number of comparisons forming the recurrence relation. *[IITG'11]*
6. Write a recursive algorithm for binary search tree and complexity. *[PTU'09, RTU'10]*
7. Write down the difference between linear search and binary search. *[WBUT'07, RTU'11]*
8. Write an algorithm for binary search and discuss its limitations. *[PTU'05, '07, '09]*
9. Design Divide-and-Conquer algorithm for computing the number of levels in a binary tree. *[JNTU'09]*
10. Write an algorithm to search an element using binary search method. Also give its recurrence relation. Analyze this for worst case, average case and best case conditions. *[RTU'08, MDU'08, '09, IGNOU'07, BPUT'05, DOEACC'08]*
11. Name three conditions under which sequential search of a list is preferable to binary search. *[PTU'08]*
12. Explain how Binary Search Algorithm finds an element in an array. Consider the following array: <11,22,33,44,55,66,77,88>. Find 50 and 77 in the above array. Show steps. *[IGNOU'07]*
13. Is the running time of mergesort depends on the value of the keys in the input file? Explain your answer. *[BPUT'05]*
14. Sort the following using mergesort method:: 75,10,20,70,80,90,100,40,30,50. *[UPTU'10]*
15. Sort the following sequence using mergesort method: <5, 2, 4, 7, 1, 3, 2, 6>. *[RTU'09]*
16. Sort the following data using mergesort. Discuss the time complexity of the algorithm if the data size is n. <15, 10, 2, 11, 17, 12, 5, 8, 9, 1, 3, 13, 6, 14, 7, 16, 4>. *[IETE ALCCS'11]*
17. Write an algorithm for mergesort and sort given list using mergesort method : < 20, 35, 18, 8, 14. 41, 3, 39 >. *[RTU'07]*

18. Suggest refinements to mergesort to make it in-place. *[JNTU'08, '10]*
19. Write an algorithm for mergesort. Show that running time is $O(n \log n)$. *[BPUT'05]*
20. Write down the recurrence relation for the work done by mergesort in the best case, explain the origin of the terms, and solve it. *[BPUT'07]*
21. Derive the complexity of mergesort. *[WBUT'07, 09, UPTU'10, PTU'05, '07]*
22. What is stable sorting? *[PTU'08, 09]*
23. A sorting algorithm is stable if equal elements are kept in the same order in the output as they occur in the input. Which of the algorithm is stable? Justify your answer with arguments or counter examples. *[IITG'11]*

    (*a*) Mergesort.

    (*b*) Quicksort.
24. Differentiate between Top-down and Bottom-up approach. *[PTU'08]*
25. Write the algorithm for the mergesort. *[UPTU'06, '10, RTU'07]*
26. How mergesort works? Explain it with suitable example. On what types of data set the method is suitable. Explain in brief. *[UPTU'08, MDU'09]*
27. Prove that the average case time-complexity of quicksort is $O(n \log n)$. You should state clearly the reasons behind the design of the recurrence relation you use for establishing this complexity. *[WBUT'07]*
28. Find the best and worst case time complexity of quicksort. *[WBUT'08]*
29. Write the algorithm of quicksort. Find the best Case, worst Ease arid average Ease time complexities of this algorithm.

    [WBUT'09, '10, UPTU'10, PTU'08, RTU'07, '08, '11, MDU'08, IGNOU'07, IETE ALCCS'11]
30. Sort the following 'array' showing the steps of quicksort algorithm : <15,.10, 13, 9, 12, 7>. [IGNOU'07]
31. Show the outcome of different passes for sorting the following sequence of data using quicksort algorithm. 8, 11, 3, 15, 6, 9, 12, 39. *[IETE ALCCS'11]*

    Choose the first element as pivot.
32. Let A [1 ... *n*] be a sorted array of n distinct integers. Give a divide-and-conquer algorithm that can find an index i such that A [i] = i (if one exists) with running time $O(\log n)$. *[WBUT'10]*
33. Find the kth smallest element of n elements with time complexity less than that of Quicksort algorithm. *[WBUT'10]*
34. Use quicksort algorithm to sort: 15, 22, 30, 10, 15, 64, 1, 3, 9, 52. Is it a stable sorting algorithm? Justify. *[UPTU'08]*
35. Why do we use Quicksort for large number of series rather than Mergesort? *[WBUT'09]*
36. Explain divide-and-conquer method and apply it on the mergesort using some example. *[UPTU'09]*
37. How can you adapt 'Quicksort' algorithm to find the m order statistics in a given list of size n? *[UPTU'05, MDU'09]*

38. On what input data does quicksort algorithm for sorting exhibit its worst-case behaviour? Justify with an example of 8 elements. *[BPUT'05]*

39. Illustrate how you would develop a divide-and-conquer approach to find the maximum and minimum elements in an array. A full algorithm is not necessary. *[BPUT'07]*

40. An array contains n random integers. It is sorted using mergesort method. Indicate the time complexity if this new sorted array is again sorted using: (i) mergesort, (ii) quicksort. *[IETE ALCCS'11]*

41. Trace quicksort(A) on the array A = [4 0 4 3 8 3 2]. What is the name of the algorithm design principle on which quicksort is based? *[DOEACC'08]*

42. Give the randomized version of Quicksort. Analyze it for finding the expected running time. *[UPTU'05]*

43. How the choice of pivot element affect the running time of quicksort algorithm? *[UPTU'08]*

44. On what kind of input does the Quicksort algorithm exhibit its worst-case behaviour? Why? *[RTU'10, DOEACC'06]*

45. In quicksort, why the pivot is chosen from the center of the list rather than from one end? *[BPUT'08]*

46. Determine the running time of Quicksort for *[JNTU'09]*
    (*a*) Sorted Input
    (*b*) Reverse-ordered input
    (*c*) Random input.

47. What is the running time of Quicksort when elements of array A have same value? *[IETE ALCCS'11]*

48. Describe a sorting algorithm that is designed based on the divide-and-conquer technique. Give a suitable example to illustrate the algorithm. Compute the worst-case computational complexity of the algorithm. *[DOEACC'09]*

❑❑❑

5 *Chapter* Dynamic Programming

Those who cannot remember the past are doomed to repeat it.

— ***George Santayana***

5.1 INTRODUCTION

The term "dynamic programming" was originally used by Richard Bellman during 1940s, to describe the solution process of problems where the best decisions one after another seem to necessary. By 1953, he had refined this term in accordance with the modern meaning. Now it specifically refers to nesting of smaller decision problems inside larger decisions.

Dynamic programming is the most powerful technique for optimization problem. It is a stage wise search method. The solutions are viewed as the result of a sequence of decisions. The most attractiveness of the approach is during the search for a solution it avoids full enumeration by reducing early partial decision solutions that cannot possibly lead to optimal solution.

Originally the word "programming" in "dynamic programming" had no connection to modern computer programming. Instead it originated from the term "mathematical programming" and is considered as a synonym for optimization. Now it is meant as "planning or a tabular method". And many optimization problems are best solved by dynamic programming paradigm.

Definition: Dynamic Programming in short DP is an algorithm design technique that can be used when the solution to a problem can be viewed as the results of a sequence of decisions.

5.2 GENERAL METHOD

The steps for achieving optimal solution using DP approach are given as:

1. ***Divide into Sub-problems:*** The original problem is divided into a number of smaller sub-problems. The solution to the problem is expressed in terms of the solution for the smaller sub-problems. Start with the smallest sub-problems.
2. ***Construction of Table for Storage:*** To avoid calculating the same sub-problem twice, usually a table is constructed to store known optimal results of sub-problems for later use.

3. ***Combining Using Bottom-up Means:*** Combining solutions of smallest sub-problems obtain the solutions to sub-problems of increasing size. The process is continued until the solution of the original problem is achieved.

5.3 TWO PRIMARY CRITERIA

For problems to be solved using DP approach, the algorithm must follow the following two essential conditions.

1. ***The Principle of Optimality:*** An optimal sequence of decisions is obtained iff each sub-sequence must be optimal. That means if the initial state and decisions are optimal then the remaining decisions must constitute an optimal sequence w.r.t. the state resulting from the first decision.
2. ***Polynomial Break up:*** The original problem is divided into several sub-problems. The division is done in such a way that the total number of sub-problems to be solved should be a polynomial or almost a polynomial number. This is done for efficient performance of dynamic programming.

5.4 USE/APPLICATION

Various types of real world problems can be solved using DP approach. Some of them are cited below.

1. Longest Increasing Sub-sequence.
2. Longest Common Sub-sequence.
3. Chain Matrix Multiplication.
4. Subset Sum.
5. Coin Change.
6. Family Graph.
7. 0/1 Knapsack Problem.
8. Optimal Binary Search Tree.
9. Flow Shop Scheduling.
10. Minimum Weighted Triangulation.
11. Travelling Salesman Problem.
12. Shortest Path Problem.
    (*a*) Single Source Shortest Path.
    (*b*) Single Destination Shortest Path.
    (*c*) Single Pair Shortest Path.
    (*d*) All Pairs Shortest Path.

5.5 MATRIX CHAIN MULTIPLICATION PROBLEM

Matrix Chain Multiplication (MCM) is an optimization problem that can be solved using dynamic programming effectively in compared to traditional approach.

Problem: Given a sequence of matrices, find the most efficient way to multiply these matrices together. The problem is not actually to perform the multiplications. It simply decides order of the multiplications.

Let A be an $n \times m$ matrix and B be an mxp matrix, then their product $C = AB$ is an $n \times p$ matrix. And $C = AB$ can be computed in $O(nmp)$ time, using traditional matrix multiplication.

Suppose we want to compute A_1, A_2, A_3, A_4. Since Matrix Multiplication is associative, so the multiplication can be done in several different orders. Consider the following example.

Example 1. Suppose, we are given three matrices

$$B = 3 \times 10$$

$$C = 10 \times 5$$

$$D = 5 \times 5$$

Then order of multiplications are either as $((BC)D)$ or as $(B(CD))$ to have 3×5 matrix after multiplication.

Now cost of multiplication (mults) for ((BC)D) can be computed as:

$$BC = 3 \times 10 \times 5 = 150 \text{ and results } 3 \times 5 \text{ matrix}$$

$$BCD = (BC)D = 3 \times 5 \times 5 = 75 \text{ and results } 3 \times 5 \text{ matrix.}$$

Thus, $((BC)D) = 150 + 75 = 225$

Similarly, cost of multiplication (mults) for $(B(CD))$ can be computed as:

$$CD = 10 \times 5 \times 5 = 250 \text{ and results } 10 \times 5 \text{ matrix.}$$

$$BCD = B(CD) = 3 \times 10 \times 5 = 150 \text{ and results } 3 \times 5 \text{ matrix.}$$

Thus, $(B(CD)) = 250 + 150 = 400$

Hence, the optimal cost of multiplication = min (225, 400) = 225 and the order of multiplication = ((BC)D).

Example 2. Suppose we are given with four matrices of the following orders:

$$A_1 = 10 \times 10$$
$$A_2 = 10 \times 5$$
$$A_3 = 5 \times 15$$
$$A_4 = 15 \times 1$$

Find the optimal order of multiplication.

$$A_1 = 10 \times 100$$
$$A_2 = 100 \times 5$$
$$A_3 = 5 \times 50$$
$$A_4 = 50 \times 1$$

Solution: It is clear that $A_1\ A_2\ A_3\ A_4$ will be a 10×1 matrix. We have 5 different orderings or parenthesisation to perform multiplication as:

(*a*) $(A_1\ (A_2\ (A_3\ A_4)))$
(*b*) $((A_1\ A_2)\ (A_3\ A_4))$
(*c*) $(((A_1\ A_2)\ A_3)\ A_4)$
(*c*) $((A_1\ (A_2\ A_3))\ A_4)$
(*e*) $(A_1\ ((A_2\ A_3)\ A_4))$

The cost of multiplication (mults) of every order can be calculated as:

Order 1: $(A_1\ (A_2\ (A_3\ A_4)))$

Now, $A_{34} = A_3\ A_4 = 5 \times 15 \times 1 = 75$ mults and results 5×1 matrix.

Thus $(A_1\ (A_2\ (A_3\ A_4)))$ becomes $(A_1\ (A_2\ A_{34}))$

Now, $A_{24} = A_2\ A_{34} = 10 \times 5 \times 1 = 50$ mults and results 10×1 matrix.

Thus $(A_1\ (A_2\ A_{34}))$ becomes $(A_1\ A_{24})$

Now, $A_{14} = A_1\ A_{24} = 10 \times 10 \times 1 = 100$ mults and results 10×1 matrix.

Total Cost = 75 + 50 + 100 = 225.

Similarly,

Order 2: $((A_1\ A_2)\ (A_3\ A_4))$

$A_{12} = A_1\ A_2 = 10 \times 10 \times 5 = 500$ mults and results 10×5 matrix.

$A_{34} = A_3\ A_4 = 5 \times 15 \times 1 = 75$ mults and results 5×1 matrix.

$A_{14} = A_{12}\ A_{34}) = 10 \times 5 \times 1 = 50$ mults and results 10×1 matrix.

Total Cost = 500 + 75 + 50 = 625

Order 3: $(((A_1\ A_2)\ A_3)\ A_4)$

$A_{12} = A_1\ A_2 = 10 \times 10 \times 5 = 500$ mults and results 10×5 matrix.

$A_{13} = A_{12}\ A_3 = 10 \times 5 \times 15 = 750$ mults and results 10×15 matrix.

$A_{14} = A_{13}\ A_4 = 10 \times 15 \times 1 = 150$ mults and results 10×1 matrix.

Total Cost = 500 + 750 + 150 = 1400

Order 4: $((A_1\ (A_2\ A_3))\ A_4)$

$A_{23} = A_2\ A_3 = 10 \times 5 \times 15 = 750$ mults and results 10×15 matrix.

$A_{13} = A_1\ A_{23} = 10 \times 10 \times 15 = 1500$ mults and results 10×15 matrix.

$A_{14} = A_{13}\ A_4 = 10 \times 15 \times 1 = 150$ mults and results 10×1 matrix.

Total Cost = 750 + 1500 + 150 = 2400

Order 5: $(A_1\ ((A_2\ A_3)\ A_4))$

$A_{23} = A_2\ A_3 = 10 \times 5 \times 15 = 750$ mults and results10×15 matrix.

$A_{24} = A_{23}\ A_4 = 10 \times 15 \times 1 = 150$ mults and results10×1 matrix.

$A_{14} = A_1\ A2_4 = 10 \times 10 \times 1 = 100$ mults and results10×1 matrix.

Total Cost $= 750 + 150 + 100 = 1000$

Hence, the optimal cost of multiplication to compute $A_1\ A_2\ A_3\ A_4$

$= \min (225, 625, 1400, 2400, 1000)$

$= 225$

And the order of multiplication $= (A_1\ (A_2\ (A_3\ A_4)))$.

From the above examples it is seen that order of operations namely scalar multiplications makes a huge difference in costs. So computation of the minimum cost of multiplication depends fully on parenthesising the chain of the matrices. Thus it minimises the number of scalar multiplications resulting minimum cost.

5.5.1 Parenthesization

A product of matrices is fully parenthesized if it is either

- A single matrix, or
- A product of two fully parenthesized matrices, surrounded by parentheses.

A sequence of n matrices requires $n - 1$ multiplications. The cost of multiplying this chain obviously depends on the order in which the $n - 1$ multiplications are carried out. That order which minimises the number of scalar multiplications can be found in many ways. In the sequence there are $n - 1$ places where each parenthesization defines a set of $n - 1$ matrix multiplications. The sequence is segmented with the outermost pair of parentheses as just after the 1st matrix, just after the 2nd matrix... and just after the $(n - 1)$th matrix. In this way we able to pick the best parenthesized matrix order. A question may arise in mind that how many parenthesization are possible? Following expression gives the answer.

Let $P(n)$ be the number of ways to parenthesize n matrices. Then it can be defined as:

$$P(n) = \begin{cases} 1, \text{ if } n = 1 \\ \sum_{k=1}^{n-1} P(k)P(n-k), \text{ if } n \geq 2 \end{cases}$$

$$= \Omega\ (4^n/n^{3/2})$$

[using Catalan numbers and Stirlings' formula]

Dynamic programming version of matrix chain multiplication problem involves the following steps.

1. Structure the parenthesization optimally.
2. Recursively define the cost of sub-problems.
3. Compute the cost of an optimal solution using bottom-up.
4. Construct an optimal solution from the computed results.

5.5.2 Structure of Optimal Parenthesization

We should remember before finding the optimal cost/parenthesization that

(*a*) The order of matrices is fixed.

(*b*) Parenthesization of the chain is required to make sub-problems.

(*c*) Break down the problem into several simpler and smaller sub-problems are similar in structure.

(*d*) The highest level of parenthesization having two matrices for multiplications is considered.

(*e*) The number of orderings is equal to the number of ways to place parentheses to multiply the n matrices in every possible way.

Let $M_1, M_2 \ldots M_i, M_{i+1} \ldots M_n$ is a given chain of matrices and M_{ij}, $1 \le i \le j \le n$, is the expected matrix. Its cost can be obtained as.

(*i*) Spilt the sequence into $M_i, M_{i+1} \ldots M_j$, $1 \le i \le j \le n$.

(*ii*) If $i < j$, split $M_i, M_{i+1} \ldots M_j$ between M_k and M_{k+1}, $i \le k < j$.

(*iii*) Compute M_{ik} from $M_i \ldots M_k$

(*iv*) Compute M_{k+1j} from $M_{k+1} \ldots M_j$.

(*v*) Multiply M_{ik} and M_{k+1j} to produce final product M_{ij}.

Thus the cost of this parenthesization = the cost of computing M_{ik} and M_{k+1j} + the cost of multiplying M_{ik}, M_{k+1j}.

The cost of the outermost parenthesization $((M_1, M_2 \ldots M_i)(M_{i+1} \ldots M_n))$ consists of solving M_{1i} and M_{i+1n} optimally and then combining the solutions.

5.5.3 Cost of Optimal Solutions to Sub-problems-Recursive Solution

The cost of an optimal solution is computed recursively in terms of the optimal solutions to sub-problems. First pick a subchain $M_i, M_{i+1} \ldots M_j$, $1 \le i \le j \le n$ of $M_i, M_{i+1} \ldots M_n$ as the problem of determining its minimum cost of parenthesization. Cost is determined in terms of number of scalar multiplications.

Suppose, $m[i]\,[j]$ = minimum number of scalar multiplications needed to compute Mij

P = A list that contains dimension values of the chain.

Thus, $M_i = P_{i-1} \times Pi$ matrix.

The matrix $m[i]\,[j]$ is recursively defined as:

(*i*) If $i = j$, $m[i]\,[j] = 0$. The problem becomes trivial. No need of scalar multiplication.

(*ii*) If $i < j$, we assume that the optimal parenthesization splits $M_i, M_{i+1} \dots M_j$ between M_k and M_{k+1}, $i \leq k < j$. then

$m[i]$ $[j]$ = The minimum cost of computing the sub problems M_{ik} and $M_k + 1j$

+ The cost of multiplying these two matrices together.

$= m[i]\ [k] + m[k+1]\ [j] + P_{i-1} P_k P_j$.

$$\text{Thus, } m[i]\ [j] = \begin{cases} 0, \text{ if } i = j \\ \min_{i \leq k < j} (m[i]\ [k] + m[k+1]\ [j] + Pi - 1PkPj), \text{ if } i < j \end{cases}$$

5.5.3.1 *Computing Optimal Costs*

Computation of $m[i]$ $[n]$ for $M_1, M_2 \dots M_n$ w.r.t $m[i]$ $[j]$, the algorithm takes $O(n^2)$ time which is not better than brute force method of checking each way of parenthesizing the product. For betterment, instead of computing $m[i]$ $[j]$ recursively, we construct a table using bottom up approach.

Suppose, $M_i = P_{i-1} \times P_i$ for $1 \leq i \leq n$.

$P = \{P_0, P_1, P_2, \dots, P_n\}$

$|P| = n + 1$

Then, we construct

$m[1 \dots n]$ $[1 \dots n]$ = Auxiliary table for storing $m[i]$ $[j]$ costs.

$S[1 \dots n]$ $[1 \dots n]$ = Auxiliary table that records k, the position of parenthesization. This table is used for finding optimal solution.

5.5.3.2 *Construction of Optimal Solution*

The optimal cost table $m[i]$ $[j]$ determines the optimal number of scalar multiplications needed to compute the chain. However, it does not directly multiply the matrices. To have optimal solution information stored in $S[1 \dots n][1 \dots n]$ are utilised.

Each entry $S[i]$ $[j]$ records the value k, position of parenthesization in $M_i, M_{i+1} \dots M_j$ splits the product between M_k and M_{k+1}. Thus computation of $M_1 \dots M_n$ optimally is given by $M_1 \dots S[1]\ [n]\ MS[1]\ [n] + 1 \dots n$.

5.5.4 MCM: Algorithm

The procedure consists of two algorithms.

1. The algorithm that optimal number of scalar multiplications $m[i]$ $[j]$.
2. The algorithm that directly shows how to multiply the chain of matrices *i.e.*, to construct an optimal solution with the help of the table $S[1 \dots n][1 \dots n]$.

5.5.4.1 *Algorithm to Construct Optimal Costs*

Algorithm: OSM_MCM(C) // OSM_MCM ⇒ Optimal Scalar Multiplication for MCM

Input C = matrix chain
n = number of matrices.
P = i/p sequence represents matrices dimension.

Step 1: Repeat for $i = 1$ to n // fill diagonal
(*a*) $m[i]\ [i] = 0$

Step 2: Initialise $x = 2$

Step 3: Repeat for $i = 1$ to $n - 1$ // fill $m[i]\ [j]$ when $i < j$
(*i*) Initialise $j = x$
(*ii*) Repeat while $j \leq n$
(*a*) $m[i]\ [j] = \infty$
(*iii*) $x = x + 1$

Step 4: Initialise $x = 2$

Step 5: Repeat while $x \leq n$ through step 7

Step 6: Initialise
(a) $i = 1$
(b) $j = x$

Step 7: Repeat while ($i \leq n$) through step 9

Step 8: If $j \leq n$
(*i*) Repeat for $k = i$ to $j - 1$ // check all possible splits
(*a*) $m[i]\ [j] = \min (m[i]\ [j], m[i]\ [k] + m[k + 1]\ [j] + P_{i-1} P_k P_j)$
(*b*) $S[i]\ [j] = k$
(*ii*) Initialize $j = j + 1$

Step 9: Output "Cost of multiplication" = $m[1]\ [n]$

Step 10: Exit.

Analysis: Every entry in the table m along the principal diagonal is 0 and unknown value (∞) in other cells. It is seen that only the upper triangular matrix is filled up. Thus steps 1 to 3 halve both time and space complexities to construct the table m. And, it is $\Theta(n^2)$. Now steps 5 and 7 iterate at most $n - 1$ and n times respectively whereas looping varies in step 8. Although the order is cubic but it would be $\Theta(n^3)$. Similarly S can be put up as well.

5.5.4.2 Algorithm for Optimal Solution

Algorithm: OS_MCM (S, i, j)

Where S : The matrix that stores optimal splits.
i : Lower subscript values of matrices.
j : Upper subscript values of matrices.

Step 1:

(*i*) if $i = j$ // for single matrix
(*a*) Output " *Mi* "
(*ii*) else
(*a*) Output " ("
(*b*) $k = S[i]\ [j]$
(*c*) call OS_MCM (S, i, k)
(*d*) call OS_MCM (S, $k + 1$, j)
(*e*) Output ") "

Step 2: Exit

Analysis: From the nested loop structures for steps 2, 3, and 5 in OC_MCM(), a simple inspection shows that the algorithm runs in $O(n^3)$ time. The matrices m and s requires $O(n^2)$ space.

Example 3. Consider a chain of matrices as:

Matrix	***Dimension***
*A*1	3 × 5
*A*2	5 × 15
*A*3	15 × 5
*A*4	5 × 10
*A*5	10 × 3

Compute *A*1 *A*2 *A*3 *A*4 *A*5 efficiently.

Solution: Dimension sequence $P = < 3, 5, 15, 5, 10, 3 >$

Compute $m[i]\ [j]$ using $m[i]\ [j] = m[i]\ [k] + m[k + 1]\ [j] + P_{i-1} P_k P_j$ for $k = i$ to $j - 1$ as

$$m[1][2] = m[1][1] + m[2][2] + P_0\, P_1\, P_2$$
$$= 0 + 0 + 3*5*15$$
$$= 225 \quad \rightarrow \quad k = 1$$ [It means min value is obtained for this k]

$$m[2][3] = m[2][2] + m[3][3] + P_1\, P_2\, P_3$$
$$= 0 + 0 + 5*15*5$$
$$= 375 \quad \rightarrow \quad k = 2$$

$$
\begin{aligned}
m[3][4] &= m[3][3] + m[4][4] + P_2 P_3 P_4 \\
&= 0 + 0 + 15*5*10 \\
&= 750 \quad \rightarrow \quad k = 3
\end{aligned}
$$

$$
\begin{aligned}
m[4][5] &= m[4][4] + m[5][5] + P_3 P_4 P_5 \\
&= 0 + 0 + 5*10*3 \\
&= 150 \quad \rightarrow \quad k = 4
\end{aligned}
$$

$$
\begin{aligned}
m[1][3] &= \min\{m[1][1] + m[2][3] \\
&\qquad + P_0 P_1 P_3,\ m[1][2] + m[3][3] + P_0 P_2 P_3\} \\
&= \min\{0 + 375 + 3*5*5,\ 225 + 0 + 3*15*5\} \\
&= \min\{375 + 75,\ 225 + 225\} \\
&= 450 \quad \rightarrow \quad k = 1 \text{ or } 2. \text{ Take } k = 1
\end{aligned}
$$

$$
\begin{aligned}
m[2][4] &= \min\{m[2][2] + m[3][4] \\
&\qquad + P_1 P_2 P_4,\ m[2][3] + m[4][4] + P_1 P_3 P_4\} \\
&= \min\{0 + 375 + 5*15*10,\ 375 + 0 + 5*5*10\} \\
&= 625 \quad \rightarrow \quad k = 3
\end{aligned}
$$

$$
\begin{aligned}
m[3][5] &= \min\{m[3][3] + m[4][5] \\
&\qquad + P_2 P_3 P_5,\ m[3][4] + m[5][5] + P_2 P_4 P_5\} \\
&= \min\{0 + 150 + 15*5*3,\ 750 + 0 + 15*10*3\} \\
&= 375 \quad \rightarrow \quad k = 3
\end{aligned}
$$

$$
\begin{aligned}
m[1][4] &= \min\{m[1][1] + m[2][4] + P_0 P_1 P_4,\ m[1][2] + m[3][4] \\
&\qquad + P_0 P_2 P_4,\ m[1][3] + m[4][4] + P_0 P_3 P_4\} \\
&= \min\{0 + 625 + 3*5*10,\ 225 + 750 \\
&\qquad + 3*15*10,\ 450 + 0 + 3*5*3\} \\
&= 495 \quad \rightarrow \quad k = 3.
\end{aligned}
$$

$$
\begin{aligned}
m[2][5] &= \min\{m[2][2] + m[3][5] + P_1 P_2 P_5,\ m[2][3] + m[4][5] \\
&\qquad + P_1 P_3 P_5,\ m[2][4] + m[5][5] + P_1 P_4 P_5\} \\
&= \min\{0 + 375 + 5*15*3,\ 375 + 150 + 5*5*3,\ 625 \\
&\qquad + 0 + 5*10*3\} \\
&= 600 \quad \rightarrow \quad k = 2 \text{ or } 3. \text{ Take } k = 3
\end{aligned}
$$

$$
\begin{aligned}
m[1][5] &= \min\{m[1][1] + m[2][5] + P_0 P_1 P_5,\ m[1][2] + m[3][5] \\
&\qquad + P_0 P_2 P_5,\ m[1][3] + m[4][5] + P_0 P_3 P_5,\ m[1][4] \\
&\qquad + m[5][5] + P_0 P_4 P_5\} \\
&= \min\{0 + 600 + 3*5*3,\ 225 + 375 + 3*15*3,\ 450 + 150 \\
&\qquad + 3*5*3,\ 495 + 0 + 3*10*3\} \\
&= 585 \quad \rightarrow \quad k = 4
\end{aligned}
$$

Therefore, the cost of scalar multiplication = 585.

The solution matrix S is constructed as:

$i \backslash j$	2	3	4	5
1	1	1	3	4
2		2	3	3
3			3	3
4				4

Now apply OS_MCM (S, i, j) to parenthesize $A_1\ A_2\ A_3\ A_4\ A_5$. Initially, $i = 1$ and $j = 5$. $i \neq j$, then $k = S[1][5] = 4$, we have OS_MCM (S, 1, 4) and OS_MCM (S, 5, 5). Similarly, for OS_MCM (S, 1, 4), $i \neq j$, $k = S[1][4] = 3$, so OS_MCM (S, 1, 3) and OS_MCM (S, 4, 4). The process is shown below.

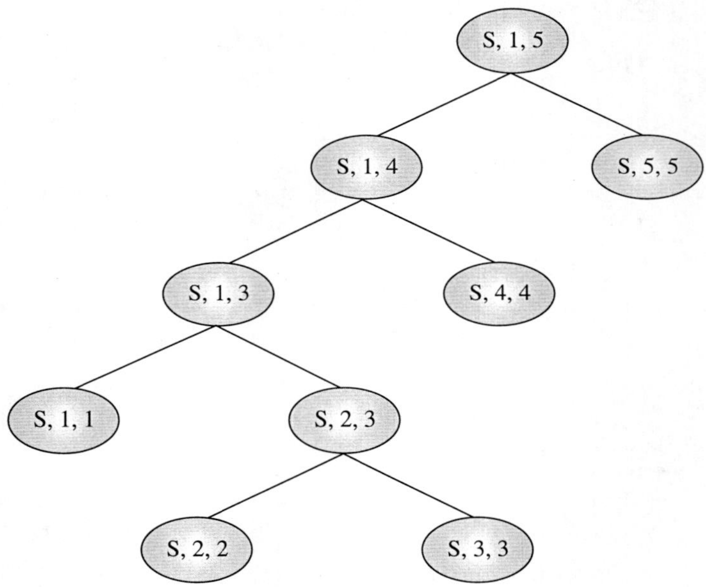

Figure 5.1 Graphical representation of matrix parenthesization.

Now start from last level, A_2 and A_3 are multiplied first. Then A_1 is multiplied with $A_2\,A_3$ *i.e.*, $(A_1\ (A_2\,A_3))$. Proceeding in the same way, order of multiplication is $(((A_1\ (A_2\ A_3))\ A_4)\ A_5)$.

Theorem: The minimum number of scalar multiplications required to multiply a chain of n matrices requires $\Theta(n^3)$ running time and $\Theta(n^2)$ space.

5.6 TRAVELLING SALESMAN PROBLEM

The Travelling Salesman Problem (TSP) is a simple combinatorial optimisation problem. It is generally studied in operation research and theoretical computer science. It can be stated very simply as: "*A salesman wants to visit n cities denoted by nodes of a graph cyclically. In a single tour he visits each city exactly once except the starting city or node, and finishes the tour up from where he started*". Our interest is that in what order the salesman should visit the cities to minimise the distance or cost travelled.

1. If $n = 2$, the problem is trivial, since only one tour is possible between the cities.
2. If $n = 3$, still the problem becomes trivial. And it is the symmetric case of TSP.
3. If all the n cities are adjacent to each other, possible different tours are $(n - 1)!$

Explanation to point 3: Pick any city from n cities as the start city of journey—then there are $n - 1$ choices for the second city to be visited, similarly $n - 2$ choices for the third, and so on. Thus, possible distinct tours are $(n - 1)!$

Again for any two cities A and B, the distance from A to B is the same as that from B to A. Exactly the same distance is travelled if we reverse the tour *i.e.,* the order of journey in which the cities are visited. Therefore, distinction between a tour and its reverse order turns into immaterial. Being symmetricity only $(n - 1)!/2$ as many distinct solutions for an n city TSP are possible.

In either case, the number of solutions becomes extremely large for large n. No efficient way is found till now to solve this problem. That's why the TSP is one of the classes of "NP Complete" combinatorial problems. However, in two famous ways we can solve it quite easily and effectively.

(*i*) DP Approach.

(*ii*) Branch and Bound Technique.

The TSP has somehow direct importance in our life. A lot of practical applications can be put in this form like.

(*i*) If we go for tour in holidays to visit different locations of India, then we always choose the tour that uses effective cost and/or time.

(*ii*) We may denote the problem as messenger or postman or courier boy problem. They take in most cases the shortest route of the recipients whose pair wise distances are known to deliver letter, drafts, etc. Of course, this problem is possibly solved by finitely many trials.

5.6.1 TSP: Classification

A TSP problem can be classified in two ways depending on its graphical representation.

(*i*) Symmetric

(***ii***) Asymmetric

In the symmetric TSP, the distance between two cities is same both in forward and reverse direction. The underlying graph is in this case an undirected graph. Each tour has the same length in both directions. This symmetric nature halves the number of feasible solutions.

In the asymmetric TSP, the distance from one city to the other need not equal the distance with the reverse direction. There may not even be a connection in the other direction. Thus, the underlying graph in this case is a digraph (directed graph). *e.g.*, One-way streets, Air-fares, etc. depend on the direction of travel.

5.6.2 TSP: Procedure

Underlying solution of the TSP involves the following graph representation.

1. Represent cities by nodes or vertices of a graph.
2. Connect a pair of vertices if there is link between them.
3. Assign a weight or cost to each edge.

Suppose $G = (V, E, W)$ be a weighted digraph with edge weight or cost between two vertices pair (v_i, v_j), $c_{ij} = W(v_i, v_j)$.

The c_{ij} can be defined as:

$$c_{ij} = \begin{cases} 0, \text{ if } i = j \\ w, \text{ if } i \neq \text{ and } (v_i, v_j) \in E \\ \propto, \text{ if } (v_i, v_j) \notin E \end{cases}$$

When $| V |= n > 2$, then a tour in G is a directed simple cycle consists of every vertex in V and the cost of this tour is the sum of the weight of the edges in the tour.

5.6.3 TSP: Computing Optimal Solution

Consider the following graph as in Figure 5.2. Suppose, the tour starts at the vertex A then some of the possible tours are:

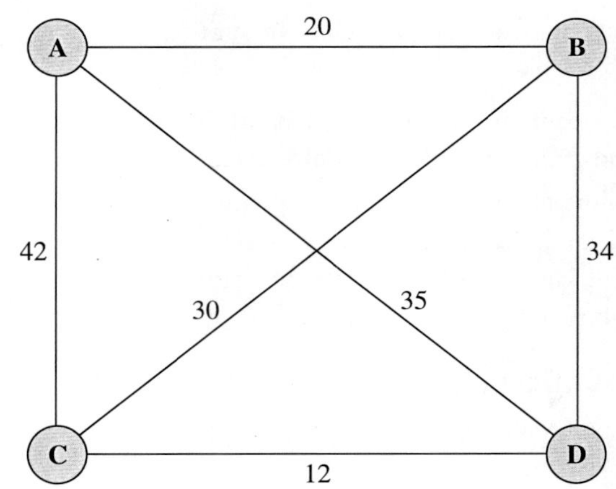

Figure 5.2 TSP with 4 cities.

(*i*) $A - B - A$
(*ii*) $A - C - A$
(*iii*) $A - D - A$
(*iv*) $A - B - C - A$
(*v*) $A - B - D - A$ and so on.

Therefore, every tour consists of an edge (A, K) for some K in $(V - K)$ and a path from K to A. This path goes through each vertex in $V - (A, K)$.

Suppose,

S = set of vertices of a tour.

OTC (i, S) = optimal tour cost starting at vertex i and going through S terminates at i.

OTC $(A, V - A)$ = optimal tour cost of the salesman.

$= \min_{2 \le k \le n}\{c_{1k} + \text{OTC}(K, V - \{1, K\})\}$ where $A = 1$, $B = 2$ and so on.

Generalising, OTC $(i, S) = \min_{j \in S}\{cij + \text{OTC}(j, S - j)\}$ for $i \notin S$

Naturally, OTC $(i, \varnothing) = c_{i1}$, $\forall\ 2 \le i \le n$.

5.6.4 TSP: Algorithm

Consider $G = (V, E, W)$, be the weighted directed graph. then suitable algorithm for the TSP is given below.

Algorithm: TSP_DP (G, C)

Where C = cost matrix. $c_{ij} = w_{ij}$ *i.e.,* $C[i][j] = W[i][j]$

Suppose vertex 1 is the starting vertex of the journey.

Step 1:

(*i*) Repeat for $i = 1$ to n // construct cost matrix.

(*ii*) Repeat for $j = 1$ to n

(*a*) If $i = j$

- $c_{ij} = 0$

Else

- $c_{ij} = w_{ij}$.

Step 2:

(*i*) Repeat for $i = 2$ to n // construct OTC$(i, \varnothing)$

(*a*) OTC $(i, \varnothing) = c_{i1}$

Step 3:

(*i*) Repeat for $i = 1$ to n // compute intermediate optimal cost.

(*ii*) Repeat for j = 1 to n

(*a*) If $i \neq j$ and $i, j \notin S$

*) OTC$(i, S) = \min(c_{ij} + \text{OTC}(j, S - \{j\}))$

Step 4:

(*i*) $x = 1$ // initialize tour

(*ii*) Tour$[x{+}{+}] = 1$

Step 5: Repeat for $k = 2$ to n // computer optimal tour cost

(*a*) OTC(1, $V - \{1\}$) = min(c_{1k} + OTC(k, $S - \{1, k\}$))

(*b*) Tour[x++] = k

Step 6: Repeat for $i = 1$ to |Tour| // display tour

(*a*) Output Tour[i]

Step 7: Exit

Analysis

Before computing OTC(1, $V - \{1\}$) we have to compute OTC(i, S) several times. Let N be that number of times. We know for each value of $|S|$ and i, there are $n - 1$ choices. Now the number of distinct stets of S of size k excluding 1 and i is $\binom{n-2}{k}$.

Thus,
$$N = \sum_{k=0}^{n-2} (n-1)\binom{n-1}{k}$$
$$= (n-1)2^{n-2}$$
$$= \Theta(n2^n)$$

Since there n vertices, total time for the TSP algorithm is given by:

$$T(n) = n\,\Theta\,(n2^n) = \Theta\,(n^2 2^n) \ll (n-1)!.$$

Example 4. Consider the following graph. Starting from vertex 1, find the optimal tour so that a person may come back to source of journey visiting all the vertices.

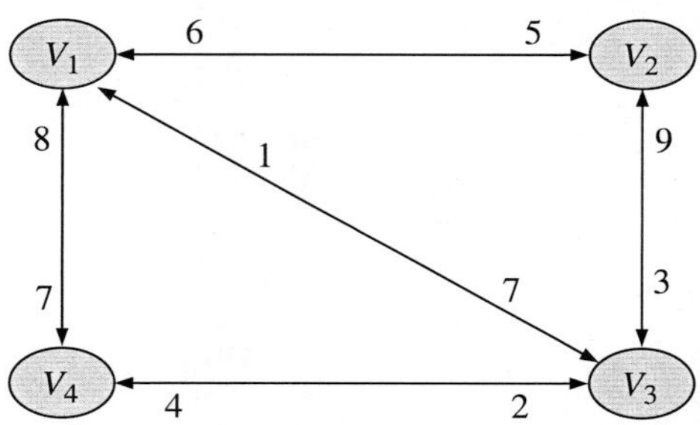

Figure 5.3 A digraph for TSP.

Solution: The cost matrix of the above graph is:

$C[i]\,[j] =$

	$V1$	$V2$	$V3$	$V4$
$V1$	0	5	7	7
$V2$	6	0	3	∞
$V3$	1	9	0	4
$V4$	8	∞	2	0

Number of vertices = 4. In the computation we represent a vertex V_i as i, *e.g.*, V_1 is 1.

Applying TSP_DP() algorithm,

$$\text{OTC}(i, \varnothing) = C_{i1},\ i \neq \text{starting vertex } 1$$

$$\text{OTC}(2, \varnothing) = C_{21} = 6$$

$$\text{OTC}(3, \varnothing) = C_{31} = 1$$

$$\text{OTC}(4, \varnothing) = C_{41} = 8$$

Using $\text{OTC}(i, S) = \min(c_{ij} + \text{OTC}(j, S - \{j\}))$ for $i \neq j$ and i, j S iteratively

For S with single vertex *i.e.*, $|S|=1$

$$\begin{aligned}\text{OTC}(2, \{3\}) &= C_{23} + \text{OTC}(3, \{3\} - \{3\})\\ &= C_{23} + \text{OTC}(3, \varnothing)\\ &= C_{23} + C_{31}\\ &= 3 + 1 = 4\end{aligned}$$

Similarly,

$$\begin{aligned}\text{OTC}(2, \{4\}) &= C_{24} + \text{OTC}(4, \varnothing)\\ &= C_{24} + C_{41}\\ &= \infty + 8 = \infty\end{aligned}$$

$$\begin{aligned}\text{OTC}(3, \{2\}) &= C_{32} + \text{OTC}(2, \varnothing)\\ &= C_{32} + C_{21}\\ &= 9 + 6\\ &= 15\end{aligned}$$

$$\begin{aligned}\text{OTC}(3, \{4\}) &= C_{34} + \text{OTC}(4, \varnothing)\\ &= C_{34} + C_{41}\\ &= 4 + 8\\ &= 12\end{aligned}$$

$$\begin{aligned}\text{OTC}(4, \{2\}) &= C_{42} + \text{OTC}(2, \varnothing)\\ &= C_{42} + C_{21}\\ &= \infty + 6\\ &= \infty\end{aligned}$$

$$\begin{aligned}\text{OTC}(4, \{3\}) &= C_{43} + \text{OTC}(3, \varnothing)\\ &= C_{43} + C_{31}\\ &= 2 + 1\\ &= 3\end{aligned}$$

For S with double vertices *i.e.*, $|S|=2$

$$\begin{aligned}\text{OTC}(2,\{3, 4\}) &= \min\{C_{23} + \text{OTC}(3, \{4\}), C_{24} + \text{OTC}(4, \{3\})\}\\ &= \min\{3 + 12, \infty + 3\}\\ &= 15\end{aligned}$$

$$\begin{aligned}\text{OTC}(3,\{2, 4\}) &= \min \{C_{32} + \text{OTC}(2, \{4\}), C_{34} + \text{OTC}(4, \{2\})\}\\ &= \min\{ 9 + \infty, 12 + \infty\}\\ &= \infty\end{aligned}$$

$$\begin{aligned}\text{OTC}(4,\{2, 3\}) &= \min \{C_{42} + \text{OTC}(2, \{3\}), C_{43} + \text{OTC}(3, \{2\})\}\\ &= \min\{\infty + 4 , 3 + 15\}\\ &= 18\end{aligned}$$

For S with triple vertices *i.e.,* $|S|=3$

$$\begin{aligned}\text{OTC}(1, \{2, 3, 4\}) &= \min\{C_{12} + \text{OTC}(2, \{3,4\}), C_{13} + \text{OTC}(3, \{2, 4\}),\\ &\qquad C_{14} + \text{OTC}(4, \{2, 3\})\}\\ &= \min\{ 5 + 15 , 7 + \infty, 7 + 18\}\\ &= 20 \qquad \text{[minimum is due to vertex 2]}\end{aligned}$$

This is the optimal tour cost and the tour can be determined with the help of the following partial optimal solution.

$$\begin{aligned}& C_{12} + \text{OTC}(2, \{3,4\})\\ &= C_{12} + C_{23} + \text{OTC}(3, \{4\})\\ &= C_{12} + C_{23} + C_{34} + \text{OTC}(4, \varnothing)\\ &= C_{12} + C_{23} + C_{34} + C_{41}\\ &= 1 - 2 - 3 - 4 - 1 \rightarrow \text{this is our tour.}\end{aligned}$$

5.7 SHORTEST PATH PROBLEM (SPP)

When you go from one place to another you generally prefer a path between two places (nodes or vertices) which is shortest one such that constituent cost/path is minimized. Shortest path problems are represented using weighted digraph $G = (V, E, W)$.

5.7.1 Variation

(*i*) **Single Pair Shortest Path Problem** — Shortest paths from a source vertex s to all other vertices in the graph.

(*ii*) **Single Destination Shortest Path Problem** — Shortest paths from all vertices in the graph to a single destination vertex d.

(*iii*) **All Pairs Shortest Path Problem** — Shortest paths between every pair of vertices v, v' in the graph.

5.7.2 Negative Weight Edges

There may be some edges whose weights are negative in a graph $G = (V, E, W)$. If it is permitted then it is confirmed that G has no negative weight cycle. Because in absence of that the shortest path weight SPL(s, v) reachable from a source vertex s to all other vertex v in V is well defined. If a negative weight cycle is reachable from s to v then no shortest path SPL(s, v) exists from s to v. And it is SPL(s, v) = ∞.

5.7.3 Negative Weight Cycle

The negative weight cycle is a cycle whose total weight/cost value is negative. In its absence shortest path consists of $n - 1$ edges in G of $|V| = n$. If shortest path has more than $n - 1$ edges, at least one vertex is repeated thus constitute a cycle. A negative cycle invalidates the notion of distance based on edge weights. So it must be eliminated.

5.7.4 Shortest Path Representation

Consider $G = (V, E, W)$ be a weighted digraph. Computation of shortest path includes vertices on the shortest paths. For the purpose a list called predecessor Pre is maintained. Pre[v] is thus either a vertex or NULL. Chain of predecessors originating at a vertex v and proceeding backward along a shortest path from s to v, results a shortest path from s to v provided Pre[v] $\neq$ NULL.

5.7.5 Computing Shortest Path

Suppose $\text{SPL}^m[v]$ is a shortest path length from s to v with at most $m = n - 1$ edges. Then in absence of negative weight cycle $\text{SPL}^1[v] = \text{cost}\,[s, v]$.

Using DP, $\text{SPL}^{n-1}[v]$ is computed in the following way.

(*i*) If the shortest path has no more than $k - 1$ edges from s to v, $\text{SPL}^k[v] = \text{SPL}^{k-1}[v]$ for $k > 1$.

(*ii*) If the shortest path has exactly $k - 1$ edges from s to v and if it is made up of a shortest path from s to some intermediate vertex x followed by the edge (x, v) then i as a candidate for x

$$\text{SPL}^k[v] = \min\{\text{SPL}^{k-1}[v],\ \text{SPL}^{k-1}[v] + \text{cost}[i, v]\}\}$$
$$k = 2, 3, \ldots, n - 1.$$

Example 5. Consider the following digraph Figure 5.4. Construct a table that shows shortest paths from a source vertex s =1.

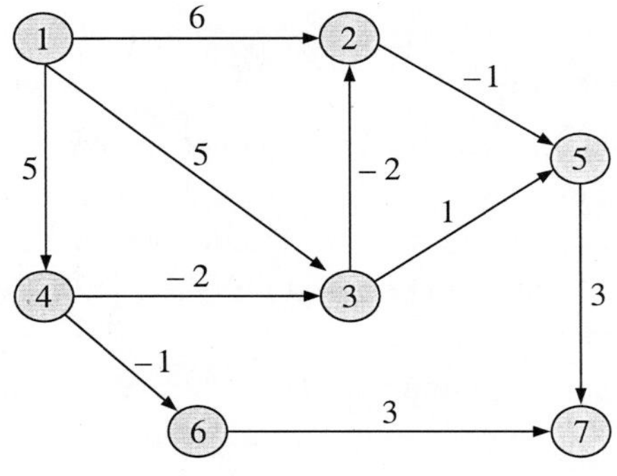

Figure 5.4 Digraph for shortest path.

Solution: Here $k = 1, 2, \ldots, 6$. Source vertex $s = 1$.

Now, $\text{SPL}^k[1] = 0, \forall k$

There are edges from vertex 1 to vertices 2, 3, 4.

So, $\text{SPL}^1[2] = 6, \text{SPL}^1[3] = 5, \text{SPL}^1[4] = 5$

There are no edges from vertex 1 to vertices 5, 6, 7.

So, $\text{SPL}^1[5] = \text{SPL}^1[6] = \text{SPL}^1[7] = \infty$

Similarly, $\text{SPL}^2[2] = \min\{\text{SPL}^1[2], \{\text{SPL}^1[i] + \text{cost}[i, 2]\}\}$

$i = 1, 2, \ldots, 7$

$= \min\{6, \min(0 + 6, 5 - 2, 5 + \infty, \infty + \infty, \infty + \infty, \infty + \infty)\}$

$= 3$

Thus, the shortest path for the problem is given in Figure 5.5.

k \ SPL^{k-1}	1	2	3	4	5	6	7
1	0	6	5	5	∞	∞	∞
2	0	3	3	5	5	4	∞
3	0	1	3	5	2	4	7
4	0	1	3	5	0	4	5
5	0	1	3	5	0	4	3
6	0	1	3	5	0	4	3

Figure 5.5 Shortest path table with negative weight cycle.

5.7.6 SPP: Applications

Application areas of shortest path problem may not be limited to the followings.

1. Automatically finds shortest path between physical locations such as driving directions on web mapping websites like MapQuest or Google Maps.
2. Shortest path algorithms can be used to find an optimal sequence of choices to reach a certain goal state.
3. In networking or telecommunications where this shortest path problem is sometimes called the min-delay path problem.
4. Other applications include operations research, robotics, transportation, and VLSI design, etc.

5.8 SINGLE SOURCE SHORTEST PATH PROBLEM

Given a directed weighted graph $G = (V, E, W)$, how we can find a shortest path from given source vertex s to each other vertex v in G. This problem can be solved easily with BFS traversal algorithm under the special case when all weights are 1. However Bellman-Ford algorithm is preferable that follows DP. We will assume either no negative-weight edges, or no reachable negative-weight cycles during computation. Algorithm will compute a shortest-path tree similar to BFS tree.

5.8.1 Bellman – Ford Algorithm

The Bellman–Ford algorithm developed by Richard Bellman, Samuel End and Lester Ford, Jr. is a label correcting algorithm that computes single-source shortest paths from a weighted digraph. The algorithm is also called Bell-End-Ford algorithm. Bellman-Ford algorithm can detect a cycle of negative weights if the graph has an edge with negative weight. The algorithm finds the shortest path repeating a vertex of a graph.

Algorithm: Bellman-Ford (G)

Where $G = (V, E, W)$ a weighted digraph.

SPL = shortest path length in terms of edge cost/weight.

Step 1: Repeat for $i = 1$ to n // construct cost/weight matrix

Repeat for $j = 1$ to n

(*i*) If $i = j$ // if self loop exists

$W[i][j] = 0$

(*ii*) Else if // if no edge between v_i, v_j exists

$W[i][j] = \infty$

(*iii*) Else // if edge between v_i, v_j exists

$W[i][j] = W(v_i, v_j)$

Step 2:

(*i*) Repeat for $i = 1$ to n // initialization

(*a*) SPL $[i] = \infty$

(*b*) Pre $[i]$ = NULL

(*ii*) SPL [1] = 0

Step 3: Repeat for $i = 1$ to $n - 1$ // relaxation/visiting of the edges

Repeat for $j = 1$ to n

If $W[i][j] \neq 0$ or ∞

(*i*) x = SPL $[i]$ + W $[i][j]$

(*ii*) If $x <$ SPL $[j]$

(*a*) SPL $[j] = x$

(*b*) Pre $[j] = i$

Step 4: Repeat for $i = 1$ to n // check negative weight cycle

Repeat for $j = 1$ to n

If $W[i][j] \neq 0$ or ∞

(*i*) x = SPL [i] + W [i] [j]

(*ii*) If x < SPL [j]

(*a*) Return False

Step 5: Return True // for no negative weight cycle.

Step 6: Exit.

Analysis: It is observed that in a shortest path at most ($|V| - 1 = n - 1$) edges can be present considering any vertex as a source vertex. Outer loop of step 3 repeats $n - 1$ times. Each iteration of this loop again iterates for all n vertices in the graph. Clearly, each iteration is called relaxation of edges. These relaxation allow minimum distances to accurately proceed throughout the graph. In absence of negative cycle, the shortest path can only visit each node at most once. Thus, the Bellman-Ford algorithm to determine shortest path from a source vertex runs in *i.e.,* $O(|V|\ |E|)$ time.

Example 6. Find the shortest path using Bellman-Ford Algorithm on the digraph in Figure 5.6.

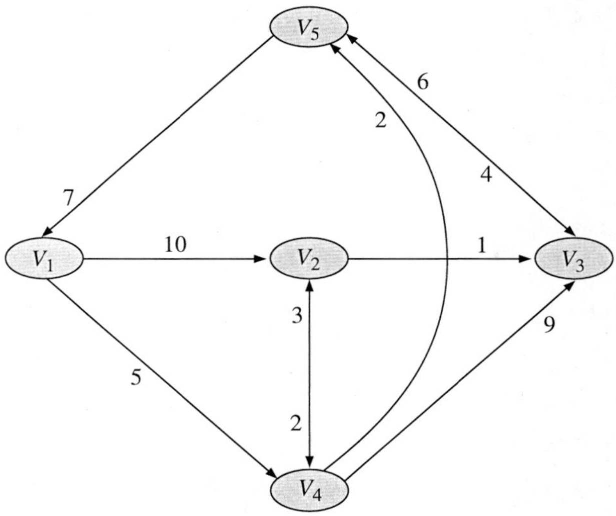

Figure 5.6 A digraph for bellman-ford algorithm.

Solution: Using steps of Bellman-Ford algorithm shortest path is found as:

Step 1: The cost/weight matrix for the graph is

$W[i][j] =$

V_i \ V_j	1	2	3	4	5
1	0	10	∞	5	∞
2	∞	0	1	2	∞
3	∞	∞	0	∞	6
4	∞	3	9	0	2
5	7	∞	4	∞	0

Step 2: Initialise SPL, Pre lists as:

$$\text{SPL}[i] = \{\infty, \forall\, i = 1 \text{ to } 5\}$$
$$\text{Pre}[i] = \text{NULL}, \forall\, i = 1 \text{ to } 5$$
$$\text{SPL}[1] = 0$$

SPL values are encircled in the corresponding vertices shown in Figure 5.7.

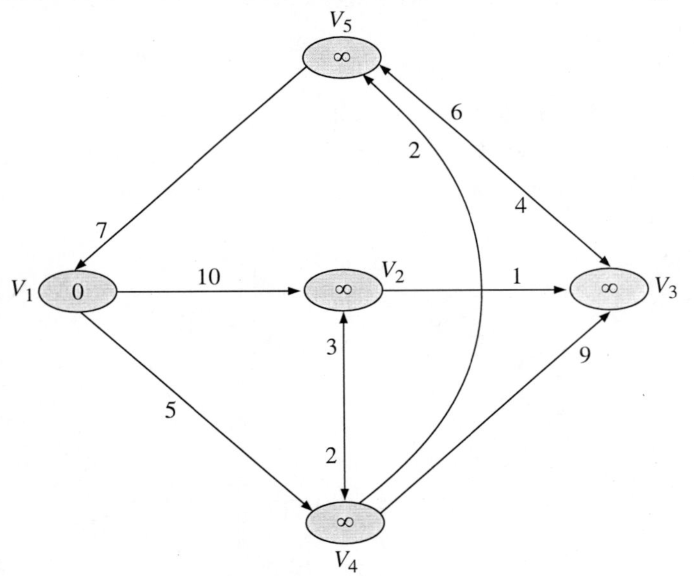

Figure 5.7 Initial Solution Graph For SSP.

Step 3: Make a list of edges $L = \{(v_i, v_j)$ in G as $(v_1, v_2), (v_1, v_4), (v_2, v_3), (v_2, v_4), (v_3, v_5), (v_4, v_2), (v_4, v_3), (v_4, v_5), (v_5, v_1), (v_5, v_3)\}$.

In first pass we relax (visit) the edges with v_1 as source. After relaxing the edges $(v_1, v_2), (v_1, v_4)$, costs are encircled in the corresponding vertices to reach v_2 and v_4 from v_1. Costs are computed as below to circle.

$$\begin{aligned}\text{SPL}[2] &= \min(\text{SPL}[2], \text{SPL}[1] + w[1][2]) \\ &= \min(\infty, 0 + 10) = 10\end{aligned}$$

$$\begin{aligned} \text{SPL }[4] &= \min(\text{SPL }[4], \text{SPL }[1] + w[1][4]) \\ &= \min(\infty, 0 + 5) = 5 \end{aligned}$$

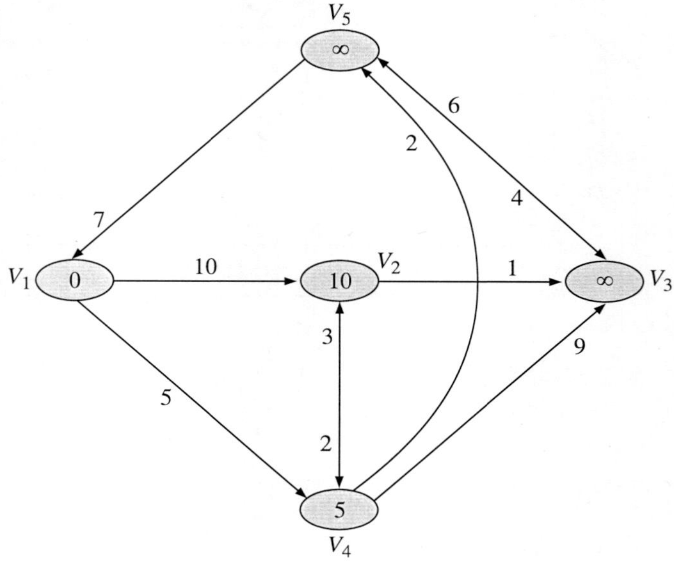

Figure 5.8 Partial solution graph for SSP.

Similarly, after relaxing the next three edges (v_2, v_3), (v_2, v_4), (v_3, v_5) and computing corresponding costs

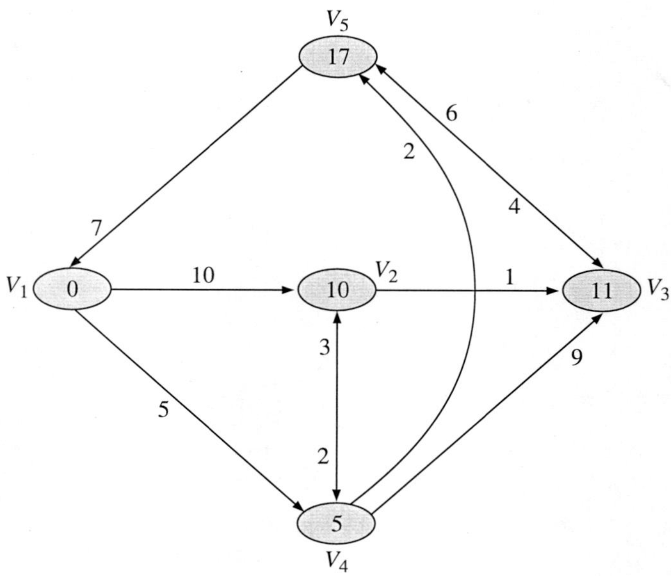

Figure 5.9 Partial solution graph for SSP.

After relaxing the remaining edges (v_4, v_2), (v_4, v_3), (v_4, v_5), (v_5, v_1), (v_5, v_3) and computing corresponding costs

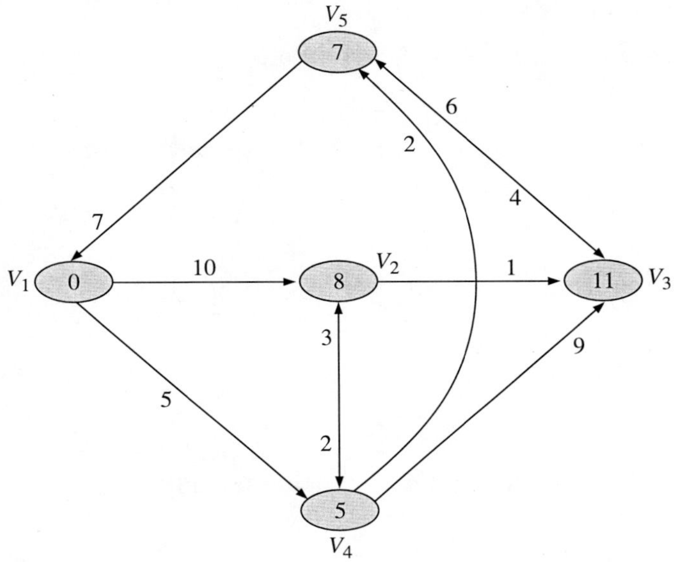

Figure 5.10 Partial solution graph for SSP.

Similarly in second pass relaxing all the edges of L, we have

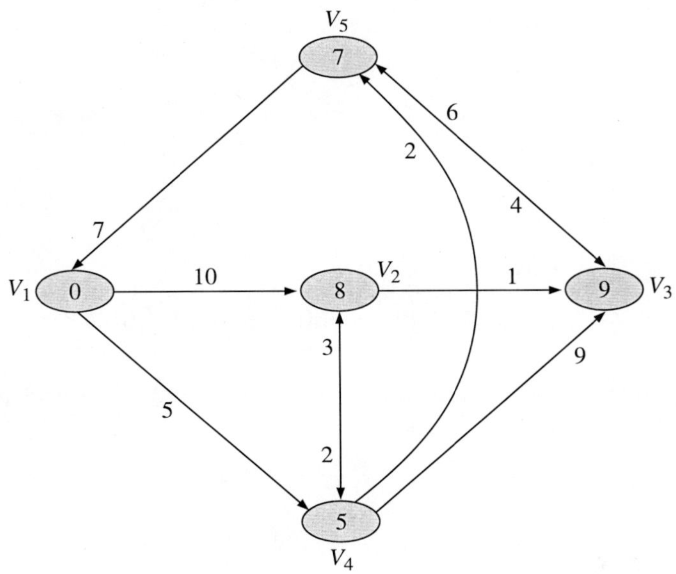

Figure 5.11 Solution graph for SSP.

In third and fourth passes there is no change in the computation. The above graph shows the shortest path length in terms of cost/weight from vertex v_1 as source vertex to all other vertices.

Example 7. Find shortest path from V_2 to all other vertices in the digraph in Figure 5.12.

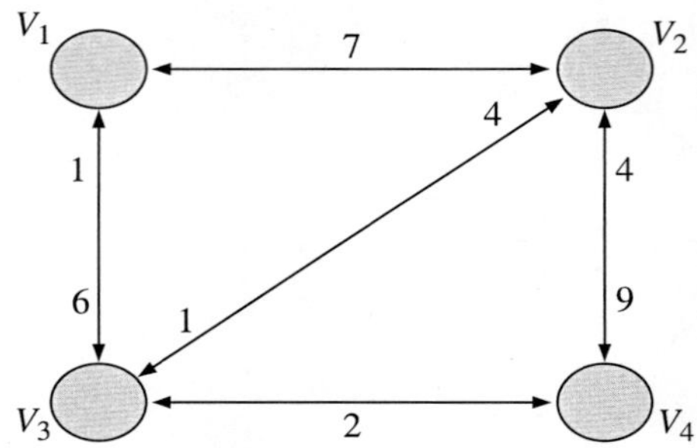

Figure 5.12 Digraph for SSP.

Solution: Existing edges are (V_1, V_2), (V_1, V_3), (V_2, V_3), (V_2, V_4), (V_3, V_1), (V_3, V_2), (V_4, V_2), (V_4, V_3). Consider initial arrangement as:

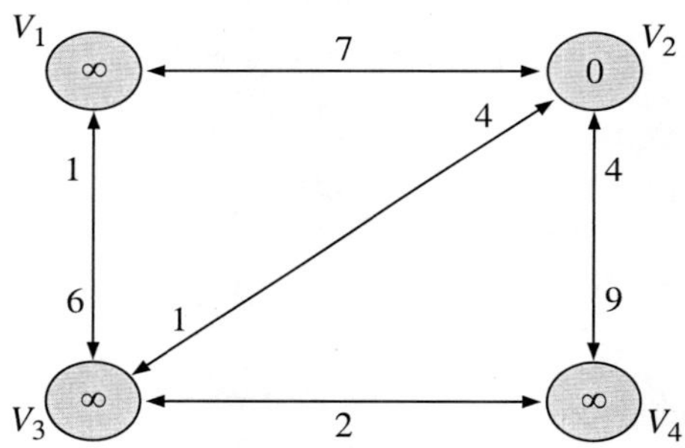

Figure 5.13 Initial solution graph for SSP.

Starting from V_2, the shortest path to all the vertices can be calculated step-by-step as follows:

Relaxing edges (V_1, V_2), (V_1, V_3)

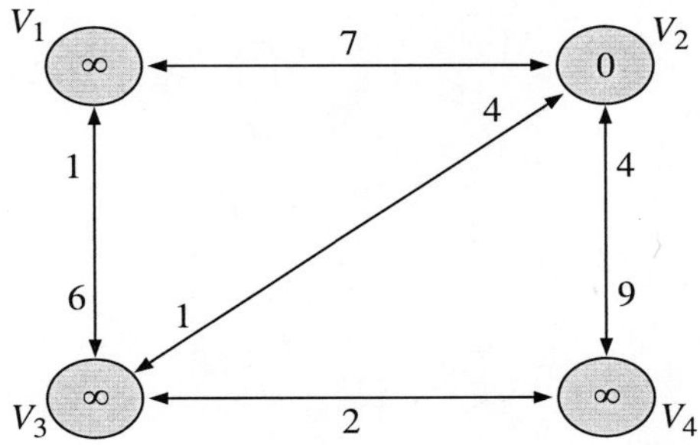

Figure 5.14 Partial solution graph for SSP.

Relaxing edges (V_2, V_3), (V_2, V_4)

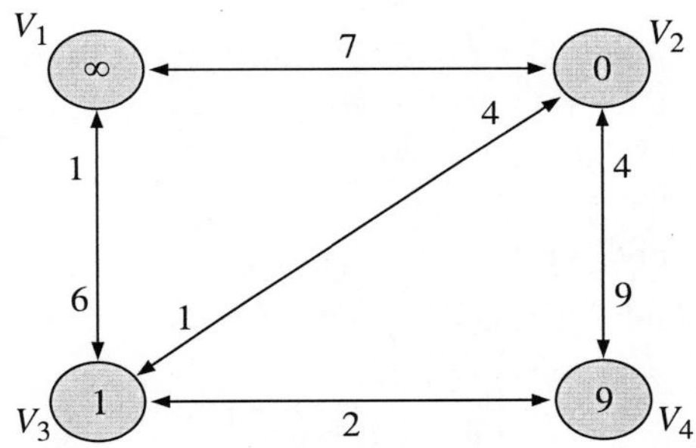

Figure 5.15 Partial solution graph for SSP.

Relaxing edges (V_3, V_1), (V_3, V_2)

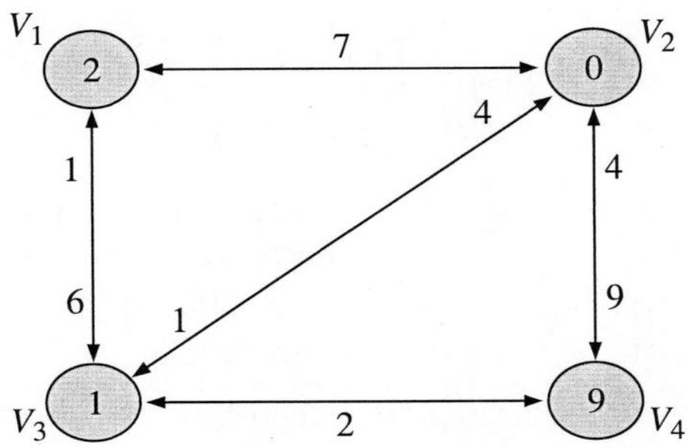

Figure 5.16 Partial solution graph for SSP.

Relaxing edges (V_4, V_2), (V_4, V_3)

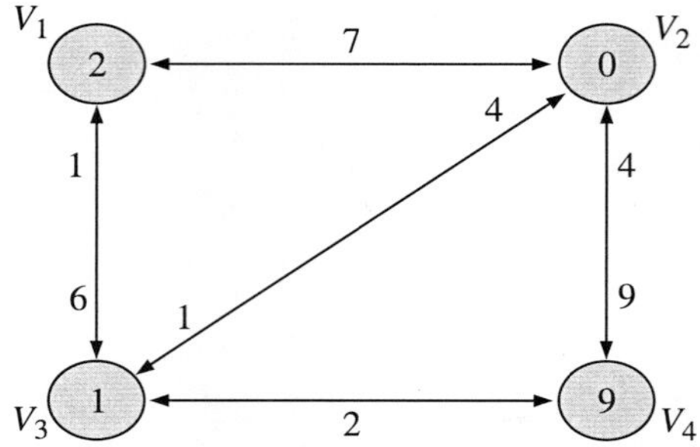

Figure 5.17 Solution graph for SSP.

No change. Thus, the shortest path obtained.

5.9 ALL PAIRS SHORTEST PATH PROBLEM

Consider a weighted digraph $G = (V, E, W)$. Solving All Pairs Shortest Path Problem (APSPP) consists of computing the shortest distance or path in terms of cost/weight between every pair of vertices or nodes (u, v) where $u, v \in V$ in the graph G. We assume absence of negative or zero weight cycle.

Common methods to solve the APSPP are:

1. Floyd-Warshall algorithm in $\Theta(n^3)$ time that uses dynamic programming. Here we will discuss only this approach.
2. Johnson's algorithm best suitable for sparse graphs but is harder to implement.
3. Repeated use of single source shortest path problem algorithm in $O(n^3)$ time.

The cost/weight matrix of G is defined as:

$$W[i]\,[j] = \begin{cases} 0, \text{ if } i = j \\ w, \text{ if } i \neq j \text{ and } (v_i, v_j) \in E \\ \infty, \text{ if } (v_i, v_j) \notin E \end{cases}$$

The matrix is of $n \times n$ size since $|V| = n$. This will be the input to the algorithm. Corresponding output will also be an $n \times n$ cost matrix. The path matrix shows shortest path cost from vertex v_i to v_j and is denoted by SPC $[i]\,[j]$. A shortest path between two vertices v_i and v_j can be obtained in two ways.

(*i*) If there is direct path from v_i to v_j then SPC $[i]\,[j]$ = W $[i]\,[j]$.

(*ii*) If the shortest path from v_i to v_j consists of some intermediate vertex v_k, SPC $[i]\,[j]$ = min (SPC $[i]\,[k]$ + SPC $[k]\,[j]$), $\forall k$.

Example 8. Find the optimal cost of the path from *V*5 to *V*6.

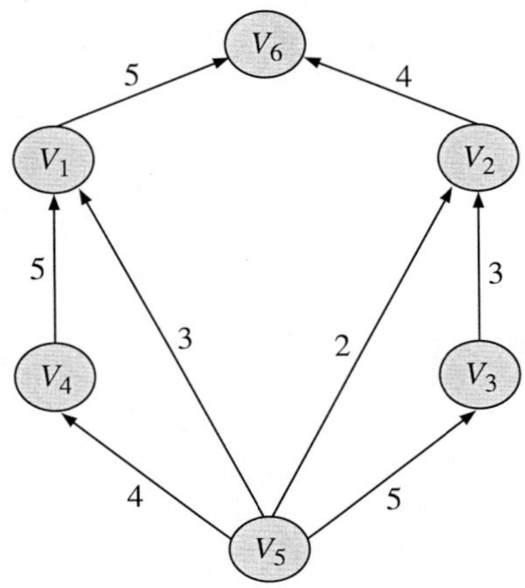

Figure 5.18 Digraph for shortest path from V_5 to V_6.

Solution: From the above figure it is seen that there is no direct path from V_5 to V_6.

So, SPC [5] [6] = ∞

Now, SPC [5] [6] = SPC [5] [1] + SPC [1] [6] = 3 + 5 = 8

SPC [5] [6] = SPC [5] [2] + SPC [2] [6] = 2 + 4 = 6

SPC [5] [6] = SPC [5] [3] + SPC [3] [2] + SPC [2] [6] = 5 + 3 + 2 = 10

SPC [5] [6] = SPC [5] [4] + SPC [4] [1] + SPC [1] [6] = 4 + 5 + 5 = 14

Thus, the minimum path length from *V*5 to *V*6 is via *V*2 and path length is 6.

5.9.1 Floyd-Warshall Algorithm

The Floyd-Warshall algorithm is named after Robert Floyd and Stephen Warshall. In a single execution of the algorithm we can find the shortest paths between all pairs of vertices of the graph. The algorithm is sometimes called WFI Algorithm or Roy-Floyd algorithm, after Bernard Roy described it in early 1960s. The algorithm is based on the principle of dynamic programming.

Another speciality of the algorithm is that it detects negative weight cycle if the path matrix contains a negative number along the diagonal. If SPC [*i*] [*i*] is negative for some vertex *i*, then this vertex *i* contains at least one negative cycle.

Algorithm: Floyd-Warshall_APSPP (*G*, SPC, *n*)

Where $G = (V, E, W)$, a weighted digraph.

SPC = A matrix for shortest path cost.

$n = |V|$, no. of vertices.

Step 1:

(*i*) Repeat for $i = 1$ to n // if there is a direct path from v_i to v_j

(*ii*) Repeat for $j = 1$ to n

(*a*) $SPC^0 [i] [j] = W [i] [j]$

Step 2:

(*i*) Repeat for $k = 1$ to n // path form v_i to v_j via intermediate vertex

(*ii*) Repeat for $i = 1$ to n

(*iii*) Repeat for $j = 1$ to n

(*a*) $SPC^k [i] [j] = \min (SPC^{k-1}[i] [j], SPC^{k-1} [i] [k] + SPC^{k-1}[k] [j])$

Step 3: Exit.

Analysis: The Floyd-Warshall algorithm compares all possible paths in the digraph G between each pair of vertices. It is possible to do this in n^3 comparisons as obtained from step 2. This is remarkable to consider that there are n^2 edges in the graph, and every combination of edges is tested n vertices. It does so incrementally updating path matrix between two vertices, until the estimate is known to be optimal. Hence the required time complexity for this algorithm is $O(n^3)$.

5.9.1.1 *Applications of Floyd-Warshalls*

The Floyd-Warshall algorithm can be used to solve the following problems, among others:

1. Shortest paths in directed graphs using Floyd's algorithm.
2. Transitive closure of directed graphs using Warshall's algorithm.
3. Finding a regular expression in a finite automaton using Kleene's algorithm.
4. Inversion of real matrices using Gauss-Jordan algorithm.
5. Optimal routing with the maximum flow between two vertices.
6. Testing an undirected graph is bipartite.
7. Maximum Bandwidth Paths in Flow Networks, etc.

Example 9. Find the shortest path among the vertices on the graph given below.

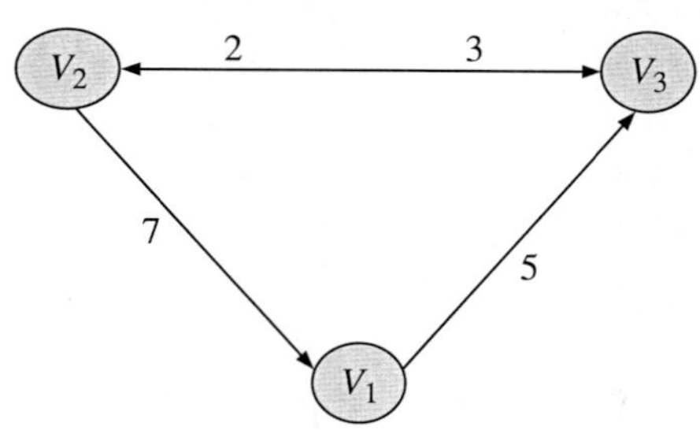

Figure 5.19 Digraph for APSPP.

Solution: The cost matrix of the above graph is

$w[i][j]=$

	V_1	V_2	V_3
V_1	0	∞	5
V_2	7	0	3
V_3	∞	2	0

$= SPC^0[i][j]$

Applying Floyd-Warshall algorithm, the problem can be solved as follows:

For $k = 1$

$$\begin{aligned}
\text{SPC}^1[1][2] &= \min\{\text{SPC}^0[1][2], \text{SPC}^0[1][1] + \text{SPC}^0[1][2]\}\\
&= \min\{\infty, 0 + \infty\}\\
&= \infty\\
\text{SPC}^1[1][3] &= \min\{\text{SPC}^0[1][3], \text{SPC}^0[1][1] + \text{SPC}^0[1][3]\}\\
&= \min\{5, 0 + 5\}\\
&= 5\\
\text{SPC}^1[2][1] &= \min\{\text{SPC}^0[2][1], \text{SPC}^0[2][1] + \text{SPC}^0[1][1]\}\\
&= \min\{7, 7 + 0\}\\
&= 7\\
\text{SPC}^1[2][3] &= \min\{\text{SPC}^0[2][3], \text{SPC}^0[2][1] + \text{SPC}^0[1][3]\}\\
&= \min\{3, 7 + 5\}\\
&= 3\\
\text{SPC}^1[3][1] &= \min\{\text{SPC}^0[3][1], \text{SPC}^0[3][1] + \text{SPC}^0[1][1]\}\\
&= \min\{\infty, \infty + 0\}\\
&= \infty\\
\text{SPC}^1[3][2] &= \min\{\text{SPC}^0[3][2], \text{SPC}^0[3][1] + \text{SPC}^0[1][2]\}\\
&= \min\{2, 2 + \infty\} = 2
\end{aligned}$$

Thus,

$\text{SPC}^1[i][j] =$

	V_1	V_2	V_3
V_1	0	∞	5
V_2	7	0	3
V_3	∞	2	0

For $k = 2$

$$\begin{aligned}
\text{SPC}^2[1][2] &= \min\{\text{SPC}^1[1][2], \text{SPC}^1[1][2] + \text{SPC}^1[2][2]\}\\
&= \min\{\infty, \infty + 0\}\\
&= \infty
\end{aligned}$$

$$SPC^2[1][3] = \min\{SPC^1[1][3], SPC^1[1][2] + SPC^1[2][3]\}$$
$$= \min\{5, \infty + 5\}$$
$$= 5$$

$$SPC^2[2][1] = \min\{SPC^1[2][1], SPC^1[2][2] + SPC^1[2][1]\}$$
$$= \min\{7, 0 + 7\}$$
$$= 7$$

$$SPC^2[2][3] = \min\{SPC^1[2][3], SPC^1[2][2] + SPC^1[2][3]\}$$
$$= \min\{3, 0 + 3\}$$
$$= 3$$

$$SPC^2[3][1] = \min\{SPC^1[3][1], SPC^1[3][2] + SPC^1[2][1]\}$$
$$= \min\{\infty, 2 + 0\}$$
$$= 9$$

$$SPC^2[3][2] = \min\{SPC^1[3][2], SPC^1[3][2] + SPC^1[2][2]\}$$
$$= \min\{2, 2 + 0\}$$
$$= 2$$

Thus,

$SPC^2[i][j] =$

	V_1	V_2	V_3
V_1	0	∞	5
V_2	7	0	9
V_3	9	2	0

For $k = 3$

$$SPC^3[1][2] = \min\{SPC^2[1][2], SPC^2[1][3] + SPC^2[3][2]\}$$
$$= \min\{\infty, 5 + 2\}$$
$$= 7$$

$$SPC^3[1][3] = \min\{SPC^2[1][3], SPC^2[1][3] + SPC^2[3][3]\}$$
$$= \min\{5, 5 + 0\}$$
$$= 5$$

$$SPC^3[2][1] = \min\{SPC^2[2][1], SPC^2[2][3] + SPC^2[3][1]\}$$
$$= \min\{7, 3 + 9\}$$
$$= 7$$

$$SPC^3[2][3] = \min\{SPC^2[2][3], SPC^2[2][3] + SPC^2[3][3]\}$$
$$= \min\{3, 3 + 0\}$$
$$= 3$$

$$SPC^3[3][1] = \min\{SPC^2[3][1], SPC^2[3][3] + SPC^2[3][1]\}$$
$$= \min\{9, 3 + 9\}$$
$$= 9$$

$$\begin{aligned}SPC^3[3][2] &= \min\{SPC^2[3][2], SPC^2[3][3] + SPC^2[3][2]\}\\ &= \min\{2, 0 + 2\}\\ &= 2\end{aligned}$$

Thus,

$SPC^3[i][j] =$

	V_1	V_2	V_3
V_1	0	7	5
V_2	7	0	3
V_3	9	2	0

This matrix shows the shortest paths between all pairs.

5.10 SOME COMMON QUESTIONS

Question 1. What is relation between TSP and Hamiltonian cycle?

Answer: Mainly two similarities are found.

1. A TSP tour is a Hamiltonian cycle in the graph.
2. An optimal TSP tour is a shortest Hamiltonian cycle.

Question 2. What are the basic characteristics of dynamic programming?

[WBUT 2009, 2010]

Answer: There are several important characteristics of dynamic programming, such as:

(*i*) The problem is divided into stages.

(*ii*) Each stage goes through a number of states.

(*iii*) The decision at a stage is forwarded to the state for the next stage.

(*iv*) The optimal decision for the remaining stages is independent of decisions made in previous states for given current state.

(*v*) Recursive relationship exists between the decision at a stage and the optimum decisions made at previous stages.

Question 3. What is meant by optimal solution?

Answer: A solution to an optimization problem which minimizes (or maximizes) the objective function is termed as an optimal solution.

Question 4. What is the difference between dynamic programming and divide and conquer?

Answer: The differences between these design techniques are given below.

Dynamic Programming	*Divide and Conquer*
1. Many decisions and sequences are guaranteed and all the overlapping subinstances are considered.	1. Divide the given problem into many subproblems. Find the individual solutions and combine them to get the solution for the main problem.
2. Follows bottom up technique.	2. Follows top down technique.
3. Split the input at every possible points rather than at a particular point.	3. Split the input only at specific points (midpoint).
4. Sub-problems are dependent on the main problem.	4. Each problem is independent.

Question 5. What are the basic elements of dynamic programming?

Answer: Three elements are–

(*a*) Sub-structure

(*b*) Table structure

(*c*) Bottom-up computations.

Question 6. What are the preconditions for the Floyd's algorithm?

Answer: The precondition for the Floyd's algorithm is that the graph should not have a cycle of negative length or cost.

Question 7. Explain why memorization is ineffective in speeding up a good divide-and-conquer algorithm such as Merge-Sort.

Answer: The Mergesort algorithm performs at most a single call to any pair of indices of the array that is being sorted. In other words, the subproblems do not overlap and therefore memorization will not improve the running time.

Exercises

1. Which of the following algorithms can be classified as dynamic programming algorithms?

   (*i*) Algorithm Insertion sort.

   (*ii*) Algorithm Mergesort.

2. Give an efficient algorithm to compute *n*th number in a Fibonacci sequence. What is the time complexity of your algorithm? Is it an exponential algorithm? Explain.

3. Write a competent algorithm to compute the binomial co-efficient $\binom{n}{r}$.

4. If the number of ways to fully parenthesise *n* matrices, $f(n) = \sum_{k}^{n-1} f(k)f(n-k)$, then prove that the solution to this recurrence is $f(n) = \frac{1}{n}\binom{2n-2}{n-1}$.

5. Multiply the following matrices: $M_1 : 4 \times 5$, $M_2 : 5 \times 3$, $M_3 : 3 \times 6$, $M_4 : 6 \times 4$, $M_5 : 4 \times 5$, and $M_6 : 5 \times 5$.

6. Give a parenthesized expression for the optimal order of multiplying the matrices in Exercise 5.5.

7. Show that the conventional matrix multiplication algorithm requires n^3 multiplications and $n^3 - n^2$ additions.

8. Consider the following five matrices:

$$M_1 : 2 \times 3, M_2 : 3 \times 6, M_3 : 6 \times 4, M_4 : 4 \times 2, M_5 : 2 \times 7.$$

   (*a*) Find the minimum number of scalar multiplications needed to multiply the five matrices.

   (*b*) Give a parenthesized expression for the order in which this optimal number of multiplications is achieved.

   (*c*) Find optimal cost of the solution.

9. Give an example of at least three matrices in which one order of their multiplication costs is at least 100 times the other order.

10. Modify the matrix chain multiplication algorithm in such a way that it produces the order of multiplications as well.

11. Let $G = (V, E)$ be a weighted directed graph, and $u, \in V$. Assume that there is at least one path from u to v.

   (*a*) If S be a shortest length from u to v that passes by another vertex x, show that the portion of the path from u to x is a shortest path from u to x.

   (*b*) Let L be a longest simple path from u to v that passes by another vertex y. Show that the portion of the path from u to y is not necessarily a longest path from u to y.

12. Find shortest path between all pairs on the weighted directed graph shown in Figure 5.20.

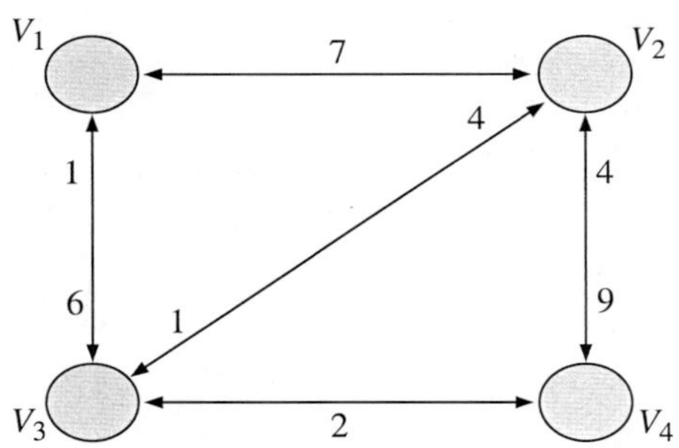

Figure 5.20 An instance of the all-pairs shortest path problem.

13. Compute the distance matrix for the directed graph with the lengths of the edges between all pairs of vertices as given by the matrices below. Use all-pairs shortest path algorithm.

(a) $\begin{bmatrix} 0 & 1 & \infty & 2 \\ 2 & 0 & \infty & 1 \\ \infty & 3 & 0 & 4 \\ 8 & 2 & 3 & 0 \end{bmatrix}$ (b) $\begin{bmatrix} 0 & 2 & 4 & 6 \\ 2 & 0 & 3 & 1 \\ 5 & 9 & 0 & 4 \\ 8 & \infty & 3 & 0 \end{bmatrix}$

14. Modify the all-pairs shortest path algorithm to detect negative-weight cycles exist in a weighted graph.
15. Consider the money change problem. We have a currency system that has n coins with values $v_1, v_2 \ldots v_n$ where $v_1 = 1$, and we want to pay change of value X in such a way that the total number of coins is minimised.

    (*a*) Give a dynamic programming algorithm to solve this problem.

    (*b*) What are the time and space complexities of your algorithm?
16. Apply the algorithm you designed in Exercise 5.17 to the instances $v_1 = 1$, $v_2 = 5$, $v_3 = 7$, $v_4 = 11$, and $X = 20$.
17. Let $G = (V, E)$ be a directed graph with n vertices and M be its adjacency matrix. The reflexive and transitive closure of M denoted by M^* is denned as follows.

$$M = \begin{cases} 1, \text{ if } u = v \text{ or } \exists \text{ a path in } G \text{ from } u \text{ to } v;\ u, v \in V \\ 0, \text{ otherwise} \end{cases}$$

    Devise a dynamic programming algorithm to compute M^* for a given directed graph. (**Hint:** slightly modify Floyd-Warshall algorithm for the all-pairs shortest path problem).
18. Let $G = (V, E)$ be a directed graph with n vertices. The $n \times n$ distance matrix D is defined as follows.

$$D = \begin{cases} d, \text{ the shortest path from } u \text{ to } v \text{ measured in the number of edges} \\ 0, \text{ otherwise} \end{cases}$$

    For example,

$$D = \begin{cases} 1, \text{ if } u = v \in V \text{ and } (u, v) \in E \\ 0, \text{ if } u = v \end{cases}$$

    Write a dynamic programming algorithm to compute D.

    (**Hint:** slightly modify Floyd-Warshall algorithm for the all-pairs shortest path problem).
19. Let $G = (V, E)$ be a directed acyclic graph (dag) with n vertices. Let u and v be two vertices in V such that the indegree and the outdegree of u and v are both 0.

    (*a*) Give a dynamic programming algorithm to compute a longest path in G from u to v.

    (*b*) What is the time complexity of your algorithm?
20. Given a set of n cities with their intercity distances. Design an algorithm to find minimum tour length. Here, a tour is a cycle that visits each city exactly once except the source of journey. What is the time complexity of your algorithm? Can this problem be solved in time $O(n^2 2^n)$?

21. Given a Knapsack/bag of some capacity and a list of items having a weight/size and cost/value associated with each item. Put the items into the Knapsack taking from the list in such a way that sum of weights of the selected items does not exceed the capacity of the Knapsack as well as value earned is optimised.

    Write a dynamic programming algorithm to earn maximum values if

    (a) There are four items of sizes 2, 3, 5, and 6 and values 3, 4, 5, and 7, and the Knapsack capacity is 11.

    (b) There are five items of sizes 3, 5, 7, 8 and 9 and values 4, 6, 7, 9 and 10, and the Knapsack capacity is 22.

22. What would happen when running the Knapsack algorithm you designed in Exercise 5.24 on an input in which one item has negative size? Explain.

23. Evaluate complexities of the algorithm you designed in Exercise 5.24.

24. Modify your algorithm designed in Exercise 5.24 so that it requires only $\Theta(C)$ space, where C is the Knapsack capacity.

25. A motorist wishes to find the shortest possible route from Kolkata to Mumbai. Given a road map of India to him on which the distance between each pair of adjacent intersections is marked. How can he determine the shortest route?

26. You have a set of n integers each in the range 0 ... K. Partition these integers into two subsets such that you may minimise $|S_1 - S_2|$, where S_1 and S_2 denote the sums of the elements in each of the two subsets.

27. You are given an ordered sequence of n cities, and the distances between every pair of cities. You must partition the cities into two sub-sequences (not necessarily contiguous) such that person A visits all cities in the first sub-sequence (in order), person B visits all cities in the second sub-sequence (in order), and such that the sum of the total distances travelled by A and B is minimised. Assume that person A and person B start initially at the first city in their respective sub-sequences.

28. You have n_1 items of size s_1, n_2 items of size s_2, and n_3 items of size s_3. Pack all of these items into bins (Knapsack) each of capacity C, such that the total number of bins used is minimised.

University Questions

1. What do you mean by dynamic programming?
   [WBUT'07, '09, '10, UPTU'05, '06, PTU'09, RTU'08]
2. List the steps for the development of a Dynamic Programming algorithm and explain each of them briefly. ***[DOEACC'08]***
3. What is the *Principle of Optimality* that is used in Dynamic Programming paradigm? Explain with an example how *use of table* is found to be efficient instead of using *recursion?* ***[BPUT'05, IGNOU'09]***
4. What are the basic characteristics of dynamic programming?
   [WBUT'09, IETE ALCCS'11]
5. When and how Dynamic programming approach is applicable? ***[UPTU'06]***
6. How does the principle of optimality reduce the number of decision sequences in dynamic programming algorithms? ***[BPUT'08]***

7. Distinguish between divide-and-conquer and dynamic programming with suitable examples. *[IGNOU'07, BPUT'05, DOEACC'07, '09]*
8. Show how A^m can be computed using dynamic programming technique, for any large positive integer value of m? *[DOEACC'09]*
9. Explain how dynamic programming is used to solve matrix-chain multiplication problem. *[BPUT'07]*
10. Derive the principle of optimality for multiplication of matrix chain. *[IGNOU'09]*
11. Compute the optimal number of scalar multiplications required to multiply the following matrices. A_1 of order 30 × 35, A_2 of order 35 × 15, A_3 of order 15 × 5. *[IGNOU'09]*
12. Give an algorithm for matrix-chain multiplication. *[WBUT'08, '09, '10, UPTU'05, RTU'08, JNTU'08]*
13. Consider the classical $O(n^3)$ matrix multiplication algorithm to multiply two $n \times n$ matrices. Observe that there are two operations *addition* (+) and *multiplication* (×). In order to determine all pairs of shortest paths for a connected graph with n vertices, this matrix multiplication algorithm can be used but one needs to replace these two operations by another two operations. Mention the name of operation to be used as a replacement of *addition* and *multiplication*. *[DOEACC'07]*
14. Find the minimum number of operations required for the following matrix chain multiplication using dynamic programming: A (10 × 20) * B (20 × 50) * C (50 × 1) * D (1 × 100). *[WBUT'08]*
15. Find the minimum number of operations required for the following chain matrix multiplication using dynamic programming. A(30, 40) * B(40, 5) * C(5, 15) * D(15, 6). *[JNTU'09]*
16. Consider the problem of chain matrix multiplication. Devise a solution for the following chain of matrices with dimensions: 30 × 35, 35 × 15, 15 × 5, 5 × 10, 10 × 20. *[IGNOU'07]*
17. Find the order of Parenthesization for the optimal chain multiplication: *[RTU'09]*

$$A_1 = 30 \times 35;\quad A_2 = 30 \times 15;\quad A_3 = 15 \times 5$$
$$A_4 = 5 \times 10;\quad A_1 = 10 \times 25;\quad A_6 = 20 \times 25$$

18. Discuss matrix chain multiplication with reference to dynamic programming technique and also apply it on the following array that contains dimensions of four matrices. *[UPTU'06]*

30	1	40	10	25

19. Find the order of parenthesization for the optimal chain multiplication: *[RTU'07]*

$$A_1 = 50 \times 20;\ A_2 = 20 \times 1;\ A_3 = 1 \times 10;\ A_4 = 10 \times 15$$

20. Find the minimum number of operations required for the following chain matrix multiplication using dynamic programming. A(30, 40) * B(40, 5) * C(5, 15) * D(15, 6). *[JNTU'08]*
21. Find lower bound for multiplying $m \times n$ matrix with $n \times 1$ vector. *[PTU'08]*
22. Solve the following 0/1 Knapsack problem using dynamic programming. $P = (11, 21, 31, 33)$, $W = (2, 11, 22, 15)$, $C = 40$, $n = 4$. *[JNTU'08]*
23. Explain the principle of ordering Matrix multiplication in the light of dynamic programming. Show that the number of ways that n matrices can be multiplied is of $O(4^n/n^{3/2})$. Hence find an optimal way to find product ABCD of matrices A(30 × 1), B(1 × 40), C(40 × 1), C(10 × 25). *[WBUT'07, BPUT'08'UPTU'07]*

24. Describe the Dynamic Programming algorithm for computing the minimum cost order of multiplying a string of n matrices $M_1 \times M_2 \times M_3 \ldots \times M_n$. and find time and space complexity of the algorithm. ***[PTU'07, DOEACC'08, '09]***

25. Let $R(i, j)$ be the number of times that table entry $m[i, j]$ is referenced while computing other table entries in a call of matrix-chain order. Show that the total number of references for the entire table is

$$\sum_{i=1}^{n}\sum_{j=1}^{n} R(i, J) = \frac{(n^3 - n)}{3}.$$

[DOEACC'06]

26. Explain the Knapsack problem. ***[RTU'09]***

27. What is 0/1 Knapsack problem? Discuss its solution. ***[IETE ALCCS'11]***

28. Consider the Knapsack instance : Number of objects $n = 3$ with corresponding weight and profit as $W_i = (2, 3, 4)$, $P_i = (1, 2, 5)$. The Knapsack capacity $m = 6$. Using dynamic approach find the solution for this instance of 0/1 Knapsack problem. ***[MDU'09]***

29. Write a dynamic programming algorithm for the 0-1 Knapsack problem. Your algorithm should be able to take as parameters two arrays: one giving the size of each of the n items and the second giving the value of each of the items, and the size of the Knapsack *W*. You are to maximize the value in your Knapsack without exceeding the capacity *W*. ***[BPUT'07]***

30. Show that the Knapsack problem does have a (suitably formulated) optimal substructure property. Can you construct (recursively) all solutions which have the property? ***[DOEACC'08]***

31. Obtain the dynamic programming recurrence equations for the 0/1 Knapsack problem of *n* instances. Assume suitable data for n instances Knapsack. ***[UPTU'10]***

32. Consider the Knapsack Instance(R_i) $n = 5$, their positive weights $(w_1, w_2, w_3, w_4, w_5) = (6, 5, 2, 1, 7)$, positive values $(v_1, v_2, v_3, v_4, v_5) = (22, 18, 6, 1, 28)$ and $W = 11$. Find the solution by Dynamic programming. ***[RTU'07, BPUT'08]***

33. Write mathematical formulation of 0-1 Knapsack problem. Use dynamic programming approach to solve the following instance of the problem. ***[RTU'10, DOEACC'06]***

Maximum capacity	:	11 units
No. of items	:	5
Weight	:	1, 2, 5, 6, 7
Profit	:	1, 6, 18, 22, 28

34. Suppose that in 0/1 Knapsack problem the order of items when sorted on increasing value is the same as their order when sorted by decreasing weight. Give an efficient algorithm to find an optimal solution to the problem. ***[IETE ALCCS'11]***

35. State Bellman's principle of optimality. Also show that how it is valid for shortest path problem with non-negative weights on the edges. ***[DOEACC'06]***

36. State an algorithm to find a minimum cost path from source to shrink in a multistage graph using backward approach using *dynamic programming* paradigm. ***[BPUT'08]***

37. Write down the reachability matrix of the graph given below by inspection. ***[BPUT'05]***

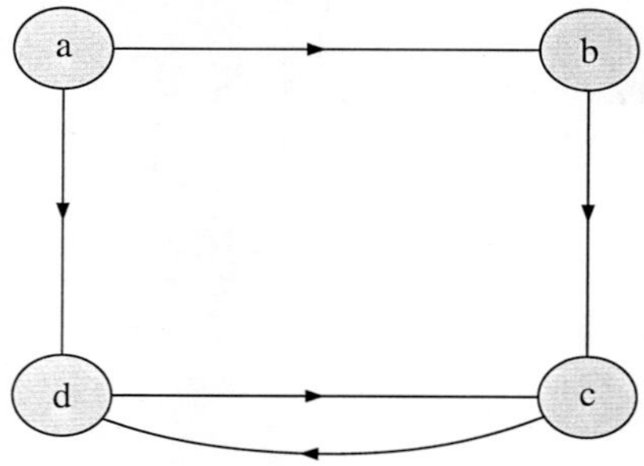

38. Write short notes on travelling salesman problem. ***[RTU'08, IGNOU'09]***
39. What is the nature of solution to the TSP problem? ***[IGNOU'09]***
40. Write the travelling salesman problem with an algorithm. ***[WBUT'10, UPTU'10]***
41. Explain Dynamic Programming method to solve a problem and use the same method to solve the following Travelling-salesperson Problem with given distance matrix: ***[MDU'08]***

$$\begin{bmatrix} 0 & 5 & 10 & 15 \\ 3 & 0 & 6 & 8 \\ 4 & 12 & 0 & 11 \\ 7 & 7 & 8 & 0 \end{bmatrix}$$

42. Explain the Floyd-Warshall algorithm with example. ***[UPTU'06, PTU'05, BPUT'05, IETE ALCCS'11, DOEACC'08]***
43. Describe Floyd-Warshall's algorithm to find all pairs shortest paths in a directed network. ***[UPTU'10]***
44. A newspaper agent daily drops the newspaper to the area assigned in such a manner that he has to cover all the houses in the respective area with minimum travel cost. Compute the minimum travel cost. The area assigned to the agent where he has to drop the newspaper is shown in the figure below: ***[WBUT'08]***

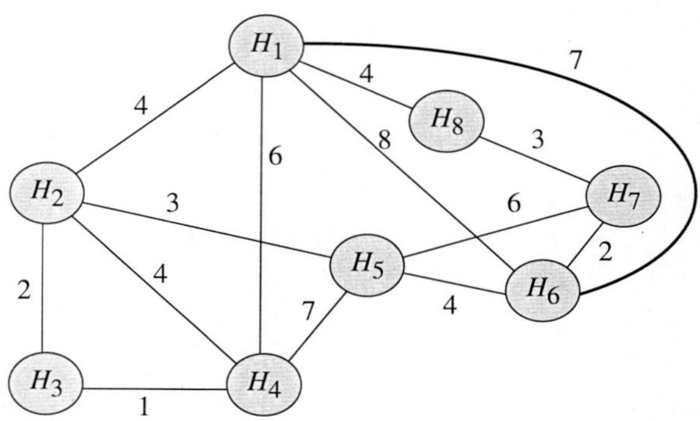

45. Write an algorithm to find all pairs shortest path in a weighted graph. Find out time complexity of the algorithm you used. ***[WBUT'09, MDU'09]***

46. For the graph (weighted, direct) apply Floyd-Warshall algorithm for constructing shortest path. *[UPTU'05]*

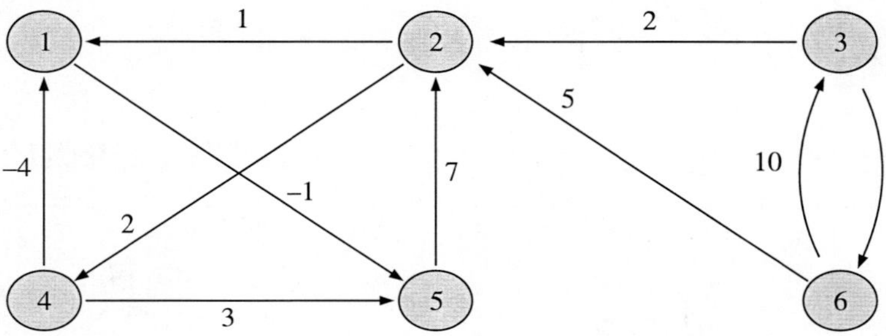

47. For the following graph G, determine the matrix C where $(i, j)^{th}$ element C_{ij} gives the cost of the shortest path from V_i to V_j, for all i and j. *[DOEACC'07]*

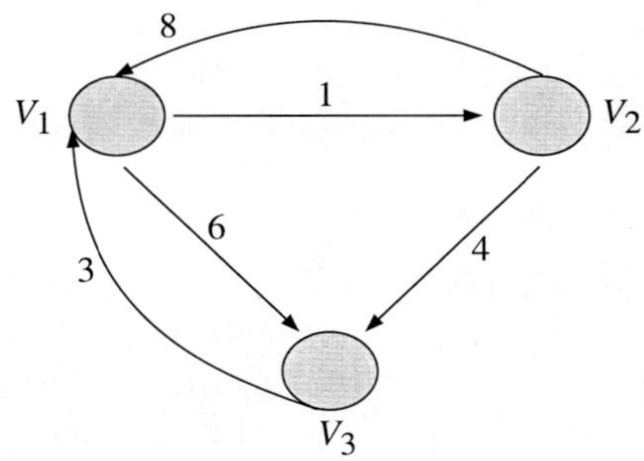

48. What do you mean by combinatorial problem? Can you find an optimal solution to this problems using dynamic programming? *[BPUT'08]*

❑❑❑

Chapter 6 Greedy Approach

6.1 INTRODUCTION

Dynamic Programming is one of the best powerful algorithm design technique for optimization problem. But it is inefficient to use in some cases since it does not always provide optimal solution to all optimization problems and often leads to a running time beyond expectation. In that places a better solution — Greedy Approach is used. The name "greedy" means to make decisions in stages based on one input at a time.

6.2 FEASIBLE SOLUTION

Greedy method is a straight forward algorithm design technique for optimization problem. In most of the cases, most of the problems use a subset that satisfies some constraints for the problem having n inputs. Any subset that satisfies these constraints is referred to as feasible solution. Our job is to find a feasible solution that optimizes *i.e.,* either maximizes or minimizes a given objective function. The feasible solution is then said to be an optimal solution. However a feasible solution may not necessarily be an optimal solution.

Example 1. A cashier adds up the largest denomination and proceeds to the smallest denomination for a given sum of money.

Example 2. Suppose we are given a set of events that have a start time and finish time. A subset with maximum number of events can be scheduled using Greedy such that no events have overlapping times.

6.3 PROCEDURE

The Greedy algorithm works in stages taking only one input at a time. At each stage an optimal solution is obtained by making decision on involvement of the input. This is done through a selection function which determines inputs in an order. The next input is only added to the partially optimal solution if it does not produce any infeasible solution. The selection itself is again based on some objective functions. For a given problem more objective functions are possible. And most of these generate sub-optimal solutions. This is called subset paradigm of the Greedy approach.

6.4 SPECIFICATIONS

In general therefore, the Greedy algorithms may have five pillars CS, SF_1, FF, OF, SF_2 where:

1. CS — candidate/input set.
2. SF_1 — selection function which selects the best candidate to add in the solution.
3. FF — feasibility function used to determine a candidate in the solution.
4. OF — objective function that assigns a value to a solution or a partial solution.
5. SF_2 — solution function that indicates complete solution.

6.5 GENERAL SOLUTION TECHNIQUES

Most problems for which this approach works well have the following three properties:

(*i*) **Feasible Solution**: We check whether all the constraints are satisfied or not so that we have an optimal solution for the problem under consideration.

(*ii*) **Local Optimal Choice**: The input is considered from the current available feasible solutions. That means we can make whatever choice seems best at the moment and then solve the sub-problems that will arise later. It is also called a greedy choice property.

(*iii*) **Unalterable**: At any subsequent stages a choice remains unaltered once it is made.

6.6 OPTIMAL SUB-STRUCTURE

A problem is said to exhibit an optimal sub-structure if the optimal solution to the problem contains optimal solutions to its sub-problems. In other words a problem has an optimal sub-structure if the next move which is assumed to be best, always leads to the optimal solution.

6.7 CLASSIFICATIONS

Greedy algorithms are ideal only for problems which have 'optimal sub-structure'. There are mainly three variations to the greedy algorithm possible:

(*i*) Pure greedy algorithms.

(*ii*) Orthogonal greedy algorithms.

(*iii*) Relaxed greedy algorithms

6.8 DYNAMIC PROGRAMMING VS. GREEDY ALGORITHMS

Greedy method is a powerful technique but not much like DP approach.

- Dynamic programming
  - Choice is made at each step.
  - The choice depends on solutions to sub-problems.
  - Decisions are based on all the decisions made in the previous stage.
  - May reconsider the previous stage's algorithmic path to solution.
  - Uses bottom up approach from smaller to larger sub-problems
  - More powerful.
  - Applicable to wide range of application.

- Greedy algorithm
  - We make the greedy choice.
  - Solve the sub-problem arising after the choice is made.
  - The choice we make may depend on previous choices
  - The choice is independent of the solutions to sub-problems.
  - Uses top down solution, problems get decreased in size.
  - Some problems can be solved effectively.

6.9 APPLICATIONS

Unlike Dynamic Programming, Greedy algorithm works well for some ranges of optimization problems.

(*i*) Constructions of MST using Prim's and Kruskal's algorithm.
(*ii*) Solving single source shortest path problem using Dijkstra's algorithm.
(*iii*) Searching set covering using Chvatal's algorithm.
(*iv*) Scheduling activity selection problem.
(*v*) Constructions of optimum Huffman trees using Huffman's algorithm.
(*vi*) Solving TSP.
(*vii*) Solving Knapsack problem.
(*viii*) Solving tree vertex splitting problem.
(*ix*) Storing optimal data on tape.
(*x*) Network routing to forward a message. Etc.

6.10 KNAPSACK PROBLEM

The Knapsack Problem (KP) also called rucksack problem is a combinatorial optimization problem. It states that: "Given a Knapsack/bag of some capacity and a list of items having a weight and cost/value associated with each item. Put the items into the Knapsack taking from the list in such a way that sum of weights of the selected items does not exceed the capacity of the Knapsack as well as the total value earned is optimised".

6.10.1 KP: Variations

A typical Knapsack problem has two variations.

(*i*) **0/1 Knapsack:** Either we select or discard an item from the list as a whole. The items are in this case indivisible *e.g.,* clothes, oil bottle, etc.
(*ii*) **Fractional Knapsack:** Any fraction of an item can be considered for the solution.

6.10.2 KP: Applications

Knapsack problem appears in the following areas.

(*i*) Resource allocation problems with financial constraints.
(*ii*) Combinatory, complexity theory, cryptography and applied mathematics.

6.10.3 KP: Greedy Criteria

Items in Knapsack are chosen based on any of the following three greedy criteria.

(*i*) **Greedy Criteria I:** Items are arranged in descending order of their values/ costs. An item having maximum value is selected first then the second largest valued item and so on. The objective is to maximise the value.

(*ii*) **Greedy Criteria II:** Items are arranged in ascending order of their weights. First a lightest item is selected then the second lightest item and so on. The process is continued unto the Knapsack capacity. The objective is to maximise the number of items into the Knapsack.

(*iii*) **Greedy Criteria III:** Items are in this case arranged in descending order of certain ratios $P_i = V_i / W_i$ where V_i is value of item I_i and W_i is weight of item I_i.

6.10.4 Fractional Knapsack Problem (FKP)

Consider, W = weight capacity of the Knapsack.

I = list of items = $< I_1, I_2, I_3..., I_n>$

n = number of elements in the list

W_i = weight of the item I_i

V_i = value associated with the item I_i

X_i = fraction of the item I_i considered. Hence $0 \leq X_i \leq 1$

The Knapsack is filled up by considering the elements from the list based on any one of the greedy criteria such that profit earned is maximized. The problem formally can be represented as:

(*i*) Maximize $\sum_{i=1}^{n} Vi\ Xi$

(*ii*) Subject to $\sum_{i=1}^{n} Wi\ Xi \leq W$

(*iii*) And $0 \leq Xi \leq 1$ for $1 \leq i \leq n$

An optimal solution will be a feasible solution if above first condition is satisfied. Obviously the feasible solution is filling the Knapsack and is any set $< X_1, X_2 ... X_n >$.

6.10.4.1 *FKP: Algorithm*

Algorithm: FracKnap_Greedy ( )

Termed used: W = weight capacity of the Knapsack

n = no. of items

W_i = weight of item I_i

V_i = value of item I_i

P = list of P_i where $P_i > P_{i+1}$ and $P_i = V_i / W_i$

I = list of items taken based on P

w = list of weights of the items taken based on P

v = list of values of items in I

$X[i]$ = fraction of ith item taken from I into Knapsack.

V = sum of values of the items of L based on P

= maximum profit.

Step 1:

(*i*) Repeat for $i = 1$ to n // initialise

(*a*) $X[i] = 0.0$

(*ii*) W = value

(*iii*) $V = 0.0$

Step 2: Repeat for $i = 1$ to n through Step 4 // fill Knapsack

Step 3: If $I[i] > W$

Go to step 7

Step 4:

(*i*) $X[i] = 1.0$ // initialize

(*ii*) $W = W - w[i]$

(*iii*) $V = V + v[i]*X[i]$

Step 5: If $i < n$

(*a*) $X[i] = W / w[i]$

(*b*) $V = V + v[i]*X[i]$

Step 6: Return V

Step 7: Exit

6.10.4.2 *FKP: Analysis*

Here we see that the items are considered based on P, the ratio value/weight in descending order. Sorting P in descending order any suitable sorting algorithm needs O(nlogn) time complexity. Again in the above algorithm we see that filling the Knapsack (step 2) requires $O(n)$ time. Now if P is already sorted then we do not need to sort twice. And our algorithm runs $O(n)$ time. Otherwise its complexity will be $O(n \log n)$ time.

Example 3: Consider the following Table that consists of some items with weights and cost values.

Items	I_1	I_2	I_3	I_4	I_5
Weight	5	10	15	22	25
Cost	30	40	45	77	90

If the Knapsack capacity W = 60 Kg, Find the feasible solution.

Solution: Using *Greedy Criteria III*

Now, cost/weight for every item are $< \frac{30}{5}, \frac{40}{10}, \frac{45}{15}, \frac{77}{22}, \frac{90}{25} >$

= < 6.0, 4.0, 3.0, 3.5, 3.6 >

cost/weight in descending order <6.0, 4.0, 3.6, 3.5, 3.0 >.

Thus the table is arranged as

Items	I_1	I_2	I_5	I_4	I_3
Weight	5	10	25	22	15
Cost	30	40	90	77	45

Now,

Set profit earned $P = 0$. Taking items sequentially until Knapsack is filled,

Take I1, then, $W = W - W_2 = 60 - 5 = 55$

$P = P + C_1 = 0 + 30 = 30$

Take I2, then, $W = W - W_2 = 55 - 10 = 45$

$P = P + C_2 = 30 + 40 = 40$

Take I5, then, $W = W - W_5 = 45 - 25 = 20$

$P = P + C_5 = 40 + 90 = 130$

Next item I_4 weight is 22 Kg. But the Knapsack requires only 20 Kg. So a part X of W_4 is to be considered and it is given by $X = \frac{20}{22}$.

Thus, take X from I_4, then

$$W = W - X*W_4 = 20 - \frac{20}{22}*22 = 20 - 20 = 0$$

$$P = P + C_4*X = 130 + 77*\frac{20}{22} = 130 + 70 = 200$$

Hence, the Knapsack has been filled up fully and the profit earned is 200.

Readers are asked to solve the above problem using greedy criteria I and II.

6.11 JOB SEQUENCING WITH DEADLINES (JSD)

The problem can be defined as: "We are given a job set JS containing n jobs. Each job i is associated with allowable time to completion called deadline d_i and cost value called profit p_i. A profit p_i is earned if ith job is completed within its deadline. Find maximum number of jobs that can be solved within a deadline. Assume that there is a single processor machine that can complete one job at a time and every job requires suppose one unit of time to completion".

The possible solution to this problem is a subset of JS without violating the above constraints. And this solution is nothing but a feasible solution set FS. The maximum profit P earned is sum of profit value of the individual job in FS. That means $P = \Sigma p_i, \forall\, i \in FS$. The optimal solution is the feasible solution with maximum profit P.

Solution of the problem depends how the next job is selecting. It requires us to choose a job from *JS* in descending order of p_i to maximise *P*.

6.11.1 JSD: Feasible Solution

Now we are ready to find out a feasible solution FS of the job sequences. First, find all possible permutations of the jobs in *JS*. Then check some permutations that can be processed and be checked within the deadline. These permutations together form *FS*. In any permutation in *FS* the jobs are generally ordered in increasing order of their deadlines.

6.11.2 JSD: Algorithm

Algorithm: JSD_Greedy(*P*, *D*, *n*, *FS*)

P = List of profits in descending order.

D = List of deadlines of the respective jobs.

FS = Set of jobs performed within deadline.

n = Number of jobs assigned.

Profit = profit earned for the feasible solution.

Initial conditions:

(*a*) $D\ [i] \geq 1$ for $1 \leq i \leq n$

(*b*) $n \geq 1$

(*c*) $FS\ [i]$ is the ith job in optimal solution for $1 \leq i \leq k \leq n$.

(*d*) At termination, $D\ [FS\ [i]] \leq D\ [FS\ [i + 1]]$ for $1 \leq i \leq k$

Step 1: Initialisation and 1st job Inclusion

(*i*) $D\ [0] = FS\ [0] = 0$

(*ii*) $FS\ [1] = 1$ // job with max profit value

(*iii*) $k = 1$

(*iv*) Profit = 0

Step 2: Repeat for $i = 2$ to n through step 5 // for feasible solution

Step 3: Initialize $r = k$

Step 4: Repeat while $D\ [FS\ [r]] > D\ [i]$ and $D\ [FS\ [r]] \neq r$

(*a*) $r = r - 1$

Step 5: If $D\ [FS\ [r]] \leq D\ [i]$ and $D\ [i] > r$ // checking deadline violation

(*a*) Repeat for $q = k$ to $(r + 1)$ down

- $FS[q + 1] = FS\ [q]$

(*b*) $FS\ [r + 1] = i$

(*c*) $k = k + 1$

Step 6: Repeat for $i = 1$ to k

(*a*) Profit = Profit + $P\ [FS\ [i]]$

Step 7: Display job sequence and profit

(*a*) Repeat for $i = 1$ to k

- Output $FS\ [i]$

(*b*) Output "maximum profit earned" = Profit

Step 8: Exit

6.11.3 JSD: Analysis

The basic parameters that affect JSD_Greedy() algorithm are mainly number of jobs assigned n and number of jobs completed k within time bound. Now step 2 executes at most $n - 1$ times whereas step 4 executes at most k times. Again step 5 may iterate $k - r$ time. Thus time complexity of the algorithm will be basically based on step 2 and 4. And the order will be $O(kn) \approx \Theta(n^2)$ since $k \leq n$.

Example 4. Find feasible solution for the following list of jobs. Also find the total profit earned. Assume that each job needs one unit time in a single machine.

Job No.	I	II	III	IV	V
Profit	50	30	10	20	15
Deadline	2	1	3	2	4

Solution: The feasible solution can be obtained step-by-step as follows.

First, sort the profit values in descending order. Accordingly place job positions. Keep their deadlines as they are.

Deadline =

Profit	50	30	20	15	10
Now Job No.	I	II	III	IV	V
Deadline	2	1	3	2	4
Old Job No.	I	II	IV	V	III

max(deadline of all jobs) = max(2, 1, 3, 2, 4) = 4. Four times slot will be there *i.e.,* at most 4 jobs will be completed within their deadlines.

Now, considering the new job numbering in the above table, jobs are scheduled as below.

Job I is allotted in [0, 1] time slot.

Time Slot	0-1	1-2	2-3	3-4	4-5
New Job No.	I				
Ordered by Deadlines	2				
Old Job No.	I				

Job II is allotted in [1, 2] time slot. But its deadline is 1. So we have to shift job I upward.

Time Slot	0-1	1-2	2-3	3-4
New Job No.	I→			
Ordered by Deadlines	2			
Old Job No.	I			

↓

Time Slot	0-1	1-2	2-3	3-4
New Job No.	II	I		
Ordered by Deadlines	1	2		
Old Job No.	II	I		

Job III is allotted in [2, 3] time slot.

Time Slot	0-1	1-2	2-3	3-4
New Job No.	II	I	III	
Ordered by Deadlines	1	2		
Old Job No.	II	I		

Job IV is allotted in [3, 4] time slot. But its deadline is 2. Already slot [1, 2] has been used. So this job IV is rejected.

Now, Job V is allotted in [3, 4] time slot.

Time Slot	0-1	1-2	2-3	3-4
New Job No.	II	I	III	V
Ordered by Deadlines	1	2	3	4
Old Job No.	II	I	IV	II

Thus, the feasible solution = <I, II, III, V> [new job no.]

Execution sequence = <II, I, III, V> [new job no.]

Profit Earned = Profit[II] + Profit[I] + Profit[III] + Profit[V] [new job no.]

= 30 + 50 + 10 + 15

= 105

6.12 MINIMUM SPANNING TREE

A Spanning Tree (ST) of a graph *G* is any tree that contains every vertex of the graph *G*. In other words, a spanning tree is a sub graph of a graph *G* which is a tree and contains all the vertices of *G*. The graph *G* must be a connected graph but may be directed/undirected or weighted/unweighted. A single graph may have many different spanning trees. Consider Figure 6.1. Some possible spanning trees are shown in Figure 6.2.

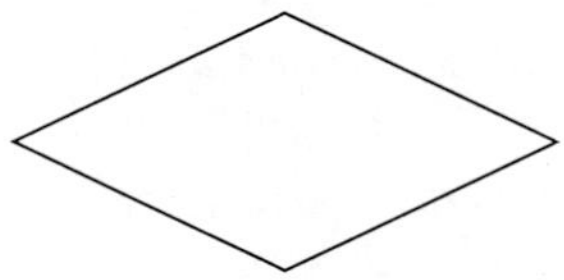

Figure 6.1 A connected graph.

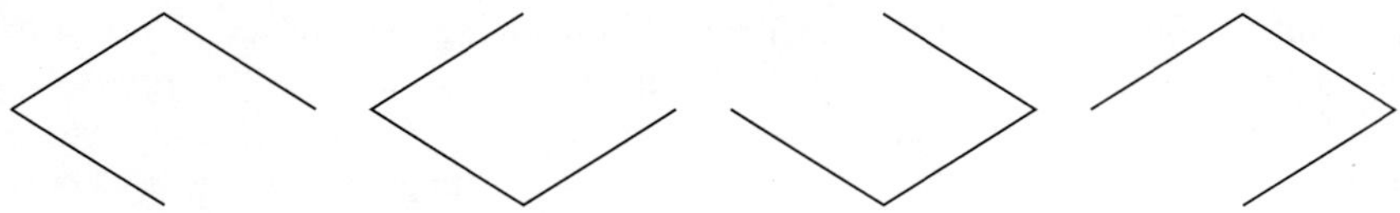

Figure 6.2 Possible spanning trees.

Every edge weight in a connected, weighted graph is assigned with a non-negative number. The sum of weights of the edges of a spanning tree of this graph when shows minimum value, the spanning tree thus obtained becomes minimum spanning tree. For a weighted graph and its minimum spanning tree is shown in Figure 6.3 and Figure 6.4 respectively.

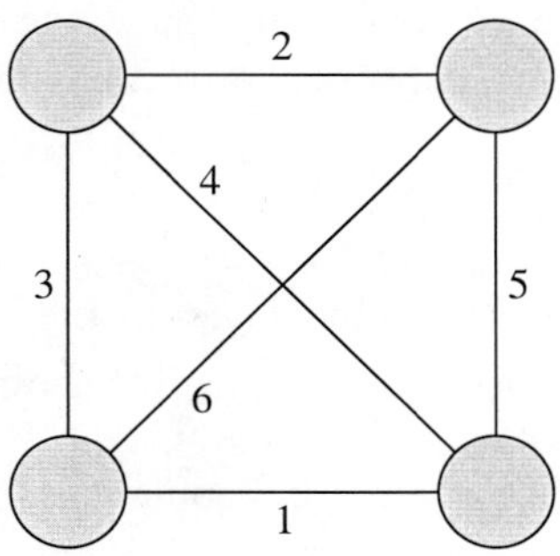

Figure 6.3 A weighted graph.

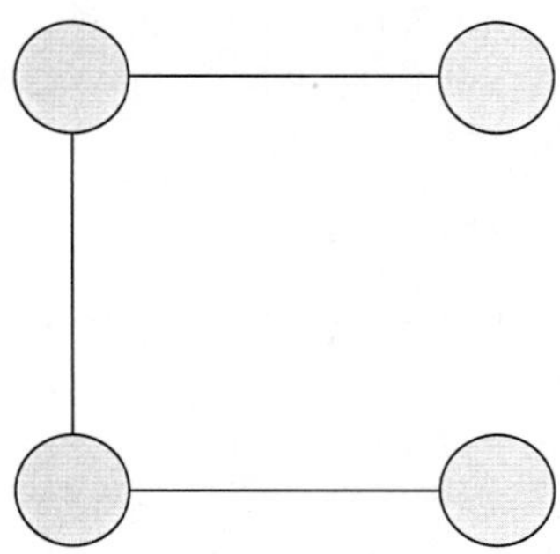

Figure 6.4 Minimum spanning tree.

A common edge of G and its spanning tree is called a branch. And an edge which belongs to *G* but not in its spanning tree is called a chord. *e.g.,* in above weighted graph and its spanning tree edges having weights 1, 2, and 3 are branches and others are chords.

6.12.1 MST: Definition

A Minimum Weighted Spanning Tree or simply Minimum Spanning Tree (MST) of a weighted, connected graph $G = (V, E, W)$ is a spanning tree of *G* in which sum of weights of the branches is minimum. Weight of a MST may be less or equal to the weight of every other spanning trees of *G*.

A MST of an undirected graph has two properties:

(*i*) It spans the graph *i.e.,* it includes all the vertices of the graph.
(*ii*) It is minimum weighted *i.e.,* the sum of weights of all the edges is least.

An undirected graph which is not connected generates minimum spanning forest instead of tree. This forest is actually a union of all MSTs for its connected components.

Example 5. Suppose a cable TV company is laying cable to a locality. All the houses are connected by different roads/paths. If there may be a constraint to bury the cable along certain paths for the houses. We can represent this locality in a graph where each node points to a house and each edge is a path connecting two houses. Some of those paths are more expensive due to longer. We represent these edges with larger weights. Here edge weight should be equivalent to cable length. A spanning tree for that graph is then a subset of those paths that does not form cycles but connects to every house. A minimum spanning tree would be one having the lowest total edge weight.

6.12.2 MST: Importance

Spanning trees are important because of the following reasons.

1. Constructing sparse sub-graph that says a lot about the original graph.
2. Designing efficient routing algorithms.
3. Solving optimisation/hard problems like Steiner tree problem, TSP, etc.
4. Designing network design, etc.

Basically, a spanning tree provides a simple, cheap even an efficient way to connect a set of terminals such as computers, cites, clustering, network construction, minimum spanning relaxation, factories, etc.

6.12.3 MST: Properties

A minimum spanning tree should in general possess the following properties.

1. **Possible MSTs:** In a graph, several spanning trees are possible. Sometimes few even all minimum spanning trees may have the same weight. If all the edges are of equal weights, every spanning tree will then be a minimum spanning tree.
2. **Uniqueness:** If each edge has a distinct weight in the graph, then there exists only one and unique minimum spanning tree. This generalises to spanning forests as well.
3. **Minimum Cost Subgraph:** When the weights are non-negative, then a minimum spanning tree is really a minimum cost sub-graph connecting all vertices.
4. **Cycle Property:** An edge having weight larger than that of other edges of any cycle in the graph can never be a branch to its any MST.
5. **Cut Property:** An edge having weight smaller than that of other edges of any cut in the graph always belongs to its all MSTs.

6.12.4 MST: Algorithms

Based on the property whether the graph is connected or disconnected, several algorithms are used to construct a MST.

1. For connected graph:

   (*i*) Borùvka's algorithm.
   (*ii*) Kruskal's algorithm.
   (*iii*) Prim's algorithm.

Though these algorithms can also be used for disconnected graph but they will be inefficient one in applying.

2. For disconnected graph—Reverse-Delete algorithm.

But keep in mind that we always deal with connected graph. Discussion on MSTs of disconnected graphs is beyond the scope.

6.12.5 Kruskal's Algorithm

In graph theory Kruskal's algorithm (KA) is a very important and easy algorithm to find a minimum spanning tree from a connected weighted graph. If the graph is not connected, then the algorithm finds a minimum spanning forest. Kruskal's algorithm is one kind of greedy algorithm.

6.12.5.1 *KA: Working Principle*

The algorithm works as follows:

1. Create a forest T, a set of trees. Initially it consists of all the vertices of the graph. Each vertex is a separate tree. Representation is initially exactly same as the graph but without any edge between vertices.
2. Create a list S containing all the edges of the graph in ascending order of their weight value.
3. Remove an edge from S with minimum weight. We call this a light edge. If more than one edge have same weight take them one after another.
4. Add the edge just removed to the forest iff it connects two different trees into a single tree. Otherwise discard that edge.
5. Repeat steps 3 and 4 till S is empty.

When the algorithm terminates, the forest will has only one component, a tree which is nothing but a minimum spanning tree of the graph.

6.12.5.2 *KA: Algorithm*

Algorithm: Kruskal_MST (G, T, S, C)

Where $G = (V, E, W)$, a weighted connected graph.

V = set of vertices in G.

E = set of edges in G.

W = weight/cost matrix for G.

$T = (V', E')$, MST of G. Initially it is a forest.

$V' = V$

E' = set of edges in T.

S = a sorted list of all the edges of G in ascending order of their weights.

C = cost of constructing T, initially zero.

Step 1:

(*i*) Consider all the vertices of V as V' in T // Initialisation

(*ii*) Consider initially $E' = \{\}$.

Step 2: (*i*) Make S.

Step 3: (*i*) Consider first edge (v_i, v_j) from S. // Starts constructing MST

(*ii*) Connect v_i, v_j in T

(*iii*) $E' = E' \, (v_i, v_j)$

(*iv*) $C = C + W(v_i, v_j)$

Step 4:

(*i*) Consider next edge (v_i, v_j) from S.

(*ii*) If connecting v_i, v_j in T does not induce any cycle

(*a*) Connect the vertices pair v_i, v_j in T.

(*b*) $E' = E' \, (v_i, v_j)$

(*c*) $C = C + W(v_i, v_j)$

Step 5: Repeat step 4 until S is fully traversed.

Step 6: (*i*) Output T // display the MST.

(*ii*) Output C // the cost of constructing T

Step 7: Exit.

6.12.5.3 *KA: Analysis*

S is made using any efficient sorting algorithm. The number of edges in G is $|E|$. Thus, step 2 requires $O(|E| \log |E|)$ time. Connecting light weight edge (v_i, v_j) that does not produce any cycle in T, needs $\Theta(\log |V|)$ time. $|V|$ is the number of vertices of G. Again there are $|E|$ edges in S. Naturally total number of iterations would be $|E|$ times. Consequently, step 4 executes in $O(|E| \log |V|)$ time. As a result, total time complexity for Kruskal algorithm to connect all the vertices pair is given by $O((|V| + |E|) \log |V|) \approx \Theta(|E| \log |V|)$ as $|E| >> |V|$.

6.12.5.4 *KA: Drawbacks*

Some common pitfalls of Kruskal's algorithm are:

1. During construction of MST we may not always have a tree rather a forest.
2. Running time complexity increases if the graph is fully connected or regular.
3. Weighted edges have to sort before construction, etc.

Example 6: Construct a MST for the following graph using Kruskal's algorithm.

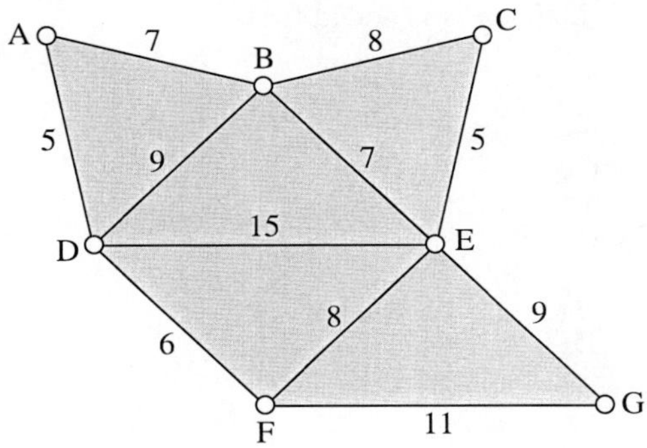

Figure 6.5 Undirected graph for Kruskal's MST.

Solution: Consider the given graph. Edges with their weights in ascending order are shown in the table below.

Weight	5	5	6	7	7	8	8	9	9	11	15
Edge	AD	CE	DF	AB	BE	BC	EF	BD	EG	FG	DE

Now, using Kruskal algorithm, we construct a MST step-by-step.

Step 1: Configure the graph as initial configuration as below.

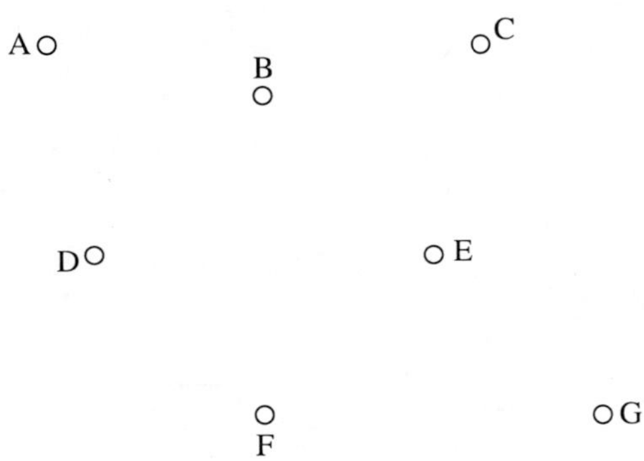

Figure 6.6 Initial configuration.

Step 2: Connect light to lighter edges in Figure 6.7 which do not form cycle. Connection is shown by labelling an edge with alphabets *a*, *b*, *c*... *e.g.*, edges having label *a*/*w* and *b*/*w* means the edge with label *a*/*w* is added first.

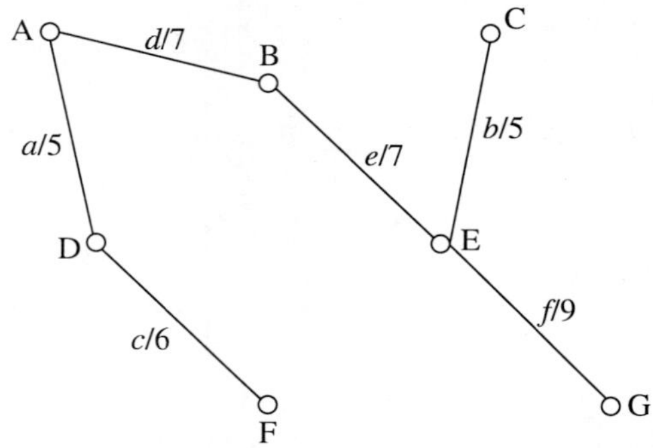

Figure 6.7 MST.

$$\text{Total cost of constructing above MST} = \text{weight } (AD + CE + DF + AB + BE + EG)$$
$$= 5 + 5 + 6 + 7 + 7 + 9$$
$$= 39$$

Example 7. Construct a MST of the graph in Figure 6.8 using Kruskal's.

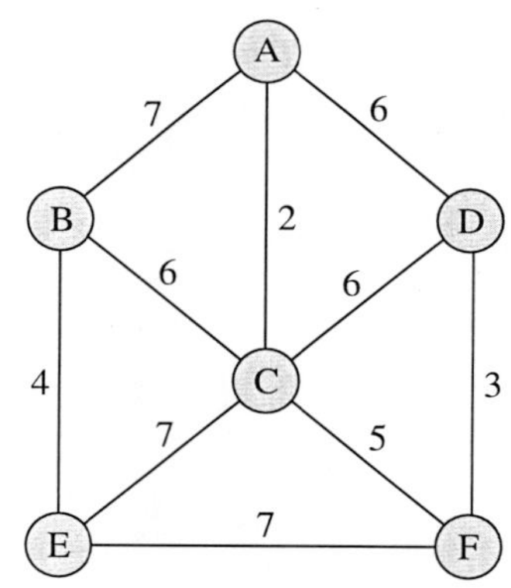

Figure 6.8 Undirected graph for Kruskal's MST.

Solution: The minimum spanning tree using Kruskal algorithm can be constructed step-wise as below.

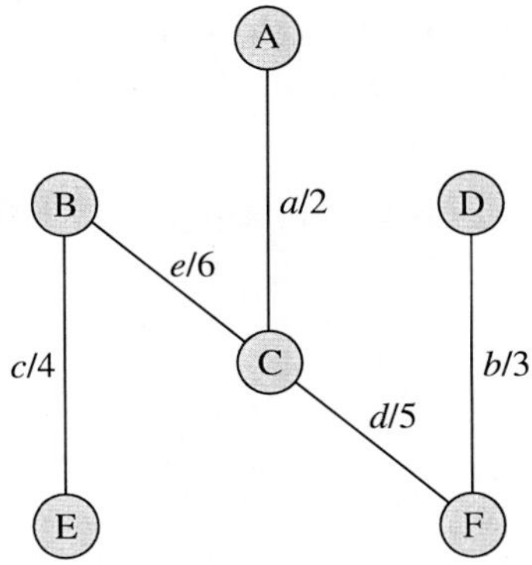

Figure 6.9 MST.

Total cost of constructing above MST= weight $(AC + DF + BE + CF + BC)$

$$= 2 + 3 + 4 + 5 + 6$$

$$= 20$$

Example 8. Find MST of the graph in Figure 6.10. Use Kruskal's Algorithm.

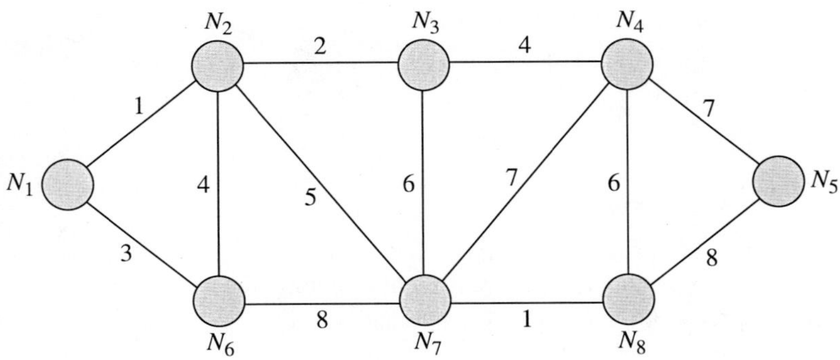

Figure 6.10 Undirected graph for Kruskal's MST.

Solution: Applying Kruskal's Algorithm on the above graph, here are the steps to the MST construction. Considering the initial configuration and edges in ascending order:

(*i*) Add edge (N_1, N_2) – cost is 1

(*ii*) Add edge (N_7, N_8) – cost = 1

(*iii*) Add edge (N_2, N_3) – cost = 2

(*iv*) Add edge (N_1, N_6) – cost = 3

(*v*) Reject edge (N_2, N_6) – cost = 4—because it forms a cycle

(*vi*) Add edge (N_3, N_4) – cost = 4

(*vii*) Add edge (N_2, N_7) – cost = 5

(*viii*) Reject edge (N_3, N_7) – cost = 6—because it forms a cycle

(*ix*) Reject edge (N_4, N_8) – cost = 6—because it forms a cycle

(*x*) Reject edge (N_4, N_7) – cost = 7—because it forms a cycle

(*xi*) Add edge (N_4, N_5) – cost = 7

Need not to proceed further as all vertices have been included. The corresponding MST is shown in Figure 6.11 below:

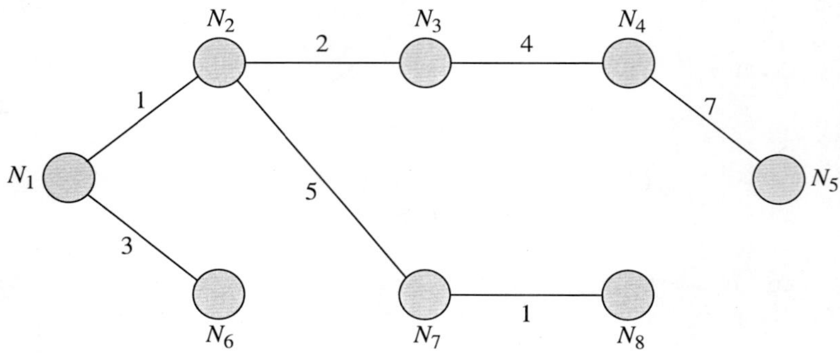

Figure 6.11 MST.

Total cost of constructing above MST

$= \text{weight } (N_1 N_2 + N_7 N_8 + N_2 N_3 + N_1 N_6 + N_3 N_4 + N_2 N_7 + N_4 N_5$

$= 1 + 1 + 2 + 3 + 4 + 5 + 7$

$= 23$

6.12.6 Prim's Algorithm

Like Kruskal's, Prim's Algorithm (*PA*) is another important greedy algorithm in graph theory that can also be used to find a minimum spanning tree for a connected weighted graph. This approach builds MST edge-by-edge.

6.12.6.1 *PA*: *Working Principle*

Consider $G = (V, E, W)$ be a connected weighted graph with vertices V and edges E and $T = (V', E)$ be the MST of G to be constructed. Here $V' = V$ and E' is a subset of E. The algorithm works by continuously increasing the size of the tree T starting with a single vertex and iterates until all the vertices are considered.

1. Set $V' = V$.
2. Set $E' = \{\emptyset\}$.
3. Find lightest edge (u, v) from G and take it into E'. If there is more than one edge having same minimum weight, then take anyone of them.

4. Now take v as source vertex u.
   (*i*) Find all the edges (v, k). $k = 1, 2, 3\ldots$
   (*ii*) Consider minimum weighted edge (v, k) just obtained.
   (*iii*) Check whether this edge if taken in E' produces a cycle or not. If not so then $E' = E' \cup (v, k)$. Otherwise discard it and take the next minimum weighted edge.
   (*iv*) $v = k$
5. Repeat step 4.
6. If it is seen that from v, the edges do not yields MST, and still there are other minimum weighted edges E' from the visited vertices except v, which will not form any cycle if included. Then we have to go back (backtrack) to the appropriate vertex as v.
7. Repeat step 4 and 5.
8. Whole process is continued until we have a MST.
9. Cost of constructing MST is sum of the weight of the edges in it.

6.12.6.2 *PA: Algorithm*

The Prim's algorithm rather Jarnik-Prim's algorithm for a MST is described below.

Algorithm: Jarník-Prim_MST (G, T, C, Near)

Where $G = (V, E, W)$, weighted connected graph

$T = (V', E')$, MST of G to be constructed. Here $V' = V$ and $E' \subseteq E$.

C = cost of constructing T. Initially zero.

Near = a list containing adjacent non-visited vertices to a current vertex.

Step 1:

(*i*) Consider the minimum weighted edge (u, v) in E'.

(*ii*) $C = W[u][v]$

(*iii*) $T[1][1] = u$

(*iv*) $T[1][2] = v$

Step 2:

(*i*) Repeat for $i = 1$ to n // Initialise Near

(*a*) If $W[i][v] < W[i][k]$

• Near $[i] = v$

Else

• Near $[i] = u$

(*ii*) Near $[u]$ = Near $[v] = 0$

Step 3: Repeat for $i = 2$ to $n - 1$ through step 5. // Include additional $n - 2$ edges in T

Step 4:

(*i*) Consider x such that Near $[x] \neq 0$ and W $[x]$ [Near $[x]$] is minimum.

(*ii*) $T[i][1] = x$

(*iii*) $T[i][2] =$ Near $[x]$

(*iv*) $C = C + W[x]$ [Near $[x]$]

(*v*) Near $[x] = 0$

Step 5: Repeat for $k = 1$ to n // Update Near

(*i*) Near $[k] \neq 0$ and $W[k][j] < W[k]$ [Near $[k]$]

(*a*) Near $[k] = j$

Step 6: Output C. // Display cost of constructing T

Step 7: Exit.

6.12.6.3 *PA: Analysis*

A simple implementation using an adjacency matrix graph representation and searching an array of weights to find the minimum weight edge to add requires $O(n^2)$ where $n = |V|$, running time. Step 2 iterates n times *i.e.,* $O(n)$. Step 3 iterates $n - 2$ times. Every statement of step 4 requires $O(1)$ time. Step 5 iterates n times independently. Due to step 3 it iterates $O(n^2)$ times. Hence Jarnik-Prim's algorithm requires $O(n^2)$ running time.

Using a simple binary heap data structure and an adjacency list representation, Prim's algorithm can be shown to run in time $O(|E| \log |V|)$. Using a more sophisticated Fibonacci heap, this can be brought down to $O(|E| + |V| \log |V|)$, which is significantly faster when the graph is dense enough.

Example 9: Construct a MST for the following graph using Jarnik-Prim's algorithm.

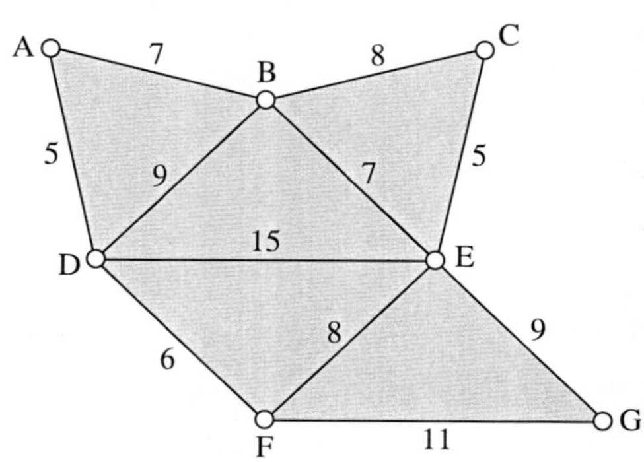

Figure 6.12 Undirected graph for Prim's MST.

Solution: Initial configuration T of the tree is

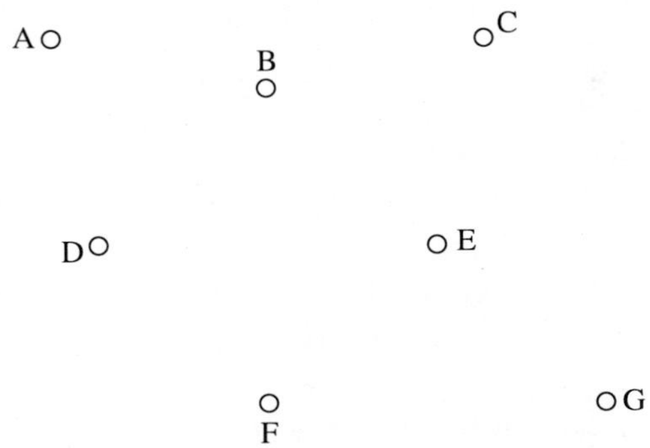

Figure 6.13 Initial configuration.

Minimum weighted edges in the graph are AD and CE. We can consider anyone of them say AD.

Let C = Cost of construction, initially 0.

S = Set of vertices traversed during construction, initially.

L = Set of edges untraversed, initially.

1. Add AD to the T.

$$C = C + w\ (AD) = 0 + 5 = 5$$
$$S = \{A, D\}$$
$$L = \emptyset$$

2. Adjacent edges not in T from

$A = AB(7)$ // number in bracket indicates its weight/cost

$D = DB(9),\ DF(6),\ DE(15)$

Light edge among these = DF and it does not form cycle in T.

Add DF.

$$C = C + w\ (DF) = 5 + 6 = 11$$
$$S = \{A, D, F\}$$
$$L = \{AB(7),\ DB(9),\ DE(15)\}$$

3. Adjacent edges not in T from

$F = FE(8),\ FG(11)$

Lightest edge among L and from $F = AB$

Add AB.

$$C = C + w\ (AB) = 11 + 7 = 18$$
$$S = \{A, D, F, B\}$$
$$L = \{DB(9),\ DE(15),\ FE(8),\ FG(11)\}$$

4. Adjacent edges not in T from

$$B = BC(8),\ BE(7),\ BD(9)$$

Lightest edge among L and from $B = BE$

Add BE.

$$C = C + w\,(BE) = 18 + 7 = 25$$
$$S = \{A, D, F, B, E\}$$
$$L = \{DB(9), DE(15), FE(8), FG(11), BC(8), BD(9)\}$$

5. Adjacent edges not in T from

$$\text{E} = \text{EB}(7), \text{EC}(5), \text{ED}(15), \text{EG}(9).$$

Lightest edge among L and from $E = EC$.

Add EC.

$$C = C + w\,(BE) = 25 + 5 = 30$$
$$S = \{A, D, F, B, E, C\}$$
$$L = \{DB(9), DE(15), FE(8), FG(11), BC(8), EG(9)\}$$

6. Adjacent edges not in T from

$$C = CB(8)$$

Lightest edge among L and from $C = CB$.

Adding CB forms a cycle. Discard it.

Next light edge in $L = FE$. For same reason discard it.

Next light edge in $L = DB$ and EG.

For same reason discard DB.

Add EG.

$$C = C + w\,(EG) = 30 + 9 = 39$$
$$S = \{A, D, F, E, C, B, G\}$$
$$L = \{DB(9), FG(11), EB(7), ED(15), CB(8)\}$$

All the vertices have been traversed. Thus the MST is constructed with cost 39.

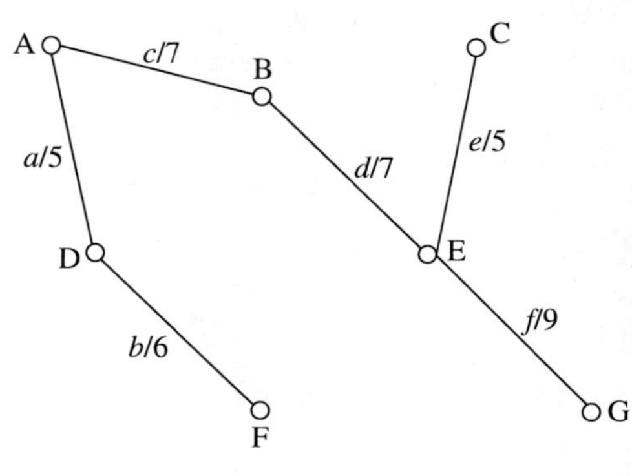

Figure 6.14 MST.

Example 10. Construct a MST of the graph in Figure 6.15 using Prim's algorithm.

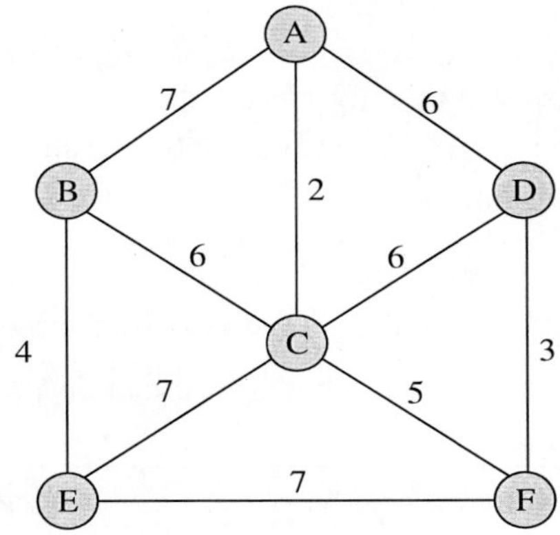

Figure 6.15 Undirected graph for Prim's MST.

Solution: The minimum spanning tree using Prim's algorithm can be constructed stepwise as below.

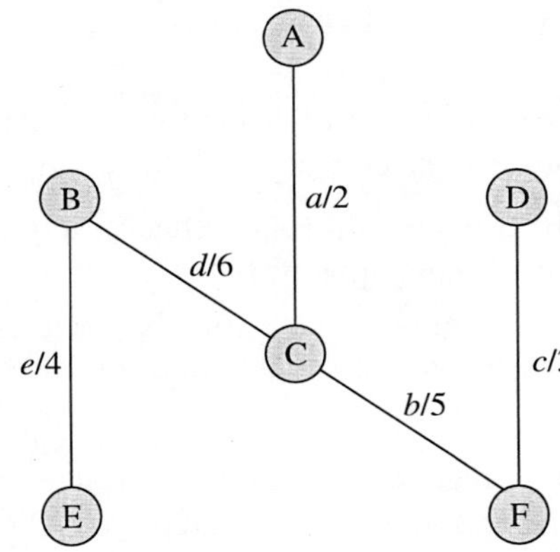

Figure 6.16 MST.

$$\begin{aligned}\text{Cost of construction} &= \text{weight } (AC + CF + DF + BC + BE)\\ &= 2 + 5 + 3 + 6 + 4\\ &= 20\end{aligned}$$

6.13 SOME COMMON QUESTIONS

Question 1. What is the difference between 0/1 Knapsack and fractional Knapsack problems? ***[WBUT 2009, 2010]***

Solution: The basic differences between 0-1 Knapsack and fractional Knapsack are –

(*i*) In 0-1 Knapsack problem each item is either taken or not taken. But in second case a fraction of each item is allowed.

(*ii*) 0-1 Knapsack can be solved by both dynamic programming and greedy approach but later can be solved by only greedy approach.

(*iii*) Optimal solution is not the usual case for 0-1 Knapsack. Fractional Knapsack provides optimal solution in most cases.

Question 2. Define feasible solution.

Answer: A feasible solution is a solution that satisfies the constraints.

Question 3. What is the difference between dynamic programming and greedy algorithm?

Dynamic Programming	*Greedy Algorithm*
1. DP computes its solution by making its choices in a serial fashion (never look backirrevocable). 2. It uses the principle of optimality.	1. GA computes the solution in bottom up technique by local choice. 2. It does not use the principle of optimality.

Question 4. Given an adjacency-list representation of a directed graph, how long does it take to compute the out-degree of every vertex? How long does it take to compute the in-degrees?

Answer: Given the adjacency-list representation of a graph the out-degree and in-degree of every node can easily be computed in O(|V| + |E|) time.

Question 5. Let (u, v) be a minimum-weight edge in a graph G. Show that (u, v) belongs to some minimum spanning tree of G.

Answer: If an edge (u, v) is a minimum weight edge then it will be safe for any cut with vertices u and v on separate sides. Therefore the edge (u, v) belongs to some minimum spanning tree.

Question 6. Show that if an edge (u, v) is contained in some minimum spanning tree, then it is a light edge crossing some cut of the graph.

Answer: Assume that an edge (u, v) belongs to some minimum spanning tree. Consider a cut with u and v on separate sides. If an edge (u, v) is not a light edge then clearly the graph would not be a minimum spanning tree.

Question 7. Let e be a maximum-weight edge on some cycle of $G = (V, E)$. Prove that there is a minimum spanning tree of G that does not include e.

Answer: Let e be the maximum-weight edge on some cycle in a graph G = (V, E) then e is not part of a minimum spanning tree of G. If e was in the minimum spanning tree then replacing it with any other edge on the cycle would yield a "better" minimum spanning tree.

Question 8. Show that for each minimum spanning tree T of $G = (V, E)$, there is a way to sort the edges of G in Kruskal's algorithm so that the algorithm returns T.

Answer: Given a minimum spanning tree T we wish to sort the edges in Kruskal's algorithm such that it produces T. For each edge e in T simply make sure that it preceeds any other edge not in T with edge weight $w(e)$.

Exercises

1. Is linear Searching technique a greedy algorithm? Explain.
2. Consider a currency system that has the following coins and their values: dollar (100 cents), quarter (25 cents), dime (10 cents), nickel (5 cents) and 1 cent coins. Give a greedy algorithm to change n cents in such a way that the total number of coins n is minimised.
3. Show that the greedy algorithm obtained in Exercise 6.2 does not always work if we instead use coins of 1 cent, 5 cents, 7 cents and 11 cents. Give a counter example.
4. Suppose in the money change problem of Exercise 6.2 the coin values are: 1, 2, 4, 8, 16... 2^k, for some positive integer k. let the value to be paid is $n < 2^{k+1}$. Write a greedy algorithm that runs in $O(\log n)$ to solve the problem.
5. A digraph may have two distinct shortest path trees. Justify the statement.
6. Find a minimum cost spanning tree using Kruskal's algorithm for the graph shown in Figure 6.17.

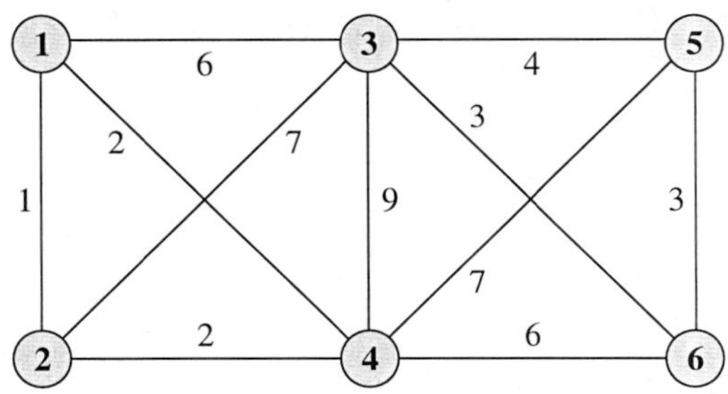

Figure 6.17 An undirected graph.

7. Find a minimum cost spanning tree using Prim's algorithm of the above graph shown in Figure 6.17.
8. Let $G = (V, E)$ be an undirected graph with $n = |V|$, $m = |E|$, and $m = O(n^{1.99})$. Find a minimum cost spanning tree for G. Choose a competent algorithm. Justify your choice.
9. Let a minimum weighted edge e G, an undirected graph. Show that e some minimum cost spanning tree of G.
10. Does Prim's algorithm work correctly if the graph has negative weights? Explain.
11. Prove that: An undirected weighted graph G has a unique minimum cost spanning tree iff no two edges have the same weight.

12. How many number of spanning trees possible in a complete undirected weighted graph G with n vertices?

13. Let T be a shortest path tree for G, a directed weighted graph with no two edges have the same weight. Let T' be a minimum spanning tree for G', an undirected graph obtained by removing the directions of the edges of G. Prove or disprove that $T = T'$.

14. Give a greedy algorithm to compute shortest path between vertices 1 and 6 on the graph shown in Figure 6.18

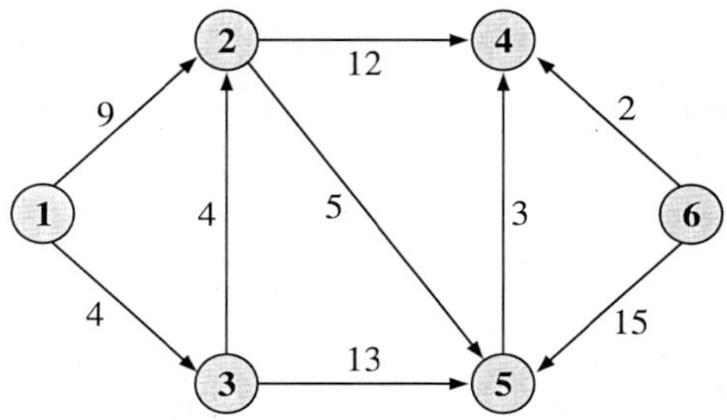

Fig 6.18 Digraph for SSSPP.

15. Find a minimum cost spanning tree using Prim's and Kruskal's algorithms of the above graph shown in Figure6.18. Consider the graph as an undirected weighted graph. Compare the costs thus obtained.

University Questions

1. What do you mean by greedy method? ***[WBUT'09, UPTU'05, DOEACC'08]***
2. Does greedy algorithm always give an optimal solution? Give arguments to support your answer. ***[PTU'09, RTU'10, IGNOU'07, DOEACC'08]***
3. What is the Greedy approach? Explain in brief. ***[RTU'08, '10, MDU'08]***
4. What are the characteristics of greedy algorithm? ***[WBUT'08, '09, PTU'08, '09, BPUT'08]***
5. What do you mean by greedy choice property? ***[BPUT'05]***
6. Explain the terms feasible solution, optimal solution and objective function ***[JNTU'08, '10]***
7. To which type of problems Greedy are applied? ***[MDU'08]***
8. Differentiate between Greedy method and Dynamic programming. ***[WBUT'07, WBUT'08, MDU'09JNTU'08, '11, BPUT'05, '07, DOEACC'06]***
9. Write a greedy algorithm to the Job sequencing with deadlines. ***[JNTU'08, '10]***
10. Discuss activity selection problem of job sequencing. ***[WBUT'08, '09]***
11. What is a feasible solution in activity selection problem? ***[BPUT'07]***

12. Write a Greedy Algorithm for Activity Selection Problem and find its time complexity. *[DOEACC'08]*
13. Give two examples in which Greedy algorithm gives optimal solution. *[RTU'10]*
14. If $p_1/w_1 \geq p_2/w_2 \geq ... \geq p_n/w_n$, prove that Knapsack generates an optimal solution to the given instance of the Knapsack problem. *[JNTU'10]*
15. If objects are selected in order of decreasing v_i/w_i then prove that the algorithm Knapsack finds an optimal solution. *[JNTU'08, '10]*
16. State the 0/1 Knapsack problem. *[WBUT'07, '10, DOEACC'08]*
17. State and write algorithm for Knapsack problem using Greedy method. *[PTU'08]*
18. Write down the algorithm for fractional Knapsack problem. *[WBUT'09]*
19. Suppose that in 0/1 Knapsack problem the order of items when sorted on increasing value is the same as their order when sorted by decreasing weight. Give an efficient algorithm to find an optimal solution to the problem. *[IETE ALCCS'11]*
20. Consider a Knapsack problem for $n = 3$, $W = 20$, $(I_1, I_2, I_3) = (25, 24, I_5)$ and $(V_1, V_2, V_3) = (18, 15, 1\ 0)$ then find the optimal solution. *[RTU'08]*
21. Find the optimal solution for the fractional Knapsack problem given as: $I = \{I_1, I_2, I_3, I_4, I_5\}$ $w = \{5, 10, 20, 30, 40\}$, $v = \{30, 20, 100, 90, 160\}$. The Knapsack capacity, $W = 60$. *[WBUT'08]*
22. Consider the Knapsack Instance $n = 5$, their positive weights $(w_1, w_2, w_3, w_4, w_5) = (10, 20, 30, 40, 50)$, positive values $(v_1, v_2, v_3, v_4, v_5) = (20, 30, 66, 40, 60)$ and $W = 100$. Find the optimal solution (by greedy algorithm). *[RTU'07]*
23. Describe greedy method. Find the optimal solution to the Knapsack instance $n = 3$, $W = [60\ 14, 10]$, $P = [20, 18, 15]$ and $C = 116$ using greedy method. *[UPTU'10]*
24. State the general Knapsack problem. Write a greedy algorithm for this problem and derive its time complexity. *[WBUT'09]*
25. Given the weight vector (2, 3, 5, 7, 1, 4, 1) and the profit vector {10, 5, 15, 7, 6, 8, 18, 3) and a Knapsack of capacity 15. Find at least three feasible solutions including optimal one for the Knapsack problem of seven objects. *[WBUT'09]*
26. Critically comment on "Greedy strategy does not work for the 0/1 Knapsack problem for all time." *[WBUT'09]*
27. Prove that the fractional Knapsack problem has a greedy-choice property. *[UPTU'05]*
28. Find the optimal solution using greedy criteria for a Knapsack having capacity 100 kg for the following list of items having values and weights as shown in the table. *[WBUT'07]*

Item	*Value*	*Weight*
I_1	10	15
I_2	20	25
I_3	30	35
I_4	40	45
I_5	50	55

29. Solve the following given instance Using Greedy Approach. Is this solution Optimal? Capacity of Knapsack is 50. *[DOEACC'08]*

Item	*Weight*	*Value*	*Value/Weight*
1	10	60	6
2	20	100	5
3	30	120	4

30. Suggest a greedy algorithm to solve 0/1 Knapsack problem. Comment on data structure to be used for implementation. What is average case time complexity of your greedy algorithm? *[BPUT'05]*
31. Write short notes on Spanning trees. *[PTU'09, BPUT'05, RTU'08, '09, DOEACC'08, '09]*
32. Write an algorithm for finding the minimum spanning tree of a graph, Discuss its time complexity. *[WBUT'09, DOEACC'06]*
33. Write short notes on Kruskal's algorithm. *[UPTU'09, PTU'09, BPUT'05]*
34. Let (u, v) be a minimum weight edge in a graph G. Show that (u, v) belongs to some minimum spanning tree of G. *[WBUT'08, JNTU'08, '11, IITG'11]*
35. Define minimum cost spanning tree. Write Prim's algorithm to generate a minimum cost spanning tree for any given weighted graph. *[UPTU'10, PTU'08, RTU'07]*
36. Describe and justify Kruskal's algorithm for finding the minimum spanning tree of an undirected graph. What is the time complexity of Kruskal's algorithm? *[BPUT''05, '07, '08, IETE ALCCS'11, DOEACC'07]*
37. Explain the reason why Kruskal's algorithm always yields optimal spanning tree. *[BPUT'05]*
38. What is the property that makes Kruskal's algorithm a unique greedy algorithm? *[DOEACC'09]*
39. Write down the difference between: Prim's algorithm and Kruskal's algorithm. *[WBUT'07, UPTU'10, JNTU'09]*
40. Write an algorithm to find a minimal spanning tree of undirected graph. Estimate the time complexity of your algorithm. *[WBUT'08]*
41. Draw a complete graph on four nodes. Also draw all of its spanning trees. *[DOEACC'06]*
42. Generate minimum cost spanning tree for the following graph using Prim's algorithm. *[UPTU'10]*

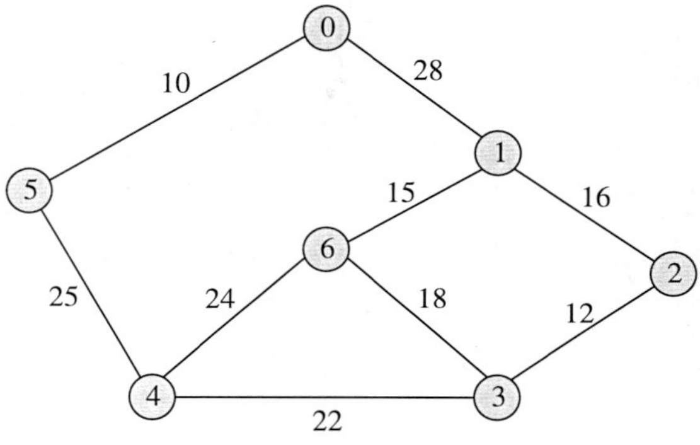

43. Trace the Kruskal's algorithm to obtain the minimum spanning tree from the following graph. ***[RTU'10]***

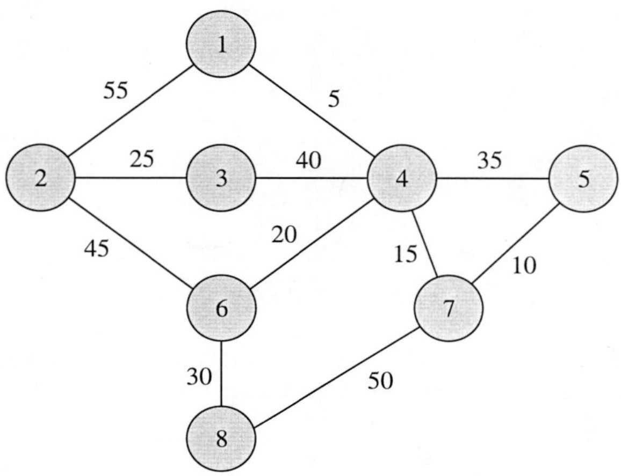

44. Find out the minimum cost spanning tree using any algorithm. ***[WBUT'07, PTU'09, MDU'08]***

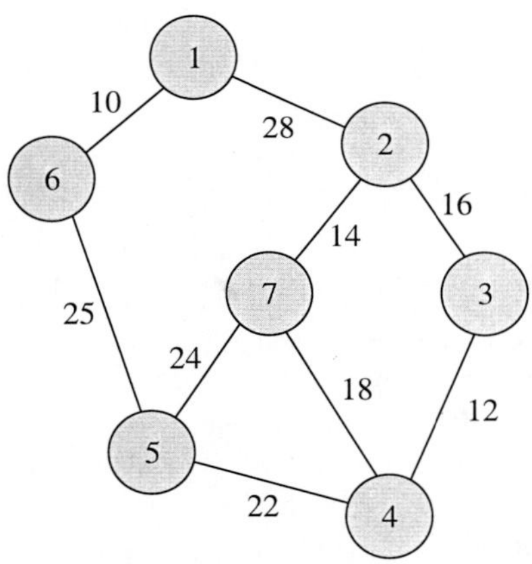

45. Using the algorithm find a minimum spanning tree of the weighted graph shown below. Will it be the unique minimum spanning tree? State reasons for your claim. ***[IETE ALCCS'11]***

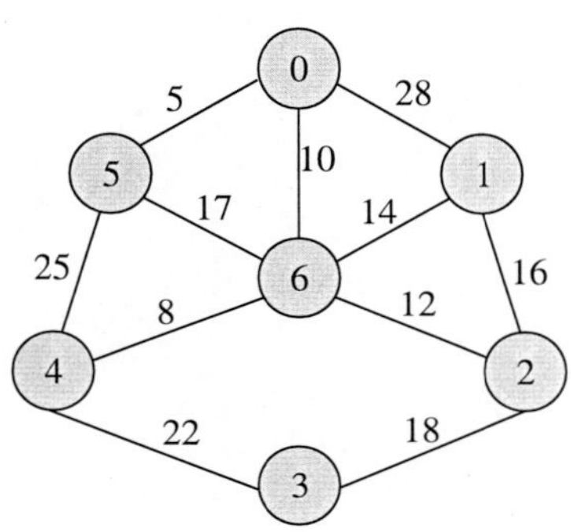

46. Find at least three spanning trees of the following graph. *[UPTU'10]*

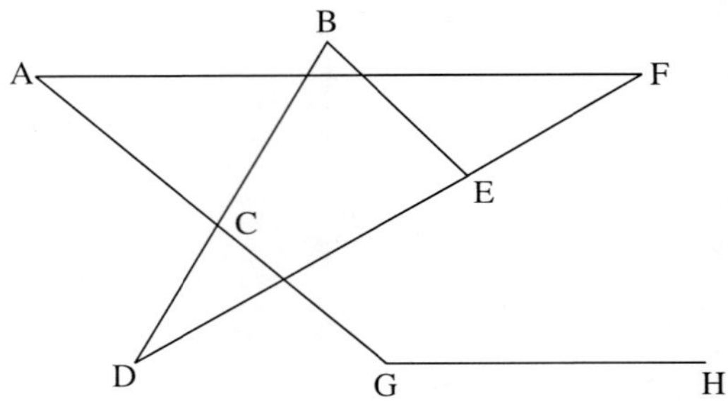

47. Obtain the minimum cost spanning tree for the following graph using Prim's algorithm. Obtain the BFS tree for the graph below also. *[IGNOU'09, DOEACC'08]*

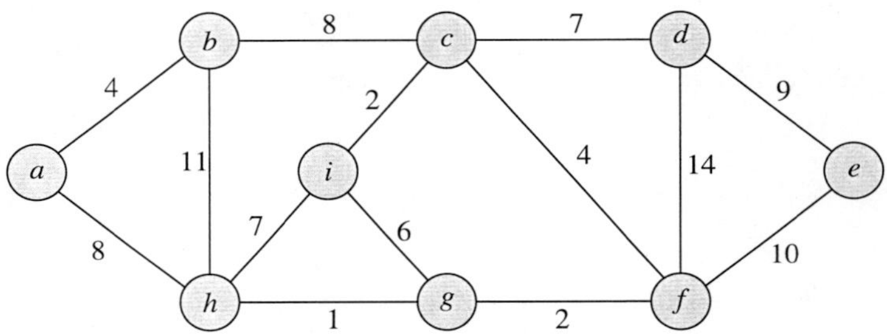

48. Prove that if the weights on the edge of the connected undirected graph are distinct then there is a unique Minimum Spanning Tree. Give an example in this regard. Also discuss Kruskal's Minimum Spanning Tree in detail. *[UPTU'06, JNTU'08]*
49. Explain Prim's algorithm to find the minimum cost spanning tree of a weighted graph and derive its time complexity. Illustrate your algorithm with an example. *[UPTU'07, MDU'09]*
50. Using Prim's Algorithm, find a minimal spanning tree for the graph given below. *[IGNOU'07]*

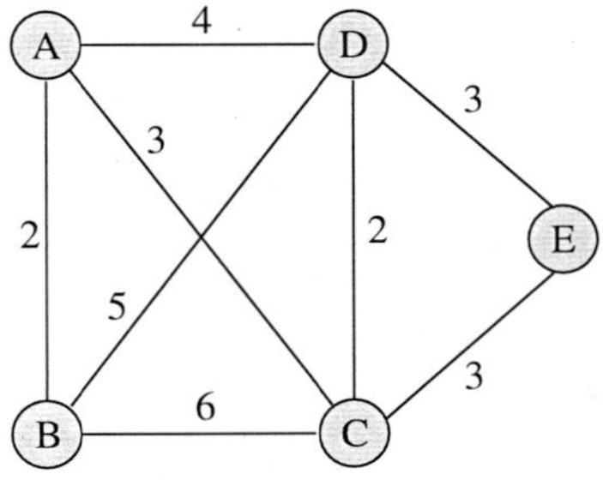

51. Apply Prim's algorithm to find a minimal spanning tree of the following graph. *[UPTU'07]*

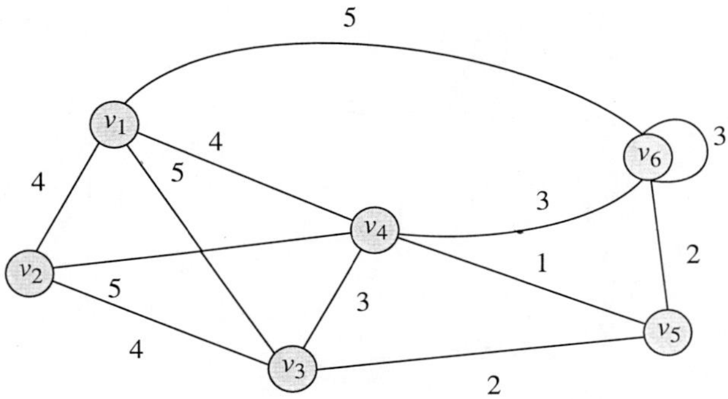

52. Show that the algorithm you have written in previous question indeed determines the optimal spanning tree. On what approach this algorithm is based on? *[DOEACC'06]*

53. Find the minimum cost spanning tree of the given graph using Prim's algorithm. *[UPTU'10]*

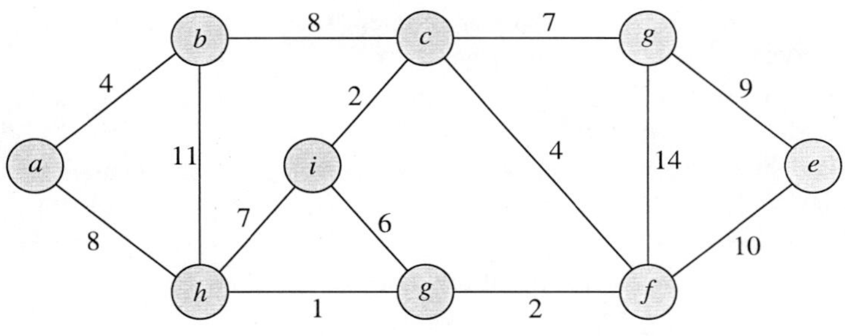

54. Find MST of the graph given below. *[BPUT'05]*

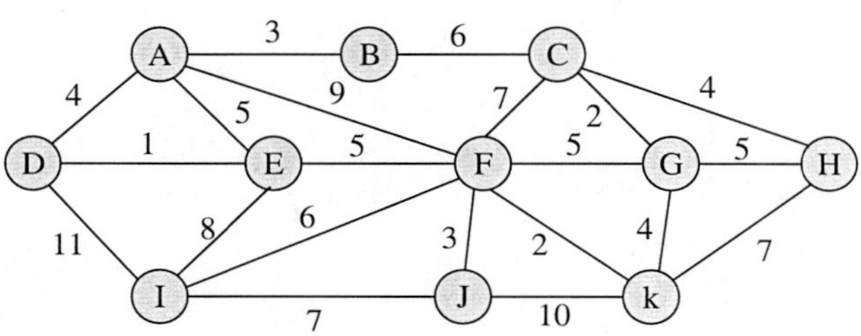

55. Find the minimum spanning tree using the Kruskal's algorithm for the following graph: ***[MDU'09]***

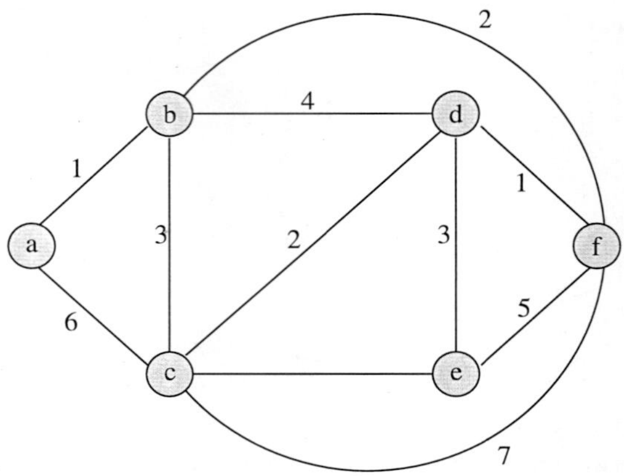

56. Show that a Hamiltonian path in a graph is a spanning tree. ***[UPTU'07]***
57. Let G be an undirected graph with edge costs $C = [Cg]$ and T is a MST of G with respect to C. Prove or disprove that if we add 1 to all edge costs $[c_e]$ then T is still a MST of G. ***[DOEACC'08]***
58. If edge weights of a connected weighted graph are not all distinct, the graph must have more than one minimum spanning tree. ***[JNTU'08]***
59. In Kruskal's method for finding a minimum spanning tree, how does the algorithm know when the addition of an edge will generate a cycle? ***[WBUT'09,JNTU'10]***
60. Consider the following graph. Using Kruskal's algorithm, calculate the minimum spanning tree, listing edges in the minimal spanning tree in the order they are added to the tree. ***[IETE ALCCS'11]***

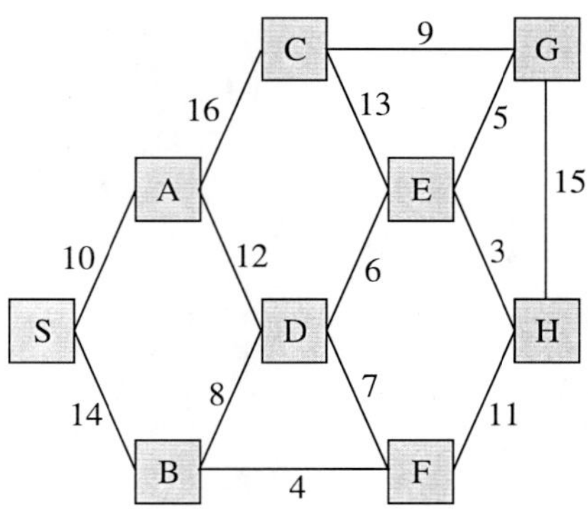

61. Trace the Kruskal's algorithm to obtain the minimum spanning tree from the following graph. ***[DOEACC'07]***

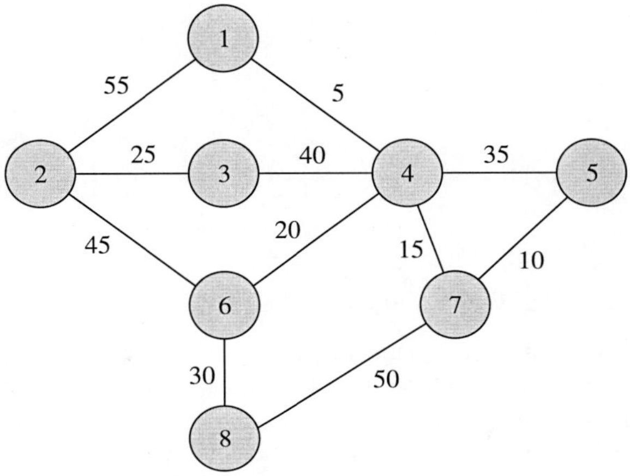

62. Trace Kruskal's algorithm to compute a minimum spanning tree for a graph *G* that contains at least 9 vertices and 13 weighted edges. ***[DOEACC'07]***

❑❑❑

Chapter 7 Backtracking

7.1 INTRODUCTION

In our journey to some unknown place, in the way there may have some location from where the road is sidetracked into at least two roads in different directions. Then we choose anyone of them. Being moved to this road if we face an obstacle, generally we come back to the location just from where we selected this path. Then follow the second path. Again if in this path there is a branching of the road the same repetition is proceed in the same way. Finally we reach to destination. This technique is called *Backtracking*.

Backtracking is a general algorithm design technique for finding solutions to some computational problem as in most of the NP-hard problems. The computational problems are mainly the optimisation problems. Many optimisation problems that search for an optimal solution through set of partial solutions can be solved using Backtracking.

These problems admit the concept of a partial candidate solution and there is a relatively quick test of whether it can possibly be completed to a valid solution, then only backtracking can be applied to them. Here a candidate is the possible path which moves towards a valid solution. Backtracking often eliminates a large number of candidates with a single test. That's why it is much faster than brute force enumeration of all complete candidates.

7.2 GENERAL METHOD

The solution to a problem using Backtracking is usually represented as an implicit graph. It is called a search tree. The algorithm performs an intelligent DFS on it so as to have all possible solutions. The valid solution is obtained by maintaining partial solutions candidates as the search proceeds. This is done by exploring the current node in the search tree.

Initially the problem has no solution. So, any node as a partial solution candidate may be considered. Then explore it to all possible candidate nodes. Consider first node of the explored nodes as obtained just like DFS. When search proceeds, a candidate is added to previous partial solution which in turn guarantees near to complete solution. The process is continued. If in this way the goal is achieved. The search is then said to be successful. Two things may happen at this stage. Either search terminates or continues for another valid solution.

If at any stage the search is unsuccessful that means from the partial solution candidate so far we achieved, a complete solution can't be defined, the candidate just added now becomes a dead state. We backtrack the search one step before. This dead candidate is removed from the partial solution. Then the next alternative candidate for the solution is chosen. Again if this candidate is dead state one then same repetition as previous is carried on.

The actual search tree in this way that is traversed by the algorithm is only a part of the desired or potential tree. The total cost of the algorithm is basically the number of nodes of the actual tree times the cost of obtaining and processing each node.

7.3 APPLICATIONS

Backtracking techniques can be applied to solve the following different kind's problems such as:

1. Knapsack Problem.
2. Hamiltonian Cycle Problem.
3. *N*-Queens Problem.
4. Subset sum Problem.
5. Graph Colouring Problem.
6. Constraint satisfaction problems, such as crosswords, verbal arithmetic, and many other puzzles.
7. Efficient technique for parsing.
8. Logic programming languages such as Icon, Planner and Prolog, etc.

7.4 *N*–QUEENS PROBLEM

The classic example of the use of backtracking is the *N*-queen's puzzle that describes all the arrangements of *N*-queens on a standard $N \times N$ chessboard so that no queen attacks any other by being in the same row, column or diagonal. A brute-force method to place *n*-queens results n^n ways. However, most of them are improper *i.e.,* not legal placements. These huge space searches can be tested in n! configurations instead of n^n simply by employing backtracking.

If $N = 1$, the problem becomes trivial. For $N = 2, 3$ there are no solutions. So N starts from 4 but our requirement is $N = 8$. Basically N is power of 2. First we will solve 4–queen's problem, then 8–queens.

In the common backtracking approach, the partial candidates are arrangements of $k < N$ queens in the first k rows of the $N \times N$ chessboard, all in different rows and columns. Any partial solution that contains two mutually attacking queens can be abandoned, since it cannot be completed to a valid solution.

7.4.1 4–Queens Problem

Consider a 4×4 chessboard as a 4×4 matrix. Rows and Columns are numbered from 1 through 4 as shown in Figure 7.1. We will place 4-queens to this matrix shown below.

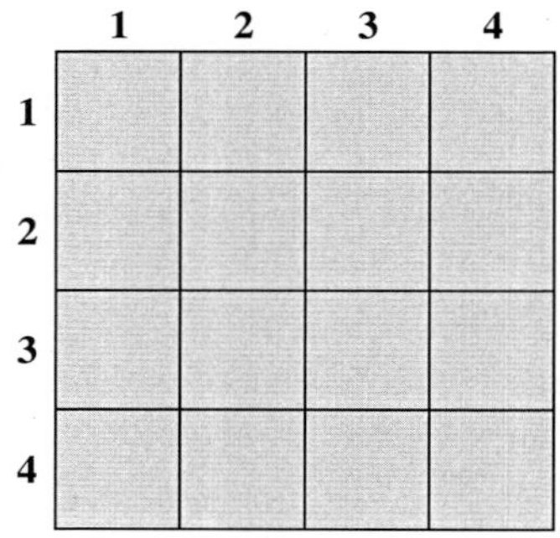

Figure 7.1 Empty chessboard for 4Qs.

There are four positions in each row. Every time a queen is placed in a cell makes a new arrangement. Maximum possible arrangements for placing these 4-queens will be 4^4. But 4-queens are to be placed in such a way that no two queens attack each other. This reduces the arrangements *i.e.,* search space surprisingly from 4^4 to 4!. Since all the queens are identical to make distinguishable them we do numbering each queen as q_1, q_2, q_3 and q_4. Also suppose that q_i is placed in row *i*. Objective is to find appropriate column of the row *i* for q_i.

Solution: Initially the chessboard rather the matrix is empty. So we can place q_1 at any column of row 1 as an acceptable position. Say we place at very first place *i.e.,* column 1. Hence the position for q_1 is (1, 1) cell. Now q_2 will be placed in (2, 3) cell because if we place it in cell (2, 1) or (2, 2) then it will be attacked by q_1 vertically or diagonally respectively. There is no suitable position at row 3 for q_3 at this moment because if we place it in cell (3, 1) then it will be attacked by q_1 vertically. If we place it in cells (3, 2), (3, 3), (3, 4) separately then it will be attacked by q_2 diagonally, vertically, diagonally respectively. As this situation has aroused due to placing q_2 in cell (2, 3) so this cell becomes dead state. So we have to backtrack to this cell and place q_2 in next possible position *i.e.,* cell (2, 4). Now we can place q_3 in cell (3, 2). For the similar reasons q_4 can't be placed in any cell of row 4. So we backtrack to cell (3, 2). We can't place q_3 in cells (3, 3) and (3, 4) due to same reasons. So we backtrack to cell (2, 4). As there is no other possible position for q_2 available, we backtrack to cell (1, 1). Now place q_1 in cell (1, 2). The q_2 will be as it is *i.e.,* in cell (2, 4). Now we place q_3 in cell (3, 1) because

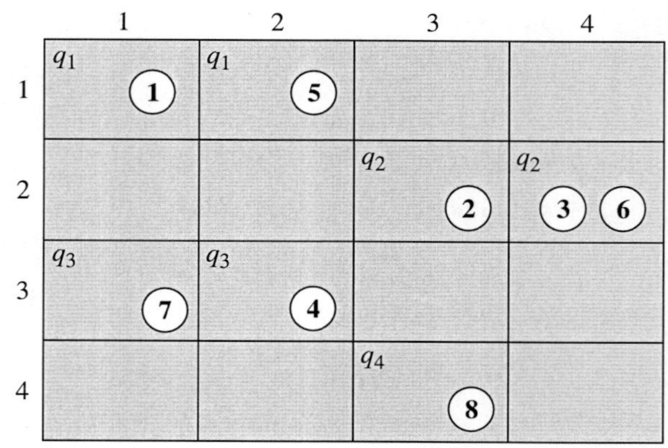

Figure 7.2 Solution space table for 4-queens.

if it is in cell (3, 2) it will be attacked by q_1 vertically. Again we can't place in cells (3, 3), (3, 4) due to q_2. Now due to q_3 in cells (4, 1) and (4, 2) q_4 can't be placed. The first possible position for q_4 is cell (4, 3).

The whole procedure is defined in the following table. The encircled number represents number of steps to solve this $4 - q_s$ problem.

Hence the solution tuple T is < 2, 4, 1, and 3 >. Position of digit in T represents row and digit itself represents column of that row.

7.4.2 8–Queens Problem

The problem is to place 8-queens on the 8 × 8 chess board so that they do not attack/check each other. The origin of this problem is probably not known because this problem is probably as old as the chess game itself. If we want to find a single solution, it is not difficult as shown below. But finding all possible solutions, then the problem becomes difficult and backtracking is the only known method then. For 8-queens, generally 92 solutions are possible. Excluding symmetry, only 12 unique solutions exist.

Solution: We can solve this problem in the same way as in 4-queens. The table given below is one of the possible solutions.

Number of queens, $N = 8$ and Queens: $Q_1, Q_2 \ldots Q_8$

	1	2	3	4	5	6	7	8
1				Q_1				
2						Q_2		
3								Q_3
4		Q_4						
5							Q_5	
6	Q_6							
7			Q_7					
8					Q_8			

Figure 7.3 Solution space table for 8-Queens.

Hence solution tuple for 8-queens is < 4, 6, 8, 2, 7, 1, 3, and 5 >.

From the above two examples it is seen that two queens will attack each other diagonally iff

(*a*) $i - j = x - y$ or

(*b*) $i + j = x + y$

From these two conditions, the general condition is thus $| i - x | = | j - y |$. Where (i, j) and (x, y) are the positions of two queens.

7.4.3 N–Queens Problem

In this section an algorithm for *N* queens has been devised and discussed. This *N* may be 4, 8, 16, ...

Algorithm: *N* queens_Backtracking (*k*, *n*)

Where k = queen k.

n = column number in which queen k is placed.

= number of queens under consideration.

Extra term:

T = solution tuple.

dead = Boolean variable indicating dead state reached. Initially it is 0.

Step 1: Repeat for $i = 1$ to n through step 3. // place n queens

Step 2: (*i*) Repeat for $j = 1$ to $k - 1$ // checking dead state

(*a*) If $T[j] = i$ or $| T[j] - i | = | j - k |$

- dead = 1 // attack occurs

(*ii*) dead = 0

Step 3: If dead = 0 // place queen k in column i

(*i*) $T[k] = i$

(*ii*) If $k = n$

(*a*) Repeat for $m = 1$ to n

- Output $T[m]$ // displaying solution tuple

Else

(*a*) Call Nqueens_Backtracking ($k + 1$, n)

Step 4: Exit

Analysis: Step 1 executes n times. Step 2 executes $k - 1$ times. Execution of Step 3 depends on the queen k. if its suitable place is found in a single movement, no need to backtrack. Since most of the placements arrives a dead state, the algorithm recursively call n times. As a result in worst case the algorithm runs in $O(n!)$ i.e. $O(n^n)$ time.

7.5 HAMILTONIAN CYCLE PROBLEM

Both the Hamiltonian path problem and the Hamiltonian cycle problem are problems of determining whether a given graph (directed or undirected) has a Hamiltonian path or a Hamiltonian Cycle. Both problems are NP-complete. In 1974 Garey and Johnson showed that the directed Hamiltonian cycle problem remains NP-complete for planar graphs and the undirected Hamiltonian cycle problem remains NP-complete for cubic planar graphs. The Hamiltonian cycle problem is a special case of the travelling salesman problem.

7.5.1 Hamiltonian Cycle

A Hamiltonian Cycle (HC) also called Hamiltonian circuit, vertex tour or graph cycle in an undirected graph is a cycle that visits each vertex exactly once except the starting vertex. A graph that contains a Hamiltonian cycle is called a Hamiltonian graph.

Similarly, Hamiltonian cycle for directed graphs, where each edge (arc) of a path or cycle can only be traced in a single direction *i.e.,* the vertices are connected with arrows and the edges traced "tail-to-head".

A Hamiltonian decomposition is an edge decomposition of a graph into Hamiltonian circuits.

Example. Let $G = (V, E)$ be an undirected graph as shown in Figure 7.4. A Hamiltonian cycle of G is $v_1 - v_6 - v_4 - v_3 - v_2 - v_5 - v_1$. The cycle goes through every vertex exactly once.

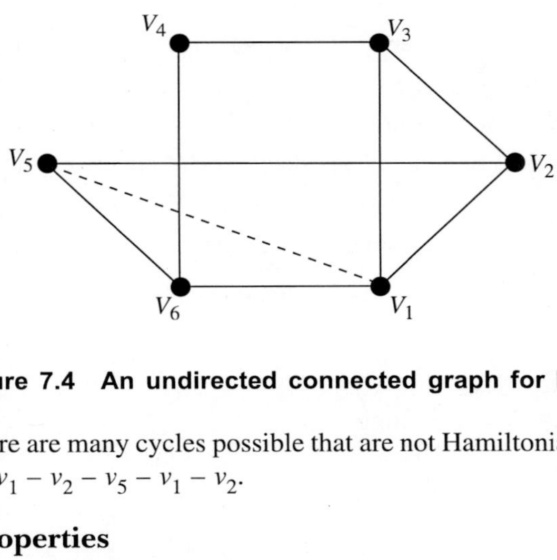

Figure 7.4 An undirected connected graph for HCP.

However, there are many cycles possible that are not Hamiltonian, e.g. $v_1 - v_6 - v_5 - v_1$ or $v_2 - v_5 - v_1 - v_2 - v_5 - v_1 - v_2$.

7.5.2 HC: Properties

A Hamiltonian cycle has some basic properties such as:

1. Any Hamiltonian cycle can be converted to a Hamiltonian path by removing one of its edges, but oppositely not true. Because a Hamiltonian path can be extended to Hamiltonian cycle only if its endpoints are adjacent.
2. A complete graph with more than two vertices is Hamiltonian.
3. Every cyclic graph is Hamiltonian.
4. A Hamiltonian cycle may be used as the basis of a zero-knowledge proof.
5. Number of different Hamiltonian cycles for a complete graph is $(n - 1)! / 2$.
6. Number of different Hamiltonian cycles for a complete Digraph = $(n - 1)!$

7.5.3 HCP: Solution Approach

The Hamiltonian Cycle Problem (HCP) asks whether a given graph *G* has a Hamiltonian cycle. Consider $G = (V, E)$ be an undirected connected graph. Backtracking finds a Hamiltonian cycle in the following way.

1. Consider any vertex *x* of *G* as a starting of searching cycle. It will be the root node of our implicit tree.
2. Explore *x*. That means consider the adjacent vertices of *x* as its children. Represent these children from left to right in alphabetical order. This node is a partial solution.
3. Take first child of *x* into partial solution if it is not visited. Explore it.
4. Repeat step 3 until all the vertices are visited.
5. If at any stage any vertex *y* makes a cycle with any vertex in partial solution other than *x* then a dead state is reached. Backtrack one step and start searching with another selection of node. After backtracking the element is removed from partial solution.
6. If a Hamiltonian cycle is obtained, the search is then successful.

7.5.4 HCP: Algorithm

The solution approach is converted in language oriented algorithm as follows.

Algorithm: HCP_Backtracking (*k*)

Where k = Arbitrary starting vertex.

Extra Terms p = Solution tuple consists of $k - 1$ distinct vertices.

g = Adjacency matrix contains the edges (v_i, v_j)

dead = Boolean variable for checking dead state. Initially it is 0.

Step 1: Repeat while dead = 0 through step 6. // generate $p\ [k]$

Step 2:

(*i*) $x = 0$

(*ii*) Repeat while $x = 0$ through step 4.

Step 3:

(*i*) $p\ [k] = (\ p[k] + 1) \bmod (n + 1)$ // find next vertex and check

(*ii*) If $p\ [k] = 0$

(*a*) $x = 1$

(*b*) Return *x*

Step 4: If $g\ [\ p\ [k - 1], p\ [k]] \neq 0$

(*i*) Repeat for $j = 1$ to $k - 1$

(*a*) If $p\ [\ j] = p[k]$

- $x = 1$
- exit

(*b*) If $j = k$
- $m = g[\,p[n], p[1]]$
- If $k < n$ or ($k = n$ and $m \neq 0$)
  - ➢ $x = 0$
  - ➢ Return x

Step 5: If $p[k] = 0$ // no vertex chosen

(*a*) dead = 1

(*b*) Return dead

Step 6:

(*i*) If $k = n$ // display Hamiltonian cycle

(*a*) Repeat for $r = 1$ to n
- Output $p\,[r]$

(*ii*) Else

(*a*) Call HCP_Backtracking ($k + 1$)

Step 7: Exit

Analysis: Step 4 requires $O(k - 1)$ time. Step 6(*i*) executes in $O(n)$ time if no dead state reached. But step 6(*ii*) requires n^{k+1} time as it recursively calls itself. Thus step 6 executes $n.n^{k+1} = n^{k+2}$ time. Total time complexity of the algorithm is given by:

$$T(n) = \sum_{k=0}^{n} n^{k+2} = \frac{n^{k+2} - 1}{n - 1} n^2 = O(n^n)$$

Example 1. Find a Hamiltonian cycle of the graph given below using backtracking.

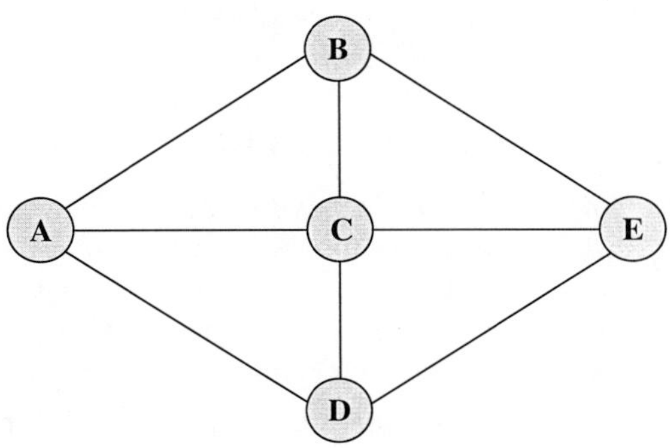

Figure 7.5 Undirected graph for HCP.

Solution: Suppose starting vertex is A. So, partial solution tuple $T = <A>$. Now the implicit tree can be drawn using backtracking with the help of above algorithm as follows.

The nodes *B*, *C*, *D* are adjacent to *A*. Next choose *B* and add it to partial solution containing *A*. Hence, $T = <A, B>$. The nodes *C*, *E* are adjacent to *B*. Next *C* is taken into partial solution. Similarly *D* and *E* are taken into partial solution. Thus,

T = <*A*, *B*, *C*, *D*, *E*>. Now all the vertices have been visited exactly once. There is however no edge between *E* and *A*. Again *E* forms cycles *E–D–C–E* and *E–B–C–E*. so, a dead state is reached. Backtrack to previous state *D* (as shown with dotted line in Figure 7.6) and remove this *E*. Now *T* = <*A*, *B*, *C*, *D*>. There is no child of *D*. again backtrack to *C* and remove *D* from *T*. Now *T* = <*A*, *B*, *C*>. Consider the next explored node *E* into partial solution. Now *T* = <*A*, *B*, *C*, *E*>. Explore *E*. it results only *D* and take it into partial solution. Now *T* = <*A*, *B*, *C*, *E*, *D*>. All the vertices have been visited exactly once and there is an edge from *D* to *A*. Thus *A–B–C–E–D–A* is a Hamiltonian cycle.

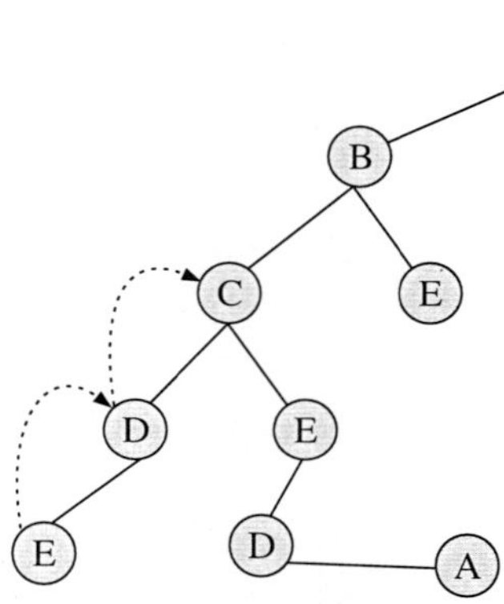

Figure 7.6 Hamiltonian cycle.

In the same way all possible Hamiltonian cycles considering other searches can be obtained.

7.6 GRAPH COLOURING PROBLEM

Graph Colouring Problem (GCP) states a way of colouring the vertices of a graph such that no two adjacent vertices share the same color. This is also called a vertex colouring. Similarly, an edge colouring assigns a color to each edge so that no two adjacent edges share the same color, and a face colouring of a planar graph assigns a color to each face or region so that no two faces that share a boundary have the same colour.

In its simplest form, vertex colouring is the starting point of this section, and other colouring problems can easily be transformed into a vertex version. For example, an edge colouring of a graph is just a vertex colouring of its line graph. The convention of using colours originates from colouring the countries of a map, where each face is literally coloured.

Graph colourings have the benefits of many practical applications as well as theoretical challenges. Different limitations are set on the graph, or on the way a colour is assigned, even on the colour itself.

7.6.1 Chromatic Number

The smallest integer number m that is required to colour a connected graph G in such a way that no two adjacent vertices have the same colour. This m is called chromatic number of G. Intention of the *GCP* is to find m.

Some observations can be drawn from the definition of this chromatic number.

(*i*) A graph is 1-chromatic if it consists of isolated vertices.

(*ii*) A graph with one or more edges without self loop is at least 2-chromatic. It is also bi-chromatic.

(*iii*) A complete graph with n vertices is n chromatic.

(*iv*) A graph is at least 3-chromatic if it has a triangle.

(*v*) A planar graph is 4-chromatic.

(*vi*) A graph with one circuit and number of vertices $n \geq 3$ is 2-chromatic iff n is even and 3-chromatic or tri-chromatic iff n is odd.

7.6.2 GCP: Solution Approach

Backtracking is an efficient approach that finds m on a connected graph $G = (V, E)$. Initially m is 0 *i.e.,* no colour is considered. With increase of m, types of colour increase. This is the partial solution tuple. The chromatic number m is established by the following steps.

1. Consider any arbitrary vertex X as starting of colouring G. This node is the root of our implicit tree.
2. Colour this X with a colour.
3. Increase m by 1 *i.e.,* $m = m + 1$
4. Explore X *i.e.,* consider next adjacent vertex Y of X on the basis of lexicographic order. We can also consider in alphabetic order.
5. Colour this Y with another colour.
6. Increase m by 1 *i.e.,* $m = m + 1$
7. Colour a vertex Z with a colour present in partial solution iff the vertices which are adjacent to Z do not share the same colour as Z, otherwise it be a dead state. So, backtrack. If this situation again happens consider a new colour in partial solution followed by $m = m + 1$.
8. Repeat step 7 until all the vertices have been coloured with distinct colour.

7.6.3 GCP: Algorithm

The above solution approach can be written in pseudocode as below.

Algorithm: GCP_Backtracking (k)

Where k = next vertex to be coloured.

Extra Terms: $G = (V, E)$ a connected graph.

g = adjacency matrix defined as

$$= \begin{cases} 0, \text{if } (i, j) \notin E \\ 1, \text{if } (i, j) \in E \end{cases}$$

m = chromatic number for G. Initially it is 0.

T = list of distinct colours = < 1, 2, 3... m >

dead = a Boolean variable.

Step 1:

(*i*) dead = 0 // find all m colours

(*ii*) Repeat while dead = 0 through step 7

Step 2:

(*i*) $x = 0$ // find next colour

(*ii*) Repeat while $x = 0$ through step 5

Step 3:

(*i*) $T[k] = (T[k] + 1) \bmod (m + 1)$ // any colour due

(*ii*) If $T[k] = 0$ // all colours are used

(*a*) $x = 1$

(*b*) Return x

Step 4: Repeat for $j = 1$ to n // checking distinctness of colours

(*a*) If $g[k][j] \neq 0$ and $T[k] = T[j]$

- $x = 1$
- Exit

Step 5: If $j = n + 1$ // consider new colour

(*a*) $x = 0$

(*b*) Return x

Step 6: If $T[k] = 0$

(*a*) dead = 0

(*b*) Return dead

Step 7: If $k = n$

(*a*) Repeat for $r = 1$ to n

- Output $T[k]$

Else

(*a*) Call GCP_Backtracking $(k + 1)$

Step 8: Exit

7.6.4 GCP: Analysis

The computing time of the algorithm is spent due to large number of nodes of the solution space tree *i.e.,* implicit tree. This number is $\sum_{i=0}^{n-1} m^i$. For each internal node $\Theta(mn)$ time is spent due to steps 2 to 5. Thus, total time required by the algorithm for n vertices is given by:

$$T(n) = n\,n \cdot \sum_{i=0}^{n-1} m^i. = \sum_{i=0}^{n-1} n \cdot m^{i+1} = \Theta(nm^n) \ = \Theta(nm^n)$$

If $m = n$ then $T(n) = O(n^n)$.

Example 2. Colour the following graph with minimum number of distinct colours using backtracking approach.

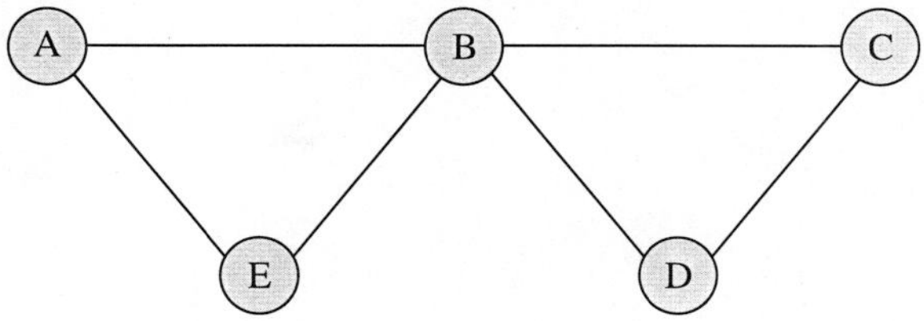

Figure 7.7 Undirected graph for GCP.

Solution: Consider the vertex A as the starting node of the implicit tree and colour the nodes in the following way.

Let C = Set of different colours used and S = set of vertices having same colour. Both are initially empty (Ø).

Step 1: Colour vertex A with a colour say, Black.

$C = C \cup \{\text{Black}\} = \{\varnothing\} \cup \{\text{Black}\} = \{\text{Black}\}$

$S = S \cup \{A\} = \{\varnothing\} \cup \{A\} = \{A\}$

Explore A.

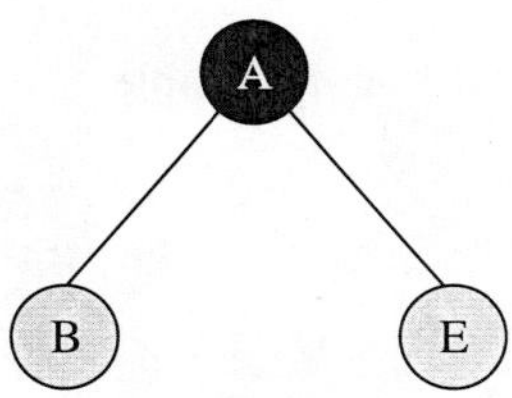

Figure 7.8 Partial solution.

Step 2: Consider vertex B.

Colour B with a new colour say, Green as it is adjacent of A and there is only one colour in C.

$C = \{\text{Black, Green}\}$

$S = \{A, B\}$

Explore B.

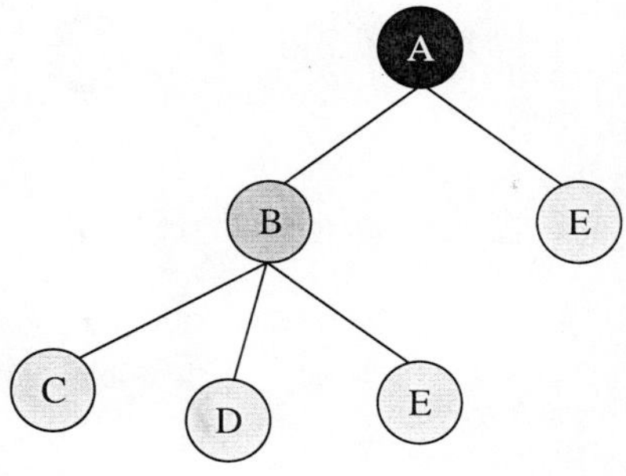

Figure 7.9 Partial solution.

Step 3: Consider vertex C.

Colour C taking from the colour set C if possible. Black is taken since it is not adjacent to A.

C = {Black, Green}. No new colour is considered. No change in set C.

$S = \{A, B, C\}$

Explore C.

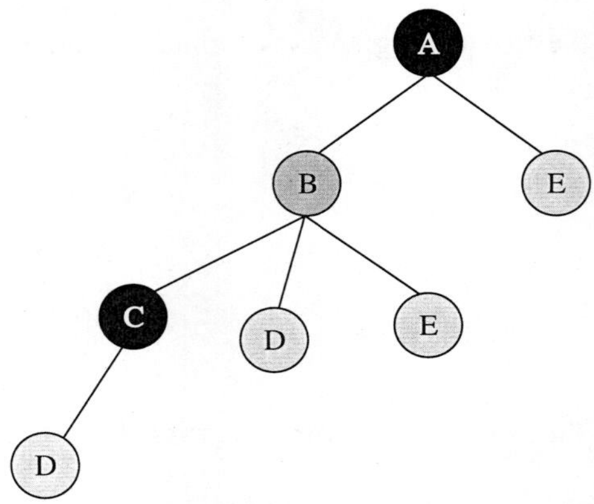

Figure 7.10 Partial solution.

Step 4: Consider vertex D.

Colour D taking from the colour set C if possible. D is adjacent to both vertices B and C. Two colours are there and they have been used for these two vertices. Take a new colour say, Red to colour D.

C = {Black, Green, Red}.

$S = \{A, B, C, D\}$

Explore D. however it is not possible.

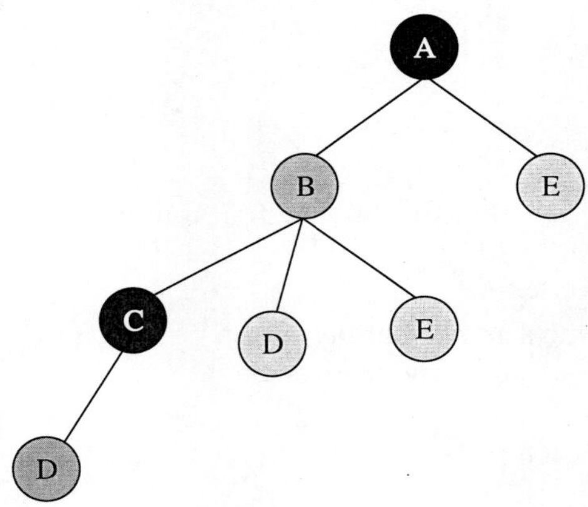

Figure 7.11 Partial solution.

All the vertices have not been traversed. *E* is remained. Backtrack to *B* to traverse *E*.

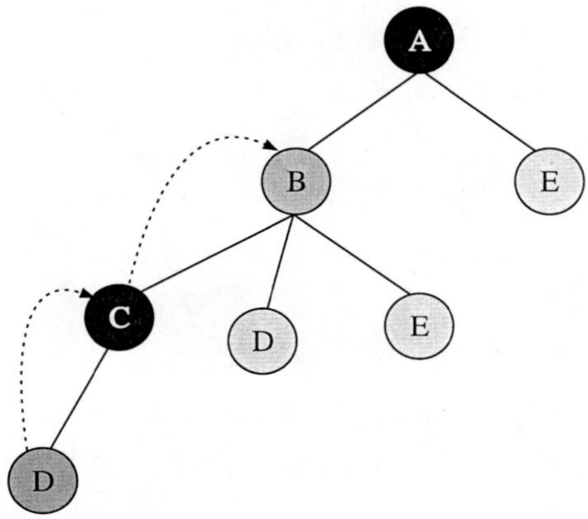

Figure 7.12 Partial solution.

Step 5: Consider vertex *E*.

Color *E* taking from the color set *C* if possible. *E* is adjacent to both vertices *A* and *B*. their colours cannot be used. But other colour Red can be cosiderd.

C = {Black, Green, Red}.

$S = \{A, B, C, D, E\}$

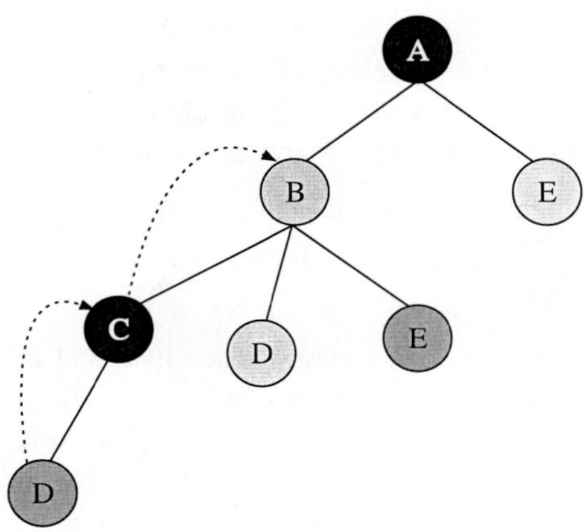

Figure 7.13 Partial solution.

Thus the given graph after colouring will be looked as below.

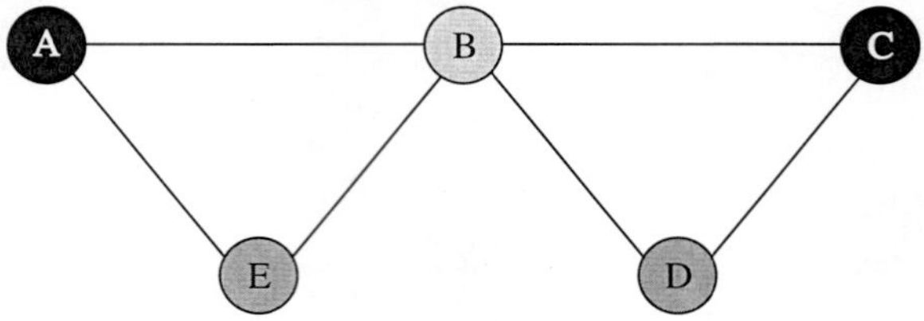

Figure 7.14 Final solution.

7.7 SOME COMMON QUESTIONS

Question 1. What are the constraints of Backtracking approach?

Answer: The valid solution is generally represented as an *N*-tuple $T = (x_i \mid 1 \geq i \geq N)$. Where x_i is a number and is chosen from a set S_i. Many problems that can be solved using backtracking require satisfying two constraints.

1. *Explicit Constraints:* The rules which restrict each x_i to take values only from a given set. *e.g., i)* $x_i \geq 0$, then S_i = {all non-negative real numbers}

   (ii) $x_i = 0/1$, then $S_i = \{0, 1\}$.

   These Explicit Constraints depend on the particular instance of the problem being solved. All tuples satisfying these constraints define a possible solution space.

2. *Implicit Constraints:* The rules which determines a particular tuple that satisfies our criteria. These constraints mainly describe a way how x_i *s* are related to each other.

Question 2. Is Hamiltonian path problem and Hamiltonian cycle problem are correlated?

Answer: There is a simple relation between these two problems. The Hamiltonian path problem for graph *G* is equivalent to the Hamiltonian cycle problem in a graph *H* obtained from *G* by adding a new vertex and connecting it to all vertices of *G*.

Question 3. Is Hamiltonian graph a Eulerian graph?

Answer: The line graph of a Hamiltonian graph is Eulerian. The line graph of an Eulerian graph is Hamiltonian.

Question 4. Find relations between a Hamiltonian path and a Tournament.

Answer: Mainly to two types of relations between them exist.

1. A tournament with more than 2 vertices is Hamiltonian if and only if it is strongly connected.
2. Every tournament has an odd number of Hamiltonian paths.

Question 5. Define exhaustive search.

Answer: Exhaustive search is simply a "brute – force" approach to combinatorial problems. It suggests generating each and every element of the problem's domain, selecting those of them that satisfy the problem's constraints, and then finding a desired element.

Question 6. Give the categories of the problem in backtracking.

Answer: Three kinds of problems are solved in backtracking.

(*a*) Decision's problem: - Whether there is any feasible solution.

(*b*) Optimization problem: - Whether there exists any best solution.

(*c*) Enumeration problem: - Finds all possible feasible solution.

Question 7. List down the examples of backtracking.

Answer: Some common examples of backtracking include-

(*a*) *n*-Queens problem

(*b*) Hamiltonian circuit problem

(*c*) Subset-Sum problem

(*d*) Constraint satisfaction problem

Question 8. How can you represent the solution for 2-queen's problem?

Answer: There is no solution for 2-queen's problem since both the queens attack themselves.

Exercises

1. Consider the graph in Figure 7.15. Find Hamiltonian circuit using backtracking.

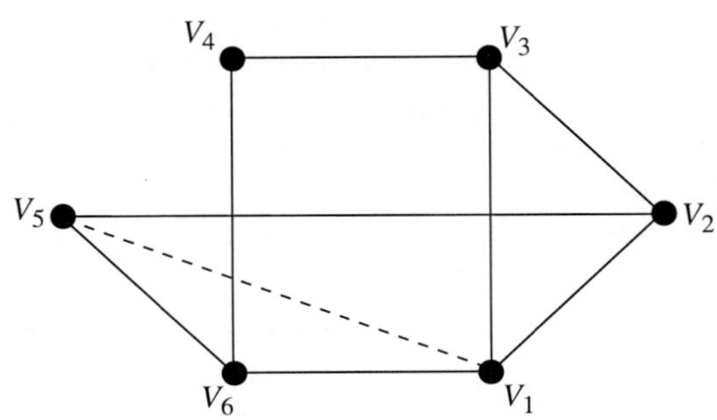

Figure 7.15 An undirected graph for Hamiltonian circuit.

2. Consider three items $< I_1, I_2, I_3 >$ with the respective weights and values as $< W_1 = 1, W_2 = 4, W_3 = 3 >$ and $< V_1 = 4, V_2 = 5, V_3 = 6>$. The Knapsack capacity $W = 6$. Apply backtracking approach to obtain optimal solution.

3. Give a suitable algorithm to solve *k*-colouring problem. What is the running time of your algorithm?

4. Applying backtracking chromatic number of the graph given in Exercise 7.1.

5. Colour the graph given below using only three colours.

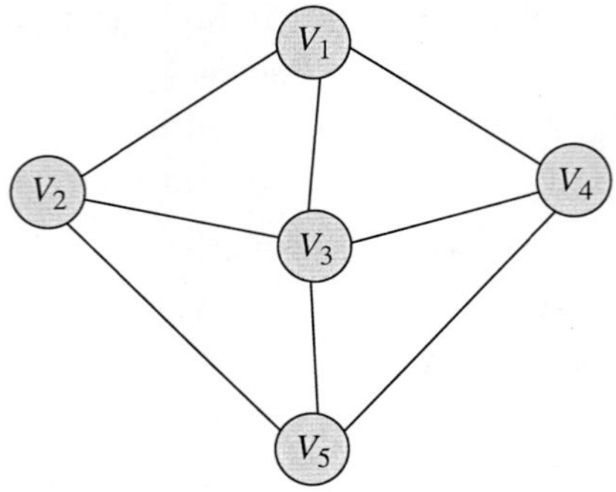

Figure 7.16 Undigraph for GCP.

6. Draw an optimum space search tree for 8-queens problem.
7. Give a suitable algorithm that can solve 4-queens problem. What is the running time of your algorithm?
8. Modify your algorithm to solve n-queens problem. Evaluate its complexity.
9. Find all possible solutions for both 4-queens and 8-queens problems in 4×4 and 8×8 chessboards respectively.
10. Determine minimum value of n in n-queens problem.
11. Find out the necessary condition that two queens will attack each other vertically or diagonally.
12. Draw some advantages and disadvantages of backtracking technique.
13. Let k-Colouring problem be a generalization of the 3-Colouring problem. How many nodes in the worst case, will be generated by its corresponding backtracking algorithm?
14. Test whether a vector corresponding to a 3-Colouring of a graph is legal or not designing an efficient algorithm. Also compute its running time.
15. How many nodes are generated by the algorithm for n-queens problem in worst case?
16. Show that two queens placed at positions X_i and X_j are in the same diagonal iff
    (*a*) $X_i - X_j = i - j$ or
    (*b*) $X_i - X_j = j - i$.
17. Write a recursive algorithm for the 8-Queens problem.
18. Does the n-Queens problem have a solution for every value of $n > 4$? Prove your answer.
19. Give a algorithm of 4-Queens so that it reduces the search space from 4^4 to 4!.
20. Design a backtracking algorithm that generates all permutations of the numbers 1, 2... n.
21. Design a backtracking algorithm to generate all 2^n subsets of the numbers 1, 2, ..., n.
22. Given an 8×8 chessboard. Is it possible to visit every square of the board, for a knight placed at a certain position of the board, exactly once and return to its start position? If yes, write an algorithm using backtracking approach.

23. Given an undirected graph $G = (V, E)$. Give an algorithm that determines whether it contains a simple cycle that visits each vertex exactly once.
[**Hint:** Backtracking algorithm of Hamiltonian Cycle problem].

24. In software firm n employees are to be assigned to n jobs such that the cost of assigning the jth job to the ith person is $C_{ij} \geq 0$, $1 \leq i, j \leq n$. Give a backtracking algorithm to find an assignment that minimises the total cost.

University Questions

1. Discuss backtracking problem solving approach. *[UPTU'06, PTU'09, RTU'09]*
2. Explain the applications of Backtracking. *[JNTU'09, '11]*
3. What are the four factors that decide the efficiency of the Backtracking Algorithm? Explain the role of rearrangement of search space in performance of Backtracking. *[BPUT'08]*
4. Write the short note on Backtracking Algorithms. *[RTU'10, JNTU'09, DOEACC'09]*
5. Describe an algorithm of some problem where backtracking is an intrinsic part of the algorithm. *[DOEACC'09]*
6. Compare general and recursive backtracking methods. *[PTU'08]*
7. Find out Hamiltonian cycle of the graph and also draw the permutation tree. *[WBUT'07]*

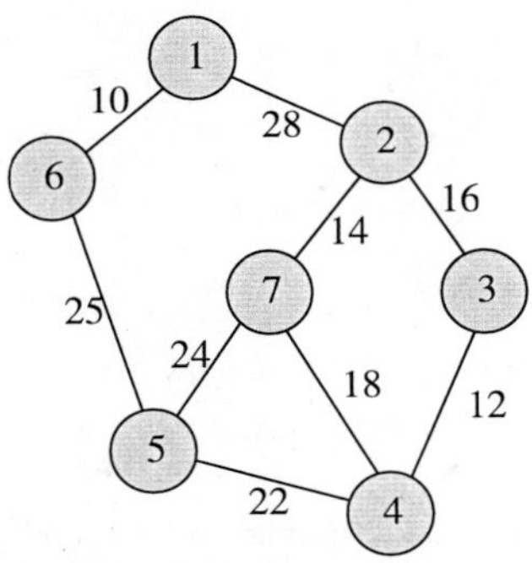

8. Design a backtracking algorithm to find all the Hamiltonian cycles in a Hamiltonian graph. What is the worst-case time complexity of the algorithm? *[WBUT'08]*
9. Give an example of a Non-Hamiltonian graph with 10 vertices such that for every pair of non adjacent vertices u and v, $d(u) + d(v) \geq 9$. The d is degree of a vertex. *[UPTU'10]*
10. Write down the steps for solving Hamiltonian circuit problem using backtracking. *[WBUT'09, PTU'08]*
11. Find Hamiltonian cycles in the following graph using backtracking. *[MDU'08]*

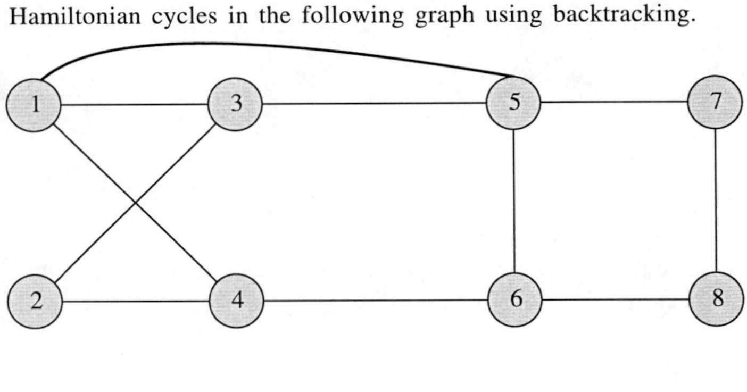

12. Give a non-deterministic graph colouring algorithm. *[WBUT'08]*
13. Consider a four vertex network as shown below. Find a least cost tour using backtracking method. *[UPTU'10]*

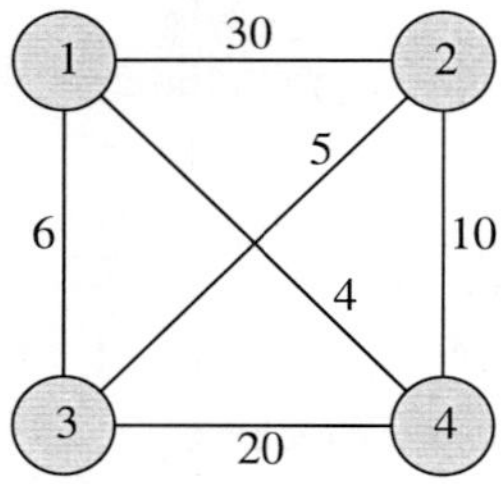

14. Apply backtracking technique to solve the 3-colouring problem for the following graph. *[WBUT'09]*

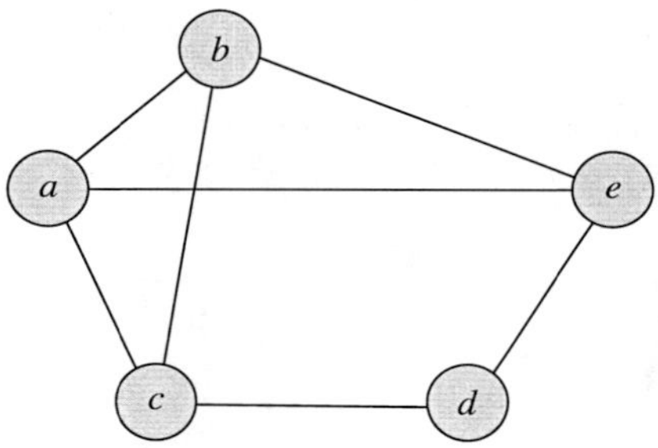

15. Explain Graph Colouring Problem. *[MDU'09]*
16. Write an algorithm of m-colouring problem. *[JNTU'08, '10]*
17. Draw the state space tree for m colouring when $n = 3$ and $m = 3$. Write a recursive backtracking algorithm. *[JNTU'08, '11]*
18. What do you mean by chromatic number of a graph? *[WBUT'09, UPTU'07, '10]*
19. Determine the chromatic number and chromatic polynomial of the following graphs. *[WBUT'06, UPTU'09]*

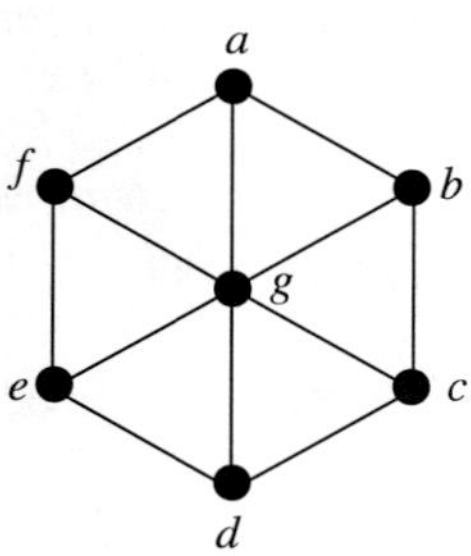

(*i*)

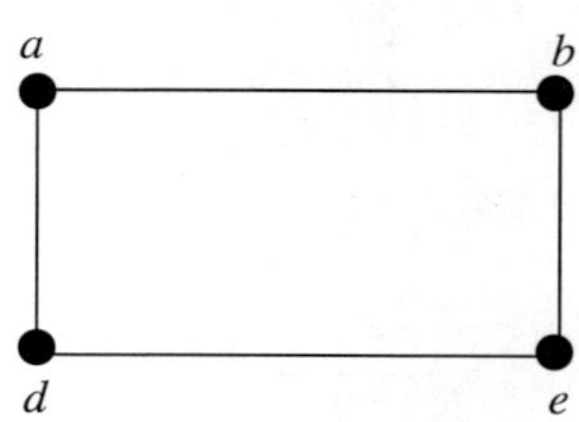

(*ii*)

20. Prove that every tree with two or more vertices is 2-chromatic. *[UPTU'10]*
21. Write backtracking algorithm to find the chromatic number of a given graph. *[MDU'08]*

22. Write short notes on *[UPTU'10]*
    (*a*) Hamiltonian cycle problem.
    (*b*) Chromatic Number.
23. Describe graph colouring problem and its time complexity. *[JNTU'08]*
24. Give an 8 × chessboard. A '*q*' is placed on the one square of the chessboard. You have to place only one '*q*' in each row and column and no two '*q*' can be placed diagonally. Write the algorithm for the above problem using backtracking method. *[RTU'10]*
25. Write an algorithm of eight-queens' problem and find time complexity of the algorithm. Explain the algorithm using an example. *[WBUT'10, JNTU'08, '11]*
26. Write an algorithm for eight-queen's problem. *[WBUT'09]*
27. Explain the backtracking problem with 4-queens on a 4 × 4 chessboard. *[PTU'08, JNTU'08, '09]*
28. Discuss *N*-Queens problem and analyze the time complexity of the same. *[MDU'09]*
29. What is the sum of subsets problem? Let $w = \{5, 7, 10, 12, 15, 18, 20\}$ and $m = 35$. Find all possible subsets of w that sum to m using recursive backtracking algorithm for it. Draw the portion of the state-space tree that is generated. *[UPTU'05]*

❑❑❑

Chapter 8 Branch and Bound

8.1 INTRODUCTION

Branch and Bound in short B&B is another general algorithm design technique that is often implemented in case of some typical optimization problems for finding the optimal solutions. These problems are mainly combinatorial and discrete global optimizations kind of problems. It consists of an orderly listing of all candidate solutions. When the domain of possible candidates is too large and all of the other algorithms fail then it is better to opt for this technique. The technique is purely based on the most elimination of the candidates where large subsets of unsuccessful candidates are discarded by using upper and lower estimated bounds of the quantity being optimised. The method was first proposed by A. H. Land and A. G. Doig in 1960 for linear programming.

B&B is composed of two main actions.

(*i*) *Branching* - Defining the tree structure from the set of candidates in a recursive manner. Basically, this is a splitting procedure over a given set *S* of candidates that returns two or more smaller sets whose union covers *S*. Nodes are the subsets of *S*.

(*ii*) *Bounding* - determining the upper and lower bounds of each node from the tree.

Selection of a node satisfying constraints leads towards optimal solution. This node is called live node or Effective node or *E*-node.

This *E*-node provides two things. One, "how far" it is located from the root of the tree. Second, "how close" it is from the optimal solution. The solution cost is basically sum of the aforesaid two distances. However, as we will see, the second distance depends on heuristics. It's just a guess.

8.2 GENERAL METHOD

Initial arrangement of the problem is a root of the search tree. Based on the conditions this root is explored to its children called candidates. Now each candidate is computed as: At every level of the tree a lower bound of every node is found out. A node having lowest bound is assigned as the *E*-node at that level. Choosing it discards others. This *E*-node is next explored to find the next *E*-node at the next level. The process is continued until the desired solution. If more than one node at any stage has the same lowest value then anyone of them is selected as E-node. But we must keep in mind that the node which seems to be nearest towards optimal solution will be judged as an *E*-node. It would however require lot of experience.

Therefore, B&B refers to all state space searching methods where a node before becoming an *E*-node, an existing *E*-node is fully explored. Two traversals namely *B*-search (BFS) and *D*-search (DFS) are mainly used. Both of these generalize to B&B strategies. And they help us to choose a node that will be live node and which path to be chosen.

The total cost for the solution of the problem is equivalent to cost of construction of the tree rather implicit graph is calculated w.r.t. live nodes. This cost is represented as a function, $f(x) = g(x) + h(x)$,

where $f(x)$ = The cost of construction of an *E*-node,

$g(x)$ = The distance between the x node and the root node,

$h(x)$ = A heuristic function because it estimates how close this node x is to the solution node (leaf). The heuristic function ought to return 0 in the case of a solution leaf since the distance is zero. The efficiency of the B&B relies on this function.

8.3 B&B *VS*. BACKTRACKING

All in all, B&B is very similar to backtracking. The main differences are that

(*i*) B&B is used only in case of optimization problems, whereas backtracking cannot be.

(*ii*) Branch-and-bound algorithms are related with maximization or minimization of a given function only. Backtracking searches for a solution or a set of solutions that suit certain properties including maximization or minimization.

(*iii*) B&B doesn't limit us to a particular way of traversing the tree. But depth-first search/preorder is used with backtracking.

(*iv*) B&B always has the entire list of successors in the queue. Backtracking always picks one single successor from the candidates.

N.B.: Out of the algorithm design techniques that we still have learned, both the backtracking and Divide-and-Conquer traverse the tree in its depth, though they take opposite routes. The greedy strategy picks a single route and forgets about the rest. Dynamic programming approaches this in a sort of Breadth-First Search variation (BFS).

8.4 APPLICATIONS

This approach is used for a number of NP-hard problems, such as:

1. 15 – Puzzle problem.
2. Knapsack problem
3. Integer programming
4. Non-linear programming
5. Travelling Salesman Problem (TSP)
6. Quadratic Assignment Problem (QAP)
7. Maximum satisfiability problem (MAX-SAT)

8. Nearest Neighbour Search (NNS)
9. Cutting stock problem
10. False Noise Analysis (FNA)

Branch-and-bound may also be a base of various heuristics. For this reason, branch-and-bound techniques are often used in game tree search algorithms.

8.5 15-PUZZLE PROBLEM

The 15-puzzle problem (15-PP) was first invented by Sam Loyd in 1878. There are 15 square tiles numbered 1 through 15. There is a square frame having 16 cells. Each cell is numbered 1 through 16 and has a size exactly same as of each tile. The task is to place all the tiles in the frame in such a way that ith tile should be in ith cell as in Figure 8.1B. Obviously, the 16th cell will remain vacant. This arrangement is called target or desire solution. Initially all the tiles are considered in the frame violating the condition Figure 8.1A. This is our initial arrangement.

1	2	3	4
5	6		8
7	9	10	11
12	13	14	15

A. Initial Arrangement of Tiles

1	2	3	4
5	6	7	8
9	10	11	12
13	14	15	

B. Final Arrangement of Tiles

Figure 8.1 Arrangements of tiles before and after solution.

In Figure 8.1A only first 6 tiles (1...6) are in proper places. Our objective is to transform this initial arrangement to target arrangement through a series of legal movements of the plates in terms of vacant cell movement. There are maximum four legal movements namely up, right, down, and left possible for vacant cell in which adjacent tile is moved to this vacant cell. It is described in Figure 8.2. On every movement a new arrangement is made. Each new arrangement made is labelled by its corresponding vacant cell movement for distinguishing them.

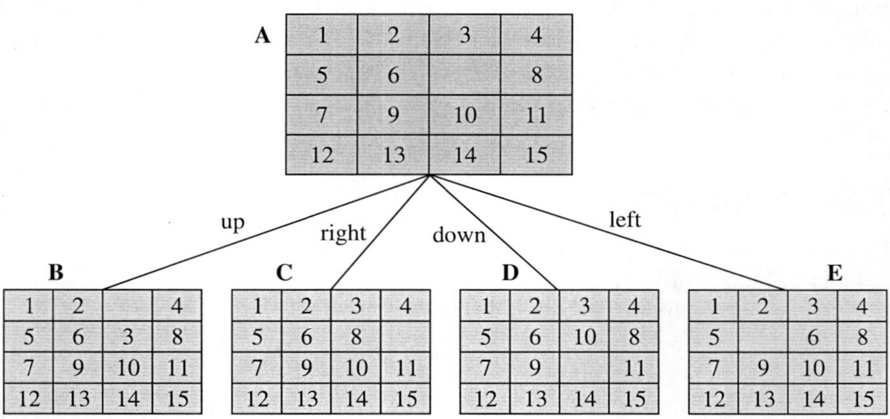

Figure 8.2 Exploring an arrangement into 4-possible legal moves (arrangements).

Every arrangement is represented as a node of the implicit tree. Obviously the initial and the target arrangements are the root and a leaf of the tree. This node is also called state. The total arrangement including initial state, internal arrangements, and final state is called solution state space.

The final state is reached through the following steps:

1. Consider the initial arrangement as root node *X*.
2. Explore *X* into legal new arrangements as its children. Never use a legal move which explores a node to one of its children which is its parent. Otherwise, a dead state comes about.
3. Find cost for each explored node.
4. Take the node having minimum cost. Select it as *E*-node.
5. If this *E*-node is not target arrangement then go to step 2.
6. Sum the costs of the *E*-nodes as the optimal cost of solution of the problem.

A state is said to be reachable from the initial state iff there is possible a sequence of legal moves to this state from initial state. The state space of a state consists of all the states that are reachable from the state. Consequently, the most straight and forward way to solve this puzzle is to search a sate space for the target state. The sequence or path from the initial state to this goal state is our optimal solution.

8.5.1 15-PP: Determination of Cost of Construction of Explored Node

15-Puzzle problem is an optimization problem which leads us to compute optimal cost for the solution space. Now the optimal solution consists of a sequence of *E*-nodes. Each *E*-node at a particular level again is found by comparing cost of every explored node. Let us see how cost of an explored node is computed.

The cost of construction *CC* [*N*] of an explored node *N* from the root *X* is given by:

$$CC\,[N] = f(N) + g(N)$$

where, $f(N)$ = The path length *i.e.*, number of edges between *X* and *N*.

$g(N)$ = The shortest path length from *N* to goal state in the sub-tree with root *X*.

= The number of no blank tiles not in proper position *i.e.*, not in goal state. This tells us that how near we are to the target node or state.

From Figure 8.2, $CC\,[B] = f(B) + g(B) = 1 + 9 = 10$. Here, $f(B)$ is 1 because number of edge between the root *A* and the current explored node *B* is 1. The tiles 3, 7, 9, 10, 11, 12, 13, 14, and 15 are not in their proper positions. Hence, g(B) = 9. Similarly, $CC\,[C] = 1 + 9 = 10$, $CC\,[D] = 1 + 8 = 9$, and $CC\,[E] = 1 + 9 = 10$. Hence, *E*-node will be *D*. The $g(N)$ excludes the blank tile position.

During finding optimal solution for the problem, cost for every explored node at the same level is computed. The nodes having minimum cost of construction will be the next *E*-node to be explored if it is not the goal state. The explored nodes having same costs of construction are generally discarded during selection of an *E*-node although they may have lowest cost value over the others at the level. This is done when other node seems near to the goal state. Sometimes anyone of the nodes having the same lowest costs may be regarded as a next *E*-node. However in this case the node as *E*-node is judged based on the guessing that this node is nearer to the goal state than the other.

8.5.2 15-PP: Algorithm

The algorithm for solving 15-puzzle problem is generally known as Least Cost Search algorithm. An effective algorithm to find optimal solution can be designed as:

Algorithm: 15PP_BB ()

Terms used:

EN = *E*-node or Effective or live node, the initial arrangement or state.

CN = Current *E*-node, initially the initial arrangement of the problem.

LN = list of *E*-nodes of the solution, initially it is empty.

GN = The goal node or target arrangement.

MC = Minimum Cost to achieve goal state.

$CC(i)$ = Cost of Constructing node i

$f(i)$ = Path length of node i from root node.

$g(i)$ = Number of tiles that are not in proper position.

Step 1:

(*i*) If $EN = GN$ // if *CN* is goal state

(*a*) $k = 1$

(*b*) $LN[k{+}{+}] = EN$

(*c*) $MC = 0$

(*d*) Go to step 4.

Step 2:

(*i*) $CN = EN$ // initialise and find sub-solution

(*ii*) Find all the child nodes i of *CN* through legal moves. If any child i is same as CN's parent, then discard this child node i.

(*iii*) Find cost of construction of all the other nodes i of *CN* as

- $CC(i) = f(i) + g(i)$

(*iv*) Set EN = child i having lowest $C(i)$ value or any one of the children having same lowest $C(i)$ value.

(*v*) $MC = MC + CC(i)$

Step 3:

(*i*) If $EN = GN$ // if goal state reached

(*a*) $LN[k{+}{+}] = EN$

(*b*) Go to step 4

(*ii*) Else If $EN = 0$ // no live node available

(*a*) Output " No goal state possible"

(*b*) Go to step 5

(*iii*) Else

(*a*) Go to step 2

Step 4:

(*i*) Repeat for $i = 1$ to k

(*a*) Output $LN [i]$

(*ii*) Output "optimal cost of solution is '*MC*'".

Step 5: Exit

Analysis: The time complexity of the algorithm basically depends on exploring a live node, computing cost of construction for each child, finding next *E*-node. Since the algorithm is based on the *B*-search *i.e.,* Breadth First Search because *B*-search always finds a goal state nearest to root node. Whatever may be the initial arrangement, the algorithm always attempts to make the same sequence of legal moves. The *B*-search always generates state space tree level wise. Therefore, the time complexity of this algorithm is $\Theta(| V | + | E |)$.

Example 1. Solve the 15-puzzle problem having the initial arrangement as given below:

1	2	3	4
5	6		8
9	10	7	11
13	14	15	12

Solution: We solve this 15-puzzle problem using implicit tree representation. The initial arrangement is the root of the tree. Then exploring this root into its children based according to the legal moves a partial solution is obtained and it is continued till the solution. Each node in the tree is labelled. First solve the problem level wise to compute cost for each node followed by finding an *E*-node. Then represent the solution in tree.

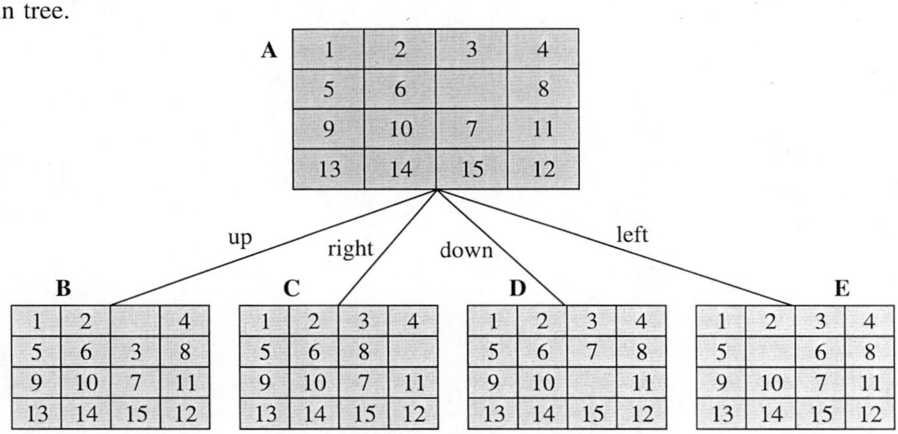

Figure 8.3 Exploring root node.

$$CC\,(B) = f\,(B) + g\,(C) = 1 + 4 = 5$$
$$CC\,(C) = f\,(C) + g\,(C) = 1 + 4 = 5$$
$$CC\,(D) = f\,(D) + g\,(D) = 1 + 2 = 3 \leftarrow E\text{-node}$$
$$CC\,(E) = f\,(E) + g\,(E) = 1 + 4 = 5$$

Explore the *E*-node *D* as it since it is not goal state.

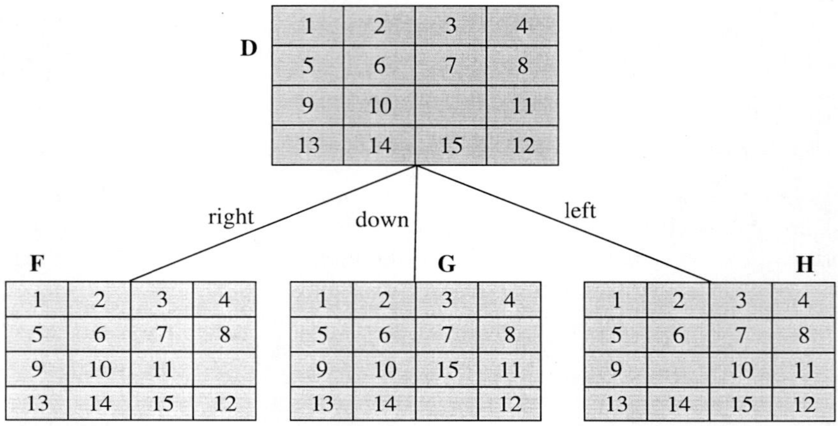

Figure 8.4 Exploring next *E*–node *D*.

$$CC\,(F) = f\,(F) + g\,(F) = 2 + 1 = 3 \leftarrow E\text{-node}$$
$$CC\,(G) = f\,(G) + g\,(G) = 2 + 3 = 5$$
$$CC\,(H) = f\,(H) + g\,(H) = 2 + 3 = 5$$

Explore the *E*-node *F*.

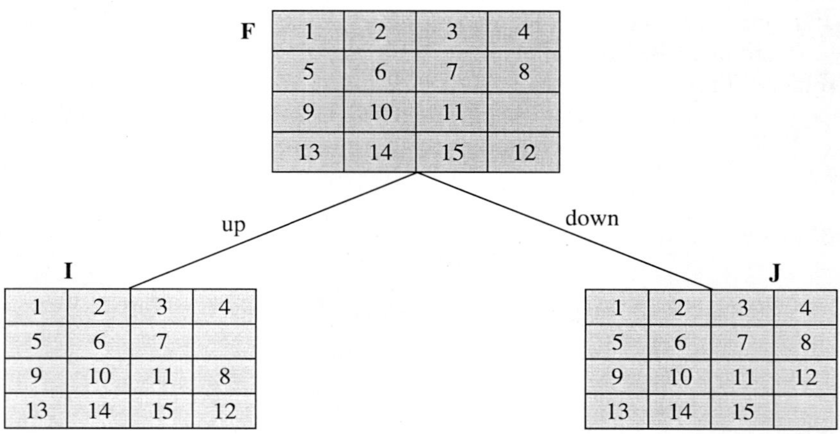

Figure 8.5 Exploring next *E*-node *F*.

$$CC\,(I) = f\,(I) + g\,(I) = 3 + 2 = 5$$
$$CC\,(J) = f\,(J) + g\,(J) = 3 + 0 = 3 \leftarrow E\text{-node}$$

Now the E-node is node J. It is seen that node J is the goal state. We thus reach to solution state and total cost of the solution rather optimal cost of the solution for the problem is $MC = CC(D) + CC(F) + CC(J) = 3 + 3 + 3 = 9$

Now combining Figure 3, 4 and 5, the solution space search tree will look as:

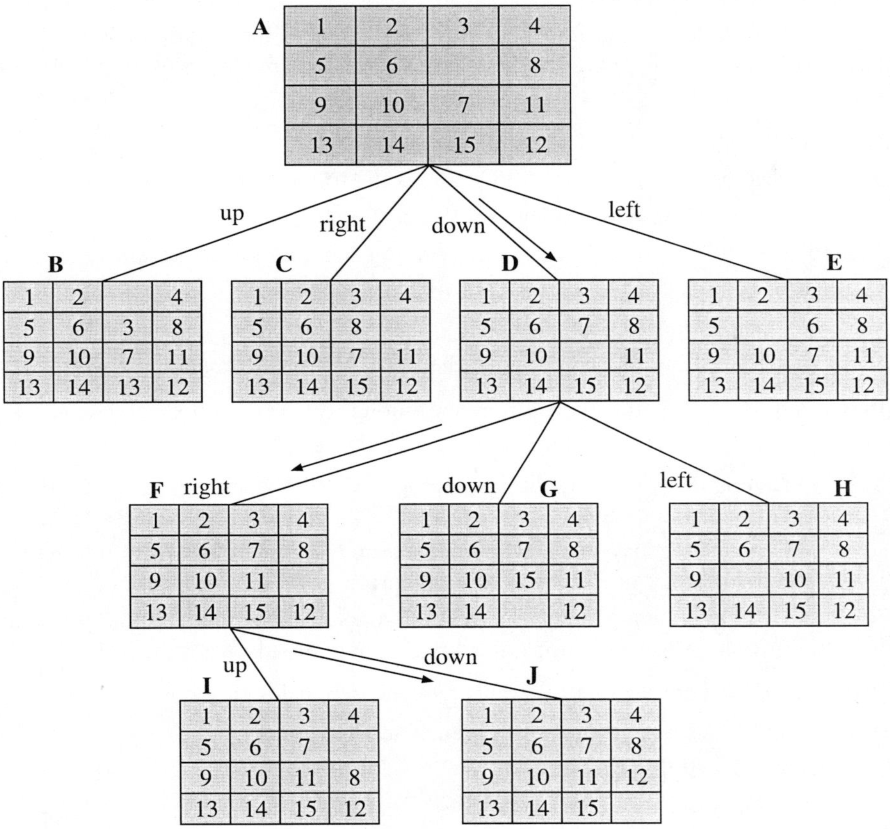

Figure 8.6 Implicit tree for 15-Puzzle problem.

8.6 SOME COMMON QUESTIONS

Question 1. Are B&B and Backtracking similar?

Answer: Branch and Bound design technique is similar to backtracking in the sense that it generates a search tree and looks for one or more solutions.

Question 2. What is state space tree?

Answer: State space tree is the tree organization of the solution space.

Question 3. Define static tree.

Answer: Static trees are the ones for which tree organizations are independent of the–

(*a*) Fixed tuple size formulation.

(*b*) Tree organization is independent of the problem instance being solved.

Question 4. What is dynamic tree?

Answer: Dynamic trees are ones for which organization is dependent on problem instances.

Question 5. Define live node.

Answer: Live node is a generated node for which all of the children have not been generated yet. An ?E-node is a live node whose children are currently being generated or explored.

Question 6. Define dead node.

Answer: Dead node is a generated node that is not to be expanded any further

(*a*) All the children of a dead node are already generated.

(*b*) Live nodes are killed using a bounding function to make them dead nodes.

Question 7. What do you mean by branch-and-bound?

Answer: Branch-and-bound refers to all state space search methods in which all children of an E-node are generated before any other live node can become the E-node.

Question 8. How do you make a difference among live node, E-node and dead node?

Answer: they could be differentiating by their definition.

- Live node is a node that has been generated but whose children have not yet been generated.
- E-node is a live node whose children are currently being explored. In other words, an E-node is a node currently being expanded.
- Dead node is a generated node that is not to be expanded or explored any further. All children of a dead node have already been expanded.

Question 9. What are the types search strategies in branch-and-bound?

Answer: the types search strategies in branch and bound are–

(*i*) FIFO search

(*ii*) LIFO search

(*iii*) Least Cost search

Question 10. Compare the FIFO and LIFO search.

Answer: They can be compared as.

FIFO search	*LIFO search*
1. BFS is an FIFO search in terms of live nodes. 2. List of live nodes is a queue	1. DFS is an LIFO search in terms of live nodes. 2. List of live nodes is a stack

Question 11. Why do we use Least Cost Search?

Answer: The next E-node is selected on the basis of this ranking function. Heuristic is based on the expected additional computational effort (cost) to reach a solution from the current live node.

Exercises

1. What is a non-trivial condition in solving 15-puzzle problem?
2. Draw an effective state space tree for 15-puzzle problem.
3. What are the application areas of B&B approach?
4. Can Traveling Salesman Problem be solved using B&B approach? If yes, give corresponding algorithm. Also apply the same on the instance given below.

$$\begin{bmatrix} \infty & 5 & 2 & 10 \\ 2 & \infty & 5 & 12 \\ 3 & 5 & \infty & 7 \\ 2 & 8 & 4 & \infty \end{bmatrix}$$

5. Using B&B with a suitable lower bound write an algorithm to solve Knapsack problem.
6. Solve the following knapsack problem.

Items	***Weights***	***Value***	***Capacity***
I_1	9	5	
I_2	6	6	14
I_3	5	7	
I_4	3	4	

7. Carry out B&B procedure to solve the following assignment problem. There are four employees and four jobs. The cost function is represented by the matrix below. In this matrix, row *i* corresponds to the ith employee, and column *j* corresponds to the *j*th job.

$$\begin{bmatrix} 3 & 5 & 2 & 4 \\ 6 & 3 & 5 & 3 \\ 3 & 7 & 4 & 5 \\ 8 & 5 & 6 & 4 \end{bmatrix}$$

8. Apply B&B to solve TSP problem on the following graph. Also find the cost of the tour.

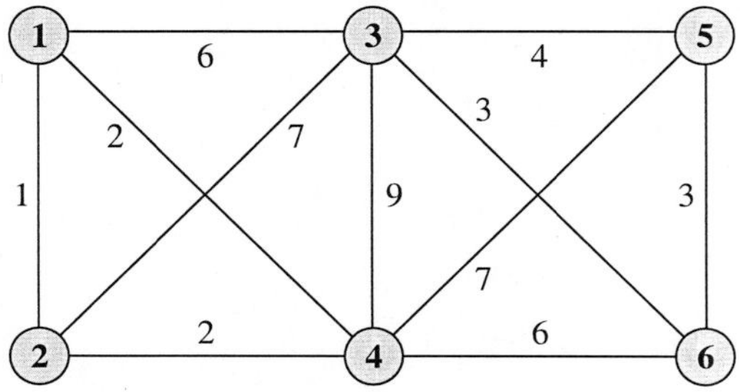

Figure 8.7 Undigraph for TSP.

University Questions

1. Write short notes on Branch-and-Bound algorithm. ***[UPTU'10, PTU'09, JNTU'08, '11]***
2. What are Explicit and Implicit constraints? ***[PTU'08]***
3. Differentiate between backtracking and branch and bound method. ***[MDU'09]***
4. Explain the general method of Branch-and-Bound. ***[JNTU'08, '09]***
5. What are the features of branch and bound algorithms? Discuss in detail. ***[PTU'09]***
6. Explain how the Knapsack problem can be solved using branch-and-bound algorithms? Analyze the time complexity for the same. ***[PTU'08, '09, MDU'09, JNTU'08, '11]***
7. Differentiate between Dynamic Knapsack and Branch-and-Bound Knapsack problem. ***[JNTU'08, '11]***
8. Discuss the working of FJFO Branch-and-Bound algorithm, considering 4-queen problem. ***[BPUT'08]***
9. Describe problem state, solution state and answer state with an example. ***[JNTU'08, 0'9]***
10. Solve the Travelling Salesman problem having the following cost-matrix using branch-and-bound technique. ***[RTU'07]***

$$\begin{bmatrix} \infty & 2 & 3 & 4 \\ 4 & \infty & 2 & 3 \\ 9 & 2 & \infty & 5 \\ 6 & 5 & 1 & \infty \end{bmatrix}$$

11. Explain the following: ***[MDU'08]***
    (*a*) LIFO Search
    (*b*) FIFO Search
    (*c*) LC - Search
12. Explain how you attempt to solve 15-puzzle problem using branch and bound strategy. Draw a portion of the state space generated by it. ***[WBUT'09]***

❑❑❑

9 Chapter — Lower Bound Theory

9.1 INTRODUCTION

In all the past chapters we have designed, analysed algorithms of some well-known problems where our main tasks were to obtain correct and efficient solutions. Now, a problem may be solved in various ways. If there are at least two algorithms discovered solving the same problem considers one algorithm having smaller order as superior over other when their time complexities differed by an order of magnitude. But we can't claim that this is the only better solution. Because we are still left with a question that continuously strikes in our mind: Does there exist at least one faster method?

Studying lower bound theory exposes to some popular techniques used to establish a given algorithm which is perhaps the most efficient one. To do this a way that discovers a function $g(n)$ as a lower bound on the time complexity of the algorithm is followed. Any algorithm must have this time bound to solve the given problem. Otherwise it is said asymptotically no better solution is found.

9.2 LOWER BOUND

In a single sentence the term lower bound is a function of growth rate below which solution of a problem is impossible. Mathematically, a lower bound of an algorithm is represented by Big Omega (Ω) notation. Definition of this notation already is given in chapter 2.

Finding lower bound of a particular problem is harder to a large extent. Indeed lower bound of numerous problems is still unknown. This is due to the fact that considering the lower bound of a problem necessitates establishment of a lower bound on all algorithms that solve this problem.

Example 1. Any searching algorithm that finds the maximum element in an unordered list of n elements must spend $\Omega(n)$ time. Searching an element always requires every element to be examined at least once.

Example 2. Any matrix multiplication algorithm to multiply two nxn matrices must compute exactly n^2 values. Each evaluation needs at least $\Omega(1)$ time. So the algorithm runs at least in $\Omega(n^2)$.

9.3 LOWER BOUND: COMPUTATIONAL MODEL

We learned deriving a good lower bound for an algorithm is more difficult than devising good efficient algorithms for a problem. Enumerating and analysing all the algorithms for a particular problem for lower boundness often become tricky. As a result, it turns into hard to obtain lower bound proofs. Still, it is possible to observe easily for many problems to have lower bound proof in terms of 1/0 size n of the problem.

Different computational models are available to determine lower bounds for some well-known problems. Some of them will be discuss here.

1. Comparison Tree.
2. Oracles and Adversary Arguments.
3. State Space Model.
4. Reductions.
5. Substitution and Linear Independence Method.

9.3.1 Comparison Tree

A very easy and popular computational model for determining lower bounds for problems is comparison tree. The problems specially include sorting and searching. The model is designed in such a way that a large number of sorting and searching algorithms can work. Decision tree for sorting and searching algorithms is based on comparisons of the elements and models all possible execution traces. Ignoring some elementary facts like control, data movement, other operations, etc. and taking count only on the comparisons about the tree, lower bounds are derived.

9.3.1.1 *Comparison Tree: Sorting Algorithm*

We assume that all the input elements are distinct in the list containing n elements. Let the input sequence is $< a_1, a_2, \ldots, a_n >$. any comparison between a_i and a_j in the list performs $a_i < a_j$, $a_i \leq a_j$, $a_i = a_j$, $a_i \geq a_j$, or $a_i > a_j$ to determine the relative order between them. the comparison tree in this case will be a binary tree. Each internal node is represented by an ellipse labelled as $i : j$ and indicates comparison of a_i with a_j. the leaf node is represented by reactangle labelled as sorted list of elements. If the problem is to sort the list in ascending order and at any internal node $a_i < a_j$ then the algorithm proceeds down the left edge otherwise the right edge of the tree from the current node. The leaves also indicate termination of the algorithm.

Analysis: Time complexity of this algorithm can be derived based on the number of unique paths. A path is the number of edges from root to a leaf node. Every path constitutes a unique permutation and elements alter along the path. Whatever the intial inputs are, the movement is same. For n items there $n!$ permutations possible. Anyone of these $n!$ permutations is a potential correct solution of the sorting prooblem. the comparison tree is ought to have atleast $n!$ leaves.

Let $T(n)$ = Minimum number of comparisons to sort n numbers in worstcase.

K = Maximum height of the the tree.

Now suppose all the internal nodes are at level $< K$. Therfore, maximum number of leaves are 2^K.

As the time complexity varies with any one path, then

$T(n) = K$ implies that $\quad 2^{T(n)} \geq n!$ [by Stirling formula]

$$\Rightarrow T(n) \geq \log n!$$

$$\Rightarrow T(n) = n \log n - \frac{n}{\log n} + \frac{1}{2} \log n + O(1)$$

$$\Rightarrow T(n) \geq n \log n$$

$$\Rightarrow T(n) = \Omega\ (n \log n).$$

Thus, any comparison based sorting algorithms for n elements run in $\Omega(n \log n)$ time.

Theorem: Any comparison-based sorting algorithm sorts n elements in $\Omega(n \log n)$ element comparisons in worst case.

Example 3. Draw a decision tree to sort a list containing 3 elements.

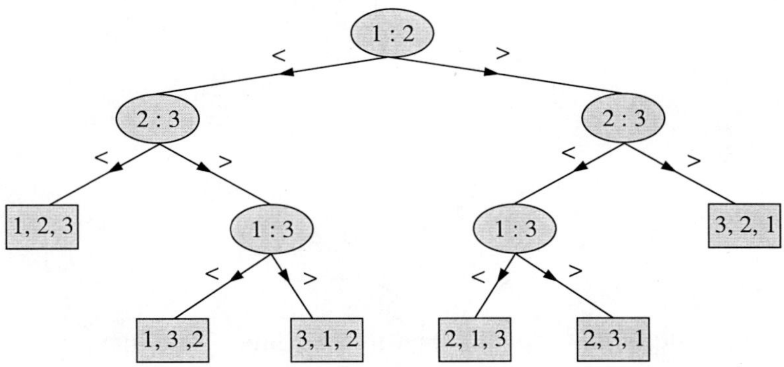

Figure 9.1 Decision tree to sort 3 elements.

9.3.1.2 *Comparison Tree: Linear Searching*

In the case of linear seraching, an unordered list L let contains n elements. If an element x is to be found in L, then compare x with every element of L *i.e.,* x is compared with $L[1], L[2], L[3], ..., L[n]$. If x is present in L, then it may be seen at any position in L. Suppose, x is present at ith positon *i.e.,* $x = L[i]$, then searching terminates after ith comparisons. Otherwise searching is continued comparisons with next elements in L. When x is found searching is successful otherwise failure occurs. The comparison tree in this case is a binary tree too. The leaves contain either failure(F) or success(S) and the internal nodes ensure the position i of x in L as $x : L[i]$.

Analysis: Depending upon number of comparisons, three cases may be derived as:

Best Case: If x is present at the root, then the time complexity is minimum and it is $\Omega(1)$.

Average Case: If x is present at any intermediate position except at the root and the leaf, then the searching algorithm returns average case time complexity. In this case time complexity $T(n)$ = average number of comparisons = $n/2 = \Omega(n)$.

Worst Case: In this case the searching is continued the entire list *i.e.,* x is present either at last position or not present in L. Number of comparisons is $n - 1$. So, time complexity in worst case is given by–

$$T(n) = \text{maximum number of comparisons} = n - 1 = O(n).$$

The tree is designed below.

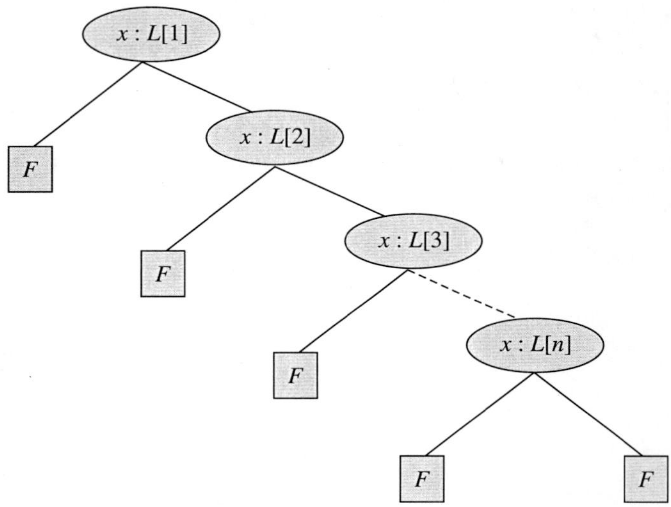

Figure 9.2 Comparison tree for linear searching.

Theorem: Any algorithm that searches a sorted sequence of n elements must perform at least ($\log n + 1$) comparisons in the worst case.

9.3.2 Oracles and Adversary Arguments

Another proof technique which is very useful for obtaining lower bounds is an Oracle. In ancient era there was the most famous oracle in history called the Delphic oracle, located in Delphi, Greece. Still this oracle can be found. But, it is now situated in some side of the hill embedded in some rocks. In those olden times, people approached the oracle and asked it a question. After elapsing some period of time, the oracle would reply. There was a caretaker who generally interpreted the oracle's answer.

Similarly we also use an oracle to establish a lower bound of a problem *i.e.,* the similar phenomenon takes place. It behaves as: suppose some model of computation such as comparison trees to oracle is considered, the oracle then returns us the outcome of each comparison. At this moment, the oracle tries at its best on the algorithm to work as hard as it can. It considers the outcome of the next test, and the result that causes the most work to determine the final answer. By keeping track of the work done, in this way a good worst case lower bound for the problem can be derived.

Example 4. Derive a lower bound for merging two sorted lists into a single sorted list.

Solution: Suppose M [1 : m] and N [1 : n] are two lists whose elements are sorted in ascending order. L is the merged list of M and N which will contain $m + n$ elements obviously in ascending order. It is possible that every element of M can be interleaved with every element of N in all possible way in L or vice versa. From elementary combinatory we know that there are $\binom{m+2}{2}$ ways to merge M and N.

Suppose, $M = < a, b >$ and $N = < c, d, e >$ then there $\binom{2+3}{2} = \binom{5}{2} = 10$ ways to merge M and N. These are shown below.

(a) a, b, c, d, e
(b) a, c, b, d, e
(c) a, c, d, b, e
(d) a, c, d, e, b
(e) c, d, e, a, b
(f) c, a, b, d, e
(g) c, a, d, b, e
(h) c, a, d, e, b
(i) c, d, a, b, e
(j) c, d, a, e, b

Now if we use comparison tree as model of computation for merging M and N, then there are $\binom{m+n}{m}$ external nodes and at least $\left\lceil \log\binom{m+n}{m} \right\rceil$ comparisons on the tree that uses any comparison based algorithms. However, the conventional merging requires $(m + n - 1)$ comparisons. Therefore, $\left\lceil \log\binom{m+n}{m} \right\rceil \leq$ merge $(m, n) \leq m + n - 1$ where *merge* (m, n) is the minimum number of comparisons required to merge m elements with n elements.

If $m = 1$, we need fewer comparisons.

If $m = n$, we have optimal number of comparisons for conventional merging and the lower bound is thus too low. Thus, conclusion is that upper and lower bound differs by far apart as $m << n$. Therefore, merge $(m, n) = 2m - 1, m \geq 1$.

Exercises

1. Give trivial lower bounds for the following problems:
   (a) Finding the inverse of an $n \times n$ matrix.
   (b) Finding transpose of an $m \times n$ matrix.
   (c) Finding the median of n elements.
   (d) Deciding whether a given list $L[1..n]$ of n elements is sorted.

2. Draw the decision tree for the following algorithms.
   (*a*) Linear search on five elements.
   (*b*) Merge sort on three elements.

3. Let A and B be two unordered lists of n elements each. Find the number of comparisons required in deciding whether the elements in A are identical to those in B.

4. Determine the minimum number of comparisons needed to test whether an array A[1...n] is a heap.

5. Let S be a list of n unsorted elements.
   (a) Construct binary search tree taking elements in S.
   (b) Show that decision tree model requires $\Omega(n \log n)$ to do this.

6. Let $S = \{x_1, x_2 \ldots . x_n\}$ be a set of n positive integers. Propose an algorithm that finds an element x greater than the median when S is sorted. Find the minimum number of element comparisons required.

7. Let A [1...n] be an array of n integers in the range [1....m], $m > n$. propose an algorithm to find an integer x in the range $\notin A$. What is the minimum number of comparisons required to solve this problem?

8. Let P and Q are two sorted lists of n elements each. Find minimum number of comparisons to merge P and Q in the worst case. [**Ans.** $2n - 1$]

9. Give an algorithm to find both the largest and second largest of an unordered list of n elements. Your algorithm should perform the least number of element comparisons.

10. If n is a power of 2, find the least number of element comparisons required in finding both the largest and second largest from an unordered list of n elements. [**Ans.** $n - 2 + \log n$]

University Questions

1. Define lower bound of a problem. *[BPUT'05,'08]*
2. Write short notes on Lower bound theory. *[RTU'07, RTU'08]*
3. Establish the theoretical minimum lower bound of time complexity for any sorting algorithm where the sorting is performed by pair wise comparison. *[WBUT'07,'09]*
4. Prove that the lower bound of sorting a sequence of n elements using comparisons based sorting algorithm is $n \log n$. *[RTU'10, DOEACC'06]*
5. What is the difference between worst case lower bound and average case lower bound? Find out the lower bound of heap sort algorithm. *[BPUT'05]*

❑❑❑

10 Chapter — Disjoint Set Manipulation

10.1 INTRODUCTION

Some applications such as tree, forest, etc, involve grouping of elements into a collection of disjoint sets. On these disjoint sets two important operations namely 'find' and 'union' are applied. These are very important operations on these sets. Designing an efficient data structure for these union-find operations is the main purpose of this chapter. We look at various methods for maintaining the data structures that support these operations. Before going in details about disjoint set and its corresponding operations first we should have little overview on common sets.

10.2 SET

A set can be defined as a well-defined collection of distinct objects or elements. *e.g.*, a group of students in subject, a number of employees in a department, etc.

10.2.1 Notation of Sets

The following conventions are generally followed.

1. Capital letters or alphabets are usually used to denote a set. *viz.*, A, B, S, X, Y, etc.
2. The elements of sets are usually represented by lower case alphabets. *viz.*, a, b, c, x, y, ...

10.2.2 Finite Set

A set S is said to be finite if it contains specific or finite number of elements.

Example 1. The following sets are finite sets.

(*i*) $S = \{x : x \in N\, x < 100\}$

(*ii*) $X = \{x : x \in I, x > 0, x \text{ is a factor of } 15625\}$

10.2.3 Infinite Set

A set S is said to be infinite if it contains infinite number of elements.

Example 2. The following are examples of an infinite set.

(*i*) Set of all points on a line segment.

(*ii*) Set of all similar triangles in a plane.

(*iii*) Set of rational numbers between the integers 1 and 2.

(*iv*) $A = \{x : x \in I \text{ and } x > 2\}$.

(*v*) $S = \{x : x \text{ is an even number}\}$.

(*vi*) Stars in the universe.

10.2.4 Set: Cardinality

The number of distinct elements in a finite set S is called its cardinality. Cardinality of a set is denoted by $n(S)$.

Example 3. The cardinality of the following sets is

(*i*) If $S = \{0, 7, 1, 2, 4, 11\}$, then $n(S) = 6$.

(*ii*) If $A = \{1, 2, 3, 2, 4, 1\}$, then $n(A) = 4$

10.2.5 Set: Size

Size of a set S is defined as the total number of elements presents in that set S and is denoted by $| S |$.

Example 4. The size of the following sets is

(*i*) If $S = \{0, 7, 1, 2, 4, 11\}$, then $| S | = 6$.

(*ii*) If $A = \{1, 2, 3, 2, 4, 1\}$, then $| A | = 6$

10.2.6 Set: Representation

There are several ways to represent a set. *A* set *S* is generally represented by the following ways:

(*a*) Roster Method.

(*b*) Set Builder Method.

(*c*) Tree.

(*d*) Linked List.

10.2.6.1 *Roster Method*

This method is also known as tabular method or object representation. In this method, a set S is represented by listing it's all the elements. Each element is separated by a comma and all the elements are enclosed within flower brackets { }.

Example 5. A set S containing 1, 2, 3, 4, 5, 6, 7, 8 is represented by Roster method as $S = \{1, 2, 3, 4, 5, 6, 7, 8\}$.

10.2.6.2 *Set Builder Method*

In this method, a set S is represented by stating all the properties which are satisfied by the elements of S only. If S contains all values of 'x' for which the condition $P(x)$ is true, then write $A = \{x : P(x)\}$. Here, x is called an object. This method is also known as Property or Object method.

Example 6. If S is a set of all positive factors of 36, then using Set Builder Method it is written as $S = \{x : x \text{ is a factor of } 36, x \in N\}$

10.2.6.3 *Tree Representation*

Another way of representing a set S is use of tree. Any arbitrary element of S is considered as root of the tree and all other nodes are linked to this root or its descendents. Root points to itself. Consider first element of S as root of the tree and it will act as a representative of S. There is no formal way to represent S using tree method meaning that height as well as type of the tree for S both varies with requirements.

Example 7. Suppose a set $S = \{1, 2, 3, 4, 5\}$. Corresponding trees for S are given below.

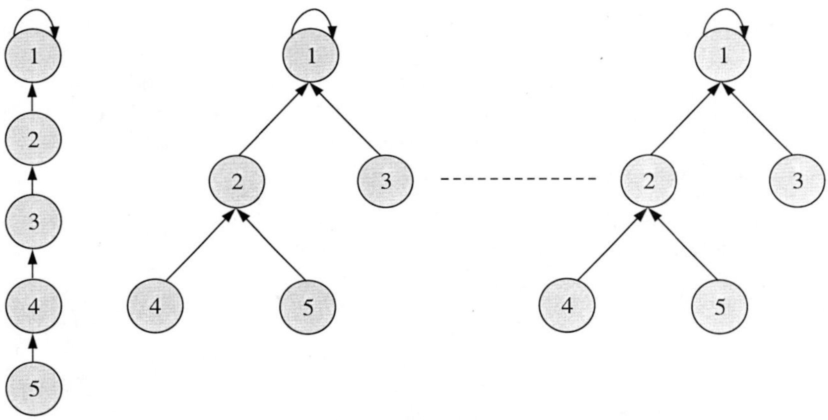

Figure 10.1 Tree representation of a set *S*.

10.2.6.4 *Linked List Representation*

Another simple way to represent a set is the linked list or simply list notation. The first element of the set serves as its representative. Each node in the liked list has three parts. Second part contains information *i.e.,* Set member and the first and third parts contain a pointer that links to representative and next node for representative and next set member or element respectively. Elements can appear in any order with the assumption that first element would be the representative in the list. For a set there may be more than one liked list. Each linked list for a set contains two extra pointer fields namely 'head' and 'tail'. The pointer 'head' and 'tail' points to the representative and the last element of the list respectively.

Example 8. Suppose the set $S = \{a, b, c, d\}$. Then one of the possible linked list representations of S will be as given below.

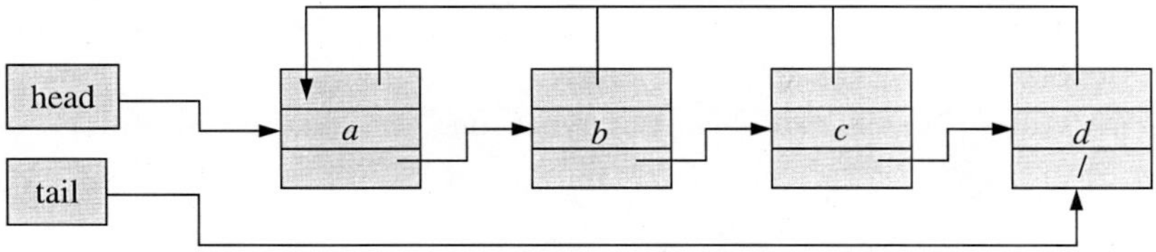

Figure 10.2 Linked list representation of set.

10.2.7 Set : Operations

There are mainly three operations that are normally performed on sets.

(*a*) *Make* (x): This function builds a new set in which x is an only member.

(*b*) *Union* (X, Y): This function unites two sets X and Y into a single set. Naturally, the new set being united contains all the elements of X and Y. Denote this operation as $X \cup Y$.

(*c*) *Find* (x): This function returns a pointer to the representative of a set which contains x. That means with this function, a set is found in which x belongs to *i.e.,* x is a member of which set.

10.3 DISJOINT SETS

Given a set of elements, it is often required and useful to break up them into a number of separate, non-overlapping sets. Non-overlapping means the partitioned sets should not have any common elements. A disjoint-set data structure is used for keeping track of such a partitioning.

Definition: Both in mathematics and computer science, two sets are said to be disjoint if they have no element in common. *e.g.,* {1, 2, 3} and {4, 5, 6} are disjoint sets. In other words, two sets A and B are said to be disjoint iff $A \cap B = \emptyset$.

This definition can be extended to any collection of sets $S = \{S_i : i \in I\}$. The family of sets S will be pair wise or mutually disjoint if any two sets S_i and S_j, $i \neq j$ and $i, \in I$ in the collection are disjoint. Obviously for each pair, $S_i \cap S_j = \emptyset$. E.g., the collection of sets {{1}, {2}, {3}, ...} is pair wise disjoint.

If S is a pair wise disjoint set containing at least two sets, then clearly its member intersection is empty that means $\cap_{i \in I} S_i = \emptyset$. However, the converse is not true *i.e.,* the intersection of the members of set say A = {{1, 2}, {2, 3}, {3, 1}} is empty, but the members are not pair wise disjoint. In fact, there are no two disjoint sets in A.

All the other terms like size, cardinality, representation, operations, etc. Are same for disjoint set as discussed earlier. Here we will discuss how the basic set operations are implementing in a suitable approach.

10.4 DISJOINT SETS: APPLICATIONS

Disjoint sets have a number utilisation in different aspects.

1. Disjoint-set data structures model the partitioning of a set. *e.g.,* this model keeps track of the connected components of an undirected graph and helps in determining belongingness of two vertices in these components. It also helps in finding cycle by adding an edge between them.
2. The Boost Graph Library uses this data structure to implement its Incremental Connected Components functionality.
3. Kruskal's algorithm can be put into practice for minimum spanning tree.

10.5 DISJOINT SETS: UNION OPERATION

The union operation unites two disjoint sets into a new set that contains all the elements of the sets under consideration. Being one member (preferably first one) in each set is considered as representative of that set then the united set should have one representative. Generally it is considered by anyone of the representatives of the sets under union operation. Union operation is preformed though the ways a set is represented.

10.5.1 Union Operation: Object Representation

The union operation requires at least two disjoint sets to complete the operation. After being united two sets say S_1 and S_2 form a new set say S_{12}. Generally all the elements of one set are first copied into S_{12}. The representative of the set S_{12} is determined based on the way the sets S_1 and S_2 are united meaning that if S_1 is united with S_2 then representative of S_2 will be the representative of S_{12} otherwise representative of S_1 will be the representative of S_{12}. If there are more than two disjoint sets are being united then continue the procedure considering sets one-by-one.

Example 9. Suppose $S_1 = \{1, 2, 3, 4\}$, $S_2 = \{5, 7, 9\}$, and $S_3 = \{6, 8, 10, 11, 12\}$ then

$$S_{12} = S_1 \cup S_2 = \{1, 2, 3, 4, 5, 7, 9\}.$$

$$S_{123} = S_{12} \cup S_3 = \{1, 2, 3, 4, 5, 7, 9, 6, 8, 10, 11, 12\}.$$

Hence the representative for the set S_{123} is 1 because S_3 is being united with S_{12} and 1 is its representative.

Algorithm of the above procedure is very simple and it is left to the reader.

10.5.2 Union Operation: Tree Representation

First represent a set by a tree. Every item of the set is in a tree. The root of the tree is the representative of the set. Union operation simply links a root of one tree of one set to the root of another tree of another set.

Consider two sets $S_1 = \{1, 2, 3, 4\}$ and $S_2 = \{5, 7, 6\}$. The corresponding trees are supposed as given below.

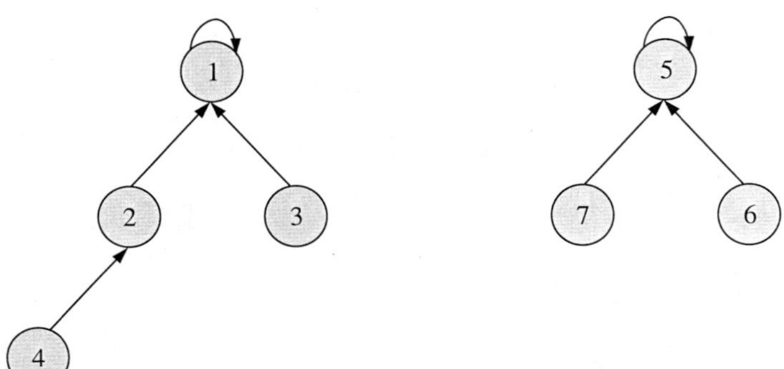

(*a*) Tree T_1 for S_1 (*b*) Tree T_2 for S_2

Figure 10.3 Tree for S_1 and S_2.

Now after union of the sets $S_1 \cup S_2$ i.e. $T_1 \cup T_2$, the united set rather the united tree is:

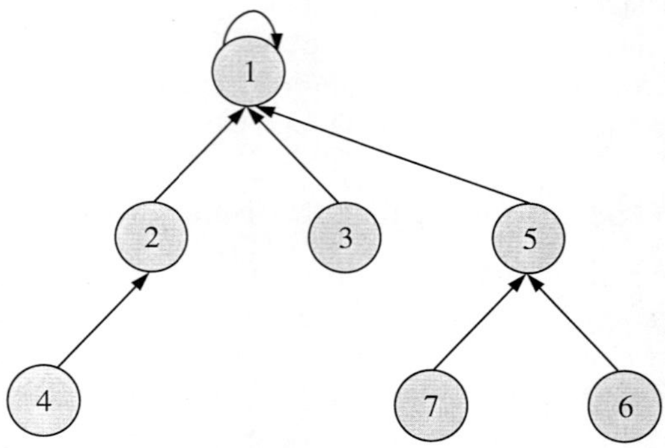

(a) $T_1 \cup T_2$

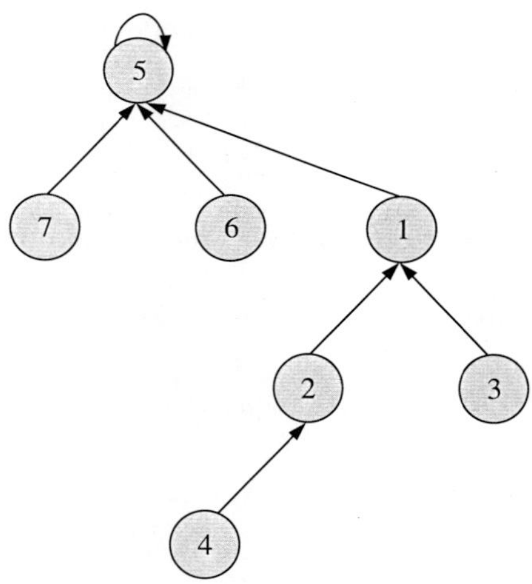

(b) $T_1 \cup T_2$

Figure 10.4 Union of S_1 and S_2 using tree.

In order to perform union operation, first make tree of the disjoint sets. Then construct an array $P[1 : n]$ where n is the total number of elements in set *i.e.,* sum of elements of disjoint sets. Assume that set elements are numbered 1 through n. The ith element of P represents the tree node that has a child i. This array element returns a parent pointer of the corresponding tree node. Since a root has no parent so represent it by –1.

Consider two sets $S_1 = \{1, 2, 3, 4\}$ and $S_2 = \{5, 7, 6\}$ and their corresponding trees from Figure 10. 3. Then P will be as:

i	1	2	3	4	5	6	7
P	–1	1	1	2	–1	5	5

Figure 10.5 Array representation of S_1 and S_2.

If we do union operation on S_1 and S_2 using Figure 10.4a then P will be as:

i	1	2	3	4	5	6	7
P	–1	1	1	2	1	5	5

Figure 10.6 Array representation of union of S_1 and S_2.

10.5.3 Union Operation: Algorithm Using Tree Representation

An algorithm for union operation of the disjoint sets with the help of their tree representations can be written as follows.

Algorithm: Union_Set_Tree ( )

Terms Used: T_i = the *i*th tree representing set S_i.
P = the array containing the parent node of the elements.

Step 1:

(*i*) Input *ns*, the number of disjoint sets. // Construct disjoint sets.
(*ii*) Repeat for $i = 1$ to *ns*
 (*a*) Input *ne*, size of set S_i
 (*b*) Repeat for $j = 1$ to *ne*
  • Add element to S_i.

Step 2: Repeat for $i = 1$ to *ns* // Construct trees
 (*a*) Construct tree T_i for S_i

Step 3:

(*i*) Repeat for all trees T_i // Construct P
(*ii*) Repeat for $j = 1$ to $|T_i|$
 (*a*) Find $pn = j \rightarrow$ parent
 (*b*) $P[j] = pn$
 (*c*) If j = root
  • $P[j] = -1$

Step 4:

(*i*) $k = 1$ // Perform union operation
(*ii*) Repeat for all trees T_i
(*iii*) If i and j be the roots of T_i and T_j respectively
 (*a*) $a[j] = i$ // T_j is united with T_i

Step 5: Repeat for $i = 1$ to $|P|$

(*a*) Output $P[i] \rightarrow$ child

Step 6: Exit

Analysis: From the above algorithm it is easily seen that the algorithm runs in $\Theta(n^2)$.

10.5.4 Union Operation: Weighting Rule

From union operations of sets using tree representation, we have seen two trees can be united in two ways. If the number of trees is more than two then there are so many ways possible in which they can be united. Most of them are alike in nature. In order to avoid the confusion that which combination is to be chosen and facilitate the union operation it is better to find the weight of individual trees. Based on the weights, unite the trees. The rule practised is called weighting rule for union. It states that:

> *If the number of nodes in a tree with x as root node is greater than the number of nodes in another tree with y as root node then make x as a parent node of y otherwise make y as a parent node of x.*

In a single sentence a tree with lower size is linked to the tree with higher size.

Example 10. Consider two sets $S_1 = \{1, 2, 3, 4\}$ and $S_2 = \{5, 7, 6\}$. The corresponding trees with union operation using weighting rule are as given below.

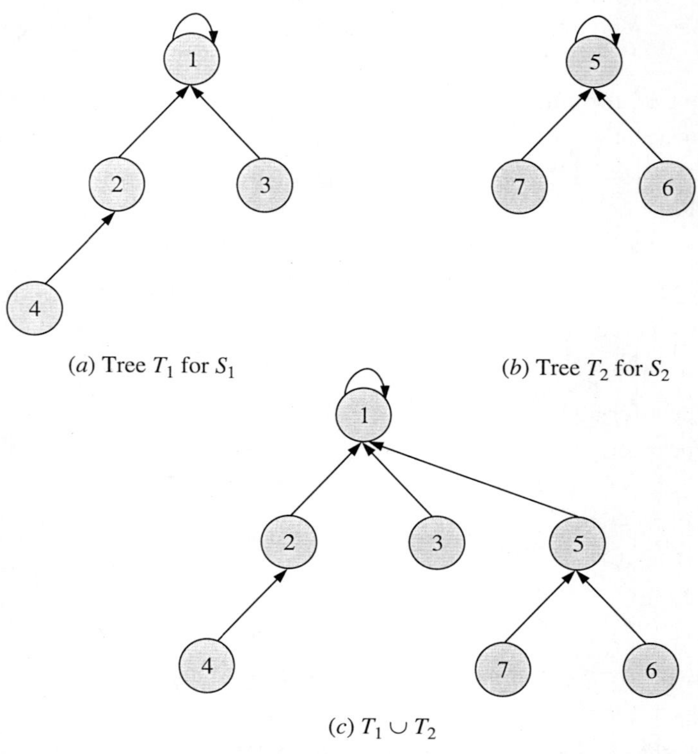

(*a*) Tree T_1 for S_1 (*b*) Tree T_2 for S_2

(*c*) $T_1 \cup T_2$

Figure 10.7 Union of sets with tree representation using weighting rule.

It is seen that T_1 has more number of nodes than T_2 meaning that according to the weighting rule root of T_2 will be linked to root of T_1.

10.5.5 Union Operation: Linked List Representation

First every set is represented in linked list form. Then union operation on sets appends one list of a set to another list of another set. Tail pointer of a list indicates the position where the second list is to be appended. Head of the second list is now linked to the head of the first list. Subsequently, head pointer of the first list and tail pointer of the second list becomes head pointer and tail pointer of the united list respectively. Obviously, the representative of the united list is the representative of the first list. But major drawback of the approach is that updating of every pointer of the second list is required. This updating takes place in linear time complexity and varies with the number of nodes in the list. Now union of two sets requires $O(1)$ time then for n sets it is $O(n - 1)$ time. Again every time ith union operation updates i elements. As a result, total time requires is $1 + 2 + \dots\dots\dots + n - 1 = \Theta(n^2)$.

Example 11. Consider two sets $S_1 = \{1, 2, 3, 4\}$ and $S_2 = \{5, 7, 6\}$. The linked lists and the union of these lists are shown below:

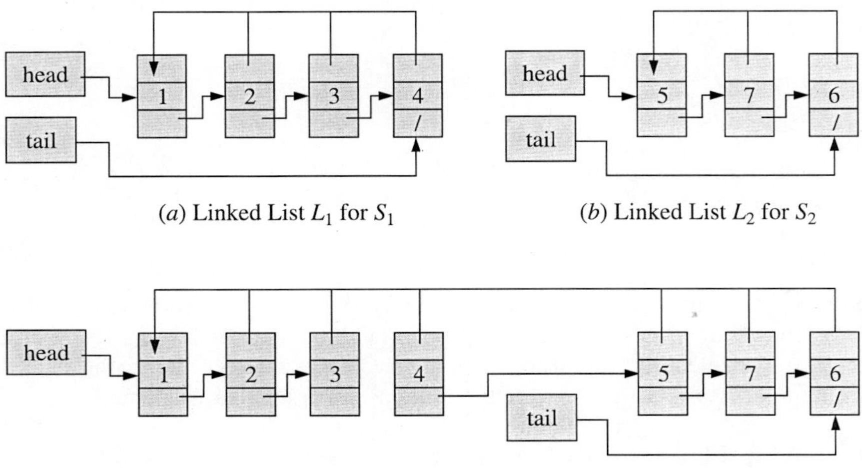

(*a*) Linked List L_1 for S_1 (*b*) Linked List L_2 for S_2

(*c*) $L_2 \cup S_2$

Figure 10.8 LL Representation of sets S_1 and S_2 with union of L_1 and L_2

10.6 DISJOINT SETS: FIND OPERATION

So far we have discussed about union operations of disjoint sets in details. Now our focus will be on find operation on these disjoint sets. It is discussed in context of the same approaches of representation as considered with union operation.

10.6.1 Find Operation: Object Representation

Find (x) operation on a set represented using Roster method returns the set that contains x. It is also helpful for decisive if two elements are in the same set.

Example 12. Suppose $S_1 = \{1, 2, 3, 4, 15\}$, $S_2 = \{5, 7, 9, 13, 14\}$, and $S_3 = \{6, 8, 10, 11, 12\}$ are three disjoint sets. Then Find (2) = S_1, Find (9) = S_2, Find (10) = S_3, and so on.

Now, the algorithm for find operation in this approach is described below.

Algorithm: Find_Set_Object (x)

Terms Used: x = The element to be found.

n = Number of sets under consideration.

ns = List of set number of the sets that contain x.

Step 1:

(*i*) Input x. // input values

(*ii*) Input n.

(*iii*) Set $m = 1$.

Step 2:

(*i*) Repeat for $i = 1$ to n // start finding

(*ii*) Repeat for $j = 1$ to $|S_i|$

(*a*) If $x = S_i [j]$

- Output x is in set S_i.
- Exit.

(*iii*) $ns[m{+}{+}] = i$

Step 3: Repeat for $i = 1$ to $| ns |$

(*a*) Output $ns[i]$

Step 4: Exit.

Analysis: From the algorithm it can be easily said that the time complexity is due to step 2. Outer loop of this step executes n times and the inner loop executes according to number of elements in current set S_i and this size varies set to set. Hence the algorithm runs in $\Theta(n| S_i |)$ *i.e.*, $\Omega(n^2)$ if every set contains equal number of elements n.

10.6.2 Find Operation: Tree Representation

The operation Find (x) on a disjoint set using tree representation helps us to determine the root of the tree for the set containing this x. A tree T_i represents a set S_i. In the operation, starting with the element j to be found, the function Find(j) tries to reach the root of the tree. If we reach –1 then the root node of the tree is reached. It is then concluded that the elements $j \in T_j$ *i.e.*, S_j. To reach –1 the array P is used.

Algorithm: Find_Set_Tree (x)

Terms Used: x = The key to be found.

S_i = Ith set.

T_i = Tree of ith set.

P = The array for all set/tree elements.

ns = Number of set.

ne = Size of S_i

pn = Parent of a node

Step 1:

(*i*) Input *ns*. // build sets

(*ii*) Repeat for $i = 1$ to *ns*

(*a*) Input ne for S_i

(*b*) Repeat for $j = 1$ to *ne*

• Add elements to S_i

Step 2: Repeat for $i = 1$ to *ns* // construct trees

(*a*) Construct T_i for S_i

Step 3:

(*i*) Repeat for $i = 1$ to *ns i.e.,* for each T_i // build *P*

(*ii*) Repeat $j = 1$ to $| T_i |$

(*a*) $pn = j \rightarrow$ parent

(*b*) $P[j] = pn$

(*c*) If j = root

• $P[j] = -1$

Step 4: Input *x*. // input key

Step 5:

(*i*) $i = x$

(*ii*) Repeat while $P\ [x] > 0$

(*a*) $y = P[x]$

(*b*) $x = y$

(*iii*) $x = i$

(*iv*) $y = abs(y)$

Step 6: Output *x* is present in set *y*.

Step 7: Exit.

Analysis: Each find operation requires a sequence from the element to be found upto the root. The time required to find an element at level *i* in a tree is $O(i)$. Thus, the total time complexity for the algorithm is $O\left(\sum_{i=1}^{n} i\right) = O(n^2)$.

10.6.3 Find Operation: Linked List Representation

With the linked list representation of disjoint sets, the find operations can be performed very easily. The operation requires $O(1)$ time because simply it returns a pointer to the node as the representative of the linked list *i.e.,* Of the set. First make linked lists of the sets. Then take any element *x* to search. The function say Find_Set_*LL*(*x*) simply returns the pointer from *x* to the representative of a linked list which indicates respective set. The algorithm is very simple and similar previous one. It is left as an exercise to the readers.

10.7 HEURISTICS TO IMPROVE RUNNING TIME OF SET OPERATIONS

The list-based data structure for disjoint set operations-union and find provides the pretty fast running time. So far it is not improved. Let's think of ways how it could be make even faster. By considering two heuristics a satisfactory running time that is almost linear can be achieved. These are namely *union by rank* and *find by path compression.*

10.7.1 Union by Rank

The first heuristic union by rank is similar to union by weighting rule. First find the rank of every node. *Rank of a node x is the longest path (maximum number of edges) from this node x to a descendant leaf in a tree. e.g.,* the following tree represents rank of every node.

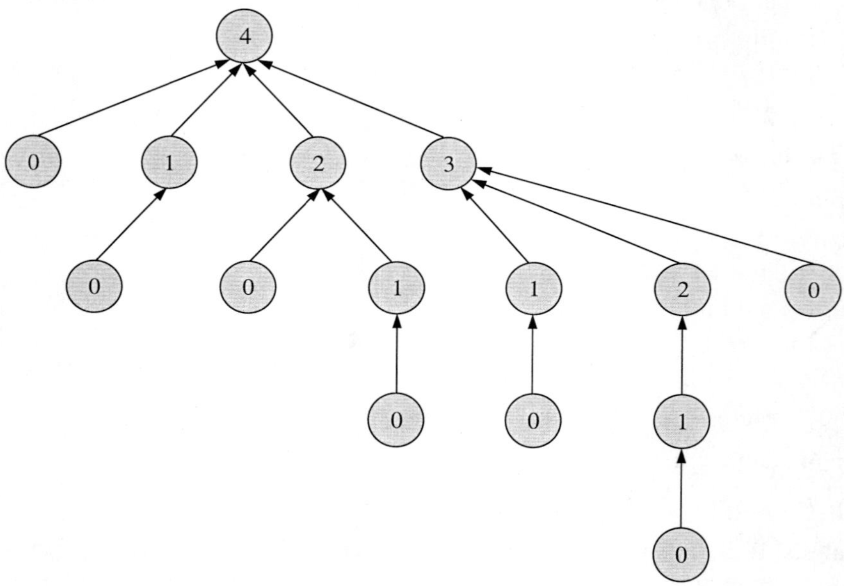

Figure 10.9 The tree with root of rank 4.

10.7.1.1 *Properties of Ranks*

To help us easy understand of the above procedure, at first some properties of ranks should be featured.

1. The rank of a node is the same as the height of its subtree without path compression.
2. Rank of x is strictly less than the rank of its parent iff x is not a root.
3. The rank of a node x can only be changed if x is a root.
4. A root node can never be root again once it becomes a non-root.
5. There are at most $n/2^r$ nodes of rank $\geq r$.

10.7.1.2 *Union by Rank: Algorithm*

The union operation by rank behaves as follows.

If one of the roots has larger rank than the other, then first one becomes the new root and the other root becomes its child in united tree. If the two roots have equal rank, then one of them arbitrarily is picked to be the new root and its rank is increased by 1. This procedure can now be converted into pseudocode as follows:

Algorithm: Union_By_Rank ()

Terms Used: x = Representative of set containing x.

y = Representative of another set containing y.

P = The array representation of trees.

Step 1: If rank (x) > rank (y)

(a) $P[y] = x$

Else

(a) $P[x] = y$

Step 2: If rank (x) = rank (y)

(a) $P[y] = x$

(b) rank (x) = rank (y) + 1

Step 3: Exit

Analysis: The running time of this union by rank algorithm depends on the find operation on the set *i.e.,* on corresponding tree. Since path compression heuristics are in use so the time complexity of this procedure will be same as the running time of find operation using path compression which is $O(\log n)$ where n is the total number of nodes in a tree.

10.7.2 Find by Path Compression

The second heuristic namely path compression used for find operation on set is also a very simple and effective one. We use a function Find_Set_PathCompression (x) to find a node x in a tree. The Find_Set_PathCompression (x) traces the path from x to the root of the tree. Let r be the root of the tree, and the path from x to r is $x\ x_1\ x_2\ x_3 \ldots x_k\ r$. Also make r linked directly to all the parent pointers of $x, x_1, x_2, \ldots, x_k$ *i.e.,* the find path nodes pointing to directly to this root r. This idea is called **path compression**. It is expected that in some *sequence for* Find-Set operation, the running time is expected to be faster. The Figure 10.10 given below describes the approach.

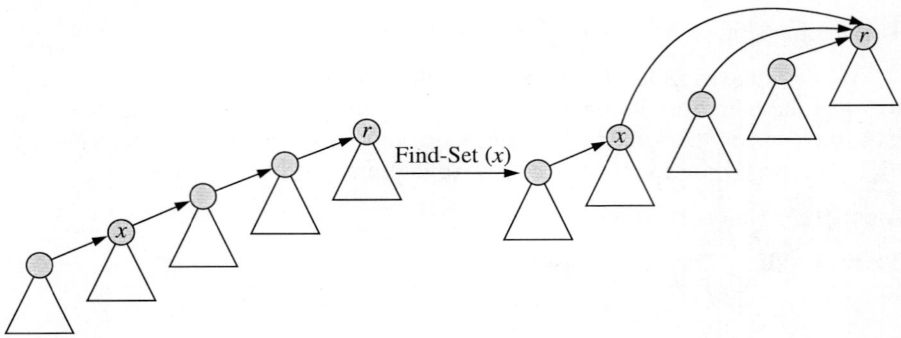

Figure 10.10 Illustration of path compression.

Algorithm: Find_Set_PathCompression (x)

Terms Used: x = the node to be found.

Step 1: If $x \neq P[x]$ // if x is not root.

(a) $P[x]$ = Call Find_Set_PathCompression ($P[x]$)

Step 2: Return $P[x]$

Step 3: Exit.

Analysis: Since the running time of Find_Set_Path Compression (x) algorithm to find a node x depends on the height h of the tree. Now for any tree, the total number nodes n iss given by the following inequality.

$$n \geq 2^h$$

$$\Rightarrow \quad h \geq \log n$$

$$\Rightarrow \quad h = O(\log n)$$

Therefore, Find_Set_Path Compression (x) algorithm runs in $O(\log n)$ time.

Exercises

1. What are the ranks of nodes 3, 7 and 5 in Figure10.11?

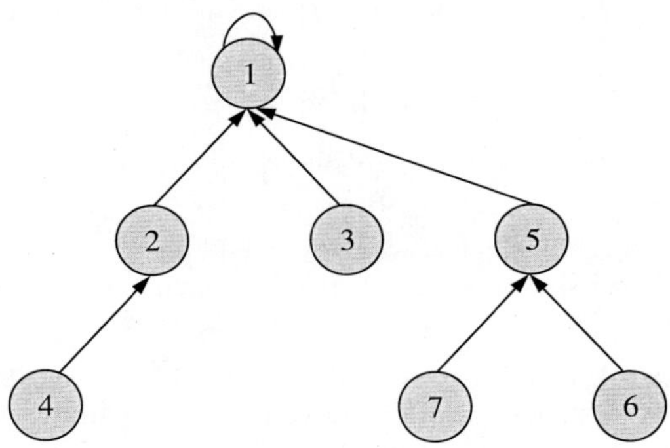

(*a*)

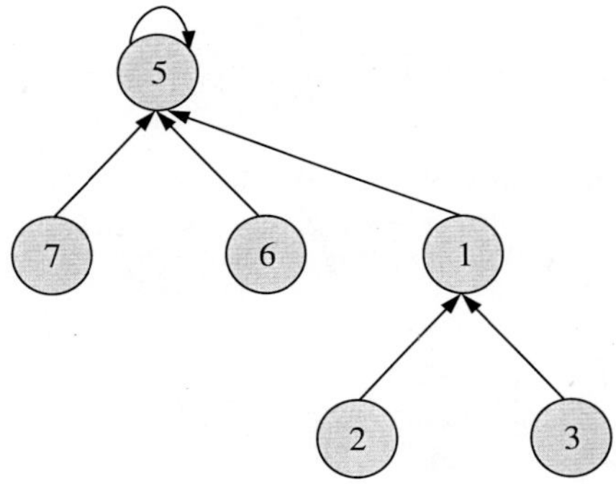

(*b*)

Figure 10.11 Trees for finding rank.

2. Let {1},{2}, {3}, ..., {8} be 8 singleton sets. Each set is represented by a tree with exactly one node. Use the union-find algorithms to find the tree representation of the set resulting from each of the followings:

   (*a*) union (1, 2)
   (*b*) union (3, 4)
   (*c*) union (5, 6)
   (*d*) union (7, 8)
   (*e*) union (1, 3)
   (*f*) union (5, 7)
   (*g*) find (1)
   (*h*) find (5)
   (*i*) find (7).

3. Do exercise 10.2 with union by rank and path compression.

4. Let x be a node in tree T resulting from a sequence of unions and finds using both the heuristics of union by rank and path compression. Prove that rank (x) is an upper bound on the height of x.

5. In a union-find all the unions occur before the finds. Let n be a sequence of union and finds. Show that, if both the heuristics-union by rank and path compression are used, the running time is linear.

6. Solve Exercise 10.2 using the weighting rule and path compression.

7. Prove that the weight-balancing or weighting rule guaran-tees $O(\log n)$ is the height of the resulting tree.

8. Let x and y be the root and a leaf in a tree T resulting from a sequence of unions and finds using the heuristics of union by rank and path compression. The ranks of the nodes on the path from y to x form a *strictly* increasing sequence. Justify.

9. Suppose two sets S_1 and S_2 represented by linked lists L_1 and L_2 respectively are to be merged. If the first element in L_1 is used as the name of the resulting set, then the pointer to the set name at each element in L_2 must be changed so that it points to the first element in L_1.

   (*a*) How do you improve this representation so that each *find* operation takes $O(1)$ time.

   (*b*) Find out the total cost of performing $n - 1$ *unions* in the worst case.

   [**Ans.** $\Theta(n^2)$]

10. When performing the union of two sets, the first element in the set with a larger number of elements is always chosen as the name of the new set, then compute the total cost of performing $n - 1$ unions. [**Ans.** $O(n \log n)$]

University Questions

1. What is union-find algorithm? *[WBUT'07]*
2. What are Union Find Problems? *[PTU'07]*
3. Explain the algorithm for Fast disjoint set union algorithm. *[PTU'07]*
4. Compute the connected components of a graph $G = (V, E)$ where $V = \{a, b, c, d, e, f, g, h, i, j, k\}$ and edges E are preprocessed in the following order: (d, i), (f, k), (g, i), (b, g), (a, h), (i, j), (d, k), (b, j), (d, f), (g, j), (a, e), (i, d). Use disjoint set data structure operations. *[IETE ALCCS'11]*

11

Chapter

Graph Traversals

11.1 INTRODUCTION

Graph Theory has gained an impulsion in the past few years in rapacious dimensions. Simultaneously, it has emerged as one of the most powerful tools for the illustration and solution of the problems that are basically chronological in nature. And many seemingly diverse problems in computer science such as finding shortest paths or minimum spanning trees have been worked out with the help of this thought.

In mathematics and computer science, graph theory is generally known as the study of graphs that models pair wise relations between objects from a set of objects. The objects are vertices (nodes) and their relations (connections/links) between every pair are edges. Diagrammatically a graph is depicted in the form of a set of dots, circles as vertices and a set of lines or curves as edges that link each pair. The Figure 11.1 gives the little bit idea of the above description.

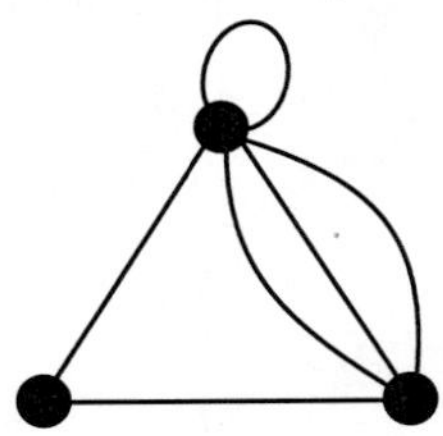

Figure 11.1 A typical graph.

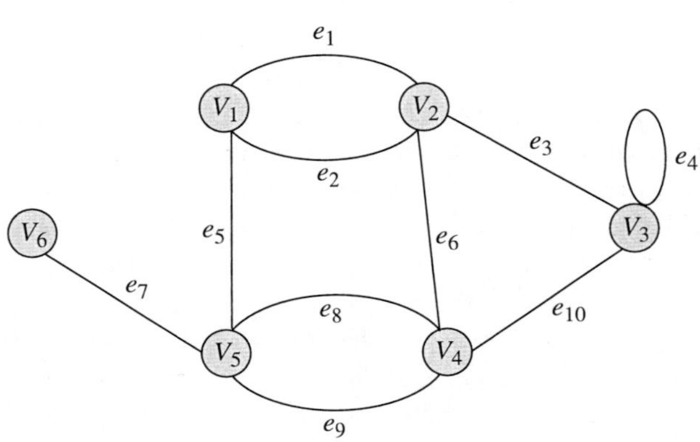

Figure 11. 2 An example of a graph.

Every vertices and edges may labelled with some well known alphabets and/or numeric values. The graph itself is named as G. The vertices and edges sets are represented by V and E respectively. A graph G is represented as $G = (V, E)$.

Definition: A linear graph or simply graph $G = (V, E)$ can be defined as a sets of $V = (v_1, v_2 \dots v_n)$ called vertices and $E = (e_1, e_2 \dots e_m)$ called edges. Here every edge $e_k \in E$ is an unordered pair (v_i, v_j) each v_i and $v_j \in V$.

11.2 GRAPH: BASIC TERMINOLOGIES

In this section we will discuss some basic terminologies related to graphs.

1. *Vertex:* The node in a graph is called vertex. E.g., $v_1, v_2 \dots v_6$ in Figure 11.2 are the vertices. They are also called ends, endpoints.
2. *Edge:* An edge e_k in a graph is a unique pair of vertices (v_i, v_j) connecting them in the graph. In Figure 11.2 $e_1 = (v_1, v_2)$ or (v_2, v_1) is the edge. An edge may be an ordered or unordered pair. Similarly $e_2, e_3 \dots$, are the edges.
3. *Digraph:* A graph $G = (V, E)$ is said to be a directed graph or digraph if V is a set of vertices and E is a set of ordered pair of vertices from V. Figure 11.3 is a digraph. In the figure $V = \{v_1, v_2, v_3\}$ and $E = \{e_1, e_2, e_3, e_4\} = \{(v_1, v_2), (v_2, v_1), (v_3, v_1), (v_3, v_2)\}$, the ordered pairs edges set. Ordered pair means (v_1, v_2) is not same as (v_2, v_1). An edge in a digraph is represented by an arrow which starts from a vertex and ends at another or same vertex.

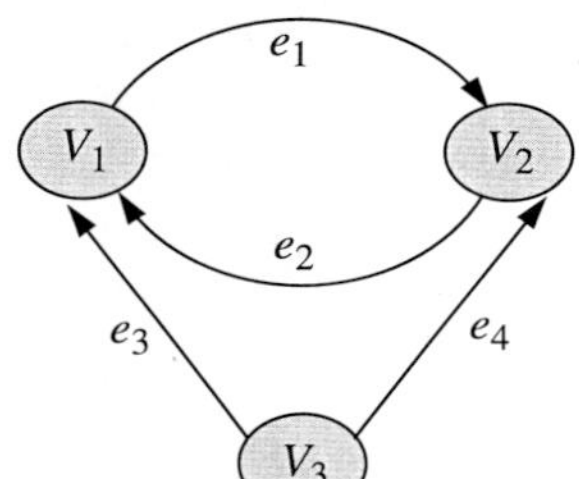

Figure 11.3 A digraph.

4. *Initial Vertex:* The vertex from which an edge is incident from in a graph is called initial vertex. *e.g.*, v_1 is the initial vertex for the edge e_1 in Figure 11.3.
5. *Terminal Vertex:* The vertex on which an edge is incident in a graph is called terminal vertex. *e.g.*, v_1 is the terminal vertex for the edge e_2 in Figure 11.3.
6. *Adjacent Vertex:* When two vertices are connected by an edge then they are said to be adjacent to each other. *e.g.*, in Figure 11.2 the vertices v_5 and v_6 are adjacent to each other as they are linked with the edge e_7.
7. *Self Loop:* If an edge in a graph is incident out and incident on the same vertex the edge is called self loop. *e.g.*, the edge e_4 is a self loop in Figure 11.2.
8. *Parallel Edge:* If two or more edges have originated from a vertex and incident on another vertex then edges are said to be parallel edges. Parallel edges have common pair of vertices. *e.g.*, in Figure 11.2 the edges pairs (e_1, e_2), (e_8, e_9) are parallel edges separately. But the edges (e_1, e_2) in Figure 11.3 are not parallel edges.

9. *Simple Graph:* A graph without any self loop or parallel edge is called a simple graph. *e.g.*, the digraph in Figure 11.3 is simple digraph.
10. *Complete Graph:* A graph is said to be complete graph if all the vertices in the graph are adjacent to each other.
11. *Connected Graph:* If there is a path between any pair of vertices in a graph then the graph is called a connected graph. *e.g.*, Figure 11.2 is a connected graph.
12. *Path:* A path in graph is a sequence of edges such that no vertices appear more than once in the sequence. *e.g.*, $e_3 - e_1$ forms a path in Figure 11.3.
13. *Cycle:* The path containing two or more edges starting and ending at the same vertex is called cycle of the graph. *e.g.*, the edges (e_1, e_2) in Figure 11.3 form a cycle.
14. *Cyclic Graph:* A graph containing at least one cycle is called cyclic graph otherwise an acyclic graph. Both the graphs in Figure 11.2 and in Figure 11.3 are cyclic graph.
15. *Sub-graph:* a graph g = (V', E') is said to be sub graph of a graph $G = (V, E)$ iff $E' \in E$ and $V' \in V$. Similarly E' and V' are said sub set of E and V respectively.
16. *Finite Graph:* A graph with finite number of vertices and edges is called a finite graph. e.g., Figure 11.2 and 11.3 are finite graphs.
17. *Degree of Vertex:* Number of edges incident on a vertex in a graph is called degree of the vertex. It is denoted for a vertex v_i as degree (v_i). *e.g.*, in Figure 11.2 degree $(v_1) = 3$. If the vertex has a self loop then it is counted twice. In Figure 11.2 degree $(v_3) = 4$.
18. *Indegree:* Number of edges incident on a vertex of a digraph is called its indegree. It is denoted for a vertex v_i as $d^-(v_i)$. *e.g.*, in Figure 11.3 $d^-(v_1) = 2$.
19. *Outdegree:* Number of edges incident out of a vertex of a digraph is called its outdegree. It is denoted for a vertex v_i as $d^+(v_i)$. Figure 11.3 $d^+(v_1) = 1$.
20. *Pendant Vertex:* A vertex with degree 1 is called pendant vertex. *e.g.*, in Figure 11.2 v_6 is a pendant vertex.

Unless stated, a graph means "undirected simple finite graph".

11.3 GRAPH: CLASSIFICATIONS

In different contexts a graph with different degrees of generalization is defined. Whenever it is necessary a strict distinction among the graphs is made. The following terms are generally used.

1. Undirected graph.
2. Directed graph.
3. Mixed graph.
4. Multi-graph.
5. Simple graph.
6. Weighted graph.
7. Regular graph.
8. Complete graph.

9. Finite graph.
10. Infinite graph.
11. Connected graph.

11.4 GRAPH: PROPERTIES

Followings are some essential properties of a graph.

1. Two edges of a graph are called adjacent or coincident if they share a common vertex.
2. Two arrows of a digraph are called consecutive if the head of the first one is at the nock of the second one.
3. Two vertices are called adjacent if they share a common edge.
4. The graph with only one vertex and no edges is called the trivial graph.
5. In a weighted graph, each edge is associated with some value called its cost/ weight.

11.5 GRAPH: REPRESENTATIONS

A graph can be represented in many ways using some standard methods like:

1. Set Representation.
2. Sequential or Matrix Representation.
   (*i*) Adjacency Matrix.
   (*ii*) Incidence Matrix.
   (*iii*) Distance Matrix.
   (*iv*) Laplacian Matrix.
3. Linked List Representation.
   (*i*) Adjacency List.
   (*ii*) Incidence List.

Here, matrix and linked list representations of the graphs briefly are discussed.

11.5.1 Adjacency Matrix

An adjacency matrix is a 2-D matrix that represents which vertices are adjacent to which other vertices in a graph. It is an $n \times n$ matrix of a finite graph $G = (V, E)$ with $|V| = n$ vertices. The matrix is denoted by Adj $(G)_{ij}$ and is defined as:

$$\text{Adj } (G)_{ij} = \begin{cases} 1, \text{ if there is an edge from vertex } i \text{ to } j. \\ 2, \text{ if there is a self loop } (i = j). \\ 0, \text{ if there is no edge from vertex } i \text{ to } j. \end{cases}$$

Summing the 1s in any row gives the degree of that vertex of the graph. Similarly, summing the 1s in any row and column gives the indegree and outdegree of that vertex of a digraph.

There exists a unique adjacency matrix for each graph meaning that adjacency matrix of one will not be same with that of the other graph. In the graph is finite, simple one, the adjacency matrix is a (0, 1) matrix with zeros on its diagonal. It is sometimes then referred as binary matrix. If the graph is undirected, the adjacency matrix is symmetric. The space required by this matrix is $O(n^2)$.

Example 1. Consider the graph G and its adjacency matrix in Figure 11.4.

	V_1	V_2	V_3	V_4
V_1	0	1	1	1
V_2	1	0	0	0
V_3	1	0	0	1
V_4	1	0	1	0

(a) Adjacency Matrix

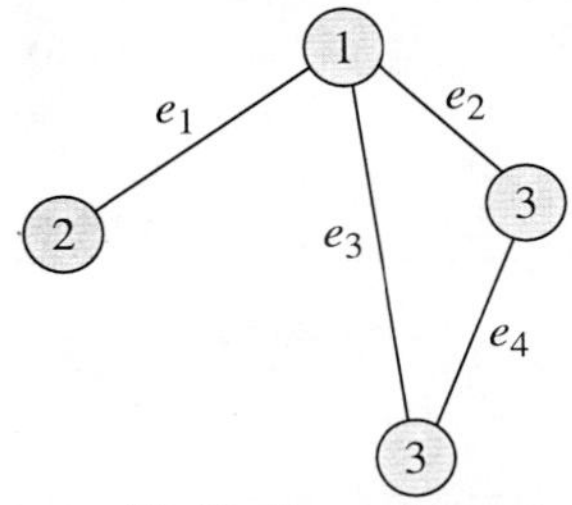

(b) Undirected Graph

Figure 11.4 An undirected graph and its adjacency matrix.

11.5.2 Adjacency List

Adjacency List is a linked list representation of a graph. It is aimed to save space. An adjacency list is the representation of all edges of a graph as a list alongwith the vertices. The list consists of nodes. Every node has two fields. First field contains a vertex. Second field contains the edge link to the next adjacent vertices. If the graph is weighted one the first field is subdivided into two parts. First sub-part contains the vertex itself. Second subpart contains weight of the adjacent edge. Typically, adjacency lists are unordered.

Vertes	Adjacent vertices links

(a) For unweighted graph.

Vertex	Weight of the linked edge	Adjacent vertices links

(*b*) For weighted graph.

Figure 11.5 General node structure for LL representation of graph.

Example 2. Consider the following digraph in Figure 11.6.

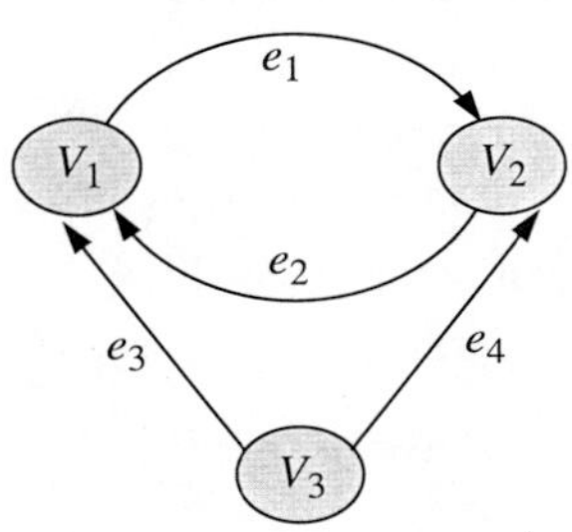

Figure 11.6 A digraph.

Now construct the following table for better understanding of adjacency list before constructing the actual adjacency list.

V_1	Adjacent to	V_2
V_2	Adjacent to	V_1
V_3	Adjacent to	V_1, V_2

The adjacent list will be as given below.

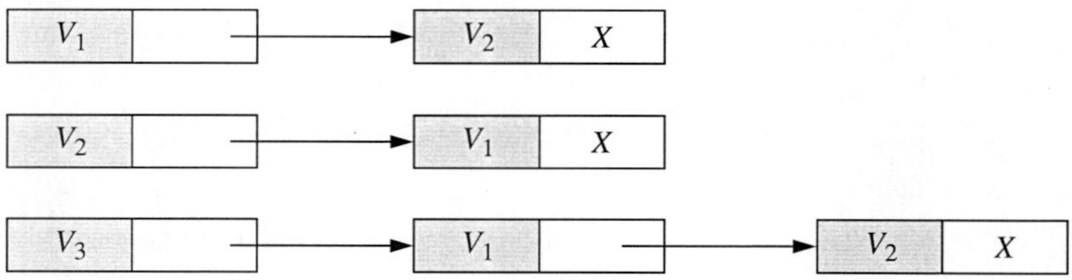

or

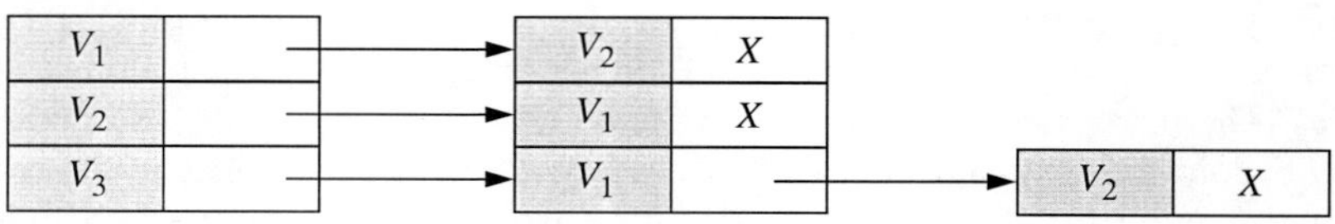

Figure 11.7 Adjacent list for the digraph in Figure 11.6.

11.6 GRAPH: APPLICATIONS

Applications of graph theory are primarily concerned with labelled graphs and various specialisations of these like.

1. *Enumeration Problem:* Large literature on graphical enumeration for the problem of counting graphs.
2. *Sub-graph Problem:* Subgraph isomorphism problem, clique problem, etc.
3. *Graph Colouring Problem:* Different kinds of colouring problems such as:
   (*i*) The four colour theorem.
   (*ii*) The strong perfect graph theorem.
   (*iii*) The total colouring conjecture (unsolved).
   (*iv*) The list colouring conjecture (unsolved).
   (*v*) The Hadwiger conjecture (graph theory) (unsolved).
4. Route Problems–these are mainly
   (*i*) Hamiltonian path and cycle problems.
   (*ii*) Minimum spanning tree.
   (*iii*) Route inspection problem.
   (*iv*) Chinese Postman Problem.
   (*v*) Seven Bridges of Königsberg.

(*vi*) Shortest path problem.

(*vii*) Steiner tree.

(*viii*) Three-cottage problem.

(*ix*) Travelling salesman problem (NP-complete).

5. *Network Flow Problems:* various notions of flows in networks *e.g.,* Max flow min cut theorem.
6. *Visibility Graph Problems:* Museum guard problem, etc.
7. *Covering Problems:* Set cover problem, Vertex cover problem.
8. Utilities Problem.
9. Electrical Network Problem.
10. Transportation Problem.
11. Shortest Path Problem.
12. Topological Sorting.
13. Binary decision Problem.
14. Seating Problem.
15. Program Flow Analysis.

11.7 GRAPH: TRAVERSALS

In finding shortest paths or minimum spanning trees, the vertices and edges are visited in a systematic order. However, the order of visiting the vertices may be unimportant. Graph traversal means visiting all the nodes in a graph in some particular manner. Tree traversal is a special case of graph traversal. Two types of traversal algorithms namely Breadth First Search (BFS) and Depth First Search (DFS) have been considered here.

11.7.1 Breadth First Search

BFS traverses a connected component of a given graph followed by defining a spanning tree. BFS label each vertex by the minimum number of edges from the start vertex. The tree thus formed is called BFS-tree.

The Breadth First Search algorithm is mainly used to solve following problems.

(*i*) Single source shortest path problem.

(*ii*) Testing whether graph is connected.

(*iii*) Computing a spanning forest of graph.

(*iv*) Computing a cycle in graph or reporting that no such cycle exists, etc.

Suppose $G = (V, E)$ be a graph and $S \in V$ is a starting vertex of this traversal. First visit this vertex S. BFS systematically traverse edges of G to explore the vertices those are reachable from S. So now explore this S. Next examine all the neighbours to S one by one. Then traverse all the neighbours of S as it is traversed. The process is continued. The linear data structure queue Q is used to keep status of traversing the neighbours. The following algorithm describes this procedure.

Algorithm: BFS(S)

Terms Used: S : starting vertex.

$G = (V, E)$, the given graph.

Q = the queue to store vertices and their neighbours.

X = the list of visited vertices.

Step 1: Push the source vertex S into Q.

Step 2:

(*i*) Delete S from Q.

(*ii*) Add the deleted vertex as visited vertex into X.

Step 3:

(*i*) Find the adjacent vertices for the most recently visited vertex.

(*ii*) Add these adjacent vertices into Q if they are unvisited and not in Q.

Step 4: If Q is not empty

(*i*) Set S = the vertex at the front of Q.

(*ii*) Go to step 2.

Step 5: Output X.

Step 6: Exit.

Analysis: We are visiting every vertex of the graph and for a vertex we are finding its neighbours *i.e.*, the linked edges. Therefore, total running time of BFS is $O(|V| + |E|)$.

Example 3. Find the BFS result on the following graph with A as starting vertex.

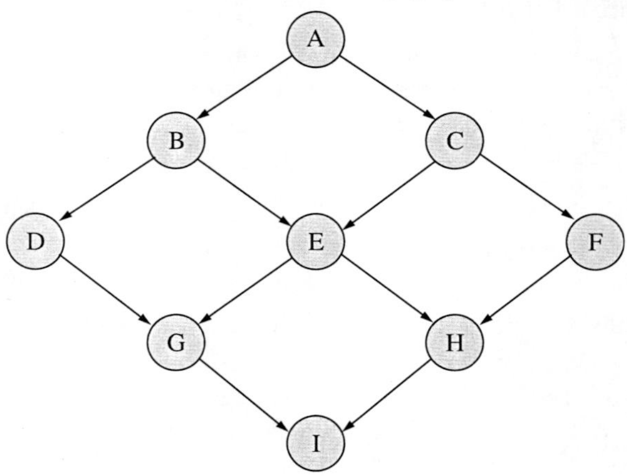

Figure 11.8 A digraph for BFS.

Solution: First construct adjacent list table for the graph in Figure 11.8 as:

Vertex	*Adjacent Vertices*
A	*B, C*
B	*D, E*
C	*E, F*
D	*G*
E	*G, H*
F	*H*
G	*I*
H	*I*
I	Ø

Here the starting vertex $S = A$. Initially the queue $Q = \{Ø\}$ and the visited or traversed vertices list $X = \{Ø\}$. Using the above table as reference for having neighbours of a node during traversing, this graph is traversed step-by-step as follows.

1. Add A into Q

   $\therefore \quad Q = \{A\}.$

   Delete a vertex from Q *i.e., A*

   $\therefore \quad Q = \{Ø\}.$

   Add recent deleted vertex A into X.

   $\therefore \quad X = \{A\}$

   Add neighbours of most recent deleted vertex *i.e., A* into Q if they are not in Q or visited already. Neighbours of A are B and C.

   $\therefore \quad Q = \{B, C\}$

2. Delete B from Q.

   $\therefore \quad Q = \{C\}$

   Add B into X

   $\therefore \quad X = \{A, B\}$

   Add neighbours B into Q if they are not in Q or not visited already.

   Neighbours of B are D and E.

   $\therefore \quad Q = \{C, D, E\}$

3. Delete C from Q.

   $\therefore \quad Q = \{D, E\}$

   Add C into X

   $\therefore \quad X = \{A, B, C\}$

   Add neighbours C into Q if they are not in Q or visited already.

Neighbours of C are E and F. Only F is added since E is already in Q.

$\therefore \quad Q = \{D, E, F\}.$

4. Delete D from Q.

$\therefore \quad Q = \{E, F\}$

Add D into X

$\therefore \quad X = \{A, B, C, D\}$

Add neighbours D into Q if they are not in Q or visited already.

Neighbour of D is G.

$\therefore \quad Q = \{E, F, G\}.$

5. Delete E from Q.

$\therefore \quad Q = \{F, G\}$

Add just deleted vertex E into X

$\therefore \quad X = \{A, B, C, D, E\}$

Add neighbours E into Q if they are not in Q or visited already.

Neighbours of E are G and H. Only H is added since G is already in Q.

$\therefore \quad Q = \{F, G, H\}.$

6. Delete F from Q.

$\therefore \quad Q = \{G, H\}$

Add just deleted vertex F into X

$\therefore \quad X = \{A, B, C, D, E, F\}$

Add neighbours F into Q if they are not in Q or visited already. No neighbour is added since H is only neighbour and it is in Q.

$\therefore \quad Q = \{G, H\}.$

7. Delete G from Q.

$\therefore \quad Q = \{H\}$

Add just deleted vertex G into X

$\therefore \quad X = \{A, B, C, D, E, F, G\}$

Add neighbours G into Q if they are not in Q or visited already. Only I is the neighbour and not in Q or visited.

$\therefore \quad Q = \{H, I\}.$

8. Delete H from Q.

$\therefore \quad Q = \{I\}$

Add just deleted vertex H into X

$$\therefore \quad X = \{A, B, C, D, E, F, G, H\}$$

Add neighbours H into Q if they are not in Q or visited already. Only I is the neighbour and it is already in Q.

$$\therefore \quad Q = \{I\}.$$

9. Delete I from Q.

$$\therefore \quad Q = \{\emptyset\}$$

Add just deleted vertex I into X

$$\therefore \quad X = \{A, B, C, D, E, F, G, H, I\}$$

Add neighbours I into Q if they are not in Q or visited already. No neighbour of I is available.

$$\therefore \quad Q = \{\emptyset\}$$

Now Q is empty so traversal is completed. Our BFS result is $X = \{A, B, C, D, E, F, G, H, I\}$ *i.e.*, A–B–C–D–E–F–G–H–I.

11.7.2 Depth First Search

It is essentially a generalisation of the preorder traversal of rooted trees Suppose $G = (V, E)$ be a graph. Traversal using DFS starts from any vertex S in G as starting vertex. The next vertex which has the highest depth value, to be visited will be the adjacent vertex of this S and then the next adjacent vertex with the next higher depth value and so on till all the adjacent nodes for that vertex are not visited. Repeat this same procedure for every visited vertex of the graph. The implementation of the DFS uses stack. DFS uses the backtracking technique. The traversal method continues the search in the forward (deeper) direction. Hence, the name depth-first search.

The DFS is useful for mainly to:

(*i*) Find a path from one vertex to another.
(*ii*) Check whether a graph is connected or not.
(*iii*) Compute a spanning tree of a connected graph.

Let's discuss the algorithm for DFS.

Algorithm: DFS(S)

Terms Used: S = Starting vertex or source vertex.
$G = (V, E)$, the given graph.
ST = The stack to store vertices and their neighbours.
X = The list of visited vertices.

Step 1: Push the source vertex S into ST.

Step 2:

(*i*) Delete S from ST.
(*ii*) Add the deleted vertex as visited vertex into X.

Step 3:

(*i*) Find the adjacent vertices for the most recently visited vertex.

(*ii*) Add these adjacent vertices into *ST* if they are not already visited and not in *ST*. Push the adjacents in increasing order of their depths.

Step 4: If *ST* is not empty

(*i*) Set S = the vertex at the top of *ST*.

(*ii*) Go to step 2.

Step 5: Output *X*.

Step 6: Exit.

Analysis: The running time of DSF is $\Theta(|V|+|E|)$.

Example 4. Find the DFS result on the following graph with *A* as starting vertex.

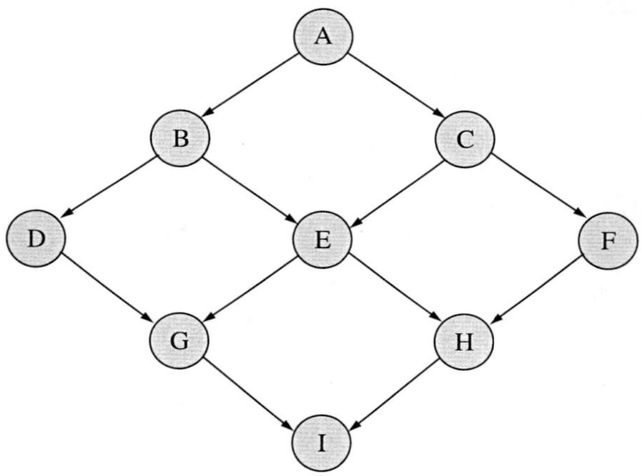

Figure 11.9 A digraph for DFS.

Solution: First construct adjacent list table for the graph in Figure 11.9 as:

Vertex	*Adjacent Vertices*
A	*B*, *C*
B	*D*, *E*
C	*E*, *F*
D	*G*
E	*G*, *H*
F	*H*
G	*I*
H	*I*
I	Ø

Here the starting vertex $S = A$. Initially the stack $ST = \{\emptyset\}$ and the visited or traversed vertices list $X = \{\emptyset\}$. Using the above table as reference for having neighbours of a node during traversing, the graph is traversed step-by-step as follows:

1. Add A into ST

$\therefore \quad ST = \{A\}$

Delete top element A from ST.

$\therefore \quad ST = \{\emptyset\}$

Add just deleted vertex A into X

$\therefore \quad X = \{A\}$

Add neighbours of A into ST if they are not in ST or visited already in increasing order of their depths. Only B and C are the neighbours.

$\therefore \quad ST = \{C, B\}$.

2. Delete top element C from ST.

$\therefore \quad ST = \{B\}$

Add just deleted vertex C into X

$\therefore \quad X = \{A, C\}$

Add neighbours of C into ST if they are not in ST or not visited already in increasing order of their depths. Only E and F are the neighbours.

$\therefore \quad ST = \{F, E, B\}$.

3. Delete top element F from ST.

$\therefore \quad ST = \{E, B\}$

Add just deleted vertex F into X

$\therefore \quad X = \{A, C, F\}$

Add neighbours of F into ST if they are not in ST or not visited already in increasing order of their depths. Only H is the neighbour.

$ST = \{H, E, B\}$.

4. Delete top element H from ST.

$\therefore \quad ST = \{E, B\}$

Add just deleted vertex H into X

$\therefore \quad X = \{A, C, F, H\}$

Add neighbours of H into ST if they are not in ST or not visited already in increasing order of their depths. Only I is the neighbour.

$\therefore \quad ST = \{I, E, B\}$.

5. Delete top element I from ST.

$\therefore$ $$ST = \{E, B\}$$

Add just deleted vertex I into X

$\therefore$ $$X = \{A, C, F, H, I\}$$

Add neighbours of I into ST if they are not in ST or not visited already in increasing order of their depths. There is no neighbour for I.

$\therefore$ $$ST = \{E, B\}.$$

6. Delete top element E from ST.

$\therefore$ $$ST = \{B\}$$

Add just deleted vertex I into X

$\therefore$ $$X = \{A, C, F, H, I, E\}$$

Add neighbours of E into ST if they are not in ST or not visited already in increasing order of their depths. These are G and H. Only G is added since H is already visited.

$\therefore$ $$ST = \{G, B\}.$$

7. Delete top element G from ST.

$\therefore$ $$ST = \{B\}$$

Add just deleted vertex G into X

$\therefore$ $$X = \{A, C, F, H, I, E, G\}$$

Add neighbours of G into ST if they are not in ST or not visited already in increasing order of their depths. It is only I and is already visited.

$\therefore$ $$ST = \{B\}.$$

8. Delete top element B from ST.

$\therefore$ $$ST = \{\emptyset\}$$

Add just deleted vertex B into X

$\therefore$ $$X = \{A, C, F, H, I, E, G, B\}$$

Add neighbours of B into ST if they are not in ST or not visited already in increasing order of their depths. These are D and E. Only D is added since E is already visited.

$\therefore$ $$ST = \{D\}.$$

9. Delete top element D from ST.

$\therefore$ $$ST = \{\emptyset\}$$

Add just deleted vertex D into X

$\therefore$ $$X = \{A, C, F, H, I, E, G, B, D\}$$

Add neighbours of D into ST if they are not in ST or not visited already in increasing order of their depths. It is only G and is already visited.

$$\therefore \qquad ST = \{\varnothing\}$$

Since the stack ST is empty hence traversing is completed. And DFS result is $X = \{A, C, F, H, I, E, G, B, D\}$. The result produces the forest as.

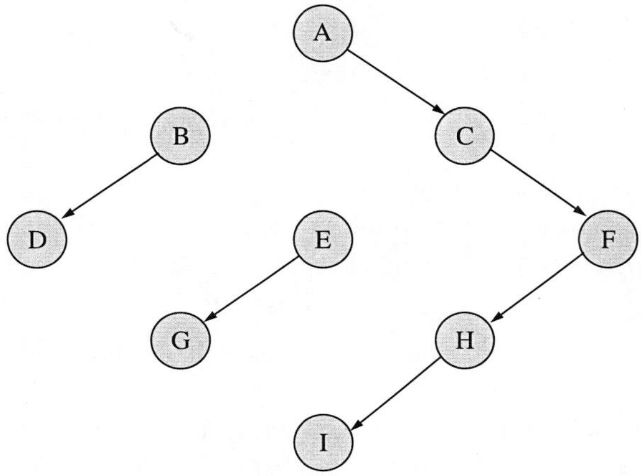

Figure 11.10 DFS Result for Figure 11.9.

11.8 SOME COMMON QUESTIONS

Question 1. Where does a Graph be applicable?

Answer: Depth-first search is a powerful tool used often in graph and geometric algorithms and has numerous applications.

(*i*) Graph acyclicity.
(*ii*) Topological sorting.
(*iii*) Finding articulation points in a graph.
(*iv*) Strongly connected components.

Question 2. Define graph traversal.

Answer: Searching a vertex in a graph can be solved by starting at vertex V and systematically searching the graph G for vertices that can be reached from V. That is traversal starts at initial vertex and visits each and every vertex exactly once and finally reaches end vertex.

Question 3. What is meant by BFS traversal?

Answer: BFS traversal means–

- It starts from the arbitrary vertex.
- It visits all vertices adjacent to starting vertex.
- Then it visits next unvisited vertex.
- This process is repeated until no more vertexes is left.

Question 4. What is the time complexity of BFS?

Answer: Time complexity of BFS traversal is based on the representation of the graph $G = (V, E)$.

- If the graph is represented by adjacency matrix, then time complexity is $O(|V|^2)$.
- If the graph is represented by adjacency List then time complexity is $O(|V|+|E|)$.

Question 5. Define depth first search traversal.

Answer: DFS traversal means–

- It starts from the arbitrary vertex.
- The exploration of the vertex V is suspended as soon as a new vertex is reached.
- At this time the exploration of new vertex U begins.
- The new vertex is explored the exploration of V continues.
- The search terminates when all reached vertices have been fully explored.

Question 6. What is the time complexity of DFS?

Answer: Time complexity of DFS traversal is based on the representation of the graph $G = (V, E)$.

- If the graph is represented by adjacency matrix: $O(n^2)$.
- If the graph is represented by adjacency List: $O(|V|+|E|)$.

Question 7. Compare DFS and BFS algorithms.

Answer: Comparison is given below:

DFS	*BFS*
1. It uses Stack representation. 2. It contains tree edge and back edge. 3. It used to identify articulation points etc.	1. It uses Queue representation. 2. It contains tree edge and cross edge. 3. It is used to identify minimum edge paths etc.

Exercises

1. Run depth-first search on the undirected graph shown in Figure 11.11. Starting at vertex A. Show the result.

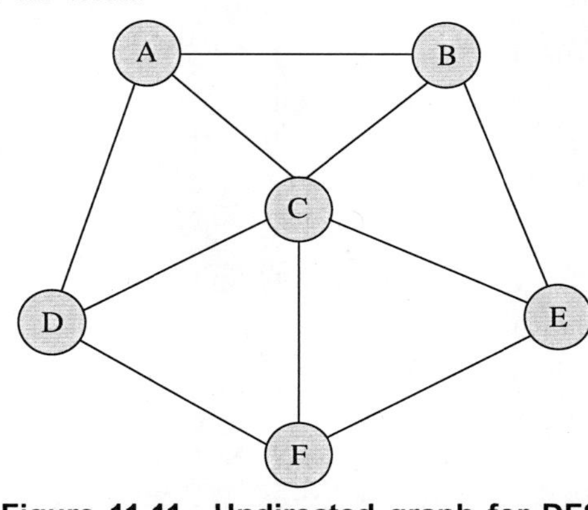

Figure 11.11 Undirected graph for DFS.

2. Run depth-first search on the directed graph shown in Figure 11.12 starting at vertex *A*. Show the result.

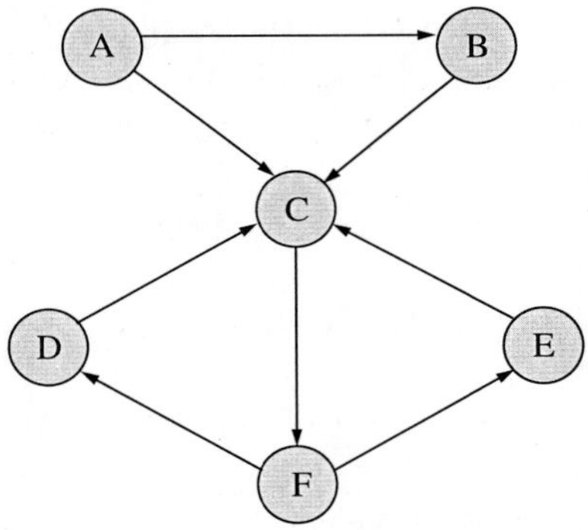

Figure 11.12 A directed graph for DFS.

3. Run depth-first search on the undirected graph shown in Figure 11.13. Starting at vertex *E*. Show the result.

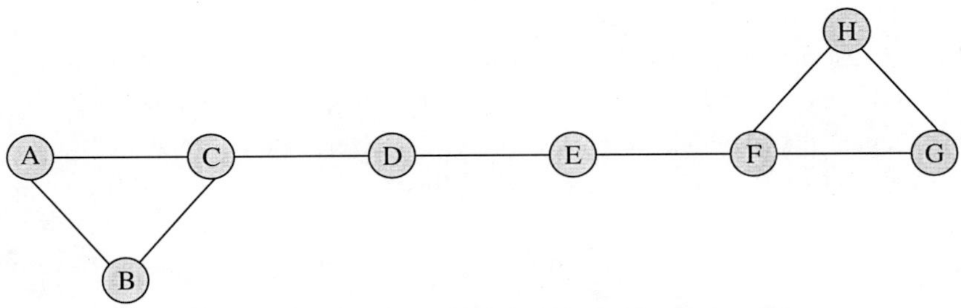

Figure 11.13 An undirected graph for DFS.

4. Run depth-first search on the undirected graph shown in Figure 11.14, starting at vertex *e*. Show the result. Give the classification of edges.

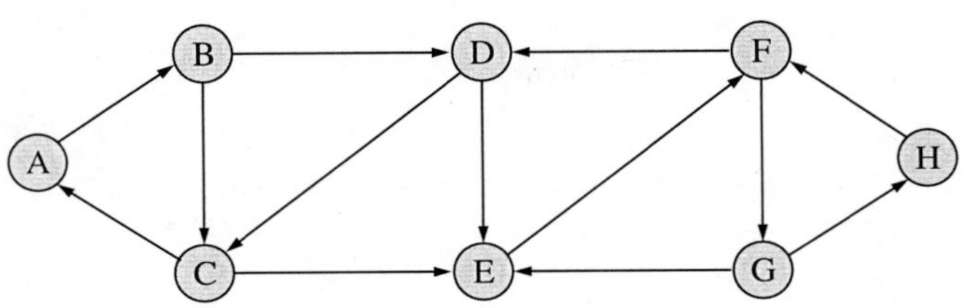

Figure 11.14 A directed graph for DFS.

5. Give an iterative version of Algorithm DFS that uses a stack to store unvisited vertices.
6. Compute the time complexity of the depth-first search algorithm when the input graph is represented by an adjacency list.
7. How many connected components are in an undirected graph possible using depth-first search or breadth-first search? Give an algorithm.

8. List the vertices separately in each connected component of an undirected graph *G* with an algorithm.
9. Detect a cycle in a connected undirected graph with *n* vertices in $O(n)$ time.
10. An edge of a connected undirected graph *G* is called a *bridge* if its deletion disconnects *G*. Modify the algorithm for finding articulation points so that it detects bridges instead of articulation points.
11. Run the breadth-first search on the undirected graph shown in Exercise 11.1. Starting vertex is *A*. Show the result.
12. Run the breadth-first search on the digraph shown in Exercise 11.2. Starting vertex is *A*. Show the result.
13. Run the breadth-first search on the undirected graph shown in Exercise 11.3. Starting vertex is *A*. Show the result.
14. Show the result of running breadth-first search on the undirected graph Exercise 11.3 starting at vertex *E*.
15. Run the breadth-first search on the undirected graph shown in Exercise 11.4. Starting vertex is E. Show the result.
16. Modify BFS algorithm so that it measures and outputs the shortest path from a vertex *v* to every other vertex in a graph.
17. Find a spanning tree for the complete graph using depth-first search.
18. Find a spanning tree for the complete graph using breadth-first search.
19. Compare the spanning trees thus obtained in Exercise 11.17 and 11.18.
20. Compute the time complexity of breadth-first search when applied on a graph with *n* vertices and *m* edges. [**Ans.** $\Theta(n + m)$]
21. Design an efficient algorithm to determine whether a given graph is bi-partite.

    [**Hint:** for definition of a bi-partite graph see any graph theory book]

University Questions

1. What is the graph? Explain the adjacency matrix to represent the directed graph. ***[UPTU'07, RTU'11]***
2. How graphs are represented in memory?

   [WBUT'09, UPTU'08, '11, MDU'08, BPUT'07, PTU'09]
3. What is adjacency list representation of a graph? Explain with the help of an example. Also write its advantages over adjacency matrix representations of graph.

   [IETE ALCCS'11]
4. Write a procedure which determines whether or not a graph G is a graph. The graph is maintained in memory by linked representation. ***[UPTU'06]***
5. Write a procedure to delete a node from the graph. ***[PTU'09]***
6. Write short notes on ***[PTU'09]***

   (*a*) Adjacency Matrix.

   (*b*) Subgraph and degree of a graph.

7. What is path matrix of a Graph? Discuss with relevant example. *[BPUT'05]*

8. Define the following with one example each: *[UPTU'07]*

   (*a*) Infinite graph

   (*b*) Hamiltonian path

   (*c*) Component of a graph

   (*d*) Euler graph

   (*e*) Spanning subgraph

9. What is meant by breadth-first traversal and breadth-first search of an undirected graph? Write down an iterative breadth-first traversal algorithm. *[IETE ALCCS'11]*

10. Define Undirected Graph with example. *[IETE ALCCS'11]*

11. Define the graph traversal and which graph traversal is better? *[RTU'11]*

12. Compare and contrast BFS vis-a-vis DFS. *[WBUT'07, '09, PTU'08]*

13. Discuss about the *time complexity* and *space complexity of B*FS algorithm with reference to the adjacency matrix and adjacency list. *[BPUT'07, DOEACC'07]*

14. Write an algorithm of DFS. Explain the algorithm with an example.

    [WBUT'09, UPTU'07, MDU'08, IGNOU'07, BPUT'05, '08]

15. Trace how DFS traverses (*i.e.,* discover and visits) the graph given below: *[DOEACC'08]*

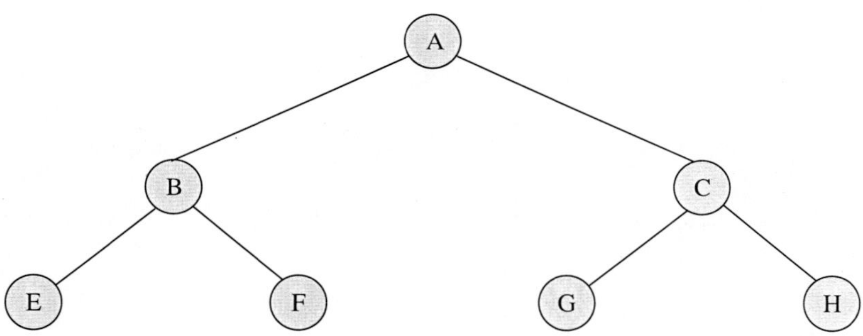

16. Write and explain the breadth first search and depth first graph traversal algorithm. What are their complexities? *[UPTU'10, '11, JNTU'08]*

17. Give an algorithm that determines whether or not a given undirected graph $G = (V, E)$ contains a cycle. Your algorithm should run in $O(|V|)$ time. *[UPTU'05]*

18. Using any method you like, compute the following subgraphs for the weighted graph. *[WBUT'10]*

    (*a*) a depth-first search tree, starting at the top vertex 6;

    (*b*) a breadth-first search tree, starting at the top vertex 6;

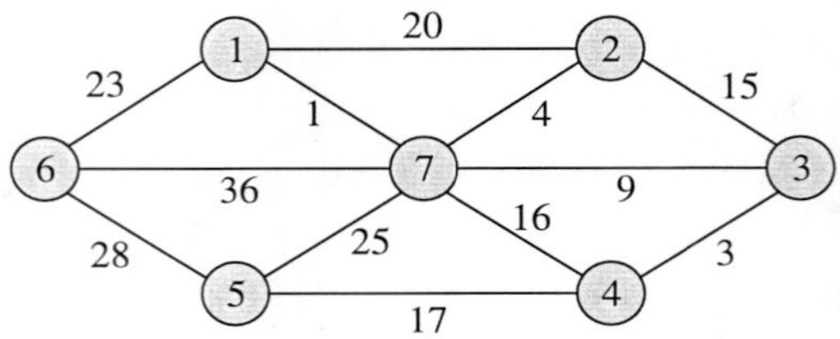

19. Discuss the algorithm to perform breadth-first traversal of a graph. Consider the following graph and list the sequence of node visited in breadth-first traversal. ***[UPTU'06, 08]***

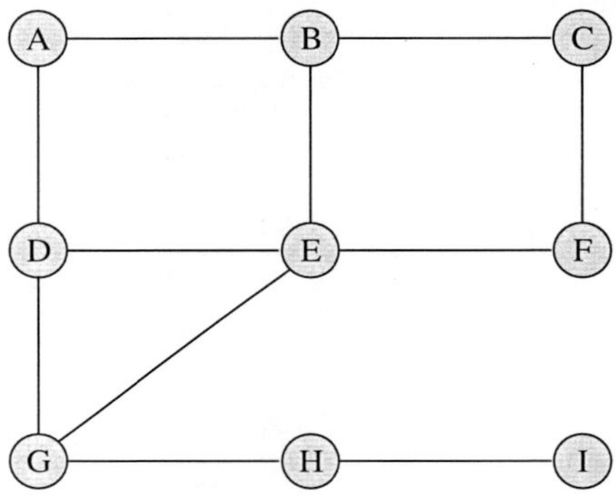

20. Define a bipartite graph. Show that the complement of a bipartite graph needs not to be a bipartite. ***[UPTU'07,'10]***
21. Define a directed graph. Given the directed graph, write down its adjacency matrix. ***[BPUT'07]***

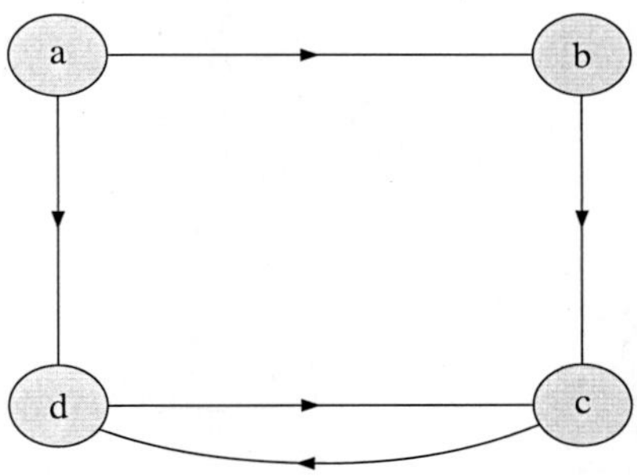

22. Find out the path from A to L using DFS. What is the complexity of your algorithm? Does this algorithm always yield optimal solution? *[BPUT'06]*

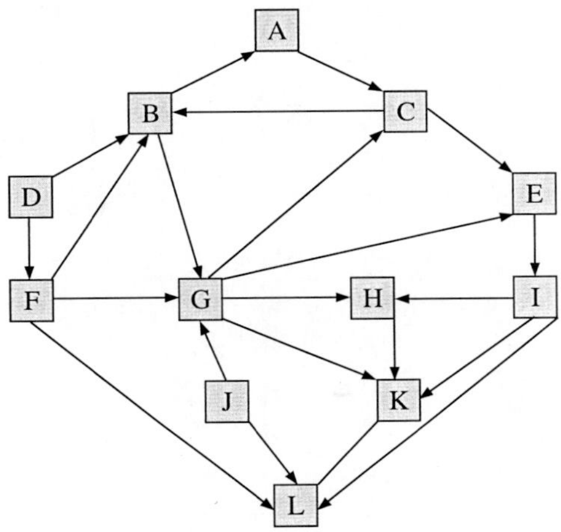

23. Differentiate, with example, a simple graph and a multigraph. Show that the maximum number of edges in a simple graph with n vertices $n(n - 1)/2$. *[UPTU'07]*

24. What is the largest number of vertices in a graph with *[UPTU'07]*

(*a*) 35 edges if all vertices are of degree at least 3.

(*b*) 24 edges and all vertices of the same degree.

25. What do you understand by isomorphic graphs? Whether the following graphs are isomorphic or not? *[BPTU'05]*

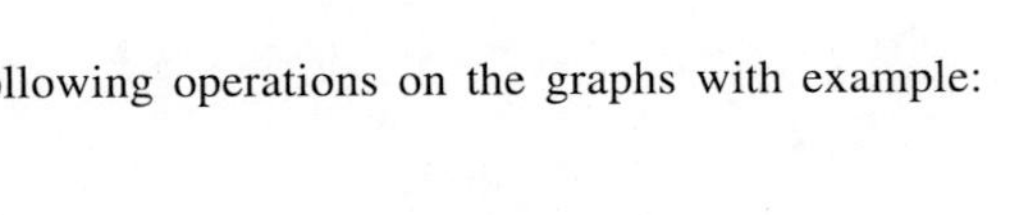

26. Define the following operations on the graphs with example: *[WBTU'07]*

(*i*) Union

(*ii*) Intersection

27. Show that: *[UPTU'10]*

(*a*) If in graph G, there is one and only one path between every pair of vertices, G is tree.

(*b*) A connected graph with n vertices and $n - 1$ edges is a tree.

28. Find the incidence matrix of the graph. *[UPTU'06]*

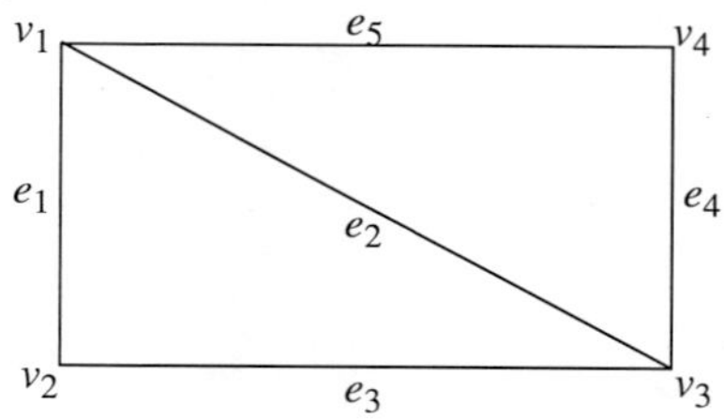

29. Given the directed graph (below], find the order in which the nodes shall be processed in a breadth-first search, starting from node a. *[BPUT'05]*

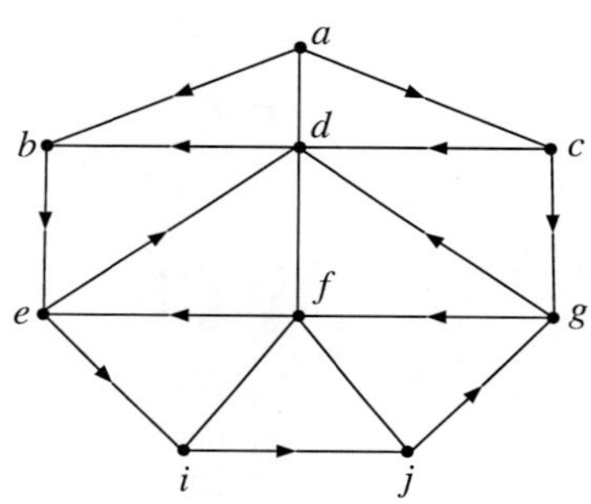

30. Use suitable data structures to represent the following directed graph in memory. *[BPUT'08]*

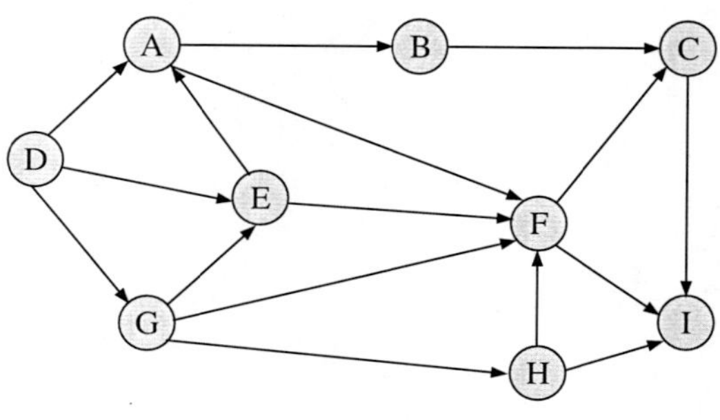

31. Find out the path from D to K using DFS. *[BPUT'08]*

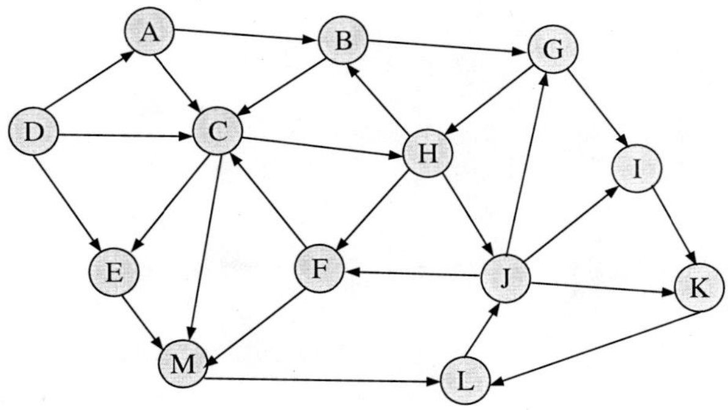

32. Consider the graph G given below. Apply depth-first search of G starting at J.

[IETE ALCCS'11]

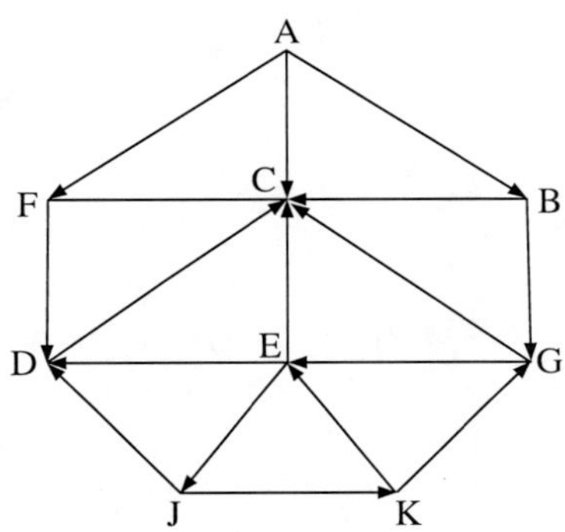

33. Given the following graph apply Depth First search and list the vertices in the order they would be visited. If nodes at same level, then do alphabetical ordering. Start with node K. *[IETE ALCCS'11]*

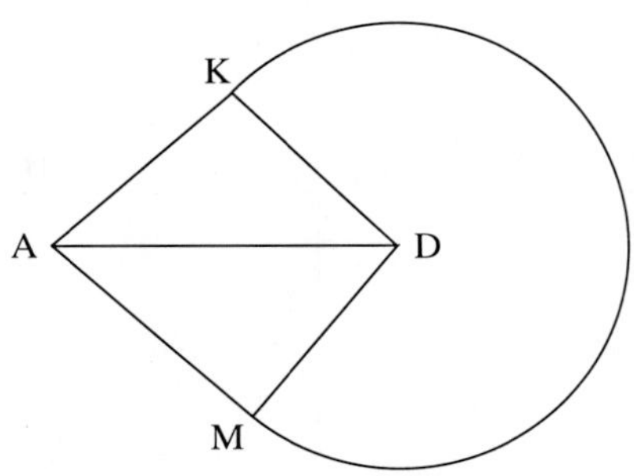

34. Draw a DFS and BFS tree for the graph starting at node A. If several nodes can be chosen at some step, pick the vertex that is alphabetically first. ***[IETE ALCCS'11]***

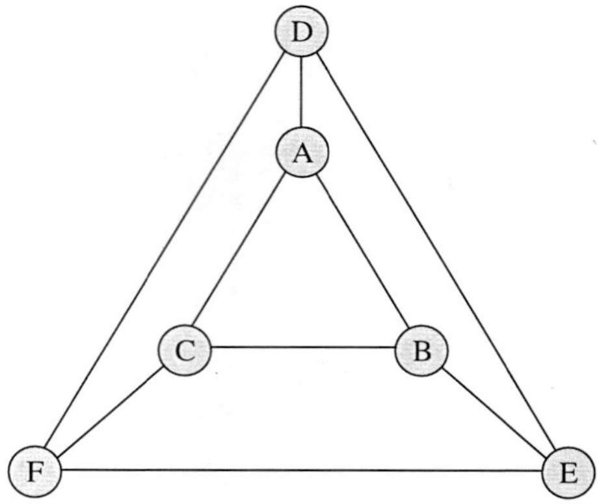

35. Let G be a graph whose vertices are the integers 1 through 8 and let the adjacent vertices of each vertex given by the table below: ***[DOEACC'06]***

Vertex	***Adjacent vertices***
1	2, 3, 4
2	1, 3, 4
3	1, 2, 4
4	1, 2, 3, 6
5	6, 7, 8
6	4, 5, 6
7	5, 6, 8
8	5, 7

(*i*) Draw G.

(*ii*) Order the vertices as they are visited in BFS traversal starting from vertex 1.

36. Give an algorithm that determines whether or not a given undirected graph $G = (V, E)$ contains a cycle. ***[DOEACC'07]***

❑❑❑

12 Chapter Matrix Manipulation Algorithms

12.1 INTRODUCTION

Most of the computational models are the mathematical symbol manipulation system. They manipulate arbitrary integers, polynomials, rational functions, etc, and are represented as mathematical expressions. This also includes much of scientific and engineering computations converted to calculations based on linear systems of equations. Again matrices are the common tool to represent these linear equations. Solving these scientific problems we first try to represent them in some mathematical equations like polynomials, simultaneous equations, linear equations, homogeneous equations, etc, so that efficiently they can be easily represented by matrices and the operations can be performed by the manipulation system.

12.2 LINEAR EQUATIONS

An algebraic equation is said to be a linear equation if each term is either a constant or the product of a constant and the first power of a single variable.

Linear equations can have one or more variables. Linear equations have great significance in applied mathematics. While modelling many phenomena, linear equations are particularly useful since many non-linear equations may be reduced to linear equations through some assumptions on the quantities of consideration that may change to only a small extent from some "background" state. The common form of two variable linear equation is $y = mx + c$.

12.3 SIMULTANEOUS EQUATIONS

Simultaneous equations in mathematics are a set of equations that contain multiple variables. Two or more equations having the same variables and the variables having same solutions are simultaneous equations. *e.g.*, $x + y = 3$ and $2x + y = 4$ are the simultaneous or linear system of equations. Common variables are 'x' and 'y', and the same solutions: $x = 1$, $y = 2$. This set is often called as a system of equations. Again every equation of this system is linear type. That's why the system of equations is known as a linear system of equations.

12.4 SIMULTANEOUS EQUATIONS: SOLUTION METHODS

Simultaneous equations are multiple equations with the same variables. The same number of equations as the number of unknown variables must be needed to get a unique solution. The equations are however mutually linearly independent. A set of particular values for all variables simultaneously satisfies all these equations. Unless a systematic approach is used it can be hard for us as a solver to solve systems of simultaneous equations with exact values. Followings are some helpful and simple methods for solution of simultaneous equations. At least one of these may be used for a solution.

(*i*) The substitution method.
(*ii*) The elimination or addition method.
(*iii*) The graphical method.
(*iv*) The matrix method.
(*v*) The least-squares method.
(*vi*) The Gauss-Jordan reduction.

12.4.1 Substitution Method

A common technique to solve simultaneous equations is the substitution method. Perhaps this is the easiest method to understand the solution approach among several algebraic techniques for simultaneous equations. The following steps demonstrate the solution method.

1. Isolate one of the variables on one side of one of the equations.
2. Substitute that isolated variable in the other equation. This yields one new equation of single variable.
3. Solve this equation for the other variable.
4. Substitute the solution value thus obtained in step 3 into the equation for the isolated variable derived in step 1.
5. If there are more than two variables, then the above steps are modified accordingly.
   (*i*) Reduce one of the equations for one unknown variable in terms of the other.
   (*ii*) Substitute that in the other equations.
   (*iii*) That will yield one equation in one unknown, which we can solve.

Example 1. Solve the following simultaneous equations using substitution method.

$$x + y = 3 \quad \ldots(i)$$

$$2x + 3y = 8 \quad \ldots(ii)$$

Solution: Variables are x and y. Now, execute the steps of substitution method.

(*i*) Isolate x of equation (*i*) as $x = 3 - y$ ...(*iii*)
(*ii*) Substitute this x of (*iii*) in other equation *i.e.*, (*ii*).

$$2(3 - y) + 3y = 8 \text{ is a new equation.}$$

(*iii*) Solve this $2(3 - y) + 3y = 8$

$$\Rightarrow \quad 6 - 2y + 3y = 8$$

$$\Rightarrow \quad y = 8 - 6 = 2$$

(*iv*) Substitute this y value in (3), $x = 3 - y = 3 - 2 = 1$

Thus, $x = 1$ and $y = 2$ are the solutions to this simultaneous equations.

Example 2. Solve the following simultaneous equations using substitution method.

$$x + y + z = 3 \quad ...(i)$$

$$2x + 3y + 3z = 8 \quad ...(ii)$$

$$x - y + z = 1 \quad ...(iii)$$

Solution: Using the substitution method we solve the equations as below.

$$x + y + z = 3$$

$$\Rightarrow \quad x = 3 - (y + z) \quad ...(iv)$$

Putting this x in equation (*ii*)

$$2(3 - (y + z)) + 3y + 3z = 8$$

$$\Rightarrow \quad 6 - 2y - 2z + 3y + 3z = 8$$

$$\Rightarrow \quad y + z = 2 \quad ...(v)$$

Putting x of equation (*iv*) in equation (*iii*)

$$3 - (y + z) - y + z = 1$$

$$\Rightarrow \quad 3 - y - z - y + z = 1$$

$$\Rightarrow \quad -2y = 1 - 3$$

$$\Rightarrow \quad y = 1$$

Putting this y in equation (*v*)

$$y + z = 2$$

$$\Rightarrow \quad 1 + z = 2$$

$$\Rightarrow \quad z = 1$$

Putting y and z in equation (*iv*)

$$x = 3 - (y + z)$$

$$\Rightarrow \quad x = 3 - (1 + 1)$$

$$\Rightarrow \quad x = 1$$

Thus, the solution is $x = y = z = 1$.

12.4.2 Elimination Method

Elimination is another commonly used method to solve simultaneous linear equations. It uses the general principles that each side of an equation remains equal the other when

(*i*) Both sides are multiplied or divided by the same quantity.

(*ii*) The same quantity is added to or subtracted from both sides.

This is done so that one or more of the variables disappear and the resulting equations with smaller number of variables are still valid representations in the system. These equations are then become easier to solve. Thus, the equations are becoming simpler through the elimination of some variables and a variable will finally appear in fully solvable form. This value can then be "back-substituted" into previously derived equations such that each "back-substitution" can solve another variable. Sometimes this method is called addition/subtraction method when elimination is done by adding or subtracting the equations.

Here is a general strategy for solving simultaneous equations having two variables: When one pair of co-efficients are negatives of one another, Add the equations vertically, and that unknown will cancel. We will then have one equation with one unknown, which we can then solve.

Example 3. Solve the simultaneous equations for x and y using elimination method.

$$2x + y = 4 \quad \text{...}(i)$$

$$x - y = -1 \quad \text{...}(ii)$$

Solution: Using elimination method we solve the above equations as below.

(*i*) Add (*i*) and (*ii*)

$$\begin{array}{r} 2x + y = 4 \\ x - y = -1 \\ \hline 3x = 3 \end{array} \Rightarrow \text{ this eliminates } y.$$

$$\Rightarrow \quad x = 1 \quad \text{...}(iii)$$

(*ii*) Putting $x = 1$ from equation (*iii*) in equation (*ii*)

$$1 - y = -1$$

$$\Rightarrow \quad y = 2$$

Hence, $x = 1$ and $y = 2$.

12.5 POLYNOMIALS

A polynomial is a representation of a function $f(x)$ of a variable x over an algebraic field F and is mathematically defined as a sum $f(x) = \sum_{i=0}^{n-1} c_i x^i$, where $c_i \in F$ are the co-efficients of the polynomial and i is integer.

Example 4. Some polynomials are given below:

(*i*) $5x^3 + 2x^2 + 7x + 3$

(*ii*) $5x^3 + 2x^2 - 7x + 3$

(*iii*) $x^3 + 2x^2 + 3$

(*iv*) $5x^4 + 2x^2 + 5$

12.5.1 Polynomials: Degree

If a polynomial $f(x)$ has highest non-zero co-efficient c_k then we say $f(x)$ is a polynomial of degree k. Any integer n just greater than k is a "degree bound" of that polynomial. The degree k of a polynomial is any integer between 0 and $n - 1$ for degree bound n of that polynomial.

12.5.2 Polynomials: Representations

By the following two ways a polynomial can be represented.

(*i*) Co-efficient Representation. (*ii*) Point Value Representation.

- ***Co-efficient Representation:*** In this case, a vector C is used to represent all the co-efficients of the polynomial $f(x) = \sum_{i=0}^{n-1} c_i x^i$ of n degree bound as $C = < c_0, c_1 \ldots c_{n-1} >$. This vector is similar to column vector or column matrix. Co-efficient representation is sometimes become convenient for some polynomial operations such as addition.

- ***Point Value Representation:*** In this case, a polynomial $f(x) = \sum_{i=0}^{n-1} c_i x^i$ of n degree bound is viewed as a set of n point value pairs $< (x_0, y_0), (x_1, y_1) \ldots (x_{n-1}, y_{n-1}) >$ such that $y_k = f(x_k)$ and x_k are distinct $k = 0, 1, \ldots. n - 1$. A polynomial may have different point value representation.

12.5.3 Polynomials: Operations

Varieties of operations can be performed on polynomials. These are listed below:

(i) Addition. (*ii*) Subtraction.
(*iii*) Multiplication. (*iv*) Division.
(*v*) Evaluation.

Here we will concentrate only on addition and multiplication operations.

- ***Polynomial Addition:*** Polynomial addition requires at least two polynomials to be performed. Coefficients of the same degree's say x are added numerically. Sum of polynomials is again a polynomial.

Suppose, $A(x) = a_{n-1}x^{n-1} + a_{n-2}x^{n-2} \ldots + a_{n-k}x^{n-k}$ and

$B(x) = b_{n-1}x^{n-1} + b_{n-2}x^{n-2} \ldots + b_{n-k}x^{n-k}$ are two polynomials of n degree.

Then their addition will result another polynomial say $C(x)$ of degree bound n.

$$C(x) = c_{n-1}x^{n-1} + c_{n-2}x^{n-2} \ldots + c_{n-k}x^{n-k}$$

where $c_k = a_k + b_k \ \forall \ k = 0, 1, \ldots n - 1$.

In general if $A(x) = \sum_{i=0}^{n-1} a_i x^i$ and $B(x) = \sum_{i=0}^{n-1} b_i x^i$ then $C(x) = A(x) + B(x) = \sum_{i=0}^{n-1} c_i x^i$ and $c_i = (a_i + b_i)$.

Example 5. Suppose $A(x) = 10x^5 + 5x^4 + 3x^2 + 20$ and $B(x) = 10x^5 - 5x^3 + 3x^2 + 20x$ then the addition will result the following polynomial.

$$10x^5 + 5x^4 + 0x^3 + 3x^2 + 0x + 20$$
$$10x^5 + 0x^4 - 5x^3 + 3x^2 + 20x + 0$$

$$(10 + 10)x^5 + (5 + 0)x^4 + (0 - 5)x^3 + (3 + 3)x^2 + (0 + 20)x + (20 + 0)$$
$$= 20x^5 + 5x^4 - 5x^3 + 6x^2 + 20x + 20$$

- ***Polynomial Multiplication:*** In polynomials multiplication, every term of a polynomial say $A(x)$ is multiplied with every term of another polynomial say $B(x)$ followed by addition of co-efficients of the same degree's x of the terms obtained after multiplication. If both $A(x)$ and $B(x)$ are of n degree bound then their product say $C(x) = A(x) \cdot B(x)$ will be of $2n - 1$ degree bound.

The degree of $C(x)$ = the degree of $A(x)$ + the degree of $B(x)$

The degree bound of $A(x)$ + the degree bound of $B(x)$

In general if $A(x) = \sum_{i=0}^{n-1} a_i x^i$ and $B(x) = \sum_{i=0}^{n-1} b_i x^i$ then $C(x) = A(x) \cdot B(x) = \sum_{i=0}^{2n-2} c_i x^i$

and $c_i = \sum_{k=0}^{i} a_k b_{i-k}$.

Example 6. Suppose, $A(x) = 5x^4 + 3x^2 + 20$ and $B(x) = 5x^3 + 3x + 10$ then $C(x) = A(x)\,B(x)$ is calculated as:

$$
\begin{array}{r}
5x^4 + 3x^2 + 20 \\
5x^3 + 3x + 10 \\
\hline
50x^4 + 30x^2 + 200 \\
15x^5 + 9x^3 + 60x \\
25x^7 + 15x^5 + 100x^3 \\
\hline
25x^7 + 30x^5 + 50x^4 + 109x^3 + 30x^2 + 60x + 200
\end{array}
$$

12.5.4 Polynomials: Evaluation

Consider a polynomial $f(x) = \sum_{i=0}^{n} c_i x^i = c_n x^n + c_{n-1} x^{n-1} + \ldots c_1 x + c_0$. Suppose, we want to compute it upto certain terms *i.e.*, $f(x_k)$ at a point x_k. For the time being two familiar methods are in hand to do this.

(*i*) Straight Forward Method. (*ii*) Horner Rule method.

- ***Straight Forward Method:*** This method is also known as Right to Left method. In this method c_0 is the initial value of the polynomial $f(x)$. Then $c_1 x_1$ is added to c_0. Now, new value is $(c_0 + c_1 x_1)$. Next $c_2 x^2$ is added with this new summed value. Now the new value is $(c_0 + c_1 x_1 + c_2 x^2)$. The process is continued till highest power of x. computation of the polynomial $f(x_k)$ involves the above procedure w.r.t. x_k instead of x.

Algorithm: Straight Forward_Polynomial_Evaluation ()

Terms Used:
$f(x)$ = The polynomial
n = Degree of $f(x)$
x_k = The point of evaluation of $f(x)$
S = The sum value initially c_0.

Step 1: Repeat for $i = 1$ to n

(*i*) $S = S + c_i * x_k^i$

Step 2: Return S.

Step 3: Exit.

Analysis: Time complexity of the algorithm is due to step 1. It absorbs both addition and multiplication operations and is repeated n times. The algorithm runs in $\Theta(n)$ time.

- ***Horner Rule Method:*** A better approach over straight forward method was first used by Sir Newton in 1971. The similar technique was also used by W. G. Horner in 1819, in order to evaluate co-efficients of a polynomial. However it suppressed the previous. Because the Horner rule helps in evaluating an nth degree polynomial optimally. Horner represented a polynomial $f(x) = c_n x^n + c_{n-1} x^{n-1} \ldots + c_0$ in the following way as.

$$f(x) = ((\ldots(c_n x + c_{n-1})x + c_{n-2})x + c_{n-3})x + \ldots + c_1)x + c_0$$

The following algorithm based on Horner rule is designed below to evaluate a polynomial.

Algorithm: Horner_Polynomial_Evaluation ()

Terms Used: $f(x)$ = The polynomial
n = Degree of $f(x)$
x_k = The point of evaluation of $f(x)$
S = The sum value initially c_n.

Step 1: Repeat for $i = n - 1$ down to 0

(*i*) $S = S{*}x_k + c_i$

Step 2: Return S

Step 3: Exit.

Analysis: Similar to straight forward method Horner's algorithm also involves n multiplications, n addition and n assignment operations. Conversely it is an improved version over straight forward method by a factor 2. However, the algorithm runs in $\Theta(n)$ time.

12.5.5 Improvement of Polynomial Evaluation

From the above discussion of two ways to evaluate a polynomial, a polynomial at n points can be evaluated in $\Theta(n^2)$ time with the help of Horner principle. However, further improvement of this evaluation using a new technique namely Fast Fourier Transform (FFT) can be possible. The FFT computes a polynomial with n values in $\Theta(n \log n)$ time. Again using complex root of unity then also polynomial evaluation needs $\Theta(n \log n)$ time.

12.6 DISCRETE FOURIER TRANSFORM

The Discrete Fourier Transform (DFT) is a specific kind of Fourier transform. It is essentially used in Fourier analysis. It transforms one function often in time domain into another function usually in frequency domain. This representation is simply the DFT of the original function. The DFT calls for an input function with discrete, non-zero values of finite or limited duration. Such inputs are often created by sampling a continuous function, like a person's voice.

Definition: The sequence of N complex numbers $x_0, x_1, ..., x_{N-1}$ called samples of a continuous function say $x(t)$ is transformed into the sequence of N complex numbers $X_0, X_1, ..., X_{N-1}$ by the DFT according to the formula:

$$X_k = \sum_{n=0}^{N-1} x_n e^{-\frac{2\pi i}{N}kn} \qquad k = 0, ..., N-1$$

Where $\qquad i$ = The imaginary unit.

$e^{\frac{2\pi i}{N}}$ = A primitive N'th root of unity.

The transform is sometimes denoted by the symbol $\mathcal{F}$.

The corresponding Inverse Discrete Fourier transform (IDFT) is given by:

$$x_n = \frac{1}{N}\sum_{n=0}^{N-1} X_k e^{-\frac{2\pi i}{N}kn} \qquad n = 0, ..., N-1$$

These equations can simply be described as: the complex numbers X_k represent the amplitude and phase of the different sinusoidal components of the input signal $x(t)$ for the input x_n. The DFT computes the X_k from the x_n. Oppositely, the IDFT computes the x_n as a sum of sinusoidal components $\left(\frac{1}{N}\right)X_k e^{-\frac{2\pi i}{N}kn}$ having frequency k/N cycles per sample.

12.6.1 DFT: Properties

The DFT have the following properties:

1. *Completeness:* An invertible, linear transformation.
2. *Orthogonality:* The vectors $e^{\frac{2\pi i}{N}kn}$ form an orthogonal basis over the set of N-dimensional complex vectors.
3. *Periodicity:* This means $X_{k+N} = X_k$.

Besides the above other features can be derived. Some of these are recorded below:

1. The shift theorem.
2. Circular convolution theorem and cross-correlation theorem.
3. Convolution theorem duality.
4. Trigonometric interpolation polynomial.
5. The unitary DFT.
6. Expressing the inverse DFT in terms of the DFT.
7. Eigen values and eigen vectors.

12.6.2 DFT: Applications

The DFT has been widely used in a large number of fields. All the applications however depend significantly on the availability of a fast algorithm that computes DFTs and their inverses.

1. Spectral analysis.
2. Data compression.
3. Partial differential equations.
4. Polynomial multiplication.
5. Multiplication of large integers.
6. Convolutions.

Example 7. Find DFT of the following polynomial of n degree-bound *i.e.,* at the n complex *nth* roots of unity ($\omega^0{}_n$, $\omega^1{}_n$, ..., $\omega^{n-1}{}_n$).

$$P(x) = \sum_{j=0}^{N-1} c_j x^j$$

Solution: Suppose, n is power of 2 and co-efficient vector of P, $c = < c_0, c_1, c_2, ..., c_{n-1} >$.

Now the results y_k for $k = 0, 1, 2, ..., n - 1$ can be defined as:

$$y_k = P(\omega^k{}_n) = \sum_{j=0}^{N-1} c_j \, \omega_n^{kj}$$

Thus, the vector $y = < y_0, y_1, ..., y_{n-1} >$ is our desired Discrete Fourier Transform (DFT) for the coefficient vector c. We may also write $y = \text{DFT}_n(c)$ for the same.

12.7 FAST FOURIER TRANSFORM

A DFT decomposes a sequence of values into components of different frequencies. So, the computation is often become too slow and it takes $O(n^2)$ arithmetical operations for n inputs. To minimise this complexity and perform the same computation more quickly it is preferable to look another approach namely the Fast Fourier Transform (FFT). This takes advantages of special properties of the complex roots of unity for the computation.

Definition: The Fast Fourier Transform or simply FFT is an algorithm that computes the DFT and its inverse efficiently. Computation of FFT takes $O(n \log n)$ operations for n inputs.

Most of the FFT algorithms involve a wide range of mathematics, from simple complex-number arithmetic to group theory and number theory. Some of the popular FFT algorithms are listed below.

(*i*) Cooley–Tukey FFT algorithm.
(*ii*) Prime-factor FFT algorithm.
(*iii*) Bruun's FFT algorithm.
(*iv*) Rader's FFT algorithm.
(*v*) Bluestein's FFT algorithm.
(*vi*) Split-radix FFT algorithm

The Fast Fourier Transform (FFT) separately defines two new polynomials $P^{[0]}(x)$ and $P^{[1]}(x)$ from a given polynomial $P(x)$ of $n/2$ degree-bound. Here first new polynomial contains all the even-index co-efficients of P, and the second one contains all the odd-index coefficients of P, these are defined as below:

$$P^{[0]}(x) = c_0 + c_2x + c_4x^2 \ldots + c_{n-2}x^{n/2-1}$$

$$P^{[1]}(x) = c_1 + c_3x + c_5x^2 \ldots + c_{n-1}x^{n/2-1}$$

Thus, $P(x) = P^{[0]}(x^2) + xP^{[1]}(x^2)$ so that its evaluation at $\omega_n^0, \omega_n^1, \ldots, \omega_n^{n-1}$ is reduced to

(*a*) Evaluation of $P^{[0]}(x)$ and $P^{[1]}(x)$ $n/2$ degree-bound polynomials at $(\omega_n^0)^2$, $(\omega_n^1)^2, \ldots, (\omega_n^{n-1})^2$ points.
(*b*) Combine the results to obtain $P(x)$.

This recursion is the basis of the following recursive algorithm for FFT to compute DFT an n element vector.

Algorithm: FFT_Rec(c)

Input: $c = n$ vector.

Output: computation of $\text{DFT}_n(c)$.

Step 1: Initialise $n = |c|$ // n is power of 2

Step 2: If $n = 1$

- Return c

Step 3: Initialise

(*a*) $\omega_n = e^{2\pi i/n}$
(*b*) $\omega = 1$
(*c*) $c^{[0]} = < c_0, c_2, c_4, \ldots, c_{n-2} >$
(*d*) $c^{[1]} = < c_1, c_3, c_5, \ldots, c_{n-1} >$
(*e*) $y^{[0]}$ = call FFT_Rec($c^{[0]}$)
(*f*) $y^{[1]}$ = call FFT_Rec($c^{[1]}$)

Step 4: Repeat for $k = 0$ to $n/2 - 1$

(*a*) $y_k = y^{[0]}{}_k + \omega\, y^{[1]}{}_k$
(*b*) $y_{(k+n/2)} = y^{[0]}{}_k - \omega\, y^{[1]}{}_k$
(*c*) $\omega = \omega\omega_n$

Step 5: Return y

Step 6: Exit.

Analysis: Step 2 of the algorithm is the basis for the recursion. Determining complexity value of the algorithm, we consider recursive calls exclusively in which each recursive call invokes $\Theta(n)$ time for an input vector of length n. since odd and even index-bound or degree-bound are under consideration. Thus the recurrence relation for the running time $T(n)$ of the FFT_Rec() algorithm is:

$$T(n) = 2T(n/2) + \Theta(n)$$

$$= \Theta(n \log n)$$

Therefore, FFT evaluates a n degree-bound polynomial in $\Theta(n \log n)$ time at the nth complex root of unity.

12.8 INTEGER MULTIPLICATION SCHEMES

A multiplication is an operation that multiplies two numbers. The algorithm used for this operation is known as multiplication algorithm. Depending on size, variation (number system), etc. different multiplication algorithms are in use. Some of the efficient multiplication algorithms have to consider here for the advent of the decimal system.

(*i*) Long multiplication method.
(*ii*) Lattice multiplication method.
(*iii*) Peasant or binary multiplication method.
(*iv*) Shift and add method.
(*v*) Gauss's complex multiplication algorithm.
(*vi*) Quarter square multiplier method.
(*vii*) Fourier transforms methods.
(*viii*) Polynomial multiplication.
(*ix*) Strassen matrix multiplication method.

12.8.1 Long Multiplication Method

This is a natural way of multiplying numbers which is generally taught in schools. Sometimes this method is known as grade-school multiplication method. The procedure says that: multiply the multiplicand (first number) by each digit of the multiplier (second number) and then add up all the properly shifted results. It thus requires memorisation of the multiplication table even for every single digit. This is done manually with base 10. Computers also use a very similar shift and add algorithm but base in this case is 2. A person doing multiplication for long integer numbers on paper will write down first all the products and then add them together.

Example 8. Suppose, multiplicand = 23958233 and multiplier = 5830 then 23958233 × 5830 can be evaluated using long multiplication method in the following way.

$$
\begin{array}{rl}
2398233 & \\
5830 & \times \\
\hline
00000000 & (=23{,}958{,}233\times \quad 0) \\
71874699 & (=23{,}658{,}233\times \quad 30) \\
191665864 & (=23{,}958{,}233\times \quad 800) \\
119791165 & (=23{,}958{,}233\times \quad 5{,}000) \\
\hline
139676498390 & (=139{,}676{,}498{,}390)
\end{array}
$$

12.8.2 Fourier Transform Method

This idea is invented by Schönhage and Strassen during 1968-71 and is known as Schönhage-Strassen algorithm. In this method the largest integer w is chosen such that it will not cause overflow during computation. Then, two numbers a and b are divided into m groups, each of w bits as.

$$a = \sum_{i=0}^{m} 2^{wi} a_i \text{ and } b = \sum_{j=0}^{m} 2^{wj} b_j .$$

Setting $a_i = b_j = 0$, $i, j > m$, $k = i + j$, and $\{c_k\}$ as convolution of $\{a_i\}$ and $\{b_j\}$.

Thus,

$$ab = \sum_{i=0}^{m}\sum_{i=0}^{m} 2^{w(i+j)} a_i b_j = \sum_{k=0}^{2m} 2^{wk} \sum_{i=0}^{k} a_i b_{k-i} = \sum_{k=0}^{2m} 2^{wk} c_k$$

Using the convolution theorem, *ab* can be computed by:

(*i*) Computing the fast Fourier transforms of $\{a_i\}$ and $\{b_j\}$.

(*ii*) Multiplying the two results entry by entry.

(*iii*) Computing the inverse Fourier transform.

(*iv*) Adding the part of c_k that is greater than 2^w to c_{k+1}.

Till 2006, this idea was the fastest known method with a time complexity of $\Theta(n \log(n) \log(\log(n)))$. In 2007, Martin Fürer improved this idea to provide a better time complexity. He achieved a time complexity of $\Theta(n \log(n) 2^{\Theta(\log^*(n))})$ using Fourier transforms over complex numbers.

12.8.3 Polynomial Multiplication Method

The fastest known algorithms to multiply large integers use the polynomial multiplication method. The method is described below.

Suppose $a(x)$ and $b(x)$ are two polynomials. Let their product $c(x) = a(x) \cdot b(x)$. Then, the ordinary product expression for the co-efficients of the polynomial *c* involves a linear, acyclic convolution. It can be rewrite as a cyclic convolution by:

- First, take the co-efficient vectors V_a and V_b for $a(x)$ and $b(x)$ respectively with constant term.
- Second, append zeros so that the resultant co-efficient vectors V_a and V_b have dimension $d > \deg(a(x)) + \deg(b(x))$.

Then, the co-efficients vector V_c for $c(x)$ is given by:

$$V_c = V_a * V_b$$

Where * is the convolution operator and is defined as:

$$(V_c)_n = \sum_{m=0}^{d-1} (V_a)_m \cdot (V_b)_{n-m} \bmod d, \qquad n = 0, 1, ..., d-1.$$

Since, the convolution becomes multiplication under the DFT, then

$$\mathcal{F}(V_c) = \mathcal{F}(V_a)\,\mathcal{F}(V_b)$$

The vector product in this case is taken element wise. Thus the co-efficients of $c(x)$ are just the terms 0... $\deg(a(x)) + \deg(b(x))$ of the co-efficient vector

$$V_c = \mathcal{F}^{-1}(\mathcal{F}(V_a)\,\mathcal{F}(V_b)).$$

The fastest known algorithms for the multiplication of very large integers use the above polynomial multiplication method. A relatively low-complexity carry-propagation step that completes the multiplication is further exploit after the polynomial multiplication is over.

12.9 MATRIX MULTIPLICATION SCHEMES

Suppose A, and B are two $n \times n$ matrices. The product $C = AB$ will also be a $n \times n$ matrix. The computation of C using conventional method will be as:

$$C(i, j) = \sum_{k=1}^{n} A(i, k)\, B(k, j), \forall\, 1 \le i, j \le n. \qquad \ldots(i)$$

Computation of above expression needs n multiplication operations. And there are n^2 elements in each matrix A, B, and C. So the conventional method for computation of C requires $\Theta(n^3)$ time.

Another approach to multiply two $n \times n$ matrices A and B is the use of Divide-and-Conquer approach. For simplicity suppose $n = 2^k$ where k is a non-negative integer. If it is seen that n is not power of 2 then we add 0s to more rows and columns as their elements to A or B or both whenever necessary such that resulting dimension becomes power of 2.

Now subdivide A and B. It results four x matrices for each of them. Now $C = AB$ is computed using the conventional method as.

$$C = AB$$

$$\begin{vmatrix} C_{11} & C_{12} \\ C_{21} & C_{22} \end{vmatrix} = \begin{vmatrix} A_{11} & A_{12} \\ A_{21} & A_{22} \end{vmatrix} \times \begin{vmatrix} B_{11} & B_{12} \\ B_{21} & B_{22} \end{vmatrix} \qquad \ldots(ii)$$

Where
$$\left.\begin{aligned} C_{11} &= A_{11}\, B_{11} + A_{12}\, B_{21} \\ C_{12} &= A_{11}\, B_{12} + A_{12}\, B_{22} \\ C_{21} &= A_{21}\, B_{11} + A_{22}\, B_{21} \\ C_{22} &= A_{21}\, B_{12} + A_{22}\, B_{22} \end{aligned}\right\} \qquad \ldots(iii)$$

If $n = 2$, then C can be computed using (*ii*) and (*iii*).

If $n > 2$, then the procedure is recursively call itself to smaller sized sub-problems until $n = 2$ to compute C.

From (*iii*) it is seen that the computation performs 4-additions and 8-multiplications. Again two $\frac{n}{2} \times \frac{n}{2}$ matrices can added in $\Theta(n^2)$ time. Total time complexity $T(n)$ for computation of $C = AB$ can be evaluated as:

$$T(n) = \begin{cases} b, n = 2 \text{ where } b \text{ is a constant.} \\ 8T(n/2) + \Theta(n^2), n > 2. \end{cases} \qquad \ldots(iv)$$

$$= \Theta(n^3).$$

It shows no improvement has been made over the conventional method of matrix multiplication. Matrix multiplication is more expensive than matrix addition. It is necessitate reformulating the equation (*iii*) by reducing multiplication operations. Volker Strassen has done this. His approach is named as Strassen's matrix multiplication scheme or algorithm. In the following section our centre of attention will be on this topic.

12.9.1 Strassen's Matrix Multiplication Schemes

Volker Strassen published his algorithm on matrix multiplication in 1969. He has discovered a new way to compute C_{ij} in equation (*iii*). He used only 7 multiplication and 18 addition/subtraction operations. He introduced 7 new $\frac{n}{2} \times \frac{n}{2}$ matrices M_1 to M_7 to compute C_{ij}. These matrices he represented in terms of A_{ij} and B_{ij} as:

$$\left.\begin{aligned} M_1 &= (A_{11} + A_{22})(B_{11} + B_{22}) \\ M_2 &= (A_{21} + A_{22})B_{11} \\ M_3 &= A_{11}(B_{12} - B_{22}) \\ M_4 &= A_{22}(B_{21} - B_{11}) \\ M_5 &= (A_{11} + A_{12})B_{22} \\ M_6 &= (A_{21} - A_{11})(B_{11} + B_{12}) \\ M_7 &= (A_{12} - A_{22})(B_{21} + B_{22}) \end{aligned}\right\} \quad \ldots(v)$$

Now he computed C_{ij} using these 7 matrices M_1 to M_7 as:

$$\left.\begin{aligned} C_{11} &= M_1 + M_4 - M_5 + M_7 \\ C_{12} &= M_3 + M_5 \\ C_{21} &= M_2 + M_4 \\ C_{22} &= M_1 - M_2 + M_3 + M_6 \end{aligned}\right\} \quad \ldots(vi)$$

Strassen's matrix multiplication algorithm uses 7 multiplications. Its time complexity can be then derived rewriting the recurrence $T(n)$ in equation (*vi*) in the following way.

$$T(n) = \begin{cases} b, \; n = 2 \text{ where } b \text{ is a constant.} \\ 7T(n/2) + \Theta(n^2), \; n > 2. \end{cases}$$

$$= \Theta(n^{2.81}).$$

Comparatively it can be said that Strassen's matrix multiplication scheme is slightly faster than the conventional or some standard methods. Yet, in practice it is applied for large matrices. Again he is the first person who has pointed out that the standard approaches for matrix multiplication are not optimal.

Exercises

1. Show that the traditional algorithm for matrix multiplication requires n^3 multiplications and $n^3 - n^2$ additions.
2. Give a traditional algorithm for matrix multiplication. What is the running time?
3. Give an algorithm that gives lower complexity than the traditional algorithm for matrix multiplication.
4. Modify Strassen's matrix multiplication algorithm independent of the matrices whose size is not necessarily a power of 2.
5. What are the limitations of Strassen's matrix multiplication algorithm?
6. State Horner's rule. Why does it necessary?
7. Evaluate the following polynomials using Horner's rule:
   (*a*) $3x^5 + 2x^4 + 4x^3 + x^2 + 2x + 5$.
   (*b*) $3x^5 + 2x^3 + 5x^2 + 3x + 7$.
8. Write short notes on:
   (*a*) DFT
   (*b*) FFT
   (*c*) Horner Rule
   (*d*) Strassen's Algorithm

University Questions

1. Discuss the procedure for Strassen's matrix multiplication to evaluate the product of n matrices. Find the resulting recurrence relation for the same and analyze its time-complexity. Is this method an Improvement over the conventional matrix multiplication method? If so, why?
   [WBUT'07, RTU'08, RTU'10, IGNOU'07, DOEACC'06]
2. Write short notes on Strassen's matrix multiplication with example.
   [WBUT'08, MDU'08, JNTU'08]
3. Use Strassen's algorithm to compute the product of two given matrices: ***[UPTU'10]***

$$\begin{bmatrix} 1 & 3 \\ 5 & 7 \end{bmatrix} \text{ and } \begin{bmatrix} 8 & 4 \\ 6 & 2 \end{bmatrix}.$$

4. Use Strassen's matrix multiplication algorithm to compute the matrix product.
   [RTU'07, 09]

$$A = \begin{bmatrix} 7 & 9 \\ 2 & 5 \end{bmatrix}, \quad B = \begin{bmatrix} 3 & 2 \\ 6 & 5 \end{bmatrix}$$

5. Show how Strassen's algorithm computes the following multiplication? ***[BPUT'07]***

$$\begin{vmatrix} 3 & 1 \\ 4 & -1 \end{vmatrix} \times \begin{vmatrix} 12 & -5 \\ 6 & -3 \end{vmatrix}$$

6. Write down the algorithm to compute Fast Fourier Transform (FFT) Hence write the recurrence relation for the work done by the Fast Fourier Transform and explain the origin of the terms.

   [BPUT'07]

7. Write down DFT algorithm and explain its method of execution. ***[WBUT'08]***
8. Write short notes on FFT algorithm. ***[WBUT'09, '10, UPTU'10]***
9. Find the computational complexity of FFT algorithm. ***[WBUT'10]***

❑❑❑

13 Chapter Notion of NP-Completeness

13.1 INTRODUCTION

So far we have discussed some of the well problems, designed and analysed their algorithms. From these discussions it has been also seen how their running times or computing times are varied. Based on their study we come to a point to cluster the problems into two groups. One group of problems behaves like polynomial bounded solution times of small degree. Some common problems of these kinds are sorting problems, searching problems, etc. The algorithms of these problems are called polynomial time algorithms. If n is the input size of a polynomial time algorithm then for a small constant k, its worst case time complexity is represented by $O(n^k)$. *e.g.,* $O(n)$, $O(n^2)$, $O(\log n)$, $O(n \log n)$, etc. As these problems are solvable in polynomial times, they are easy *i.e.,* tractable.

The second group of problems are obviously made up of non-polynomial time bounded. Because there are many problems that have been already studied, for which no polynomial time algorithms is known. Some of these problems are travelling salesman problem, graph colouring problem, the Knapsack problem, Hamiltonian cycle's problem, Boolean satisfiability problem, etc. Similarly their algorithms are called non-polynomial time algorithms. For input size n, worst case time complexity value of any problem of this kind may be $O(n^2 2^n)$, $O(2^n)$, $O(n^n)$, etc. Naturally these values are so high. Unfortunately no one till now however able to develop a polynomial time bounded algorithms for these kinds of problems. Thus these problems are intractable as they require super polynomial time. The reason is that these are hard to solve. A polynomial time algorithm verifies the validity of any supposed solution of these problems. *e.g.,* determining a Hamiltonian Cycle in a graph is hard. However, whether a sequence of nodes in that graph defines a Hamiltonian cycle or not that can be easy to verify.

In this chapter, we are neither going to discuss how to obtain polynomial time algorithms these intractable problems nor we don't say that their polynomiality does not exist. Instead we are going to show that no polynomial time algorithms for some of these problems are computationally related.

13.2 CLASSIFICATION OF PROBLEMS

Different classes of problems available such as:

(*i*) Decision problem.

(*ii*) Optimization problem.

(*iii*) Search problem.

(*iv*) Counting problem.

(*v*) Function problem, etc.

13.2.1 Decision Problems

Any problem for which the answer is either *yes* or *no* depending on the values of some input parameters is called a decision problem. These input parameters can be natural numbers as well as strings of a formal language. Instead of yes/no sometimes it also uses 1/0, success/failure, or true/false as output. Outputs of decision problems are Boolean.

An algorithm for solving a decision problem is termed as decision algorithm or procedure for that problem. And the problem is then called decidable or effectively solvable. A decision problem is decidable if it is a recursive set.

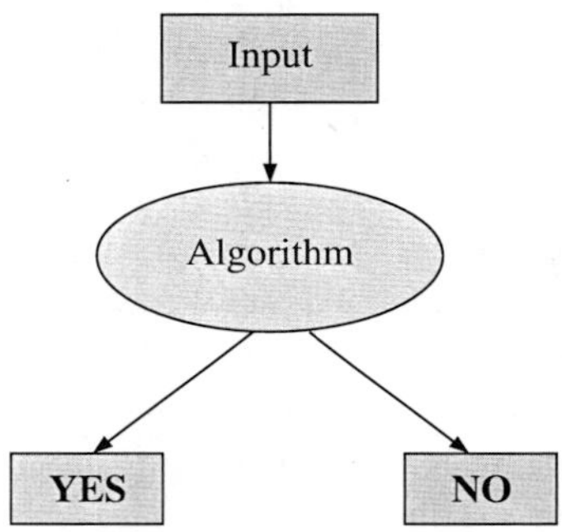

Figure 13.1 A decision algorithm for any input.

Example 1. The problem, "given two numbers x and y, does x evenly divide y?" is a decision problem. Because the answer can be either 'yes' or 'no' depending upon the values of x and y.

Example 2. Consider the problem "Find a Hamiltonian cycle in graph G". The problem is however not a decision problem, but if problem "Is the graph G Hamiltonian?" then it is a decision problem.

Example 3. The 0/1 Knapsack problem is a decision problem since the problem is to determine whether an item is to be considered or not *i.e.,* 0/1 assignments of the values.

Example 4. A classic example of a decidable decision problem is to effectively decide whether a given natural number is prime or not.

Example 5. Let S be a sequence of real numbers. Are all the numbers in S distinct? This is a decision problem.

Example 6. Consider a graph $G = (V, E)$. The decision problem is: Is it possible to colour G using a specified number of colours? This problem is intractable.

Example 7. Consider a graph $G = (V, E)$. "Does there exist a clique of size k (some positive integer) in G?" is decision problem.

13.2.2 Optimization Problems

The problem of finding the best solution from all feasible solutions is an optimization problem. Optimization problem involves the identification of an optimal solution *i.e.*, either maximum or minimum. Similarly, an algorithm which is used to solve an optimization problem is called optimization algorithm. A cost function is used for the purpose.

Optimization problems have corresponding decision problems meaning that many optimization problems can be recast into decision problems that ask whether there is a feasible solution. However, a decision problem can be solved in polynomial time iff the optimization problem can.

Sometimes some hard problems are time consuming for their optimum solution. Using approximation algorithms in these cases, algorithms that find near-optimal solutions to these problems are worked out.

Example 8. Suppose a graph G which contains vertices u and v. "find a path from u to v that uses the fewest edges" is then an optimization problem. This problem might have an answer say 5. A corresponding decision problem of this optimization problem would be "is there a path from u to v that uses 10 or fewer edges?" Then the answer can be simply 'yes' or 'no'.

Example 9. The max clique problem in a graph is an optimization problem which has to determine largest clique size in the graph.

Example 10. Let S be a sequence of real numbers. Finding an element with maximum frequency is an optimization problem.

Example 11. Consider a graph $G = (V, E)$. Finding chromatic number to colour G is an optimization problem.

Example 12. Consider a graph $G = (V, E)$. Finding max-clique of size k (some positive integer) of in G is an optimization problem.

13.3 CLASSIFICATION OF ALGORITHMS

Results retuned by an algorithm may be defined or may not be defined uniquely. Based on this factor algorithms are normally categorised into two groups.

(*i*) Deterministic Algorithms.

(*ii*) Non-deterministic Algorithms.

13.3.1 Deterministic Algorithms

A deterministic algorithm behaves as expected about the result of every operation of the algorithm. For a particular given input, deterministic algorithm always produces the same output, and the underlying machine will always pass through the same sequence of states.

A mathematical function always produces the same output for given an input, and then this might be considered one simple model for deterministic algorithms. Any deterministic algorithm's complexity can be reduced to $O(1)$.

Deterministic algorithms are related with some major problems.

(*i*) Finding deterministic algorithms of some problems are hard.

e.g., the algorithm that determines whether a given number is prime is considerably slower in practice.

(*ii*) Another major problem is that sometimes we don't want the results to be predictable.

e.g., in cryptography private keys are often generated using pseudo-random number generator. Sometimes, a hacker able to guess this generated private key. It can be avoided by using a cryptographically secure pseudo-random number generator.

13.3.2 Non-deterministic Algorithms

A non-deterministic algorithm is an algorithm which has one or more choices and without any specification any one of them can be chosen. So, non-deterministic algorithms produce an answer by a series of "correct guesses".

Specification of non-deterministic algorithms in general requires introduction of three new functions:

(*i*) Choice(S): arbitrarily chooses one of the elements of set S.

(*ii*) Failure(): indicates an unsuccessful completion.

(*iii*) Success(): indicates a successful completion.

Running time of each of the above three functions is $O(1)$. The set of choices is always considered. This set leads to a successful completion of the operation and helps in termination of the algorithm. A non-deterministic algorithm terminates unsuccessfully iff no set of choices whenever leading to a successful indication is on hand.

Some problems that can be solved by non-deterministic algorithms are:

(*i*) N-Queens Problem.

(*ii*) Vertex Set Covering Problem.

(*iii*) Construction of Spanning Tree.

(*iv*) Shopping List Problem.

(*v*) Searching Problem.

(*vi*) Merge Sorting, etc.

13.4 CLASSES OF PROBLEMS

Based on the above descriptions, the problems on the basis of their behaviour, techniques for solving them, running times, etc. are formed to groups. There are mainly four classes of problems exist.

13.4.1 P Class

The term *P* means polynomial time (*P*-time) also known as deterministic time (*D*-time) is one of the most fundamental complexity classes. The class of problems that can be solved in polynomial time is called *P* class. This class of problems are solved in $O(n^k)$ time for a constant k of lower value and input size *n* by some deterministic algorithms. These are basically tractable. A set of decision problems with yes/no answer including previous examples belongs to this *P* class such as:

(*i*) Decision versions of linear programming.
(*ii*) Calculating the greatest common divisor.
(*iii*) Finding a maximum matching.

13.4.2 NP Class

NP is one of the most fundamental classes of problems in computational complexity theory. The abbreviation *NP* refers to non-deterministic polynomial time (*NP*-time or *N*-time). The class of problems that are verifiable in polynomial time is called *NP* class problems. The statement means that if we able to certify a solution anyhow then it becomes ease of proving its correctness using polynomial time algorithms.

Example 13. Hamiltonian cycle problem is in *NP* class problem. In this problem a certificate will be the sequence of vertices $\{v_1, v_2, ..., v_n\}$ for the undigraph $G = (V, E)$ where $v_i \in V$ for $i = 1, 2, ..., n$. It is possible to verify in polynomial time whether the edges (v_i, v_{i+1}) E, $i = 1, 2, ..., n - 1$ and (v_n, v_1) take part to form the cycle or not.

Example 14. Graph colouring problem is an *NP* class problem.

Equivalently, many natural computer science problems like the set of decision problems on searching and optimization problems that can be solved in polynomial time on a non-deterministic. Turing machine or by a non-deterministic algorithm are in the *NP* class problems. Here is a list of problems that are in *NP*.

(*i*) All problems in *P*.
(*ii*) The class of decision problems for which there exists a non-deterministic algorithm that runs in polynomial time.
(*iii*) Integer factorization problem.
(*iv*) The graph isomorphism problem.
(*v*) All *NP*-complete problems *e.g.,* travelling salesman problem.

13.4.3 NP–Complete Class

The class of problems "*NP*-complete" stands for the sub-class of decision problems in *NP* that are hardest. The class *NP*-Complete is abbreviated as *NP-C* or *NPC* and comes from:

– *Non-deterministic Polynomial*
– *Complete — "Solve one, Solve them all.*

A decision problem *L* is said to be *NP*-complete if:

(*i*) *L* is in *NP* that means any given solution to this problem can be verified quickly in polynomial time.

(*ii*) Every problem in *NP* is reducible to *L* in polynomial time.

It means that if a solution to this *L* can be verified in polynomial time then it can be shown to be in *NP*. A problem that satisfies second condition is said to be *NP*-hard that will be examined in recent. Informally it is believed that if a NPC problem has a solution in polynomial time then all other NPC problems can be solved in polynomial time.

The list given below is the examples of some well-known problems that are *NP*-complete when expressed as decision problems.

(*i*) Boolean Circuit Satisfiability problem (C-SAT).
(*ii*) *N*-puzzle problem.
(*iii*) Knapsack problem.
(*iv*) Hamiltonian path problem.
(*v*) Travelling salesman problem.
(*vi*) Subgraph isomorphism problem.
(*vii*) Subset sum problem.
(*viii*) Clique Decision Problem (CDP).
(*ix*) Vertex cover problem.
(*x*) Graph colouring problem.

Obviously a question may arise in our mind how to prove an unknown problem as NPC problem. By the following easiest way we may able to prove that some new problem is NPC.

(*i*) First prove that it is in *NP*.
(*ii*) Assume it is not *NP* complete.
(*iii*) Reduce some known *NP*-complete problem to it.

The following techniques can be applied to solve NPC problems and they often give rise to substantially faster algorithms:

(*i*) Approximation Approach
(*ii*) Randomisation
(*iii*) Restriction
(*iv*) Parameterisation
(*v*) Heuristics

13.4.4 NP-Hard Class

The class *NP*-hard written as NPH or *NP-H* stands for non-deterministic polynomial time hard. The class can be defined as:

(*i*) NPH is a class of problems that are "at least as hard as the hardest problems in *NP*".
(*ii*) A problem *H* is *NP*-hard iff there is a NPC problem *L* that is polynomial time reducible to *H* (*i.e.*, *L H*).
(*iii*) A problem *H* is said to be NPH if satisfiability reduces to *H*.
(*iv*) If a NPC problem *L* can be solved in polynomial time by an oracle machine with an oracle for *H*.

NP-hard problems are generally of the following types-decision problems, search problems, or optimization problems. To prove a problem *H* is *NP*-hard, reduce a known *NP*-hard problem to *H*. If a NPH problem can be solved in polynomial time, then all NPC problems can also be solved in polynomial time. As a result, all NPC problems are NPH, but all NPH problems are not NPC.

Example 15. Given a set of integers. Does any non-empty subset exist that sum up to zero? This is a yes/no question. And it happens to be *NP-H*.

Suppose a set of some integers as {–7, –3, –2, 5, 8} is given and we wish to decide whether a subset of this set sum up to zero. The answer is 'yes' since the subset of integers {–3, –2, 5} corresponds to the sum (–3) + (–2) + 5 = 0.

Example 16. The Travelling Salesman Problem is another example of an optimization problem of NPH type.

Example 17. The Turing machine halting problem is decision problem that is *NP*-hard but not *NP*-complete. The problem asks "given a program and its input, will it run forever?" This is a yes/no question, so this is a decision problem.

Consider the Boolean circuit satisfiability problem which can be reduced to the halting problem by transforming it to the description of a Turing machine that tries all truth values. When it finds one that satisfies the formula it halts and otherwise it goes into an infinite loop.

Example 18. Subgraph Isomorphism problem is sometimes thought to be *NP*-hard only but isn't thought to be *NP*-complete. Though, it is in *NP* class.

13.5 RELATIONSHIP AMONG CLASSES OF PROBLEMS

From the discussion in previous section we have learned that:

(*i*) *P* problems are decision problems that are solvable in polynomial time by deterministic algorithms.

(*ii*) *NP* problems are decision problems that are solvable in polynomial time by non-deterministic algorithms and their solution is verifiable using a deterministic algorithm. There are certain problems which are only *NP* types not *P* types.

(*iii*) *NP*-Complete problems are only decision problems which are both *NP* and *NP*-Hard types.

(*iv*) *NP*-Hard problems are both optimization and decision problems. Even there are certain decision problems which are only *NP*-Hard types not *NP*-Complete types.

Relationship among these four classes of problems can be diagrammatically represented as–

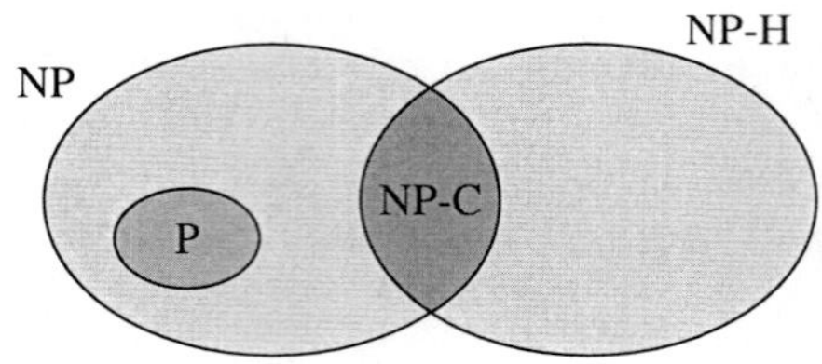

Figure 13.2 Relationship between *P*, *NP*, *NP*-complete, and *NP*-hard set of problems.

13.6 REDUCIBILITY

Solving a problem similar to another problem already solved, often we transform the instances of first problem into instances of the second problem. Using our existing solutions our desired solution is consequently obtained. In computational complexity theory, this procedure is called as reduction.

Reduction is a transformation of one problem into another problem. This is called the "many to one" or many-one reduction. The algorithm is called reduction algorithm. Depending on the reductions used complexity classes are defined on a set of problems.

A problem L_1 is reducible to another problem L_2 written as $L_1 \preccurlyeq L_2$ if L_2 has solutions and give these solutions to L_1 whenever L_1 has solutions. Thus, solving L_1 can be no harder than solving L_2. The symbol $\preccurlyeq$ is used to indicate the type of reduction being used.

Let α = The instances of a new decision problem $L_{1.}$
β = The instances of an old decision problem $L_{2.}$

In order to solve this L_1 in polynomial time using reduction algorithm, go through the following steps.

(*i*) Transform instances α of L_1 by the reduction algorithm to an instance β of $L_{2.}$
(*ii*) Run the polynomial time bounded decision algorithm on β for L_2.
(*iii*) Use the answer thus obtained for β as the answer for α.

Diagrammatically the above steps can be represented as shown below:

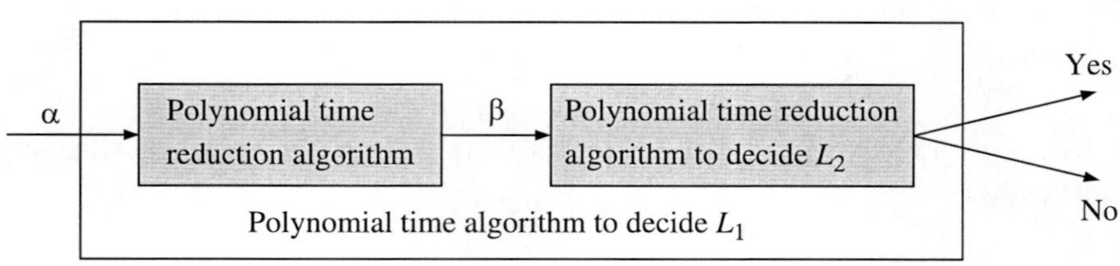

Figure 13.3 Reduction of $L_1 \preccurlyeq L_2$.

13.7 COOK'S THEOREM

Stephen Cook has published his paper "The complexity of theorem proving procedures" for the *NP*-completeness in 1971. He said that any problem in *NP* can be reduced in polynomial time to a problem of determining whether a Boolean formula is satisfiable using a deterministic Turing machine. He with Levin proposed a statement which is known as the Cook–Levin theorem or simply Cook's theorem. The theorem can be stated by any one of the followings.

(*i*) Any *NP* problem can be converted to SAT in polynomial time.

(*ii*) Circuit-Satisfiability (C-SAT) is *NP*-complete.

(*iii*) Satisfiability is in *P* iff $P = NP$.

(*iv*) Satisfiability is *NP*-complete.

13.8 SATISFIABILITY

Satisfiability is the problem of determining whether a Boolean formula evaluates True or 1 for the variables of the given Boolean formula assigned in a particular way. Similarly, if no such assignments exist, the function expressed by the formula is identically False or 0 for all possible variable assignments. In this case it is said the function is unsatisfiable. Due to its binary/dual nature, this problem is commonly referred a Boolean or propositional satisfiability problem. The abbreviation "SAT" is usually used to denote it. So, the SAT problem is to determine whether an input Boolean or propositional formula is satisfiable. Boolean circuit is a combinational circuit formed of AND, OR, and NOT logic gates. The Boolean circuit satisfiability problem (C-SAT) or propositional satisfiability problem (P-SAT) is a decision problem. Instance of this problem is then obviously a Boolean expression or formula written using only logic gates (AND, OR, NOT), variables, and parentheses.

Most important thing regarding SAT problems is that it covers wide area of applications such as computer science as well as theoretical computer science, hardware design, artificial intelligence, algorithms, automation, and verification, etc.

Example 19. Consider a Boolean formula $f(x, y) = xy$. Now the circuit and truth table for this Boolean formula are given below.

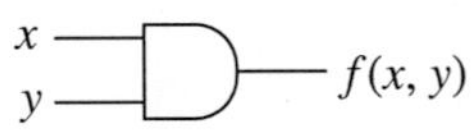

Input		Output
x	y	$f(x, y)$
0	0	0
0	1	0
1	0	0
1	1	1

(*a*) Boolean Circuit for $f(x, y)$ (*b*) Truth Table for $f(x, y)$

Figure 13.4 Boolean formula satisfiability.

From the above truth table it is seen that the assignments $x = 1$, $y = 1$ makes $f(x, y) = 1$. So, this Boolean formula is satisfiable.

Example 20. Write a pseudocode to find sum of two numbers. Consider the following Boolean circuit as shown in Figure 13.5.

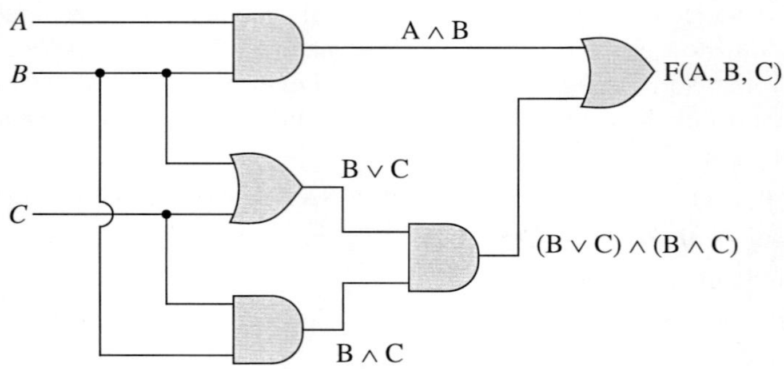

Figure 13.5 A boolean circuit of 3 variables.

The Boolean formula $F(A, B, C) = (A \wedge C) \vee (B \vee C) \wedge (B \wedge C)$. The corresponding truth table is:

INPUT			*OUTPUT*			
A	B	C	$(A \wedge B)$	$(B \vee C)$	$(B \wedge C)$	$F(A, B, C)$
0	0	0	0	0	0	0
0	0	1	0	1	0	0
0	1	0	0	1	0	0
0	1	1	0	1	1	1
1	0	0	0	0	0	0
1	0	1	0	1	0	0
1	1	0	1	1	0	0
1	1	1	1	1	1	1

From the truth table it is seen that the assignments $A = B = C = 1$ make $f(x, y) = 1$. So, this Boolean formula $F(A, B, C) = (A \wedge B) \vee (B \vee C) \wedge (B \wedge C)$ is satisfiable.

13.9 C-SAT IS NP-C

A combinational Boolean circuit is satisfiable *i.e.,* C-SAT iff for a given set of input the circuit always returns 1 or True as output. To prove C-SAT is *NP*-C, show the followings.

(*i*) C-SAT is *NP*
(*ii*) C-SAT is *NP-H*

- **C-SAT is NP**

Let, A = Two input non-deterministic polynomial time algorithm used to verify C-SAT.

C = Combinational Boolean circuit.

First guess the input values as well as corresponding output values of each logic gate of C. Represent them in a graph. Each logic gate is a vertex and connection is an edge. Then visit each logic gate *i.e.,* each vertex in C with at least one incoming edge or input to verify the guessed values. This evaluation process is verified using the non-deterministic algorithm A in polynomial time. If the verification for every gate succeeds and the output is 1 *i.e.,* the algorithm returns 'yes', there is a satisfying assignment of the input variables in the circuit C. Thus the algorithm shows 'yes' for selective statements of a possible collection of outcomes in polynomial time. Similarly there must be a possible satisfying assignment of input values for C. Thus, C-SAT is *NP*.

- **C-SAT is NP-H**

The second part is to show C-SAT is *NP*-hard meaning that every problem in *NP* is polynomial reducible to C-SAT.

Suppose L = A language in *NP* that defines some decision problem.

C = A Boolean circuit.

D = Another polynomial algorithm to be designed.

F = Reduction function computed by D and maps any string x in L to $C = F(x)$ iff C is in C-SAT.

Y = A certificate.

Because of x in L, there exists a deterministic algorithm A that verifies any x in L in polynomial time $p(n)$ where n is size of x. The algorithm D will use this algorithm A to compute the reduction function F. Consider the worst case time complexity to compute x is $p(n) = O(n^k)$ for some constant $k \geq 1$ and the length of the certificate Y is $O(n^k)$. Therefore, the algorithm A computes through a sequence of configurations or reductions.

The reduction algorithm D computes the initial configuration of the circuit C *i.e.,* computes $C = F(x)$ that is satisfiable for the string x in L for the certificate Y such that $A(x, Y) = 1$. To do that D first computes $n = |x|$ followed by constructing a circuit C'. Thus $C = F(x)$ will be obtained by slightly modifying C'. First the input to C' is corresponding to A *i.e.,* all inputs except x to the circuit correspond to Y. Second, ignore all the intermediate outputs except last one that corresponds to the output of A. Thus the circuit computes $C(Y) = A(x, Y)$ for any Y of length $O(n^k)$. Therefore the reduction algorithm D computes this circuit and outputs for x. since Y is certificate then the output of C *i.e.,* $C(Y) = A(x, Y) = 1$. So C is satisfiable. Oppositely, if C is satisfiable then for any input Y to C, $C(Y) = 1$ that concludes that $A(x, Y) = 1$. Thus D correctly computes reduction function F. thus C-SAT is at least hard as any language L in *NP*. Therefore, C-SAT is *NP*-hard.

Thus, it is concluded that C-SAT is *NP*-Complete.

13.10 CLIQUE DECISION PROBLEM IS NP-C

A clique is a complete subgraph in a graph $G = (V, E)$. Size of a clique means the number of vertices $k \leq |V|$ in it. The clique problem is the problem of determining whether a graph contains a clique of at least a given size k. The corresponding optimization problem *i.e.,* the maximum clique problem is to find the largest clique in a graph. The decision version of the problem is that whether a graph contains a clique of size k or not. It is seen that the number of possible cliques of size k on a graph G of size $n = |V|$ is equal to

$$\binom{n}{k} = \frac{n!}{k!(n-k)!}$$

The clique decision problem (CDP) is *NP*-C. To prove this, prove the followings:

(*i*) CDP is *NP*.

(*ii*) CDP is *NP-H*.

- **CDP is NP**

Suppose, $g = (V', E')$ be clique of a graph $G = (V, E)$. Then, $V' \subseteq V$ and use this set V' as a certificate for G. Checking each pair of vertices $(u, v) \in V'$ and the edge $(u, v) \in E$, it can be verified whether g is a clique or not. This can be accomplished in polynomial time by a non-deterministic algorithm. So, conclusion is that CDP is *NP*.

- **CDP is NP-H**

We know that conjunctive Normal Form or CNF is *NP-H* type. Now, we will show that CDP can be reduced to CNF.

Suppose, $F = C_1 \vee C_2 \vee ... \vee C_n$ be a Boolean formula in CNF. And consider x_i, $i = 1, 2, ...$ are the variables in F. Construct the graph $G = (V, E)$ from F such that $g = (V', E')$ be a clique of G. It is only possible iff F is satisfiable. Now this G can easily be computed that means if there is a polynomial time algorithm for CDP, then there is also a polynomial time algorithm for this construction. And this time complexity varies with the length of F. therefore, CNF-satisfiability CDP. Thus, CDP *NP-H*.

And so, CDP is *NP-C*.

13.11 SOME COMMON QUESTIONS

Question 1. What makes algorithms non-deterministic?

Answer: A variety of factors can cause an algorithm to behave as non-deterministic if:

(*i*) It uses more external states such as user input, a global variable, a hardware time, or other data except the input.

(*ii*) It operates in timing-sensitive way, *e.g.,* when multiple processors write the same data at the same time, the precise order of writing data by each processor affects the result.

(*iii*) A hardware error occurs then it may cause its states to change in an unexpected way, etc.

Question 2. Is P = NP? Justify your answer.

Answer: The class of P problems is a sub-class of NP problems. Because a problem in P can be solvable in polynomial time without giving any certificate and NP type problem is trying to solve (verifiable) in polynomial time. Thus, $P \subseteq NP$.

Question 3. Show that 2-SAT is P.

Answer: To show that 2-SAT belongs to P, we should consider a polynomial time algorithm for solving 2-SAT. The idea is to construct a digraph $G(F) = (V, E)$ where

(*a*) $V = \{x, x' \mid \text{the literal } x \in F\}$

(*b*) $E = \{(x', y), (y', x) \mid (x \vee y) \in F\} \subseteq (V \times V)$

Here edges of G are interpreted as implications. Because, $x \vee y = (x' \to y)$ $(y' \to x)$. Thus, an assignment of F corresponds to an assignment of G and G is satisfiable iff for no x there exists a path from x to x' or vice-versa in G. It means x must not imply x', and x' must not imply x.

Thus, for input of size $n = |F|$, 2-SAT can be solved by the following steps:

(*i*) Construct the graph $G(F)$, which has $O(n)$ vertices and edges

(*ii*) search a path from x to $x¢$ for each vertex x. it reueires $O(n^2)$ time.

(*iii*) If no such paths exist, return 'yes', otherwise return 'no'

This algorithm has complexity $O(n^2)$.

Question 4. What are the two classes of non-polynomial time problems?

Answer: These are (1) NP-hard and (2) NP-complete.

Question 5. State whether true or false.

Answer: 1. All NP-complete problems are NP-hard – TRUE.

2. All NP-hard problems are not NP-complete – TRUE.

Question 6. What do you mean by Satisfiability?

Answer: Let $x_1, x_2, x_3 ..., x_n$ denotes Boolean variables. Let x_i denotes the relation of x_i.

A literal is either a variable or its negation. A formula in the prepositional calculus is an expression that can be constructed using literals and the operators and $\wedge$ or $\vee$. A clause is a formula with at least one positive literal. The satisfiability problem is to determine if a formula is true for some assignment of truth.

Question 7. What is non-deterministic algorithm?

Answer: Algorithm that contains operations whose outcomes are not uniquely defined but are limited to specified set of possibilities.

Question 8. What is deterministic algorithm?

Answer: Algorithm that contains whose outcomes are uniquely defined.

Question 9. What do you mean by optimization problem?

Answer: Computational problem where we try to maximize or minimize the objective function.

Question 10. What do you mean by decision problem?

Answer: Computational problem with intended output of yes or no *i.e.,* binary 1 or 0.

Question 11. Define Cook's theorem.

Answer: Satisfiability is in *P* if and only if *P* = *NP*.

Question 12. What is a relation between *NP* hard and *NP* complete problem?

Answer: Relation could be as.

- **NP-Hard Problem:** A problem *L* is *NP*-hard if any only if satisfiability reduces to *L*.
- **NP-Complete:** A problem *L* is *NP*-complete if and only if *L* is *NP*-hard and $L \in NP$.
- There are NP-hard problems that are not *NP*-complete. E.g., *Halting problem is NP-hard decision problem, but it is not NP-complete.*

Question 13. Is the dynamic-programming algorithm for the 0-1 Knapsack problem a polynomial-time algorithm?

Answer: The dynamic programming algorithm for the Knapsack problem runs in time $O(nW)$ where *n* is the number of items and *W* is the maximum weight of the items. Since *W* does not depend on the size of the input the algorithm is not polynomial.

Question 14. Show that any language in NP can be decided by an algorithm running in time $2^{o(n^k)}$ for some constant *k*.

Answer: Any NP complete language can be decided by an algorithm running in time $2^{o(n^k)}$ for some constant *k* simply by trying all the possible certificates.

Exercises

1. Given a list of *n* integers. Give an efficient algorithm to decide whether the list is in non-decreasing order? What is the time complexity of your algorithm?
2. Given two sets of integers. Give an efficient algorithm to decide whether their intersection is empty. What is the time complexity of your algorithm?
3. Given an undirected graph *G*, is it 2-colourable. Decide with a polynomial time algorithm. [**Hint:** Colour the first vertex white, all adjacent vertices black]
4. Design a polynomial time algorithm for 2-SAT problem.
5. Design a non-deterministic algorithm to solve Satisfiability problem.
6. Design a non-deterministic algorithm to solve the TSP problem.
7. $P \subseteq NP$. Justify your answer.
8. *P* is a superset of *NP*. Justify your answer.
9. Let *P* and *Q* be two problems such that $P \propto_{\text{poly}} Q$. Let *Q* can be solved in $O(n^q)$ time and the reduction needs $O(n^r)$ time. Show that *P* can be solved in $O(n^{qr})$ time.
10. Show that Hamiltonian cycle problem for undirected graphs, is *NP*-complete.
11. Show that Hamiltonian cycle problem for digraphs, is *NP*-complete.

12. Find a clique of a given undirected graph $G = (V,E)$ with n vertices. Give a suitable algorithm too. Your algorithm should be polynomial time bounded.
13. Show that the Clique problem is *NP*-complete.
14. Consider the following formula f :

$$(\overline{X}_1 \vee X_2 \vee \overline{X}_3) \wedge (\overline{X}_1 \vee \overline{X}_2) \wedge (\overline{X}_1 \vee X_3) \wedge (\overline{X}_2 \vee X_3)$$

   (*a*) Transform the above formula into a Clique for which the answer is *yes* iff the above formula is satisfiable.
   (*b*) Find a clique of size 4 in your graph and convert it into a satisfying assignment for the formula given above.
15. Any cover of a clique of size n must have exactly $n - 1$ vertices. Justify.
16. If $P = NP$, a polynomial time algorithm for the Satisfiability problem.
17. Knapsack problem with Knapsack capacity C and n items, can be solved in time $\Theta(nC)$. Again it is *NP*-complete. Is there any contradiction? Explain.
18. If an *NP*-complete problem Q is solved in polynomial time, then prove that $NP = P$.
19. Prove that $NP = P$ iff some *NP*-complete problem $L \in P$.
20. Finding a shortest simple path between two vertices p and q in a graph with negative edge weights is *NP*-complete.
21. Show that 3-SAT problem is *NP*-complete.
22. Show that 3-Colouring problem is *NP*-complete.

University Questions

1. Define classes *P*, *NP* and *NP*-complete, *NP*-Hard.
   [WBUT'08, '09, PTU'07, '09, RTU'08, '09, JNTU'08, BPUT'05, DOEACC'07, 08]
2. Give example of each of the above classes. ***[RTU'09].***
3. Discuss problem class in detail. ***[UPTU'06]***
4. Discuss the relationship between the classes *P*, *NP*, *NP*-complete and *NP*-Hard problem with suitable example.
   [WBUT'09, BPUT'05, 07, UPTU'05, PTU'07, 08, 09, MDU'08, IETE ALCCS'11]
5. Differentiate among *P*, *NP*, *NP*-complete and *NP*-hard class of problems with suitable examples.
   [PTU'07, 08, 09, MDU'08, BPUT'05, IETE ALCCS'11, DOEACC'09]
6. State which of the sentences given below are correct, false or unknown: ***[DOEACC'08]***
   (*i*) If a problem is in *P*, it must also be in *NP*.
   (*ii*) If a problem is in *NP*, it must also be in *P*.
   (*iii*) If a problem is *NP*-complete, it must also be in *NP*.
   (*iv*) If a problem is not in *P*, it must be *NP*-complete.
7. What is polynomial time reducibility? Give example(s). ***[DOEACC'09]***

8. What are the three properties of *NP*-complete problems? ***[BPUT'08]***
9. Write a note on *NP*-hard problems. ***[IGNOU'09]***
10. Write steps through which we can prove that a particular problem is *NP*-Complete. ***[IGNOU'07]***
11. Prove that $P \subseteq NP$. ***[RTU'09, JNTU'08]***
12. Under what circumstances the condition "$P = NP$" is true? ***[DOEACC'09]***
13. State the independent-set problem. Formulate a related decision problem for this problem and prove that it is *NP*-complete. ***[DOEACC'08]***
14. Describe circuit satisfiability problem and prove that circuit-SAT is in *NP*. ***[WBUT'08]***
15. Discuss what you mean by polynomial reductions. ***[WBUT'09]***
16. Explain vertex cover and set cover problem. ***[RTU'09, 10]***
17. Explain the satisfiability problem. ***[JNTU'09, 11]***
18. What is meant by Halting problem explain with an example. ***[JNTU'08]***
19. Prove that vertex cover is *NP* complete. ***[RTU'09]***
20. What is Cook's theorem? Explain. ***[RTU'09, '10, MDU'08, JNTU'09, DOEACC'09]***
21. Describe Clique Decision Problem (CDP). ***[WBUT'09, JNTU'08]***
22. Write the algorithm for the CDP. ***[JNTU'08]***
23. Prove that CDP is *NP*-complete. ***[WBUT'09, IGNOU'07, DOEACC'06]***
24. Show that 2SAT is in *P* but 3SAT is *NP*-complete. ***[WBUT'10]***
25. Show that 3-colouring problem is *NP*-complete. ***[WBUT'10]***
26. Mention a strategy that can be adopted to show that a problem X is *NP*-hard. ***[DOEACC'07]***
27. Assuming 3-CNF satisfiability problem to be *NP*-Complete, prove clique problem is also *NP*-Complete. ***[RTU'10]***
28. Write short notes on Polynomial time algorithm. ***[PTU'08]***
29. Write short notes on *NP*-Completeness. ***[UPTU'10, PTU'07]***
30. Differentiate between deterministic and non-deterministic algorithms? ***[PTU'09, MDU'08, IGNOU'07]***
31. Show that travelling salesman problem is *NP*-complete. ***[RTU'10, DOEACC'06]***
32. Write a non-deterministic Knapsack algorithm. ***[JNTU'08]***
33. Explain Decidable and Undecidable problems. Give example for each. ***[IGNOU'09]***
34. For problem *P*, if we are given an input *I* and a possible answer *A*, and we find way to verify whether or not A really is a valid answers no *P* given *I*, then what kind of problem is *P*? ***[BPUT'08]***
35. What do you mean by non-deterministic algorithms? ***[WBUT'09, PTU'09]***

❑❑❑

14 Chapter Approximation Algorithms

14.1 INTRODUCTION

There are many combinatorial optimization problems that are intractable. In the previous chapter various *NP*-hard combinatorial optimization problems have been studied. Again some of them are *NP-C* type as well. An optimal solution of many of these *NP*-complete problems cannot be solved efficiently *i.e.,* finding the solutions in reasonable amount of time are too hard. As these problems are very much related in our practical life and can be solved in polynomial time therefore we should have some efficient algorithmic approach which might not provide the optimal solution but most acceptable solution "reasonable" (near to optimal solution). Three approaches are generally followed to solve these problems.

(*i*) *General Optimization Methods:* Backtracking, branch and bound, neural networks, genetic algorithms, etc. are some techniques.

(*ii*) *Exponential Algorithms:* The best known algorithms for *NP-H* problems.

(*iii*) *Approximation Algorithms:* Solve *NP-H* problems in polynomial time either in worst case or average case finding a near optimal solution.

Ideally, an approximation algorithm is the best choice because it approximates the optimal solution. Its performance bound guarantees that this approximate solution will not be far away from the neighbourhood of the exact solution. However, approximation algorithms are increasingly becoming too expensive finding an optimal solution in polynomial time.

An approximate solution of an optimization problem *P* using approximation algorithm can be obtained as: first forget that the algorithm that solves *P* does not always result an optimal solution. Instead it has a feasible solution (approximate value) which is near to the optimal solution. Select the most appropriate one.

14.2 APPROXIMATION ALGORITHMS: APPLICATIONS

Since approximation approach is related strictly with *NP-H* problems, in this chapter some common *NP-C* problems not all have been explored with their polynomial time approximation algorithmic approach. Some application areas where approximation approaches are preferred over others are stated below.

(*i*) Bin packing problem.

(*ii*) Set covering problem.

(*iii*) 0/1 Knapsack problem.
(*iv*) Vertex covering problem.
(*v*) Travelling salesman problem.
(*vi*) Planar graph colouring problem.
(*vii*) Maximum programs stored problem
(*viii*) Scheduling independent tasks problem, etc.

14.3 APPROXIMATION ALGORITHMS: TYPES

Several kinds of approximation algorithms can be defined. Some of them have been defined here with the following considerations.

Suppose, P = A *NP-H* type optimization problem.
I = Instances of P.
A = An approximation algorithm for P.
$O_s(I)$ = The value of optimal solution to I for P.
$F_s(I)$ = The value of feasible solution to I returned by A.

1. *Absolute Approximation Algorithm:* If $| O_s(I) - F_s(I) | \leq k$ for some constant k, then A is said as an absolute approximation algorithm for P.
2. *F(n)-Approximation Algorithm:* When $\frac{|O_s(I) - F_s(I)|}{O_s(I)} \leq F(n)$ for the problem size $n = |I|$ and $| O_s(I) | > 0$, then A is said to be a $F(n)$-approximation algorithm for P.
3. *ε-approximation algorithm:* The above $F(n)$-approximation algorithm is said to be an - approximation algorithm when $F(n) \leq \varepsilon$, where is a constant and it is normally greater than 1 if P is a maximization problem.
4. *Approximation scheme:* If for every non-negative constant ε, if $\frac{|O_s(I) - F_s(I)|}{O_s(I)}$ for the input instance I of P, assuming $| O_s(I) | > 0$, the algorithm A for represented as $A(\varepsilon)$ results feasible solutions. And then $A(\varepsilon)$ is called approximation scheme.
5. *Polynomial Time Approximation Scheme:* For every non-negative fixed constant ε, if the computing time an approximation scheme is polynomial time bounded for the input instance I of the problem P, then that scheme is called polynomial time approximation scheme.
6. *Fully Polynomial Time Approximation Scheme:* If the computing time of an approximation scheme is fully polynomial for both I and $1/\varepsilon$, then the scheme is called fully polynomial time approximation scheme.

Naturally, our desired approximation algorithms should be absolute approximation algorithms. Unfortunately, we don't have it for most of the NP-H problems.

14.4 PERFORMANCE GUARANTEES

In this chapter we would restrict ourselves to those approximation algorithms that run in time polynomial time for any input size n and return a solution which is a member of feasible solution region F.

Definition: A feasible solution value in F of an approximation algorithm is thus guaranteed to be approximate or near-optimal in some well-defined sense. Such a guarantee is called the performance guarantee.

Performance guarantees may be of two kinds:

(*i*) *Absolute Performance Guarantee:* It is meaning that the additive difference between the optimal and the approximate value is bounded.

(*ii*) *Relative Performance Guarantee:* The approximate value remains within a multiplicative factor of the optimal value.

Before an approximation algorithm with a performance guarantee returns a solution for a problem, we do implicitly have a bound on the optimal solution for the problem. Designing approximation algorithms perhaps is the most fundamental challenging tasks to obtain such bounds. Because if one unable to compute the optimal value, then how it can be expected to prove that the approximate value is near-optimal solution.

Three common techniques what are known to us as:

(*i*) Witnesses,

(*ii*) Relaxation, and

(*iii*) Coarsening.

- ***Witnesses:*** Intuitively, a witness constructs a short and easily verified proof for the optimal solution providing a dual role (maximal or minimal) to feasible solutions.
- ***Relaxation:*** A new optimization problem formulated from a given optimization problem is then called a relaxation of the original problem. *e.g.*, both Linear programming and Randomised Rounding use this approach.
- ***Coarsening:*** This means to restrict a problem instance in to a less complex feasible region so that the resulted problem can be solved efficiently. Dynamic programming is used for this in general.

14.5 PERFORMANCE RATIO

Sometimes it is required as well as possible to prove certain properties about the approximations of the optimum results of some approximation algorithms.

Suppose, P = A NP-H type optimization problem.

I = Instances of P.

A = An approximation algorithm for P.

$C_o(I)$ = The cost of optimal solution to I for P.

$C_f(I)$ = The cost of feasible solution to I produced by A.

We are working with optimization problems, so our interests are in finding a ratio $C_o(I)/C_f(I)$ or $C_f(I)/C_o(I)$ as small as possible for a solution of maximisation or minimisation problems respectively for a given problem instance I in the set of feasible solutions.

Definition: The approximation ratio denoted by $\rho(n)$ of an approximation algorithm for any input size n of the given problem instance I is defined by the following inequality as

$$\rho(n) \leq \max(C_o(I)/C_f(I),\ C_f(I)/C_o(I))$$

This means that the algorithm A has a ratio bound of $\rho(n)$ if the feasible cost $C_f(I)$ of the feasible solution produced by A is within a factor of $\rho(n)$ of the optimal cost $C_o(I)$ of an optimal solution. This definition applies for both minimization and maximization problems. In this case the algorithm A is called ρ-approximation algorithm.

If solution produced by approximation algorithm is true optimal solution *i.e.,* $C_f(I) = C_o(I)$ then clearly $\rho(n) = 1$. Otherwise, $\rho(n)$ is always greater than 1.

If $C_f(I) \geq C_o(I)$, the problem is minimization problem, and the cost of the approximate solution $C_f(I)$ is larger than the cost of an optimal solution $C_o(I)$ by the factor $C_f(I)/C_o(I)$. Similarly, if $C_o(I) \geq C_f(I)$, the problem is a maximization problem, and the cost of an optimal solution $C_o(I)$ is larger than the cost of the approximate solution $C_f(I)$ by the factor $C_o(I)/C_f(I)$.

14.6 RELATIVE ERROR

The factor ρ is also called the relative performance guarantee. An approximation algorithm has an absolute performance guarantee or bounded error or relative error $\varepsilon(n)$ if it has a proof for every instance I.

The relative error denoted by $\varepsilon(n)$ of an approximate algorithm for any input size n can be defined as $\varepsilon(n) = |C_f(I) - C_o(I)| / C_o(I)$.

And if $\varepsilon(n) \geq |C_f(I) - C_o(I)| / C_o(I)$ then the approximation algorithm has relative bound of $\varepsilon(n)$.

14.7 0/1 KNAPSACK PROBLEM: APPROXIMATION ALGORITHM

The Knapsack problem sometimes called rucksack problem is a combinatorial optimization problem. The problem states that: "Given a set I of n items, each with a weight $w_i > 0$ and a value $v_i > 0$, $i = 1, 2, ..., n$. Determine the number of items from I so that sum of the weight of each item thus considered is less $\leq W$, the capacity of the Knapsack and the total value is as large as possible". Hence, the problem formally can be represented as:

(*i*) Maximise $\sum_{i=1}^{n} V_i x_i$

(*ii*) Subject to $\sum_{i=1}^{n} w_i x_i \leq W$

(*iii*) $x_i = \begin{cases} 0, \text{ if an item is not selected} \\ 1, \text{ if an item is selected} \end{cases}$

An approximation solution will be a feasible solution if the first condition mentioned above is satisfied. Obviously the feasible solution is filling the Knapsack and is any set $S \subseteq I$. This problem is NP-hard. It needs to devise a fully polynomial time approximation scheme for this problem. Obviously this approximation algorithm is for maximization problem. Consider any real constant $\varepsilon > 0$ that will return a feasible/approximate solution. And it is at least $(1 - \varepsilon)$ times the optimal value in polynomial in the input size and in $1/\varepsilon$.

Algorithm: 0/1_Knapsack_Approximation()

Terms Used: w = List of weights of n items = $< w_1, w_2, ..., w_n >$
v = List of values of n items = $< v_1, v_2, ..., v_n >$
W = Capacity of the Knapsack.
S = Subset of items, initially empty.
ε = Any real number > 0

Step 1: Find maximum value v_{max} from v.

Step 2: Repeat for $i = 1$ to n

- compute $v_i' = (nv_i) / (\varepsilon v_{max})$

Step 3: Run dynamic programming approach using v_i', w_i, W.

Step 4: Return the subset as a feasible solution.

Step 5: Exit.

Analysis: Already it is known to us that the running time for 0/1 Knapsack problem using dynamic programming is $O(n^2)$. Now, total time is due to the step 3. And in this case it is given by $O(n^2 v'_{max})$ *i.e.,* $O\left(\frac{n^3}{\varepsilon}\right)$.

14.8 SOME COMMON QUESTIONS

Question 1. Write three characteristics of approximation approach.

Answer: Most of the approximation algorithms are fast due to greedy heuristics. Besides they have marking characteristics such as:

(*i*) An approximation algorithm is a way of dealing with *NP*-completeness for optimization problem.

(*ii*) Approximation algorithm does not guarantee the best solution.

(*iii*) The goal of an approximation algorithm is to find as close as possible to the optimum value in polynomial time.

Exercises

1. Show that the performance ratio of the *MST* heuristic is achievable.
2. Give an instance of the Euclidean TSP problem on which the *MST* heuristic results in a performance ratio of 2.
3. Prove the theorem: number of vertices of odd degree in a graph is even.
4. Show that the performance ratio of the Knapsack algorithm using greedy is 2.
5. Evaluate $\sum_{k=0}^{m} \binom{n}{k}$. [**Ans.** $O(mn^m)$]
6. Show that Greedy approach in the following algorithm does not always result in a minimum vertex cover in an undirected graph.

   Algorithm: Find_Vertex_Cover()

   1. Sort the vertices by decreasing order of degree.
   2. Repeat until all edges are covered.
      (*a*) Pick a vertex of highest degree that is incident to at least one edge in the remaining graph.
      (*b*) Add it to the cover.
      (*c*) Delete all edges incident to that vertex.
7. Show that the performance ratio of the approximation algorithm in Exercise 14.6 is unbounded.
8. Give an approximation algorithm using greedy approach that results a clique of maximum size.
9. Consider the following approximation algorithm.

   1. Delete a vertex from *G* which is not connected to every other vertex in it.
   2. Repeat step 1 until the resulting graph is a clique.

   Show that this greedy approach does not always result in a clique of maximum size. Find performance ratio of the algorithm. [**Ans.** Unbounded]
10. Consider the following approximation algorithm.

    1. Set $C = \{\}$
    2. Repeat for all vertices v in G that is not in C or connected to every other vertex in C.
       (a) $C = C \cup \{v\}$

    Show that the performance ratio of the heuristic algorithm is unbounded.
11. Give an approximation algorithm to find chromatic number of an undirected graph. Check whether its performance ratio is bounded.
12. Give an approximation algorithm to find maximum number of vertices that are mutually disconnected from each other. Prove that its performance ratio is bounded.
13. Give an approximation algorithm that finds a vertex cover in a tree in $O(n)$ time.

University Questions

1. Write short notes on Approximation algorithms. *[WBUT'08, UPTU'10, PTU'09]*
2. What are approximation algorithms? *[UPTU'05, PTU'08, BPUT'07, '08, DOEACC'08]*
3. Define absolute approximation. *[PTU'08]*
4. Define E-approximation with example. *[PTU'08]*
5. Write short notes on Approximation schemes. *[WBUT'09]*
6. Explain how can we modify almost any algorithm to have a good best case running time? *[UPTU'05]*
7. How do you characterize approximation algorithms? *[BPUT'08]*
8. Explain approximation vertex cover. *[PTU'09]*
9. Write approximation algorithm for vertex cover problem. *[DOEACC'08]*
10. What is approximation ratio? *[UPTU'05, BPUT'07, '08]*
11. What are the key characteristic of a approximation algorithm? *[BPUT'08, DOEACC'08]*
12. Explain Circuit Satisfiability Problem. *[MDU'09]*
13. What do you mean by relaxation in the process of designing an approximation algorithm for NP-Hard problem? *[BPUT'08]*
14. What are the differences between heuristic and approximation algorithms? *[DOEACC'07, 09]*
15. Discuss the approximate and hard problems for Travelling Salesman Problem. *[UPTU'06]*
16. Approximate the Travelling Salesman problem with triangle Inequality. *[UPTU'05, BPUT'07,'08]*
17. *Travelling Salesman Problem (TSP)* finds a minimal cost tour in a connected weighted graph that touches every vertex of the graph exactly once and comes back to the initial vertex; show that the approximate solution of *TSP* is twice the *MST* of the graph. *[BPUT'07, DOEACC'09]*
18. Design an approximation algorithm for colouring a planar graph. *[DOEACC'07]*

❑❑❑

15

Chapter Large Integers Multiplication

15.1 INTRODUCTION

The ***divide-and-conquer*** approach discussed in chapter 4 employs here the same strategy on an instance of a problem. The problem which is perhaps initially larger in size is divided into two smaller instances. The smaller instances are themselves instances of the original problem. If the smaller instances are still too large to be solved, they can be enthusiastically divided into still smaller instances as like previous. This process of dividing the instances continues until they are small enough that a solution is readily obtainable. Solution to these smaller instances is combined to obtain the solution of the previous larger instance. The process is continued. The solution to the original instance can be obtained by combining these solutions. Therefore, the divide-and-conquer approach is a ***top-down*** approach. Meaning that, the solution to a *top-level* instance of a problem is obtained by going *down,* obtaining solutions to smaller instances, and moving up.

One great example to designing algorithms using divide-and-conquer, is most probably drawn from after the brilliant strategy employed by the French emperor Napoleon in the Battle of Austerlitz on December 2, 1805. A combined army of Austrians and Russians outnumbered Napoleon's army by about 15,000 soldiers. The Austro-Russian army launched a gigantic attack against the French right flank. Anticipating their attack, Napoleon drove against their center and split their forces in two. Because the two smaller armies were individually no match for Napoleon, they each suffered heavy losses and were compelled to retreat. By *dividing* the large army into two smaller armies and individually conquering these two smaller armies, Napoleon was able to *conquer* the large army.

We now introduce the divide-and-conquer approach with finding efficient way to multiplying large integers.

15.2 REPRESENTATION OF LARGE INTEGERS

One way to represent large integers in a straightforward way is to use an array of integers, in which each array cell stores one digit but in reverse order. For example, the integer 123456 can be represented in the array $S[0 \ldots n]$ or $S[n \ldots 0]$ as follows: first digit (LSB) 6 is stored in cell 1 *i.e.,* $S[1]$, second digit is stored in $S[2]$, and so on. The cell $S[0]$ is used for sign bit. If $S[0] = 0$ then the number is positive. If $S[0]=1$, then the number is negative.

$S[0 \ldots 5]$	0	1	2	3	4	5
Number	0	5	4	3	2	1

or

$S[5 \ldots 0]$	5	4	3	2	1	0
Number	1	2	3	4	5	0

Figure 15.1 Representation of Non-negative Integers in array.

Similarly, the number - 543,127 can be represented as below.

$S[0 \ldots 5]$	0	1	2	3	4	5	6
Number	1	7	2	1	3	4	5

or

$S[6 \ldots 0]$	6	5	4	3	2	1	0
Number	5	4	3	1	2	7	1

Figure 15.2 Representation of Negative Integers in array.

Another way to represent a large integer N of n digits is to use divide-and-conquer. First split N into two sub-integers U and V of m digits each in the following way.

$U = M/10^m$ and $V = M\%10^m$ where $m = n/2$.

Thus, $N = U \times 10^m + V$

Following are two examples of such splits.

$$\underbrace{567{,}832}_{\text{6 digits}} = \underbrace{567}_{\text{3 digits}} \times 10^3 + \underbrace{832}_{\text{3 digits}}$$

$$\underbrace{9{,}423{,}723}_{\text{7 digits}} = \underbrace{9423}_{\text{4 digits}} \times 10^3 + \underbrace{723}_{\text{3 digits}}$$

Figure 15.3 Multiplication of Two Integers.

In general, if n is the number of digits in the integer N, we will split the integer into two integers, one with $\lfloor n/2 \rfloor$ and the other with $\lfloor n/2 \rfloor$ if n is odd number otherwise $n/2$.

15.3 MULTIPLICATION OF LARGE INTEGERS

In chapter 12 section 8, we were introduced with some integer multiplication schemes. A simple quadratic-time algorithm for multiplying large integers is one that we learned in our school days. We will here develop one algorithm that is better than quadratic running time. Our algorithm is based on using divide-and-conquer to split an n-digit integer into two integers of approximately $n/2$ digits as mentioned in previous section.

Suppose, A, and B are two large integers each of n-digits length. We need to compute $C = A \times B$ *i.e.,* we want to multiply A, and B in a simplest way. The simplest approach based on divide-and-conquer is stated below.

(*a*) Divide each of A, and B into two integers of $m = n/2$ digits.

(*b*) Suppose, A is divided into U and V such that $A = U \times 10^m + V$.

(*c*) Suppose, B is divided into X and Y such that $B = X \times 10^m + Y$.

(*d*) Multiply A, and B as–

$$\begin{aligned} C &= A \times B \\ &= (U \times 10^m + V)(X \times 10^m + Y) \\ &= U \times X \times 10^{2m} + U \times Y \times 10^m + VXXX\ 10^m + V \times Y \\ &= U \times X \times 10^{2m} + (U \times Y + V \times X) \times 10^m + V \times Y \end{aligned}$$

We can multiply A and B by doing four multiplications on integers with about half as many digits and performing linear-time operations. Recursively, This division process is continued until a smaller integer simply called *threshold* value is reached, at which the multiplication can be done in the standard way.

Although we illustrate the method in this section using integers A, and B with about the same number of digits, it is still applicable when the number of digits is different. In that case we simply use $m = \lfloor n/2 \rfloor$ to split both of them, where n is the number of digits in the larger integer A or B. The algorithm now follows. We keep dividing in the same way as mentioned above until one of the integers A, and B is 0 or we reach some *threshold* value for the larger integer, at which time the multiplication is done in the usual way.

15.3.1 Algorithm for Large Integers Multiplication

In this section we are going to represent pseudocode of the above approach to multiply two large integers. Consider the situation stated below.

Problem: Multiply two large integers A and B. Each of them is n-digits long.

So, input to this algorithm are A, and B. And say *mult* is the final result obtained from this multiplication.

Algorithm: *MLI(A, B)* // MLI stands for Multiplication of Large Integers

Where A, B are two large integers of size p digits and q digits respectively. Assume that the threshold value is set to 2. The algorithm returns an integer which may be large.

Step 1: Consider four large integers U, V, X, and Y.

Step 2: Initialize $n = 0$, $m = 0$, $mult = 0$

Step 3: Find $n = \max(p, q)$

Step 4: If $A = 0$ or $B = 0$

Return *mult*

Step 5: (*i*) If $n \leq 2$ //checking threshold

(*a*) $mult = A \times B$

(*b*) Return *mult*

(*ii*) Else

(*a*) $m = \lfloor \boldsymbol{n/2} \rfloor$

(*b*) $U = A/10^m$ //integer division operation

(*c*) $V = A\%10^m$ // integer remainder operation

(*d*) $X = B/10^m$

(*e*) $Y = B\%10^m$

(*f*) $P = MLI(U, X) \times 10^{2m}$

(*g*) $Q = MLI(U, Y) \times 10^m$

(*h*) $R = MLI(V, X) \times 10^m$

(*i*) $S = MLI(X, Y)$

(*j*) $mult = (P + Q + R + S)$

(*k*) Return *mult*

Step 6: Exit

Complexity Analysis: We analyze how long the algorithm *MLI*() takes to multiply two *n*-digit integers. Basic operations here are mainly addition, multiplication, and division. Now, the manipulation of one decimal digit in a large integer when adding 10^m digits, dividing by 10^m, or multiplying by 10^m, each of these basic operations is done *m* times. Suppose *n* is a power of 2. Then *U*, *V*, *X*, and *Y* all have exactly *n*/2 digits. It means that the input size to each of the four recursive calls to *P*, *Q*, *R*, and *S* *i.e.*, to *MLI*() is $m = n/2$. So, addition, division by 10^m, modulo-division (remainder) by 10^m, and multiplication by 10^m all have linear-time complexities in terms of $\Theta(n)$. Our recurrence relation for the *MLI*() procedure will be then,

$$\begin{aligned} T(n) &= 4T\left(\frac{n}{2}\right) + \Theta(n) \\ &= \Theta(n^2) \end{aligned}$$

However, exact time complexity is not straightforward if *n* is not restricted to being a power of 2. But, it is possible to establish a recurrence like the previous one but involving floors and ceilings values. Our recurrence is then

$$\begin{aligned} T(n) &= 2T(\lfloor \boldsymbol{n/2} \rfloor) + 2T(\lfloor \boldsymbol{n/2} \rfloor) + \Theta(n) \\ &= 2T\left(\frac{n}{2}\right) + 2T\left(\frac{n}{2}\right) + \Theta(n) \\ &= \Theta(n^2) \end{aligned}$$

15.3.2 Improved Algorithm for Large Integers Multiplication

Our algorithm *MLI*() for multiplying large integers is still quadratic. The reason is that the complexity of the algorithm is mainly due to basic multiplication operations. We see the above procedure does four multiplications with half as many digits as the original integers. If we are able to reduce the number of these multiplications operations, we can possibly achieve a complexity better than quadratic. We do this in the following way recalling *A*, *B*, *C*, *U*, *V*, *X*, and *Y* from previous section.

$$\begin{aligned} C &= A \times B \\ &= (U + V)(X + Y) \\ &= U \times X + U \times Y + V \times X + V \times Y \\ \Rightarrow \quad U \times Y + V \times X &= C - U \times X - V \times Y \\ &= (U + V)(X + Y) - U \times X - V \times Y \end{aligned}$$

Thus we get only three multiplications, while doing some additional linear-time additions and subtractions. Now we rewrite the procedure *MLI*() in the following way.

Algorithm: ***IMLI(A, B)*** // *IMLI* stands for Improved Multiplication of Large Integers

Where A, B are two large integers of size p digits and q digits respectively. Assume that the threshold value is set to 2. The algorithm returns an integer which may be large.

Step 1: Consider four large integers U, V, X, and Y.

Step 2: Initialize $n = 0$, $m = 0$, *mult*

Step 3: Find $n = \max(p, q)$

Step 4: If $A = 0$ or $B = 0$

Return *mult*

Step 5: (*i*) If $n \leq 2$ //checking threshold

(*c*) $mult = A \times B$

(d) Return *mult*

(*ii*) Else

(*a*) $m = \lfloor n/2 \rfloor$

(*b*) $U = A/10^m$ //integer division operation

(*c*) $V = A\%10^m$ // integer remainder operation

(*d*) $X = B/10^m$

(*e*) $Y = B\%10^m$

(*f*) $P = IMLI\ (U + V, X + Y)$

(*g*) $Q = IMLI\ (U, X)$

(*h*) $R = IMLI\ (V, Y)$

(*i*) $mult = (Q \times 10^{2m} + (P - Q - R) \times 10^m + R)$

(*j*) Return *mult*

Step 6: Exit

Complexity Analysis: Basic operations here are mainly addition, subtraction, multiplication, and division. Now, the manipulation of one decimal digit in a large integer when adding 10^m digits, subtracting 10^m digits, dividing by 10^m, or multiplying by 10^m, each of these basic operations is done m times. Our recurrence relation for the *IMLI*() procedure will be then,

$$
\begin{aligned}
T(n) &= 3T\left(\frac{n}{2}\right) + \Theta(n) \\
&= \Theta(n^{1.58})
\end{aligned}
$$

Using Fast Fourier Transforms, Borodin and Munro (1975) developed a $\Theta(n(\log n)^2)$ algorithm for multiplying large integers.

Example 1. Multiply the numbers $A = 78$ and $B = 21$.

Solution: Here $n = 2$

$$\Rightarrow \quad m = n/2 = 2/2 = 1$$

$$U = A/10 = 78/10 = 7$$

$$V = A\%10 = 78\%10 = 8$$

$$X = B/10 = 21/10 = 2$$

$$Y = B\%10 = 21\%10 = 1$$

We can now rewrite the product $A \times B$ in terms of the digits as:

$$\begin{aligned} A \times B &= (U \times 10 + V) \times (X \times 10 + Y) \\ &= (U \times X) \times 100 + (U \times Y + V \times X) \times 10 + V \times Y. \\ &= (7 \times 2) \times 100 + (7 \times 1 + 8 \times 2) \times 10 + 8 \times 1 \\ &= 1400 + 230 + 8 \\ &= 1638. \end{aligned}$$

Example 2. Multiply the numbers $A = 78$ and $B = 21$ efficiently.

Solution: Here, $n = 2$

$$m = n/2 = 2/2 = 1$$

$$U = A/10 = 78/10 = 7$$

$$V = \mathrm{A}\%10 = 78\%10 = 8$$

$$X = B/10 = 21/10 = 2$$

$$Y = \mathrm{B}\%10 = 21\%10 = 1$$

$$\begin{aligned} P &= (U + V)(X + Y) \\ &= (7 + 8)(2 + 1) \\ &= 15 \times 3 \\ &= 45 \end{aligned}$$

$$\begin{aligned} Q &= U \times X \\ &= 7 \times 2 \\ &= 14 \end{aligned}$$

$$\begin{aligned} R &= V \times Y \\ &= 8 \times 1 \\ &= 8 \end{aligned}$$

$$\begin{aligned} A \times B &= (Q \times 10^{2m} + (P - Q - R) \times 10^{2m} + R) \\ &= 14{\times}100 + (45 - 14 - 8) \times 10 + 8 \\ &= 1400 + 230 + 8 \\ &= 1638 \end{aligned}$$

Example 3. Multiply the numbers $A = 5678$ and $B = 4321$ efficiently.

Solution: Here, $n = 4$

$$m = n/2 = 4/2 = 2$$
$$U = A/10^m = 5678/100 = 56$$
$$V = A\%10^m = 5678\%100 = 78$$
$$X = B/10^m = 4321/100 = 43$$
$$Y = B\%10^m = 4321\%100 = 21$$
$$\begin{aligned} P &= (U + V)(X + Y) \\ &= (56 + 78)(43 + 21) \\ &= 134 \times 64 \\ &= 8576 \end{aligned}$$
$$\begin{aligned} Q &= U \times X \\ &= 56 \times 43 \\ &= 2408 \end{aligned}$$
$$\begin{aligned} R &= V \times Y \\ &= 78 \times 21 \\ &= 1638 \end{aligned}$$
$$\begin{aligned} A \times B &= (Q \times 10^{2m} + (P - Q - R) \times 10^m + R) \\ &= 2408 \times 10000 + (8576 - 2408 - 1638) \times 100 + 1638 \\ &= 24080000 + 453000 + 1638 \\ &= 24534638 \end{aligned}$$

Contemporary P, Q, R, and $A \times B$ have to be computed recursively when $n \geq 4$ to obtain final result for $A \times B$.

15.4 DETERMINATION OF THRESHOLDS

Divide-and-Conquer approach for Large Integers Multiplication Problem is a based on recursion. One of the major drawbacks of recursion is that it requires a fair amount of overhead in terms of computer time. For example, is it really worth this overhead to sort n elements so that we can use $\Theta(n \log n)$ running time based algorithm instead of a $\Theta(n^2)$ time based algorithm? We then have to develop a method that determines for what values of n this algorithm is at least as fast to call an alternative algorithm as it is to divide the instance further. These values are called ***threshold values.*** These *threshold values* depend on–

(*a*) The divide-and-conquer algorithm,

(*b*) The alternative algorithm, and

(*c*) The computer on which they are implemented.

Ideally, we would like to find an ***optimal threshold value*** of n. However, an optimal threshold value does not always exist. If the analysis does not yield an optimal threshold value, we can then select a threshold value from the analysis results. We then modify the divide-and-conquer algorithm accordingly so that the instance is no longer divided once n reaches that threshold value.

15.5 SOME COMMON QUESTIONS

Question 1. When we do not use Divide-and-Conquer Approach.

Answer: In two cases primarily we should avoid to follow divide-and-conquer methodology. These two cases are:

(*a*) A problem of size n is divided into two or more sub-problems each almost of size n.

(*b*) A problem of size n is divided into almost n problems of size n/c, where c is a constant.

Question 2. Why do we not use Divide-and-Conquer Approach against Question 1?

Answer: The first partitioning leads to an exponential-time algorithm. E.g., find nth Fibonacci Term Recursively. Computing nth term we divide it into two instances that are the $(n-1)$st term and the $(n-2)$nd term. Their sizes are nearly same as the original instance.

The second partitioning leads to an $n^{\Theta(\log n)}$ algorithm. E.g. the Towers of Hanoi problem which requires an exponentially large number of moves in terms of n.

However, neither of these values is acceptable for large n. Such partitionings lead to poor performance.

Exercises

1. Use Algorithm IMLI() for Multiplication Large Integer to find the product of 12534 and 233,103.
2. How many multiplications are needed to find the product of the two integers in previous question?
3. Use Algorithm MLI() for Multiplication Large Integer to find the product of 12534 and 233,103.
4. How many multiplications are needed to find the product of the two integers in previous question?
5. Write algorithms that perform the following operations.

   (*a*) $U \times 10^m$

   (*b*) $U / 10^m$

   (*c*) $U \% 10^m$.

   Where U represents a large integer, m is a non-negative integer. Analyze your algorithms. Can these operations be done in linear time?

6. Modify *MLI*() for Multiplication Large Integer so that it divides each n-digit integer into
   (*a*) Three smaller integers of $n/3$ digits (*i.e.*, assume that $n = 3^k$).
   (*b*) Four smaller integers of $n/4$ digits (*i.e.*, assume that $n = 4^k$).
   (*c*) Five smaller integers of $n/5$ digits (*i.e.*, assume that $n = 5^k$).
   (*d*) Four smaller integers of $n/4$ digits (*i.e.*, assume that $n = 5^k$).
7. Analyze your algorithms, and show their time complexities in order notation as you developed in previous question.
8. Modify *MLI*() for Multiplication Large Integer so that it divides each n-digit integer into
   (*a*) Three smaller integers of $n/3$ digits (*i.e.*, assume that $n = 3^k$).
   (*b*) Four smaller integers of $n/4$ digits (*i.e.*, assume that $n = 4^k$).
   (*c*) Five smaller integers of $n/5$ digits (*i.e.*, assume that $n = 5^k$).
   (*d*) Four smaller integers of $n/4$ digits (*i.e.*, assume that $n = 5^k$).
9. Analyze your algorithms, and show their time complexities in order notation as you developed in previous question.
10. How do you determine Thresholds for obtaining product of two large integers?
11. Suppose M, and N are two integers. Number of digits in M, and N are same as size of 64×64 matrix. How many multiplications would be performed in finding the product of M and N.
12. Write a recurrence equation for the modified *MLI*() algorithm.
13. Solve the recurrence equation derived in question 13, and verify your answer.
14. When do we not use Divide-and-Conquer approach?
15. Write an iterative version of the *MLI*() procedure. Analyze your algorithm.
16. Write an iterative version of the *IMLI*() procedure. What is its running time?

University Questions

1. Let A and B be two n-bit numbers where n is a power of 2. A divide-and-conquer based algorithm divides the numbers into two equal parts to compute the product as $AB = A_1B_1\ 2^n + (A_1B_2 + A_2B_1)\ 2^{n/2} + A_2B_2$ where $A = (A_1\ 2^{n/2} + A_2)$ and $B = (B_1\ 2^{n/2} + B_2)$ and A_1B_1, A_1B_2, A_2B_1 and A_2B_2 are computed recursively. Then,
   (*i*) Determine the time complexity.
   (*ii*) If $(A_1B_2 + A_2B_1)$ is computed as $(A_1 + A_2)(B_1 + B_2) - A_1B_1 - A_2B_2$ then will there be any change in the time complexity? Justify. ***[DOEACC'07]***
2. Explain how the Divide-and-Conquer strategy is used to multiply large integers. Apply the same to compute 2101 * 1130. ***[VTU, 2008]***

❑❑❑

Chapter 16 File Compression

16.1 INTRODUCTION

Compression both in computer science and in information theory (also known as coding theory) is useful because it helps to reduce the consumption of expensive resources, such as hard disk space or transmission bandwidth. Compressed data are stored data in a format that requires less space than usual. Compressed data are decompressed to be used at the receiver side, and this extra processing may sometimes be disadvantageous to some applications. For instance, a compression scheme for video data may require sometimes an expensive hardware device for the compressed video data to be decompressed fast enough. Ironically, the techniques that are used to save space are "coding" methods.

Definition: Data Compression or simply File Compression is the process of encoding information using fewer bits than the original representation would use. It is sometimes called source coding or bit-rate reduction.

The design of data compression schemes therefore involves trade-offs among various factors, including the degree of compression, the amount of distortion introduced and the computational resources.

16.2 CLASSIFICATION OF DATA COMPRESSION

Data can be compressed mainly in two ways. One is called *Lossless Compression*, and second one is called *Lossy Compression*.

Lossless Compression refers to data compression techniques in which no data is lost. The PKZIP compression technology is an example of lossless compression. In most cases different types of data, lossless compression techniques can reduce the space needed by 50%. *Lossy compression* is used for greater compression. *Algorithms used for Lossless Compression,* exploit statistical redundancy such that these represent data more concisely without error. Since most real-world data has statistical redundancy naturally *Lossless* compression is possible. For example, in English text, the letter '*e*' is much more common than the letter '*z*', and the probability that the letter '*q*' will be followed by the letter '*z*' is very small.

Lossy Data Compression (or Perceptual Coding) refers to data compression techniques in which some amount of data is lost. It is possible if some loss of fidelity is acceptable. And data in this scheme are called lossy data. Generally, a Lossy data compression is based on how people perceive data. For example, the human eye is more sensitive to subtle variations in luminance than it is to variations in colour. Lossy data compression provides a way to obtain the best fidelity for a given amount of compression. Lossy compression technologies attempt to eliminate redundant or unnecessary information. Most video compression technologies, such as MPEG, use a lossy technique.

16.3 TYPES OF CODES

One of the objectives of Encoding is always to make it more simple and efficient. Computer system handles data in binary form. It will be more specific to use *Binary Character Code* or simply *Code* to encode data. In *Binary Character Code* each character is represented by a unique character string (codeword). Encoding is always simple for any binary character code. Therefore we just concatenate the codewords used to represent each character of a word in the file.

We consider here only those codes in which no codeword is a prefix of some other codeword. Such codes are called *Prefix Codes*. It is possibly seen that the optimal data compression is achievable by a *Binary Character Code* representation with a prefix code. Prefix codes are desirable because they simplify decoding of the encoded text. Reason is that as long as a codeword is not a prefix to any other codeword the codeword that begins an encoded file as a result till then remains unambiguous. Simply the initial codeword is identified, and is then translate it back to the original character. The decoding process is repeated on the rest of the encoded texts in the file.

16.4 TYPES OF ENCODING

Suppose we are given a file, which consists of string of characters. We wish to compress this file as much as possible. Compression is done in such a way that the original file can easily be reconstructed. Let the set of characters in the file be $C = \{C_1, C_2, ..., C_i \; ..... \; C_n\}$, $1 \le i \le n$. Each C_i be the frequency of character say x in the set C in the file *i.e.* the number of times x appears in the file. Using a fixed number of bits to represent each character, called the *encoding* of the character, the size of the file depends only on the number of characters.

There are many ways to represent data in a file. Suppose we have to encode a text that comprises characters from some word of n-characters. Each letter of the text is assigned by some sequence of hits. This is called *codeword.* Two types of encoding are generally used to represent this codeword. One is *Fixed Length Encoding*, and another is *Variable Length Encoding*.

16.4.1 Fixed Length Encoding

In *a fixed-length encoding,* each character is assigned with a bit string of length m ($m = \lceil \log_2 n \rceil$). This is exactly what the standard ASCII code does. One way of getting a *fixed-length encoding* scheme that yields a shorter codeword (bit string) is based on a simple idea. The idea is that assigns smaller (shorter) codewords to more appeared (frequent) characters and assign larger (longer) codewords to less appeared characters. This idea was used by Samuel Morse during the mid-19th century and in particular, is known as the *telegraph code*.

Example 1. Store a 100,000-character data file of six distinct alphabets "*A through F*" compactly using *fixed-length encoding.*

Solution: Number distinct characters $n = 6$. So, codeword or bit length $m = \lceil \log_2 6 \rceil = 3$. We need then 3 bits to represent each of six characters: $A = 000$, $B = 001$, $C = 010$, $D = 011$, $E = 100$ and $F = 101$.

Thus total number of bits required to encode the file

$$= (\text{Number of characters in the file}) \times (\text{codeword length})$$
$$= 100000 \times 3$$
$$= 300,000$$

Example 2. Encode normal ASCII character set using *fixed-length encoding.*

Solution: The normal ASCII character set consists of roughly 100 "printable" (distinguish) characters. So, codeword $m = \lceil \log_2 100 \rceil = 7$. Now, seven bits allow the representation of at most 128 characters, so the ASCII character set adds some "nonprintable" characters such as "\n", "\t", " blank space" etc.

Example 3. Find out codeword for a file with the size of the character set is *C.*

Solution: The size of the character set is *C,* therefore, the codeword $m = \lceil \log_2 C \rceil$.

16.4.2 Variable Length Encoding

In real life, files can be quite large. Many of the very large files are output of some program and there is usually a big disparity between the most frequent and least frequent characters. Then *fixed-length encoding* will not probably be a better approach. Alternatively a *variable-length Encoding* can do considerably better than a fixed-length code. *Variable-length encoding* assigns codewords of different lengths to different characters. Generally frequent characters are assigned with short codewords and infrequent characters are assigned with long codewords.

Example 4. Consider Example 1 and the table below. Calculate the size of the file after encoding using variable codeword.

Characters	A	B	C	D	E	F
Frequency (in thousands)	50	13	12	15	10	5
Codeword	0	011	101	111	1101	11110

Solution: Number of characters in the file = 100,000.

Number of distinct characters $n = 6$

Therefore, size of the encoded file

$$= \sum_{i=1}^{n} (\text{frequency} \times \text{codeword})$$

$= (50 \times 1 + 13 \times 3 + 12 \times 3 + 15 \times 3 + 10 \times 4 + 5 \times 5) \times 1{,}000$

$= (50 + 39 + 36 + 45 + 40 + 25)$

$= 235{,}000$ bits.

Example 5. Consider Examples 1 and 4. Calculate savings of storage spaces.

Solution: Using fixed length codeword, encoded file size is 300,000 bits. Using variable length codeword, encoded file size is 235,000 bits.

Therefore, savings of storage spaces

$= 300{,}000 - 235{,}000$

$= 65{,}000$ bits.

Thus, % of savings disk spaces $= \dfrac{65{,}000}{300{,}000} \times 100 \;= 21.67$

Another way to determine number of bits (codeword) of an encoded text that represents character is to use *prefix-free* (or simply *prefix)* codes. In a prefix code as mentioned earlier no codeword is a prefix of a codeword of another character. We simply scan a bit string with such an encoding until the first group of bits that is a codeword for some character. This character replaces these bits. This operation is repeated till the end of bit string is reached.

16.5 APPLICATION OF DATA COMPRESSION

File compression is the practice of packaging a file or files to use less disk space. The File Compression also includes software programs that will archive our files and extract archived files such as ZIP and RAR files. Some common application areas include-

(*a*) Communication Theory.
(*b*) Computer Networks.
(*c*) Backup Utilities.
(*d*) Spreadsheet Applications.
(*e*) Database Management Systems.
(*f*) Bit-Mapped Graphics.
(*g*) Data Warehousing.
(*h*) Data Mining, etc.

Data compression reduces storage cost by storing more logical data per unit of physical capacity. Performance is improved because there is less physical data to retrieve during scan-oriented queries. Performance is further enhanced since data remains compressed in memory.

16.6 HUFFMAN CODING

Even though the capacity of secondary storage devices are getting larger comparatively with their cost is getting smaller, the devices continue to fill up due to increased storage demands. It is desirable to find an efficient way to store data in the file as efficiently as possible. The problem of *data compression* is to find an efficient method for encoding a data file. Here, we discuss the greedy based encoding method called *Huffman encoding* for a given file. Huffman codes are a widely used and very effective technique for compressing data. **Huffman coding** is an entropy encoding algorithm used for lossless data compression. Depending on the characteristics of the data being compressed, typical savings of storage spaces ranges from 20% to 90%.

A binary prefix code for some alphabet is determined by the association of the alphabet's characters with leaves of a binary tree. Each left edge is labelled by 0 and every right edge is labelled by 1. The codeword of a character can then be obtained by reading the edge labels on the simple path from the root to the leaf containing the character. Any such tree is called *prefix coded tree or coding tree*. More specifically it is well known as *Huffman tree.* One aspect of this tree is that if there is no simple path from the root to a leaf containing a character and that continues to another leaf, no codeword can be a prefix of another codeword. Construction of Huffman tree exploiting the occurrences (frequencies) of character was first invented by David A. Huffman in 1951 during his graduate study at MIT. The tree assigns shorter bit strings to the character with high-frequencies.

16.6.1 Properties of Huffman Coding

An optimal code for a file is always obtained from Huffman tree which is represented by a *full* binary tree in which every non-leaf node has two children. Basic properties can be derived for a Huffman tree as–

(*a*) Huffman coding is a form of statistical coding.

(*b*) Code word lengths are no longer fixed like ASCII.

(*c*) All characters may not occur with the same frequency.

(*d*) All characters may be allocated the same amount of space.

(*e*) Prefix codes tend to have inefficiency on small alphabets.

(*f*) Code word lengths vary and will be shorter for the more frequently used characters.

(*g*) The term refers to the use of a variable-length codeword for encoding a source symbol.

(*h*) Huffman coding uses a specific method to represent each symbol resulting in a prefix code.

(*i*) Huffman coding is optimal when the frequency (probability) of each input symbol is a negative power of 2.

(*j*) The worst case for Huffman coding can happen when the probability of a symbol exceeds $2^{-1} = 0.5$.

16.6.2 Variations of Huffman Coding

Several variations of the Huffman coding are possible. Some of these codings are simply Huffman-like algorithms. Some codings find optimal prefix codes. These methods need not be Huffman-like, and need not be polynomial time bounded.

(*a*) *n-ary Huffman coding* – it builds an n-ary tree and uses an alphabet in $\{0, 1, \ldots, n-1\}$ to encode message.

(b) *Adaptive Huffman coding* – the probabilities are calculated dynamically and are based on recent actual frequencies in the sequence of source symbols. The coding tree is updated according to new probability estimates.

(c) *Huffman template algorithm* – any kind of weights (costs, frequencies etc.) and any combining methods are used to encode.

(d) *Length-limited Huffman coding* – the goal is to achieve a minimum weighted path length but each codeword length is less than a given constant.

Other variations include Huffman coding with unequal letter costs, the canonical Huffman code Optimal alphabetic binary trees (well known as Hu-Tucker coding) etc.

16.6.3 Applications of Huffman Coding

Applications of Huffman coding are pervasive throughout computer science. This coding scheme is not limited to encoding messages in English text. Today it is often used as a "back-end" to compression methods.

(*a*) *Arithmetic coding* – every symbol has a probability of the form: $1/2^k$.

(*b*) DEFLATE (PKZIP's algorithm) .

(*c*) Multimedia codec such as JPEG and MP3.

(*d*) Quantization.

(*e*) Facsimile compression.

(*f*) Huffman coding can be used effectively anywhere there is a need for a compact code to represent a long series of a relatively small number of distinct bytes.

16.6.4 Huffman Encoding

Huffman's encoding is one of the most important and efficient file compression techniques. It yields an optimal (minimal-length) encoding. In addition, it is simple and versatile. The simplest version of Huffman encoding, replace each character by its codeword in a file. Other versions are use of fixed length encoding and variable length encoding. All these we discussed before. In this section we will discuss how a Huffman tree is constructed followed by an achievement to optimal file compression. A Huffman tree is a minimum weighted binary tree that can be constructed using Huffman algorithm. However the Huffman tree constructed is not unique in spite of minimum weighted length. The Huffman tree can be constructed either employing any standard Sorting algorithm or Min-Heap. But the construction procedure is same in both the cases. We follow the convention that all the leaves and internal nodes in the Huffman tree are represented by square and circles respectively.

16.6.4.1 Huffman Tree Using Sorting Algorithm

Suppose frequency of each character in a file is assumed as weight of corroding character. Say there are n characters and weight of each character is w_i, $1 \leq i \leq n$. Now a Huffman tree is constructed in the following way.

Algorithm: HuffmanTree_Sorting(L, n)

Where: L = weight of each character.
n = total number of characters.

Step 1: Sort the weights of the characters in ascending order using standard sorting algorithm.

Step 2: Choose two minimum weights from the sorted list.

Step 3: Construct a subtree with these two minimum weights as the weights of external node. Root's weight is sum of weights of its children.

Step 4: Add Root's weight in proper position in and remove its children weights from the list.

Step 5: Repeat above steps till the list contains only one element which nothing but the Root of the Huffman tree.

Complexity Analysis: Since the above procedure is based on the basically on sorting algorithm. Complexity of this algorithm depends on selection of the sorting algorithms. Say Merge Sorting is considered. So step 1 runs in $O(n \log n)$ times. Selection of two elements and adding their sum in the list needs constant time. The step 5 iterates $n - 1$ times *i.e.* $O(n)$ times. Thus time complexity of this algorithm is $O(n)$.

Example 6. Consider a message consists of six alphabets and their weights. Construct a Huffman tree. Also compute optimal code length.

Alphabet	A	B	C	D	E	F
Weight	3	5	2	4	1	7

Solution: Sorted List L = 1, 2, 3, 4, 5, 7

(*a*) Choose 1 and 2. Construct subtree as Figure 16.1(*a*). Return root to list *L*.

(*b*) *L* = **3**, 3, 4, 5, 7. Choose **3** and 3. Construct subtree as Figure 16.1(*b*). Return root to list *L*.

(*c*) L = 4, 5, **6**, 7. Choose 4 and 5. Construct subtree as Figure 16.2(*a*). Return root to list L.

(*d*) L = **6**, 7, **9**. Choose **6** and 7. Construct subtree as Figure 16.2(*b*). Return root to list *L*.

(*e*) *L* = **9, 13.** Choose **9** and **13.** Construct subtree as Figure 16.3(*a*). Return root to list *L*.

(*f*) *L* = **22.** Thus we get single node. This is the Root of the Huffman tree. Thus the Huffman tree is Figure 16.3(*b*).

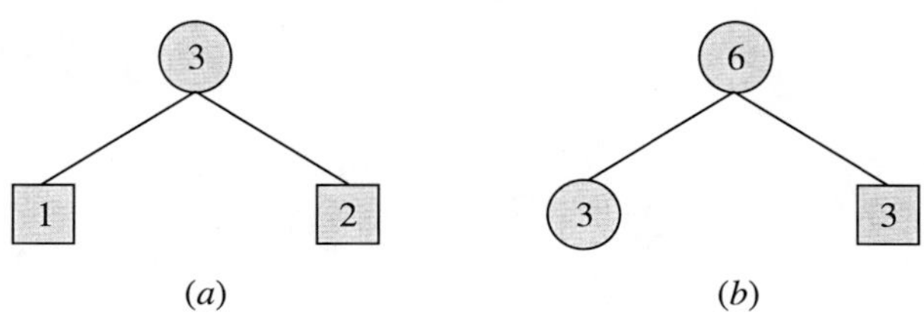

Figure 16.1 Partial Huffman Tree.

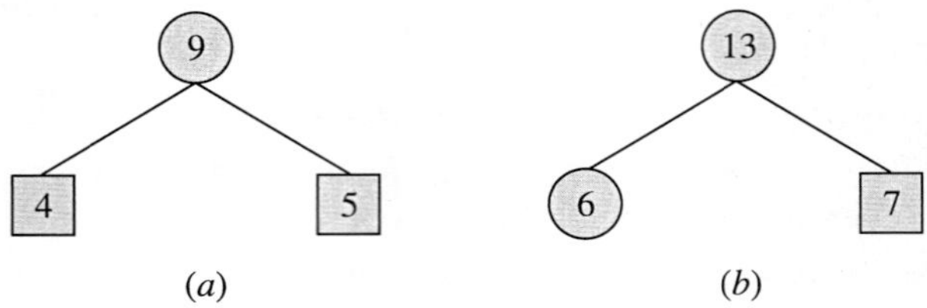

Figure 16.2 Partial Huffman Tree.

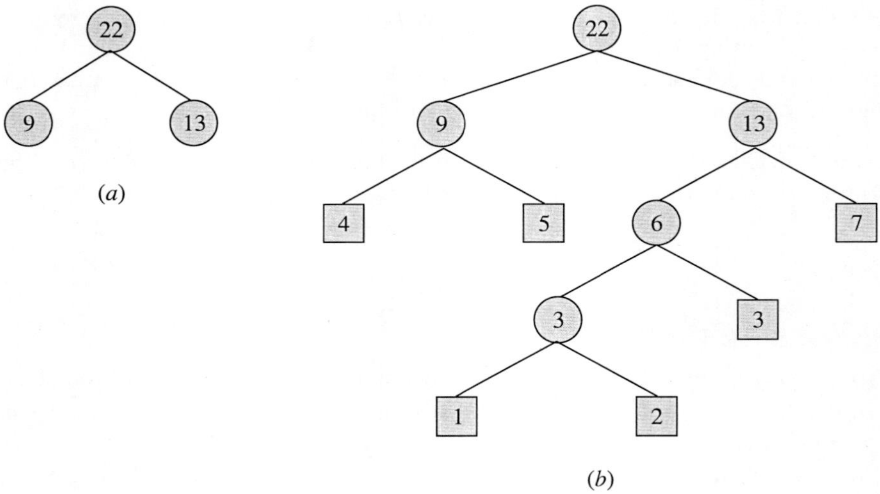

Figure 16.3 Partial as well as Huffman Tree.

The boxes are the weight of each alphabet. Sum of products of weight and path length from root of each leaf node yields the optimal code *i.e.,* total number of bits *i.e.,* encoded message size. Total bits $= 4\times2 + 5\times2 + 1\times\mathbf{4} + 2\times\mathbf{4} + 3\times\mathbf{3} + 7\times2 = 53$ bits. Prefix code for each alphabet can be obtained assigning 0 to each left edge and 1 to each right edge.

Thus, Prefix code for 4 *i.e.,* $D = 00$

Prefix code for 5 *i.e.,* $B = 01$

Prefix code for 1 *i.e.,* $E = 10000$

Prefix code for 2 *i.e.,* $A = 1001$

Prefix code for 3 *i.e.,* $C = 101$

Prefix code for 7 *i.e.,* $F = 11$

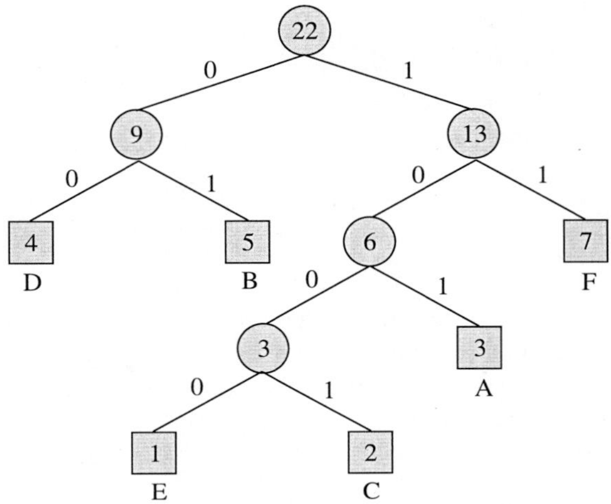

Figure 16.4 Binary Coded Huffman Tree.

It is seen that path length of each alphabet is nothing but the prefix code length.

16.6.4.2 Huffman Tree Using Min-Heap

Consider S is a set of n characters and each character $c \in S$. Assume that weight or frequency of each c is $f[c]$. The algorithm builds the contemporary tree T corresponding to the optimal code in bottom-up manner. The algorithm begins with a set of n leaves rather a forest of n trees. Each tree consists of each character node and performs a sequence of $n - 1$ "*merging*" operations to create the Huffman tree. Say, Q is a minimum priority queue. Each key of the Q is each $f[c]$. Two least-frequent characters are identified from Q and are then merged together to build T. That means two smallest weighted trees are merged. Weight of a tree is its root value. The result of the merge operation of two characters (trees) yields a new character (tree) whose frequency (root value of the tree) is the sum of the frequencies of the two characters that were used to merge. At the end of the algorithm there is one tree, and this is the *optimal Huffman coding tree*.

Algorithm: HuffmanTree_MinHeap(S, Q)

Where, S = set of n characters

Q = priority queue constructed using minimum heap.

R = weight/frequency of Root node

Lchild = weight/frequency of Left child to R

Rchild = weight/frequency of Right child to R.

Step 1: Enqueue every frequency from S to Q.

Step 2: Repeat for $i = 1$ to $n - 1$

  (i) Lchild = Extract_Min(Q)

  (ii) Rchild = Extract_Min(Q)

  (iii) R = Lchild + Rchild

  (iv) Enqueue(Q, R)

Step 3: Return Extract_Min(Q) // Return the root of the tree.

Step 4: Exit.

Complexity Analysis: We can construct a heap rather min-heap to maintain the weight/frequency of each character (and intermediate trees). It is done in step 2 (iv) to enqueue operation. Now a min-heap construction algorithm requires $O(\log n)$ time. But the loop in step 2 iterates times *i.e.*, there are $O(n)$ iterations. Thus the time complexity of the Huffman algorithm is $O(n \log n)$.

Example 7. Consider a message consists of eight alphabets A ... H and their weights. Construct a Huffman tree. Also compute optimal code length.

Alphabet	A	B	C	D	E	F	G	H
Weight	12	6	7	9	3	7	15	6

Solution: The Huffman tree is constructed stepwise below.

(a) Construct initial arrangement of the forest.

Figure 16.5 Initial forest.

(b) Extract two minimum weighted trees. Merge them.

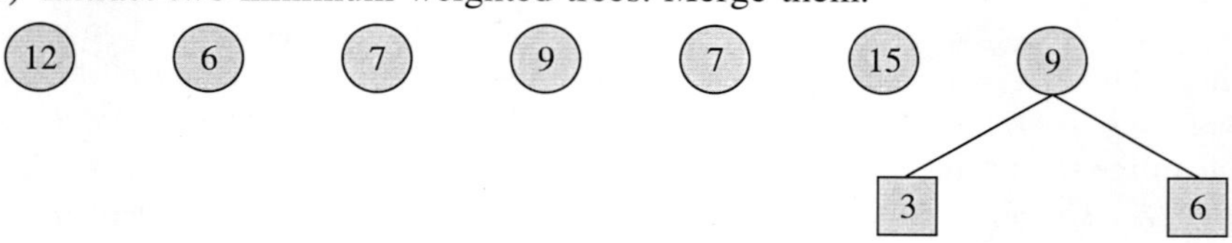

Figure 16.6 Merge tree with merge of 3 and 6.

(c) Repeat previous step till one node remained.

Figure 16.7 Merge tree with merge of 6 and 7.

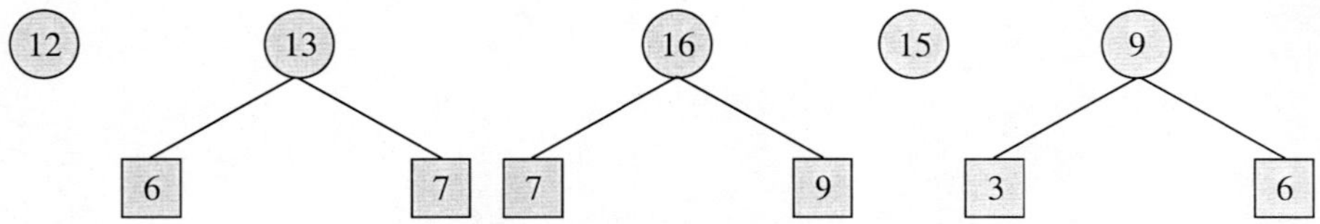

Figure 16.8 Merge tree with merge of 7 and 9.

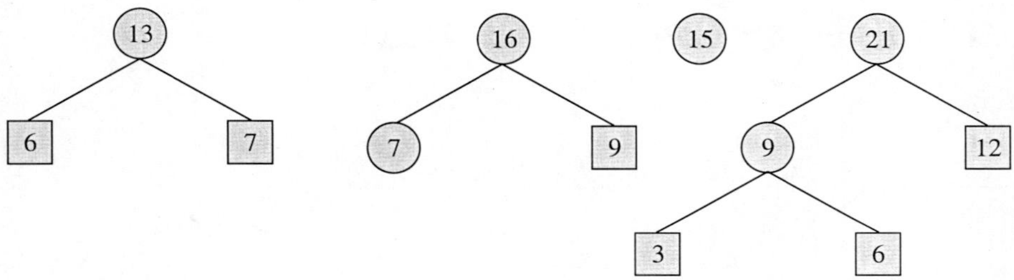

Figure 16.9 Merge tree with merge of 9 and 12.

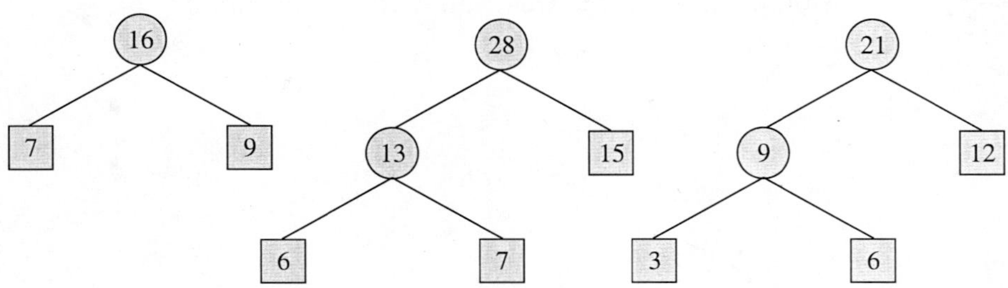

Figure 16.10 Merge tree with merge of 13 and 15.

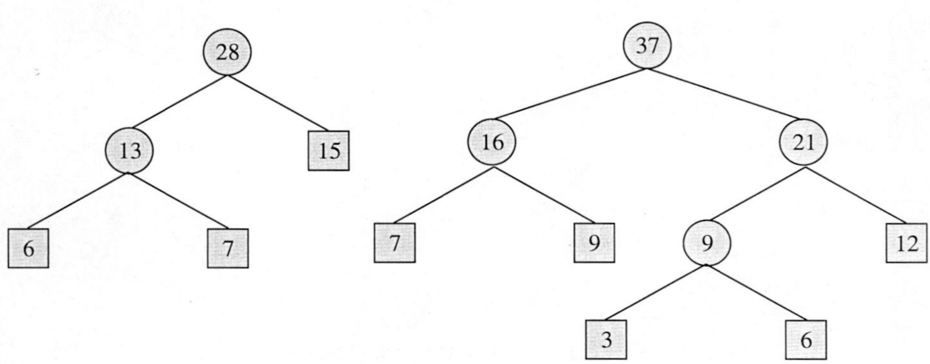

Figure 16.11 Merge tree with merge of 16 and 21.

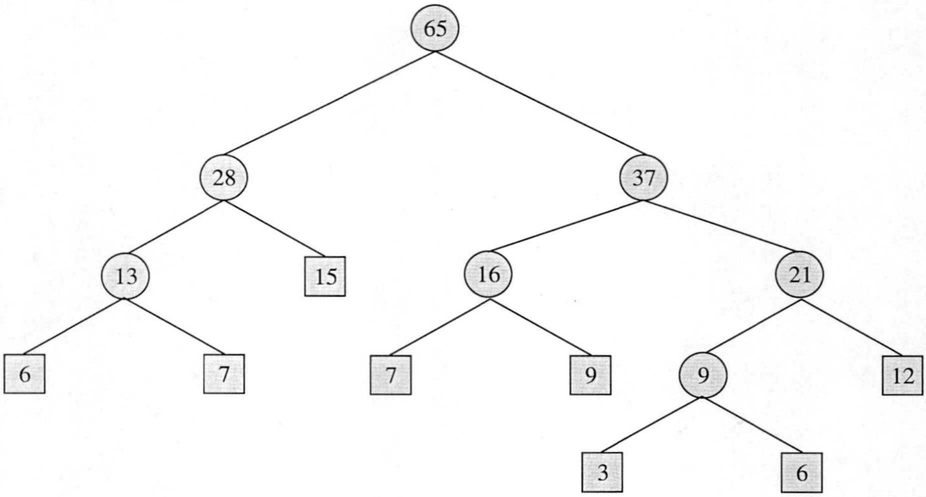

Figure 16.12 Merge tree with merge of 28 and 37.

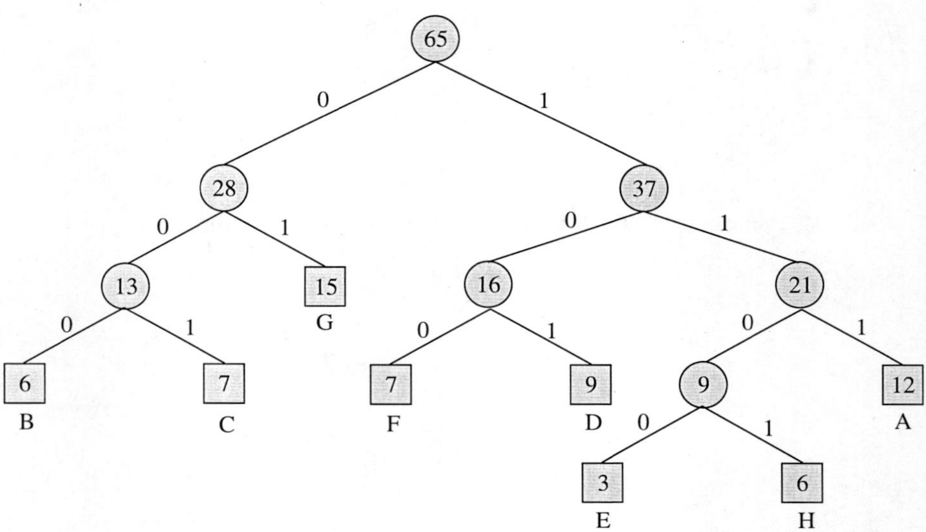

Figure 16.13 Binary Coded Huffman Tree.

Optimal code is shown in the table below.

Alphabet	*Weight*	*Codeword*	*Number of bits*
A	12	111	12 × 3 = 36
B	8	000	8 × 3 = 24
C	7	001	7 × 3 =21
D	9	101	9 × 3 = 27
E	3	1100	3 × 4 = 12
F	7	100	7 × 3 = 21
G	15	01	15 × 2 = 30
H	6	1101	6 × 4 = 24
Total Number of bits =			195

Example 8. Consider a message consists of five alphabets E, A, M, N, T. Construct Huffman tree of this message. Assume character's frequency in decreasing order of their use.

Solution: We assign shorter code to the character with higher frequency. E has highest frequency so it will have shortest code say 1, 00 to A. This is illustrated in the Huffman tree.

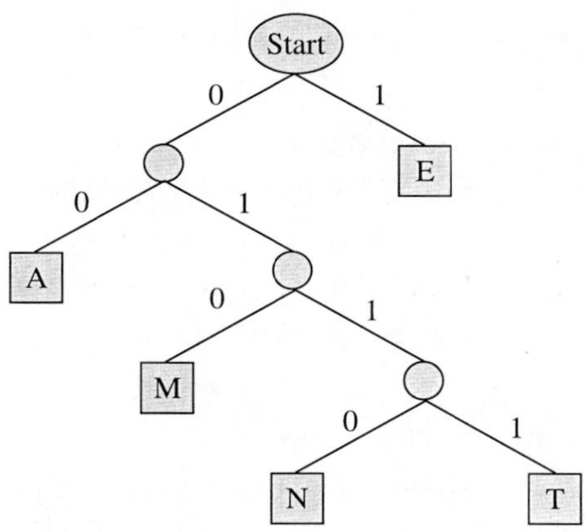

Figure 16.14 Binary Coded Huffman Tree.

Code to each character is as in table below.

Letter	*code*	*Lenght*
E	1	1
A	00	2
M	010	3
N	0110	4
T	0111	4

Therefore the Huffman tree in Figure 16.14 may be the best illustration for the message consists of E, A, M, N, T if frequencies of N and T are same.

16.7 SOME COMMON QUESTIONS

Question 1. For any optimal prefix code the tree must be "full binary tree". Why?

Answer: If some internal node had only one child then we could simply get rid of this node and replace it with its unique child. This would decrease the total cost of the encoding.

Question 2. What are basic advantages of Huffman coding?

Answer: Potential advantages offered by Huffman coding are–

(*a*) The Huffman coding has minimum average length so encoded data needs low transmission time and lower storage space.

(*b*) Because of unique prefix property, the Huffman coding does not require any end character delimiter.

Question 3. What are basic disadvantages of Huffman coding?

Answer: Potential disadvantages offered by Huffman coding are –

(*a*) Main drawback is that the code is variable length code.

(*b*) Processing time for both encoding and decoding is required.

(*c*) If a bit is changed due to noise, it may be unable to recognize remaining part of the message.

Question 4. Why is Huffman Coding Greedy?

Answer: Huffman's algorithm is an example of a greedy algorithm. It's called greedy because the two smallest nodes are chosen at each step, and this local decision results in a globally optimal encoding tree.

Exercises

1. For the given problem show that a Huffman tree can be constructed in linear time if the alphabets are given in a sorted order of their frequencies.

character	A	B	C	D	–
probability/frequency	0.1	0.15	0.15	0.2	0.4

2. Write a program that constructs a Huffman code for a given English text and encode it.
3. Write a program for decoding an English text that has been encoded with a Huffman code.
4. Construct a Huffman tree of the following table. Suppose frequencies correspond to code length. Then what will be optimal Huffman code?

Char	*Code*
E	0000
i	0001
y	0010
l	0011
k	0100
.	0101
space	011
e	10
r	1100
s	1101
n	1110
a	1111

5. Experiment with your encoding program to find a range of typical compression ratios for Huffman's encoding of English texts of 1000 words.
6. The following set of weights is known to the letters – L to S as 1, 2, 1, 2, 3, 4, 6, and 7 respectively. Show that there are a number of different Huffman tree possible.
7. Write an algorithm to calculate the total external weighted path length for a given Huffman tree.
8. Consider a document file. Count the frequency of various characters occurring in the file. Construct a Huffman tree taking the frequencies of characters as weights. Store the document file using the Huffman encoded bit string. Decode the compressed file so obtained.
9. Experiment with your encoding program to find out how sensitive the compression ratios arc to using standard estimates of frequencies instead of actual frequencies of character occurrences in English texts.

10. Generalize Huffman's algorithm to ternary codewords (*i.e.*, codewords using the symbols 0, 1, and 2), and prove that it yields optimal ternary codes.
11. Draw a Huffman coding tree for the string "ABRACADABRA." How many bits does the encoded message require?
12. What is the Huffman code for a binary file? Give an example showing the maximum number of bits that could be used in a Huffman code for a N-character ternary (three-valued) file.
13. Suppose that the frequencies of the occurrence of all the characters to be encoded are different. Is the Huffman encoding tree unique?
14. Huffman coding could be extended in a straightforward way to encode in two-bit characters (using 4-way trees). What would be the main advantage and the main disadvantage of doing so?
15. What would be the result of breaking up a Huffman-encoded string into five-bit characters and Huffman encoding that string?
16. Construct a Huffman tree for the following data and obtain its Huffman code.

character	A	B	C	D	E	–
probability/frequency	0.5	0.35	0.5	0.1	0.4	0.2

17. Indicate whether each of the following properties is true for every Huffman code.
    (*a*) The codewords of the two least frequent characters have the same length.
    (*b*) The codeword's length of a more frequent character is always smaller than or equal to the codeword's length of a less frequent one.
18. Write a program to calculate the maximal length of a codeword possible using Huffman encoding in a word of n alphabets.
19. Prove that Huffman Codes are Optimal.
20. Prove that a binary tree that is not full cannot correspond to an optimal prefix code.
21. What are the optimal codes of the following Huffman trees shown below?

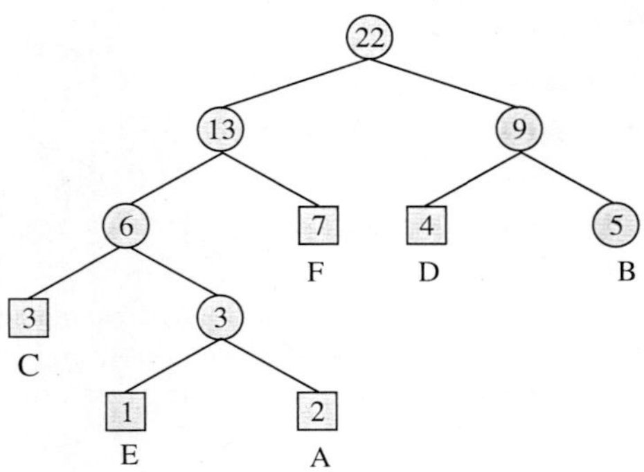

(*a*)

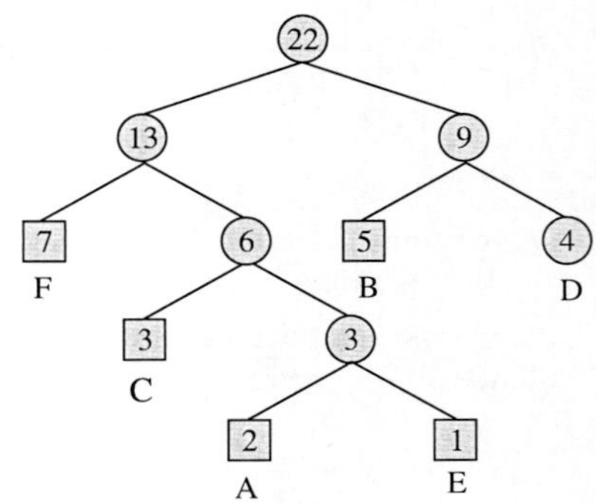

(*b*)

22. Write a pseudocode for the Huffman-tree construction algorithm. What is the time efficiency class of the algorithm for constructing a Huffman tree as a function of the alphabet's size?

University Questions

1. What is data compression and why we need it? Explain compression and reconstruction with the help of block diagram. *[UPTU'06]*
2. What is an optimal Huffman code for the following set of frequencies, based on the first 8 Fibonacci numbers?

   a:1, b:1, c:2, d:3, e:5, f:8, g:13, h:21

   Can you generalize your answer to find the optimal code when the frequencies are the first n Fibonacci numbers? *[IITG'07, '11]*
3. Determine the cost and structure of an optimal binary search tree for a set of $n = 7$ keys with the following probabilities. *[JNTU'08]*

   i - 0, 1, 2, 3, 4, 5, 6, 7.

   p_i - 0.04, 0.06, 0.08, 0.02, 0.1, 0.12, 0.14

   q_i - 0.06, 0.06, 0.06, 0.06, 0.05, 0.05, 0.05, 0.05
4. Differentiate between static length and variable length coding schemes. Explain with the help of examples. *[UPTU'07]*
5. What are the measures of performance of data compression algorithms? *[UPTU'07]*
6. What do you understand by length of Huffman code and how it is defined? *[UPTU'07]*
7. Compare Huffman and Arithmetic coding. *[UPTU'06]*
8. Design Huffman code for the following character: *[RTU'07]*

Character	*a*	*b*	*c*	*d*	*e*
Frequencies	8	2	6	3	9

9. Develop a Huffman code for the input string "a fast runner need never be afraid of the dark". Explain how the greedy-choice property used in this algorithm? *[BPUT'05]*
10. Given the following set of characters with their corresponding probabilities

a	0.32	*e*	0.08
b	0.24	*m*	0.08
c	0.16	*o*	0.02
d	0.09	*n*	0.01

   Construct a Huffman code for the same source, and compute its weighted average code length. What is time complexity of the algorithm used? *[BPUT'07]*
11. Design Huffman codes for the following symbols: *a*, *b*, *c*, *d*, *e*, *f*, and *g* having relative frequency 2, 4, 6, 8, 10, 16 respectively. Discuss on the time complexity of the Huffman algorithm with binary heap. *[WBUT'05, BPUT'06]*
12. What is an optimal Huffman code for the following set of frequencies based on first 8 Fibonacci series: 1, 1, 2, 3, 5, 8, 13, 21 *[IETE ALCCS'11]*
13. How Huffman coding algorithm is used to design Huffman code for a source that takes latter from an alphabet set $A = \{a_1, a_2, a_3, a_4, a_5\}$. Subscripts are the frequencies. *[UPTU'07]*

14. Write short note on Huffman codes. *[IETE ALCCS'11]*

15. Write short note on Huffman's algorithm. *[UPTU'06, 07'10]*

16. Suppose we build the Huffman code tree for the set of letters and frequencies given below:

    Character: A, B, C, D, E, F
    Frequency: 1, 5, 20, 30, 40, 50

    What will be the length of the code for the character *B*? *[IETE ALCCS'11]*

17. Let there be a set of four identifiers $\{a_1, a_2, a_3, a_4\} = \{do, end, stop, while\}$. Let us define five classes of identifiers E_0, E_1, E_2, E_3, and E_4 where E_i contains all identifiers X satisfying $X < a_i$, $i = 0, 1, 2, 3$ and E_4 consists of all identifiers X satisfying $X > a_4$. Let p_i be the probability for a successful search of the identifier a_i, $i = 1, 2, 3, 4$ and q_i be the probability that the identifier being searched is in class E_i, $i = 0, 1, 2, 3, 4$. Obtain the Optimal Binary Search Tree for the identifier set $\{a_1, a_2, a_3, a_4\}$ having $p_1 = 0.1, p_2 = 0.2, p_3 = 0.2, p_4 = 0.1, q_0 = 0.1, q_1 = 0.1, q_2 = 0.1, q_3 = 0.05$ and $q_4 = 0.05$. *[DOEACC'07]*

❑❑❑

17 Chapter Biconnectivity

17.1 INTRODUCTION

Depth-first search on a graph $G = (V, E)$ is a generalization of traversing a tree similar to Preorder tree traversal. Starting at some vertex v, we visit v and then recursively traverse all vertices adjacent to v. On each iteration, the algorithm proceeds to an unvisited vertex that is adjacent to the current one. If there are several adjacent vertices to current vertex, a tie can be resolved arbitrarily. DFS algorithm has been discussed in chapter 11. DFS traversal produces a tree called *Depth First Spanning Tree or simply Depth First Tree* in $O(|E|+|V|)$ time.

All vertices of a tree with vertices V are systematically visited in a total of $O(|V|)$ time when Preorder is performed on the tree. But we should take care when we perform this process on an arbitrary graph because we need to be careful to avoid cycles. To do this, when we visit a vertex v, we mark it *visited*, and recursively call depth-first search on all adjacent vertices that are till unmarked. If the graph is undirected but not connected or directed but not strongly connected, this strategy might fail. Because some vertices till remain unvisited. We apply a depth-first traversal in such a case to search for an unmarked vertex, and continue this process until there are no unmarked vertices. Instead of depth first tree a *Depth First Forest* (set of trees) is generated. It is observed that in DFS strategy every edge is encountered only once.

17.2 CLASSIFICATION OF EDGES

Another interesting property of depth-first search is that the search can be used to classify the edges of the input graph $G = (V, E)$. We classify edges mainly into four categories. This edge classification can be used to gain additional information about a graph. Remember that these edges are the edges of the depth-first forest (sometimes tree) produced after a depth-first search on G. Consider an edge (u, v) in G.

1. ***Tree Edges:*** An edge (u, v) in depth first forest is called a *Tree Edge* if v is first discovered by exploring edge (u, v). Simply an edge in a DFS tree is a tree edge.
2. ***Back Edges:*** An edge (u, v) that connects a vertex u to an ancestor v in a depth-first tree. Obviously the edge (v, u) belongs to G. This is a non-tree edge from a vertex u to a proper ancestor of u in a DFS tree. Self-loops in digraphs are supposed to be back edges.

3. ***Forward Edges:*** An edge (*u*, *v*) that connects a vertex *u* to a proper descendant *v* in a depth-first tree.
4. ***Cross Edges:*** All edges (*u*, *v*) except above three categories are *Cross Edges*. This is a non-tree edge that connects vertices in two different DFS-trees or two vertices in the same DFS-tree neither of which is the ancestor nor the descendant of the other.

However, the way we classify the edges is not unique. It depends on starting node and the order the algorithm happens to select successors to visit. Remember that both Forward- and Cross-Edges are only applied to directed graphs. Therefore, every edge of DFS tree of an undirected graph *G*, is either a tree edge or a back edge.

Consider the digraph shown in Figure 17.1 and its DFS tree in Figure 17.2. DFS is performed with *s* as staring vertex. The edges labelled with B are *Back Edges.* The edges labelled with F are forward edges. The edges labelled with C are *Cross Edges.* The edges without label are *Tree Edges.*

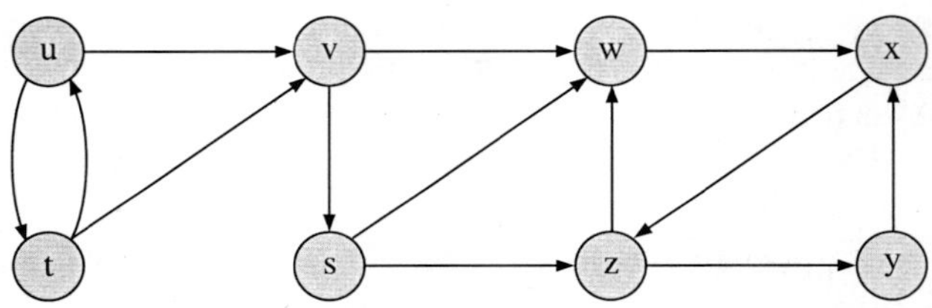

Figure 17.1 A digraph.

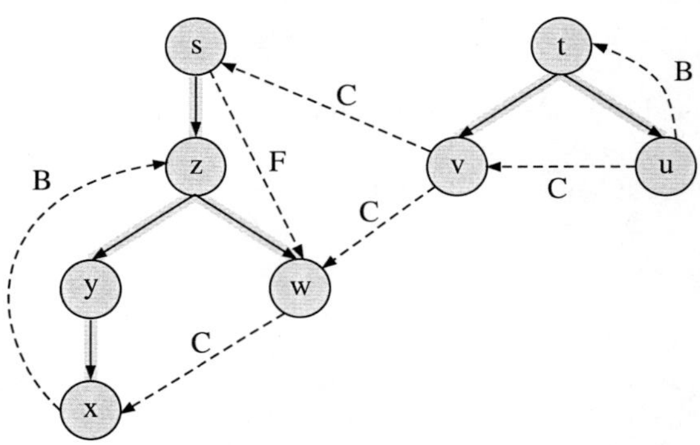

Figure 17.2 DFS Tree with labelled edges.

We graphically illustrate above four kinds of edges in sequentially with a *depth-first tree* shown in Figure 17.4 of an undirected connected graph shown in Figure 17.3. We start DFS traversal from the vertex A. The root of the tree is *A*, the first vertex is marked visited. Each edge (*u, v*) of the graph in Figure 17.3 is present in its DFS tree. When we process an edge (*u, v*), we will see that *v* is unmarked (not visited), or when we process an edge (*v, u*), we will see that *u* is unmarked, so this edge (*u, v*) is a tree edge.

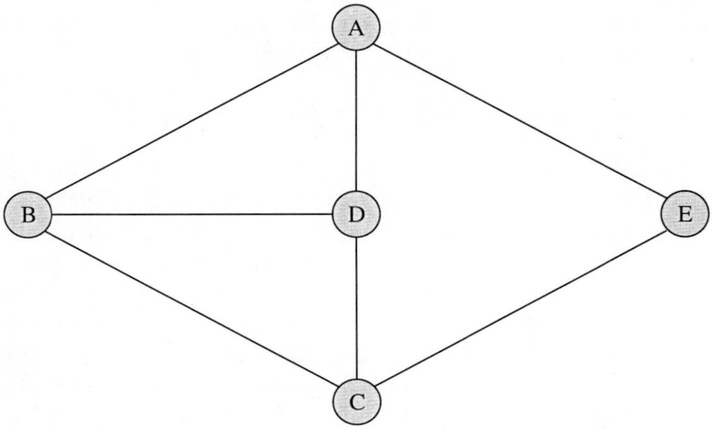

Figure 17.3 An Undirected Connected Graph.

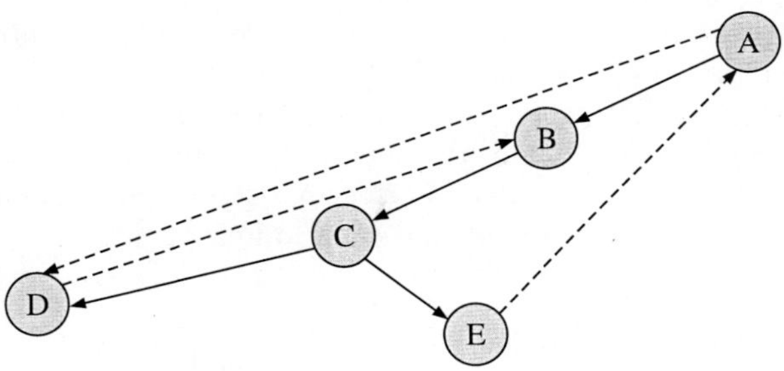

Figure 17.4 DFS Tree of Figure 17.3.

When we process an edge (u, v), we may see that v is already marked *visited*, and processing the edge (v, u), if we find that u is already marked, we then draw a dashed line from v to u. We call this edge (v, u) a *back edge*. This back edge indicates that this "edge" does not belong to the DFS tree.

17.3 APPLICATION OF DEPTH FIRST TRAVERSAL

DFS is typically used to traverse an entire graph. Algorithms that use depth-first search as a building block on a graph include:

(*i*) Finding Connected Components.
(*ii*) Topological Sorting.
(*iii*) Finding Bridges.
(*iv*) Finding Strongly Connected Components.
(*v*) Graph Planarity Testing.
(*vi*) Finding Biconnectivity.

We discuss three important application areas of Depth First Search approach. In this chapter we will discuss *Biconnectivity*. Strongly Connected Components and Topological Sorting are discussed in the subsequent chapters.

17.4 BICONNECTIVITY

An undirected graph is connected if every vertex is reachable to other. That means starting from any vertex if we can visit all other vertices, then the graph is connected. In other words, if a depth first search starting from any vertex completes visiting every vertex without any backtracking. We assume that the graphs considered are connected. If they are not connected then there is at least one vertex that may discrete (when removed the graph) the graph into two or more components. We can find these vertices connected components using DFS. The algorithm is easy to apply and efficient at the moment. As a first example of a non-trivial graph algorithm in this section we will look at a generalization of connectivity called *biconnectivity.*

Definition: A connected undirected graph $G = (V, E)$ is said to be *biconnected* if there are no vertices whose removal disconnects the rest of the graph.

If a graph is not biconnected and the vertices if exist whose removal would disconnect it (the graph) are known as *articulation points or cut vertices*. If an edge when removed from G and disconnects it then the edge is called a *bridge*. The vertex C in Figure 17.5(*a*) is an articulation point or cut vertex. The edge (C, D) in Figure 17.5(*b*) is a bridge. Cut vertices are critical in many applications. Any edge in a graph that does not lie on a cycle is a bridge. A bridge has at least one articulation point at its end. It is not necessary for an articulation point to link a bridge.

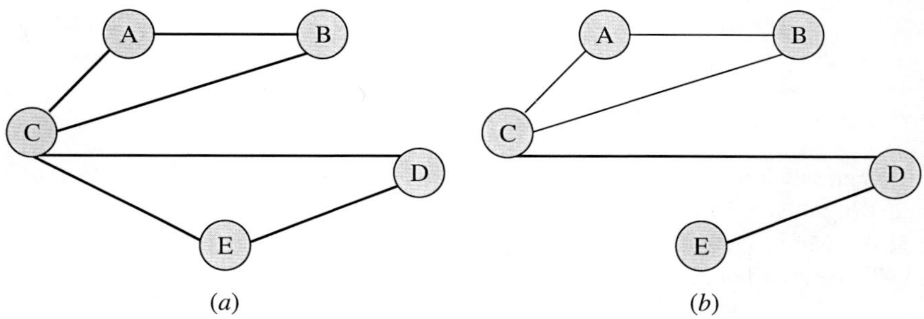

Figure 17.5 A graph with articulation point.

Consider three distinct vertices x, y, and z in an undirected connected graph G. If every path from x to y contains z then z is an articulation point of G. Naturally the G is no longer biconnected. In general, an undirected connected graph G without any articulation point is called biconnected graph. The graphs in Figure 17.6 are biconnected.

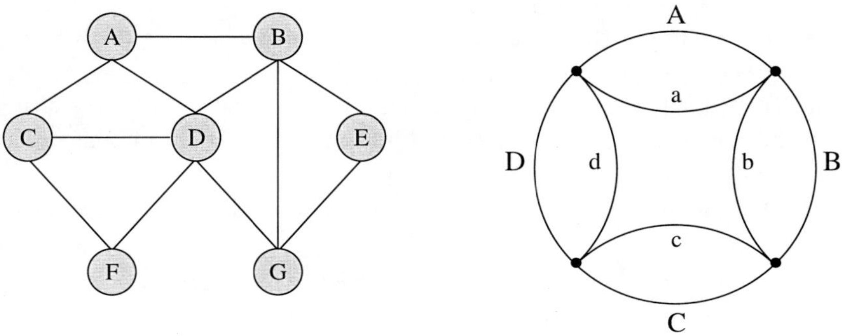

(a) (b)

Figure 17.6 Biconnected graphs.

17.4.1 Applications of Biconnectivity

Connectivity problem which particularly arises frequently involves a dynamic situation when edges are added to the graph one by one and is interspersed with queries as to whether two particular vertices belong to the same connected component. Biconnectivity has great importance especially in computer networks and communication theory to identify link failure or network failure because of fault happened eventually at some linked device. For instance, consider a network (as an undirected graph) built on set of nodes. Suppose the nodes are computers and the edges are links. Then if any computer goes down (or gets fault) then network mail is unaffected except at the down computer. This is probably possible only if the network is biconnected. It would therefore be reasonable to design a network more than one route between linked points (linked device) so as to handle possible failures at the connection points.

Another application includes Circuit Theory. An integrated circuit is designed such that the main communications lines remain biconnected so that the rest of the circuit still can function if one component fails.

Another application which is sometimes considered as nonrealistic is followed at wartime. In a wartime situation an enemy must bomb at least two stations in order to make the communication by rail disconnected.

It is therefore important in all the above cases that a graph *be* connected for some application, it might also be important that it *stay* connected on failure.

17.4.2 Two Important Properties for Biconnectivity

Let us consider T be a DFS tree of an undirected connected graph $G = (V, E)$. To be confirmed the G is biconnected by checking whether it has an articulation point, following two properties are very helpful. One is *Numvalue* and another is *Lowvalue* of the vertices of T.

(*a*) *Numvalue (v)* of a vertex *v* in *T* is the number assigned to *v* as it appeared during Preorder traversal on *T*. We can numbering the vertices for their Numvalues during DFS traversal on *G* as they are visited.

(*b*) *Lowvalue (v)* is the lowest-numbered vertex *i.e., v* has minimum Numvalue that is reachable from *v* by taking zero or more tree edges and possibly at most one back edge in that order. We can find Low value for *v* using Postorder tree traversal on *T*.

By the definition of *Lowvalue* of a vertex *v, Lowvalue* (*v*) is the minimum of

(*a*) *Numvalue* (*v*).

(*b*) The lowest *Numvalue* (*w*) among all *back edges* (*v, w*), or

(*c*) The lowest *Lowvalue* (*w*) among all *tree edges* (*v, w*).

Only third condition allows us to choose some tree edges and possibly one back edge to find *Lowvalue* (*v*). We succinctly describe this condition with a recursive call that evaluates *Lowvalue* for all the children of *v* before evaluating its (*v*) *Lowvalue.* As stated earlier this is a Postorder traversal. Looking at a glance on *Lowvalue* (*v*) and *Lowvalue* (*w*), we can say whether any edge (*v, w*) is a tree edge or a back edge.

17.4.3 Algorithms for Biconnectivity

Here we are interested in knowing if there is more than one way to get from one vertex to another in the graph. That is even if one vertex and all its adjacent edges are removed, the graph remains still connected. We will look at a method for testing whether a graph is biconnected using Depth First Search. Basically the algorithm finds all articulation points in a connected graph.

Algorithm: Biconnectivity (*G*)

Where, *G* = (*V, E*), an undirected connected graph.

Step 1: Traverse the graph *G* using DFS.

Step 2: Find number of adjacent edges *N* (or children) of the Root node of the DFS tree *T* generated in Step 1. // root property

Step 3: If $N \geq 2$

- Output "the graph is not biconnected."
- Exit.

Step 4: Repeat for each vertex *v* (except root) in DFS tree T

- Test a vertex *w* (child to v) if there is a tree edge (*v, w*) such that Lowvalue(*w*) ≥ Numvalue(*v*). // complicated property
- Output "vertex *v* is an articulation point".

Step 5: Output "the graph is not biconnected."

Step 6: Exit.

Complexity Analysis: The running time complexity of the graph depends on the DFS traversal algorithm as well as finding *Numvalues* and *Lowvalues* of the vertices in DFS tree. *Numvalue* for a vertex is found by Preorder traversal algorithm on the DFS tree

T where as *Lowvalue* for the vertices is found by Postorder traversal on *T*. Both these tree traversals run in $O(|V|)$ explicitly. On the other hand, running time of DFS algorithm is $O(|V|+|E|)$. Therefore, time complexity of the algorithm to test biconnectivity on a graph is $O(|V| + |E|)$. However, this complexity may be $O(|V|)$ if *T* is the input instead of *G*. So, Depth-first search provides a linear-time algorithm to find all articulation points in a connected graph.

The algorithm says that the Root node of the DFS tree is always an *articulation point* if and only if it has at least two children. Because it is when removed, disconnects the tree into subtrees. It is merely true for the Root if it has only one child. Any other vertex *v* is an *articulation point* iff *Lowvalue(w)* ≥ *Numvalue(v)* for some *w* as child *v*. This condition is always for the Root. So, we overlook this condition for this node. This condition also says that vertex v will not be an articulation point if every descendant *w* has some node lower in the tree connected (via a dotted link) to a node higher in the tree than *v*, thus providing an alternate path from *v* to *w*.

Example 1. Find articulation point(s) of the graph shown in Figure 17.7. Is the graph biconnected?

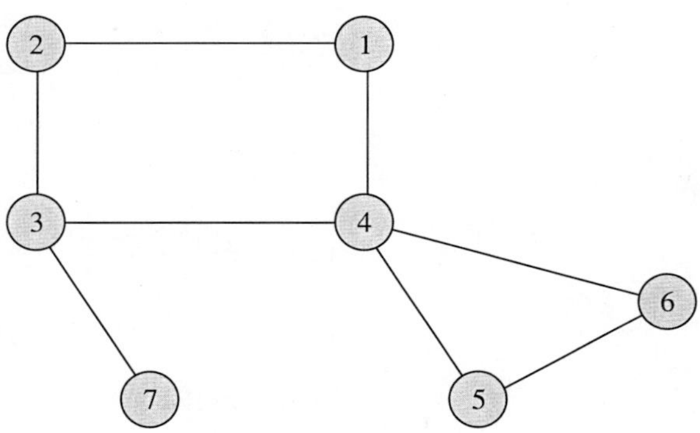

Figure 17.7 Undirected Connected Graph.

Solution: Articulation points are found stepwise as below.

(*a*) Find DFS tree with back edges of the graph of Figure17.7. Dashed lines are back edges.

(*b*) Do numbering for every vertex *v* to get *Numvalue(v)* using Preorder traversal starting from vertex A of DFS tree in Figure 17.8. the number associated with each node in Figure 17.9 is its Numvalue.

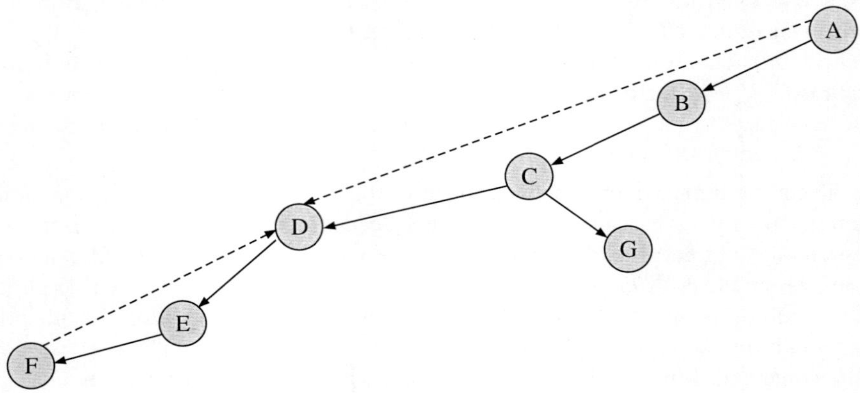

Figure 17.8 DFS tree with back edges.

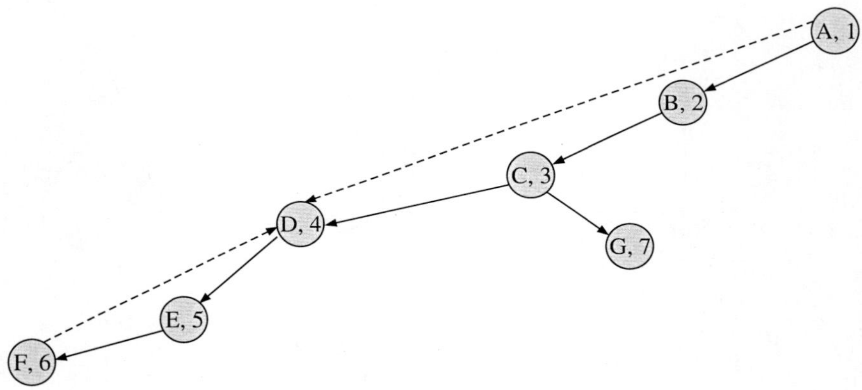

Figure 17.9 DFS tree with Numvalue for each vertex.

(*c*) Do numbering for every vertex *w* to get *Lowvalue(w)* using Postorder traversal starting from vertex *A* of DFS tree in Figure 17.9.

Vertex *A* is root node. So its Lowvalue(*A*) = Numvalue(*A*) = 1.

Lowvalue(*B*) = Numbvalue(*A*) = 1, since *B* is a child of *A*.

Lowvalue(*C*) = Numvalue(*B*) = Numvalue(*A*) = 1. Since *C* is a child of *B* and there is a path from *C* to *B*, and *C* to *A* using back edge *DA* in the cycle (or component) *A-B-C-D-A*.

Lowvalue(*D*) = Numvalue(*C*) = Numvalue(*B*) = Numvalue(*A*) = 1 due to similar reason.

Lowvalue(*E*) = Numvalue(*D*) = 4. Since *E* is a child of *D* and there is a path from *E* to *D* using back edge *FD*. Also *D* belongs to the cycle *D-E-A-D*. So, *D* is current root to the component subtree *D-E-A*.

Lowvalue(F) = Numvalue(E) = Numvalue(D) = 4. Reason is similar,

Lowvalue(G) = Numvalue(G) = 7. Although G is child of C but there is no path from G to C. So, it is single node component subtree G is its root.

Lowvalue for each vertex is shown in Figure 17.10. Here I/J for a vertex is treated as that I is it's Numvalue and J is it's Lowvalue.

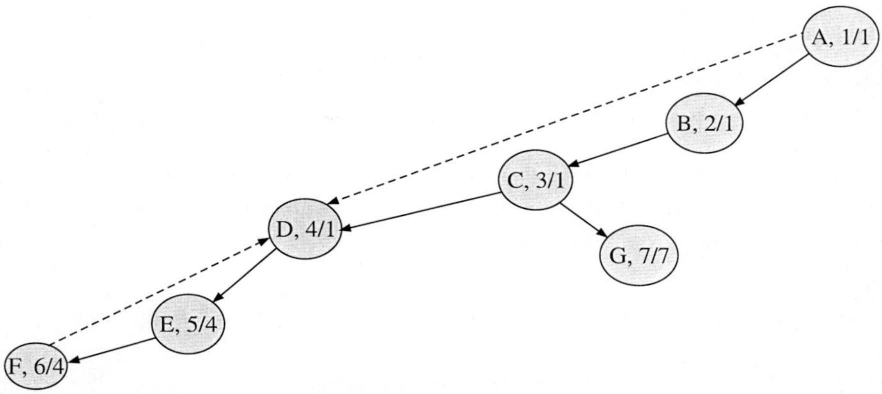

Figure 17.10 DFS tree with Numvalue and Lowvalue for each vertex.

(*d*) Compare for every child w to v whether Lowvalue(w) $\geq$ Numvalue(v) for each vertex pair (v, w) in Figure 17.10 except $v = A$.

Consider $v = B$, $w = C$, then Lowvalue(C) $\geq$ Numvalue(B) $\Rightarrow 1 \geq 1 \Rightarrow$ true. But child D (lower node to C) is linked with the ancestor A (higher node to B) by the back edge DA. So B is not articulation point.

Consider $v = C$, $w = D$, then Lowvalue(D) $\geq$ Numvalue(C) $\Rightarrow 1 \geq 1 \Rightarrow$ true. But no child (lower node to D) exists. So C may be an articulation point.

Consider $v = C$, $w = G$, then Lowvalue(G) $\geq$ Numvalue(C) $\Rightarrow 7 \geq 1 \Rightarrow$ true. There is no child (lower node to G) exists. So C may be an articulation point.

No other child for C exists. So C is an articulation point.

Consider $v = E$, $w = F$, then Lowvalue(E) $\geq$ Numvalue(D) $\Rightarrow 4 \geq 5 \Rightarrow$ false. So E may be an articulation point.

No other child for E exists. So E is an articulation point.

No child for both F, and G exists. Therefore they are not articulation points.

Therefore, the graph has two articulation points – C, and D.

$\Rightarrow$ *The graph is not biconnected.*

17.5 SOME COMMON QUESTIONS

Question 1. Define biconnected graph and biconnected component.

Answer: A biconnected graph is a connected graph that has no articulation points. A biconnected component of a connected undirected graph is a maximal biconnected sub graph of G.

Question 2. What is meant by articulation point?

Answer: An articulation point is a vertex v of G such that the deletion of v, together with all edges incident on v, produces a graph, G', that has at least two connected components.

Question 3. What are the methods to identify the articulation point?

Answer: Three methods are –

(*a*) Common vertex in b_i connected component.
(*b*) Depth first spanning tree.
(*c*) Lowest depth first number.

Question 4. Define connected component.

Answer: A graph is said to be connected if for every pair of its vertices U and V, there is path from U to V. A connected component is the maximal sub graph of a given graph.

Question 5. How do identify connected component using breath first search?

Answer: BFS can be used to determine whether G is connected. Using BFS all newly visited vertices are put on the list. Subgraph formed by the vertices on this list make up a connected component.

Question 6. What is tree edge and cross edge?

Answer: In the breath first search forest, whenever a new unvisited vertex is reached for the first time, it is attached as a child to the vertex from which it is being reached. Such an edge is called tree edge. If an edge is leading to a previously visited vertex other than its immediate predecessor, that is noted as a cross edge.

Exercises

1. Is it possible to reconstruct the original graph (directed or undirected) from its DFS tree? If yes, explain how.
2. Check if the graph in Figure 17.11 is biconnected.

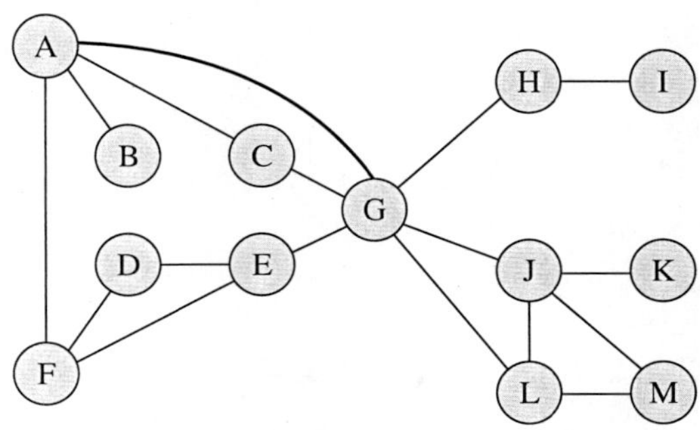

Figure 17.11 Undirected connected graph.

3. Is the DFS tree in Figure 17.12 biconnected? If not Determine its articulation point(s).

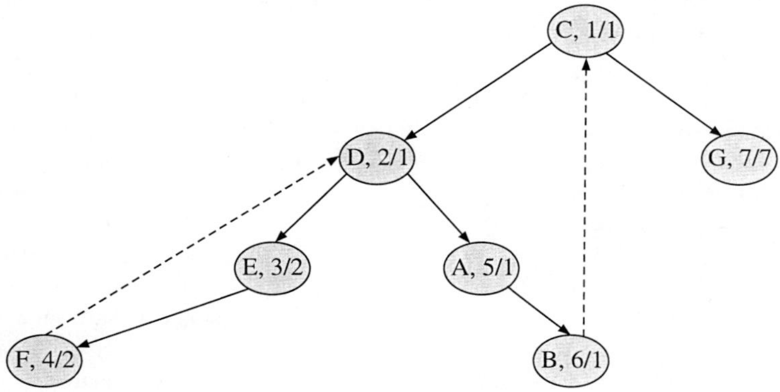

Figure 17.12 DFS tree for Question 2.

4. Apply your algorithm as mentioned for Question 3 to the DFS tree in Figure 17.12. Analyze time complexity of your algorithm.
5. Find whether the graph with its DFS tree in Figure 17.13 has articulation poin(s).

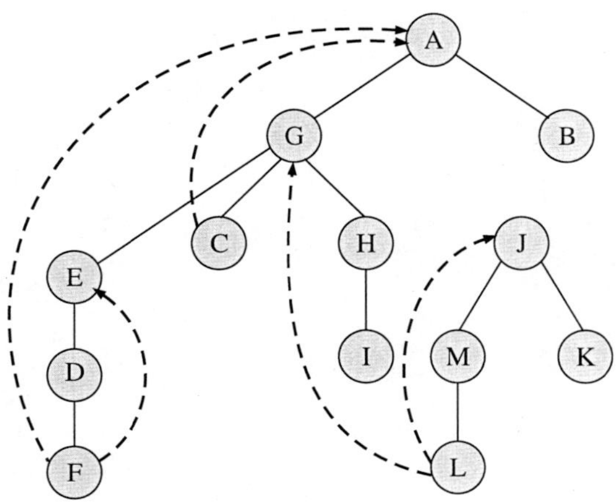

Figure 17.13 DFS tree for Question 5.

6. Write a pseudocode to print out the biconnected components of a graph.
7. Rewrite the algorithm for DFS using a stack. How does it affect to check whether a graph is biconnected?
8. A digraph $G = (V, E)$ is called *Singly Connected* iff at most only one simple path exists from u to v $u, v \in V$. Give an efficient algorithm to determine whether a digraph is singly connected.

University Questions

1. Write a suitable algorithm to find the components of a graph. *[UPTU'08]*
2. Write an algorithm which counts the number of connected components in a graph. *[UPTU'06]*
3. Define a connected graph. For a connected graph of n vertices, how many minimum and how many maximum edges should be in the graph? Prove it. *[IETE ALCCS'11]*
4. Devise a $O(n + m)$ time algorithm for computing a component graph of a directed graph $G = (V, E)$, where $|V| = n$ and $|E| = m$. Make sure that your algorithm produces at most one edge between any pair of vertices in the component graph. *[DOEACC'09]*
5. Prove or disprove an undirected graph $G = (V, E)$ is biconnected if and only if for each pair of distinct vertices u and v there are two distinct paths from u to v that have no vertex in common except u and v. *[JNTU'08, '09]*
6. Write an algorithm of Biconnected components and also analyze its time complexity. *[JNTU'08]*
7. Find a necessary and sufficient condition for the root of a depth first search for a connected graph to be an articulation point. Prove it. *[JNTU'08]*

❑❑❑

18 Chapter Strongly Connected Components

18.1 INTRODUCTION

Connectivity in undirected graphs is pretty straightforward. An undirected graph is said to be connected if there exists a path between every pair of vertices. Otherwise, the graph is called disconnected. In directed graphs, connectivity is more subtle. A directed graph or digraph is called connected if every pair of distinct vertices in the graph is connected. In particular, a path from vertex a to vertex b in a digraph if exists and also a path from b to a if exists then this digraph is called *Strongly Connected Graph*. However, if the digraph is not *Strongly Connected* but if the undirected representation (just ignoring direction of an edge) of this digraph is connected then this digraph is called *Weakly Connected Graph*. The *Connected Components* of an undirected graph are the separate "pieces" (subgraph) of the graph such that there is no connection between the pieces. That is each piece contains disjoint set of vertices. In case of a directed graph it is called *Strongly Connected Component (SCC)* that means a maximal set of vertices such that for every pair of vertices u and v in the component, vertices u and v are reachable from each other. A *Connected Component* is a maximal connected subgraph of a graph. In other words, the strong components are the maximal strongly connected subgraphs. Each vertex belongs to exactly one connected component, as does each edge. In previous chapter we have discussed one application of Depth First Search Approach. We now in this chapter consider another classic application of depth-first search: decomposing a digraph into its strongly connected components.

18.2 FEATURES OF SCC

The *Strongly Connected Components* of a directed graph $G = (V, E)$ are its maximal strongly connected subgraphs. But why do we decompose a graph into its components. Some basic characteristics of *Strongly Connected Component* include.

(*a*) If each strongly connected component is *contracted* to a single vertex, the resulting digraph is a *directed acyclic graph* (*DAG*.) this is the *condensation* of G. That means a digraph is acyclic if and only if it has no (nontrivial) strongly connected subgraphs.

(*b*) Algorithms for finding strongly connected components may be used to solve 2-satisfiability problems.

(*c*) An undirected graph may be oriented in such a way that it becomes strongly connected, if and only if it is 2-edge-connected.

(*d*) Many algorithms that work with directed graphs begin with such a decomposition etc.

This section shows how to do this decomposition using two depth-first searches. After decomposition, the algorithm is run separately on each strongly connected component. The solutions are then combined according to the structure of connections between components.

18.3 APPLICATIONS OF SCC

The graph G is strongly connected if every pair (u, v) of vertices in G is reachable from one another. This property is important and utilized in many areas.

(*a*) Networking - Strongly connected components define network nodes connected by multiple, therefore redundant paths.

(*b*) Communications

(*c*) Web - To gather pages for search engines, Web crawlers move from one vertex (page) to another along edges defined by hyperlinks.

18.4 DECOMPOSITION INTO SCC

The decomposing a directed graph into its strongly connected components has a great importance. And this decomposition of a directed graph into its strongly connected components is very informative and useful. Many algorithms that work with digraphs begin with such decomposition. The algorithm is run separately after decomposition, on each strongly connected component. The solutions are then combined as per the structure of connections between components. This section shows how to do this decomposition. Mainly we dot it using depth-first search calling two times. By using these two depth-first search traversals, we can test whether a directed graph is strongly connected. If it is case that the digraph is not strongly connected, we can actually then produce the subsets of vertices that are strongly connected to themselves. Instead of using DFS twice, this can also be done in only one depth-first search in that the method would be little bit trickier. Here the method used to find SCC is much simpler to understand.

Algorithm: SCC(G)

Where $G = (V, E)$, a connected digraph

Step 1: Perform depth-first search on the input graph G. The result produces a depth-first spanning forest T.

Step 2: Do numbering of the vertices of G by a Postorder traversal of the T.

Step 3: Reverse the direction of all the edges in G a new graph called *Reverse Graph* is generated. Say, this reverse graph is G_a.

Step 4: Perform depth-first search on G_r always starting from the vertex with higher number assigned to it in step 2.

Step 5: The trees in the new depth-first spanning forest are the *strongly connected components* of G.

Step 6: End.

Complexity Analysis: Time complexity of the algorithm SCC() can be easily be analyzed. As it is seen from the steps of the algorithm that both step 1 and step 4 has the same time complexity and it is $O(|V|+|E|)$. Step 2 runs in $O(|V|)$ time. Reversing the direction of each edge in G needs $O(1)$ time. So, this process completes in $O(|E|)$ time. Therefore, total time complexity of the algorithm is twice of $O(|V|+|E|)$. That means complexity of the SCC() algorithm belongs to $O(|V|+|E|)$.

Example 1. Decompose the digraph G shown in Figure 18.1 into its *strongly connected components.*

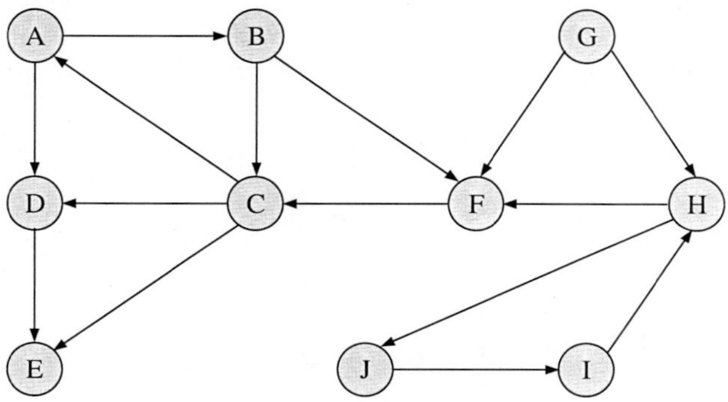

Figure 18.1 A connected digraph.

Solution: Suppose we start DFS first on the graph at the vertex A, then we have the *dfs* tree as in Figure 18.2(*a*)

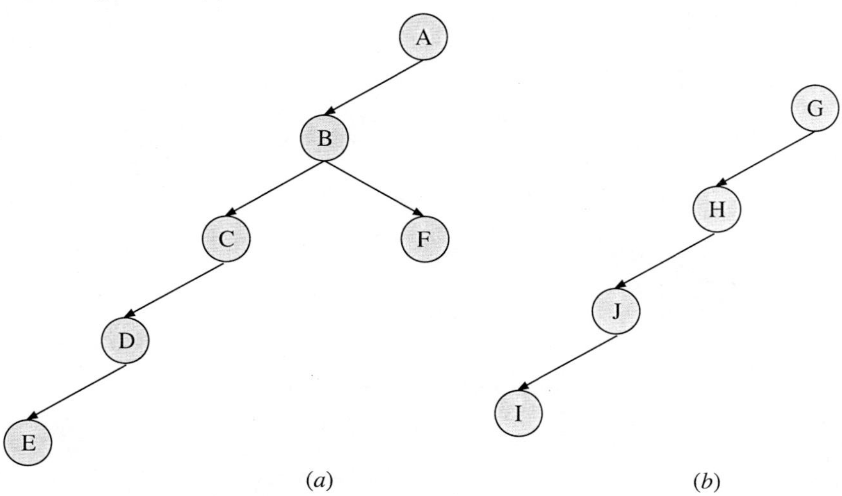

Figure 18.2 DFS forest.

Then start DFS from the vertex *G*. It will generate the *dfs* tree as in Figure 18.2(*b*). You can add *cross edges* and *back edges* to above forest.

Now do numbering the vertices of the graph in Figure 18.1 using Postorder. First take the first *dfs* tree *i.e.,* Figure 18.2(*a*). The number to each vertex is given below.

E = 1, D = 2, C = 3, F = 4, B = 5, A = 6, I = 7, J = 8, H = 9, G = 10.

After numbering the G would look like as in Figure 18.3.

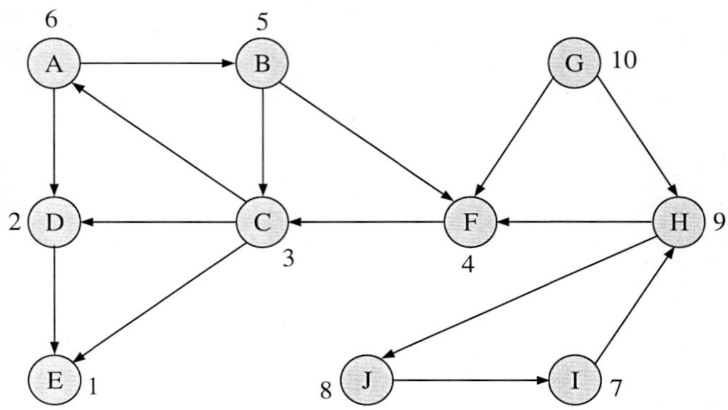

Figure 18.3 Postorder numbering.

Now reverse directions of the edges of the graph in Figure 18.3 to get G_r.

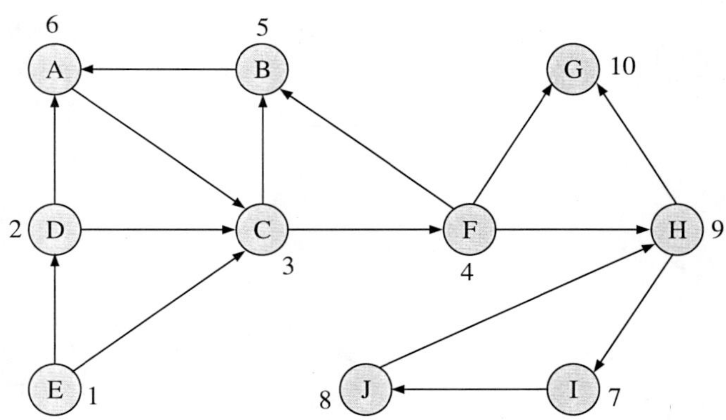

Figure 18.4 Reverse graph.

Apply DFS on Reverse graph starting at the vertex *G* in Figure 18.4, we have the *dfs* tree which is the vertex *G* itself because no other vertex can be reachable from this vertex. Next we start DFS at *H*. The *dfs* trees are shown in Figure 18.5 below. You can add *cross edges* and *back edges* to the forest below.

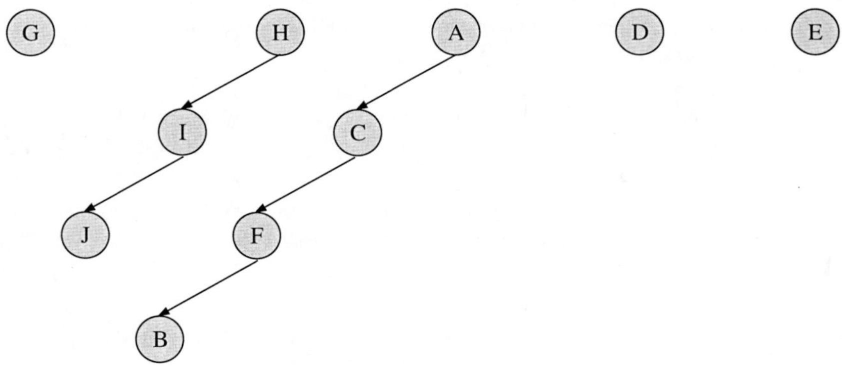

Figure 18.5 DFS forest.

Individual trees $\{G\}$, $\{H, I, J\}$, $\{A, B, C, F\}$, $\{D\}$, $\{E\}$ are the *strongly connected components.*

18.5 SOME COMMON QUESTIONS

Question 1. Why the algorithm SCC() works?

Answer: If two vertices u and v in a connected digraph G are in the same strongly connected component, then u and v are reachable to each other. Hence it is also true for the reverse graph G_r. Now, if two vertices u and v are not in the same depth-first spanning tree of G_r, clearly they cannot be in the same strongly connected component of G.

Question 2. How do you prove that the algorithm SCC() works?

Answer: To prove that this algorithm works, we must show that if two vertices u and v are in the same depth-first tree of the reverse graph G_r of the graph G, then from u and v are reachable to each other. Equivalently, we can show that if x is the root of the depth-first spanning tree of G_r and u is vertex in it then there is a path from x to u and from u to x. Similarly it is true for the vertex v. These paths imply that u and v are reachable to each other.

Exercises

1. Apply an algorithm to the following digraph to determine its strongly connected components.

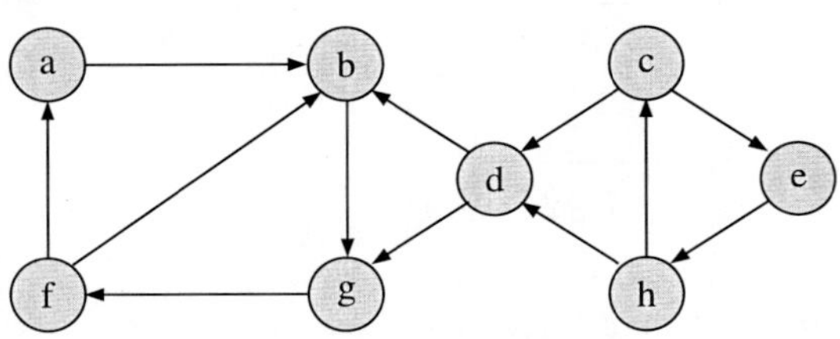

Figure 18.6 A Digraph.

2. What is the time efficiency of the algorithm you used in question 1? Give separate answers for the adjacency matrix representation and adjacency list representation of an input graph.
3. A legend among a group of n people is a person who is top rated in all aspects. The task is to identify a legend by asking a single question of the form "Who deserves it?" Design an efficient algorithm to identify a legend or determine if the group has no such person. How many questions does your algorithm need in the worst-case?
4. Run the strongly connected components algorithm on the following directed graphs. Whenever there is a choice of vertices to explore, always pick the one that is alphabetically nearest.

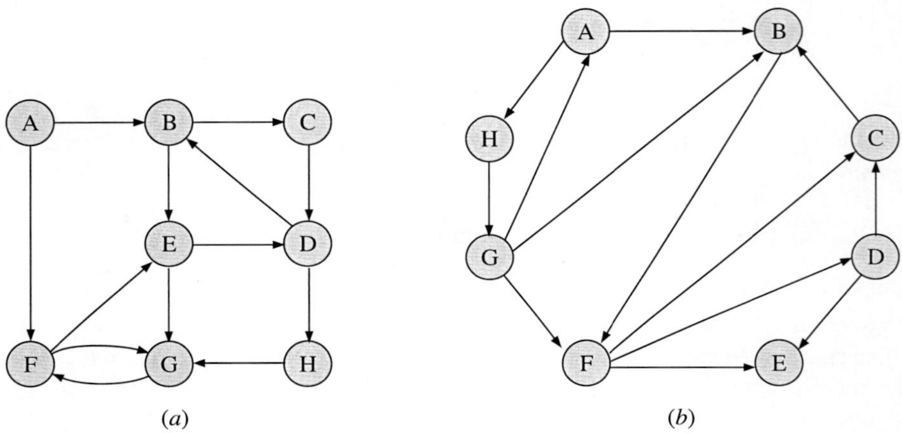

Figure 18.7 Two Digraphs.

5. Consider a digraph that consists of garments as below. Find strongly connected components of this graph.

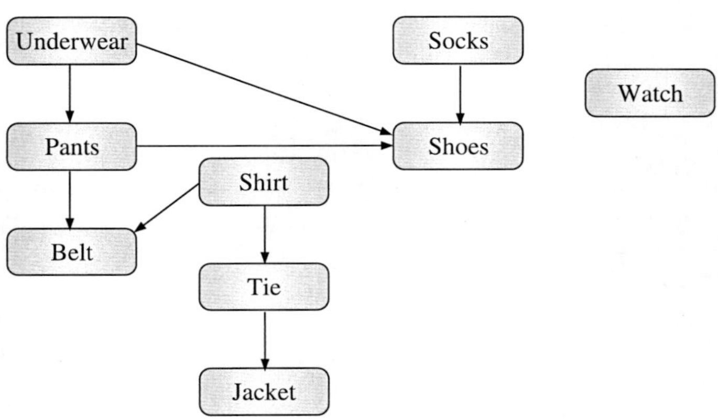

Figure 18.8 A typical digraph for garments.

6. Prove that SCC(G) correctly computes the strongly connected components of a directed graph G.
7. How can the number of strongly connected components of a graph change if a new edge is added?
8. Find strongly connected components of the graph given below. Can you determine how many components are there?

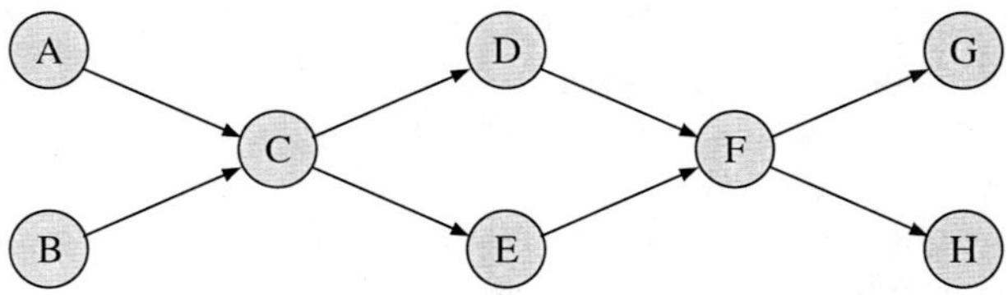

Figure 18.9 A Digraph.

9. Run the SCC(G) algorithm for the undirected connected graph given below. How many strongly connected components does a dag have?

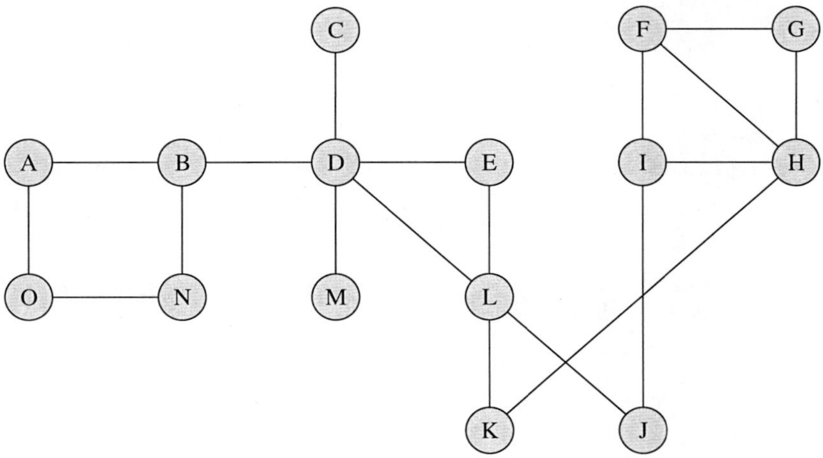

Figure 18.9 An Undirected graph.

10. Could you make any difference on apply SCC(G_1) and SCC(G_2). Here G_1 and G_2 are digraph and undirected graph. But both are connected.
11. Professor Beans claims that the algorithm for strongly connected components can be simplified by using the original (instead of the reverse) graph in the second depth-first search and traversing the vertices in order of *increasing* finishing times (preordered assigned number). Is the Prof. Beans correct?
12. Show how depth-first search can be used to find the connected components of an undirected graph.

University Questions

1. Explain the properties of strongly connected components. ***[JNTU'08, 10]***
2. Define Euler diagraph with example. Prove that every Euler diagraph without isolated verities is strongly connected. Also, show by constructing a counter example, that converse is not true. ***[UPTU'07]***
3. Discuss strongly connected components of a graph with example. ***[IETE ALCCS'11]***
4. Show that a depth-first search of an undirected graph G can be used to identify the connected component of G and that the depth-first forest contains as many trees as G has connected components. ***[DOEACC'06]***
5. Show how to modify depth-first search so that each vertex v is assigned an integer label $cc[v]$ between 1 and k, where k is the number of connected components of G, such that $cc[u] = cc[v]$ if and only u and v are in the same connected component. ***[DOEACC'06]***
6. Devise a $O(n + m)$ time algorithm for computing a component graph of a directed graph $G = (V, E)$, where $|V| = n$ and $|E| = m$. Make sure that your algorithm produces at most one edge between any pair of vertices in the component graph. ***[DOEACC'06, '07, 09]***
7. Define the term *Strongly Connected.* ***[PTU'08]***

19 Chapter Topological Sorting

19.1 INTRODUCTION

In this section, we discuss an important problem for directed graphs. There are many problems involving a set of tasks in which some of the tasks must be done before others. For example, consider the problem of taking a course by students of IIT, Guwahati only after taking its prerequisites. If the subjects are to be taken based on a systematic way of linearly arranging the courses, then what is the order that they should be taken? The answer to this question is simply to choose *Topological sorting*. Before going to topological sort we must have an overview about Directed Acyclic Graphs. A directed graph or digraph is said to be acyclic if it does not contain any cycle. Directed Acyclic Graphs (DAGs) are used in many applications to indicate precedence among events. They can also be used to represent a number of interesting relations. Though this class includes trees, but is less in general than the class of all digraphs. For example, a directed acyclic graph in Figure 19.1 is used to represent an arithmetic expression $\{(a + b)(c + d) + (a + b)(c - d)\}$ with repeated subexpression $(a + b)$.

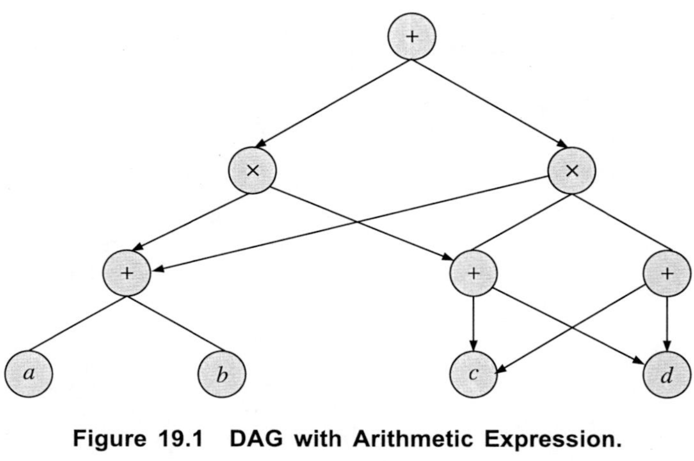

Figure 19.1 DAG with Arithmetic Expression.

These graphs also offer a natural representation for partial orderings, such as the relation of set-inclusion. For example, partial ordering defined on the divisors (positive integers) of 24 can be represented by the Figure 19.2. Here, an edge from node i to node j is said to exist iff i is a proper divisor of j.

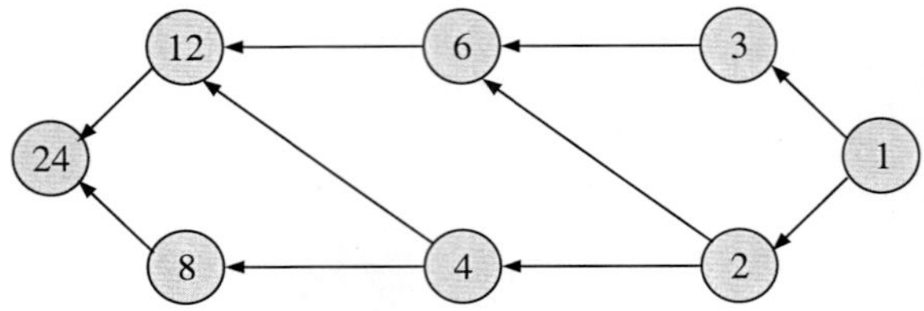

Figure 19.2 Partial Ordering for Divisor of 24.

The adjacency matrix and adjacency lists are still two key ways of representing our digraphs. Two notable differences between undirected and directed graphs represented by the above means can be made as: (1) the adjacency matrix of a digraph does not need to be symmetric, and (2) an edge in a directed graph appears only once in its adjacency lists. Depth-first search and breadth-first search are the traversal algorithms for traversing digraphs.

From the DFS forest obtained after DFS traversal on a directed graph, a remarkable feature can be drawn that a back edge in the forest can connect a vertex to its parent. The presence of a back edge in the forest indicates that the digraph has a directed cycle. Conversely, it can be said that the digraph is a dag if a DFS forest of the digraph has no back edges.

19.2 TOPOLOGICAL SORTING

A *topological sorting or sort* of a dag $G = (V, E)$ is a linear ordering of all its vertices such that if G contains an edge (u, v), then u appears before v in the ordering. In this kind of ordering the nodes of the graph are listed in such a way that if there exists an edge (u, v), then node i precedes node j in the list. For example, Figure 19.4 represents topological ordering of Figure 19.3.

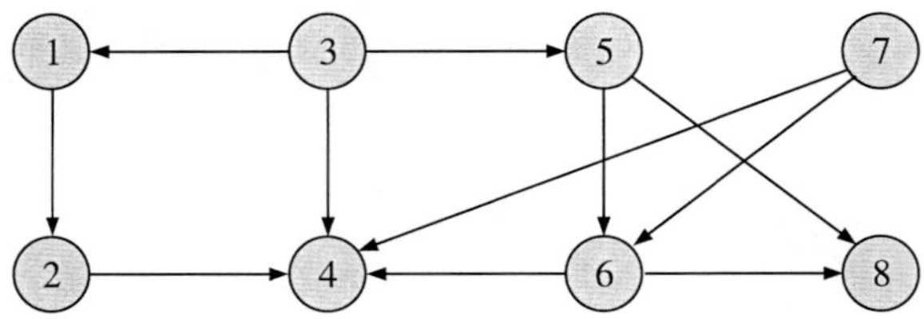

Figure 19.3 A DAG.

If the graph is not acyclic, then no linear ordering is possible. Being a dag is one of the necessary conditions for topological sorting to be possible. It makes sure us that being a dag is not only necessary but also sufficient for topological sorting to be possible.

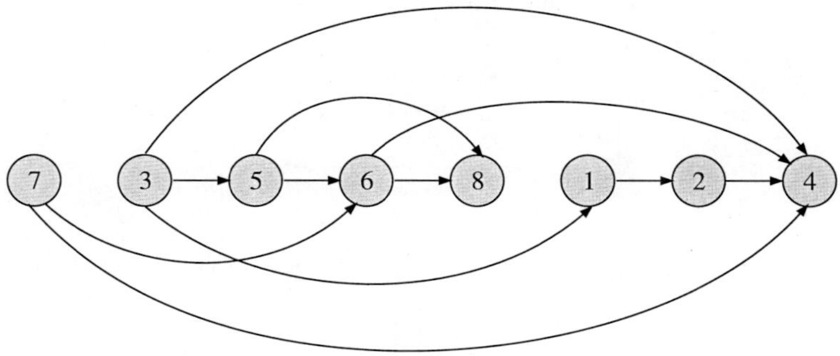

Figure 19.4 Topological Sorting for Figure 19.3.

19.2.1 Topological Sorting Algorithms

A topological sort of a DAG can be viewed alternatively as an ordering of its vertices along a horizontal line so that all directed edges go from left to right. That means vertices of a DAG can be ordered linearly with the intention that starting vertex for every edge is listed before its ending vertex. Furthermore, the ordering is not necessarily unique; any legal ordering will do. Topological sorting is thus different from the usual kind of "sorting". The following simple algorithms topologically sort a DAG.

19.2.1.1 First Topological Sorting Algorithm

A simple algorithm to find a topological ordering has been described below. Although the algorithm assumed to be faster still it is not optimal one.

Algorithm: TopSort(G) // TopSort indicates Topological Sorting

Where $G = (V, E)$, a DAG.

Q = a queue of size $|V|$.

Step 1: Find any vertex u in G with no incoming edges.

Step 2: Remove u and its adjacent outgoing edges (u, v) from G. Now G is new G without u and its adjacent outgoing edges.

Step 3: Enqueue (insert) u in Q.

Step 4: Choose any v (obtained in step 2) as u with no incoming edges. Go to step 2.

Step 5: Q is the topological sort of the vertices of input G.

Step 6: End.

Complexity Analysis: Searching a vertex u with no incoming edges involves at most $O(|V|)$ time. Removal of u and its outgoing edges can be possible in constant time. Similarly inserting u into queue needs linear time. Step 4 actually iterates $|V| - 1$ times previous three steps. Therefore the running time complexity of the algorithm is $O(n^2)$ if the DAG is represented by adjacency list and is $O(n^3)$ if the DAG is represented by adjacency matrix.

Example 1. Find the topological sorting of the following graph given in Figure 19.5.

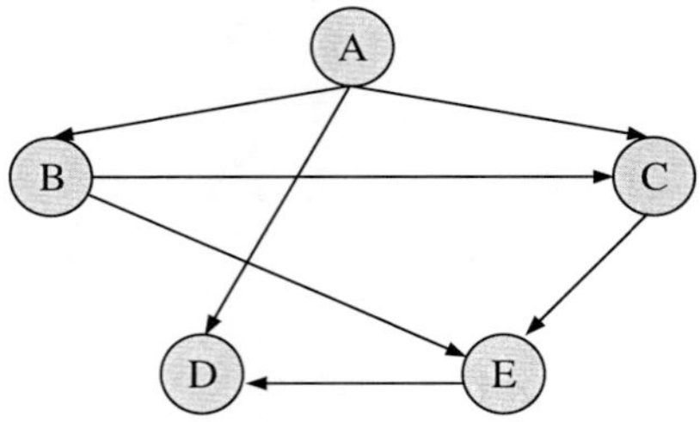

Figure 19.5 A DAG.

Solution: Suppose, Q be the queue that will contain the result. It is initially empty. From the graph it is seen that vertex A has no incoming edges. Applying the above algorithm, remove A and its adjacents outgoing edges. Include $Q = \{A\}$. The graph is reduced to as Figure 19.6.

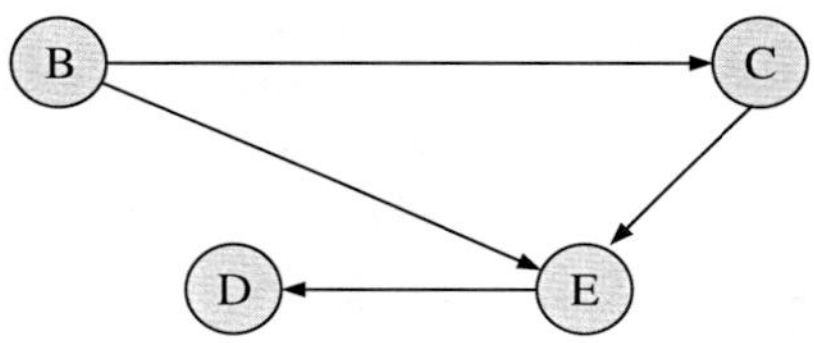

Figure 19.6 Reduced DAG.

Proceeding in the same way, we reach to the solution as below.

Remove B and its adjacent outgoing edges from Figure 19.6. $Q = \{A, B\}$. Reduced G is

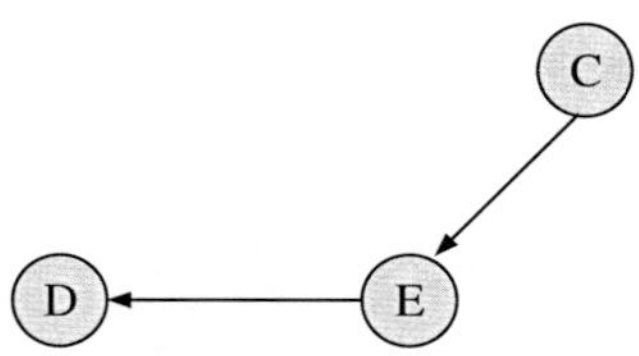

Figure 19.7 Reduced DAG.

Remove C and its adjacent outgoing edges from Figure 19.7. $Q = \{A, B, C\}$. Reduced G is

Figure 19.8 Reduced DAG.

Remove E and its adjacent outgoing edges from Figure 19.8. $Q = \{A, B, C, E\}$. Reduced G is

Figure 19.9 Reduced DAG.

Now we have just a vertex. Thus $Q = \{A, B, C, E, D\}$is the topological sorting of the algorithm.

Example 2. Find the topological sorting of the following graph given in Figure 19.10.

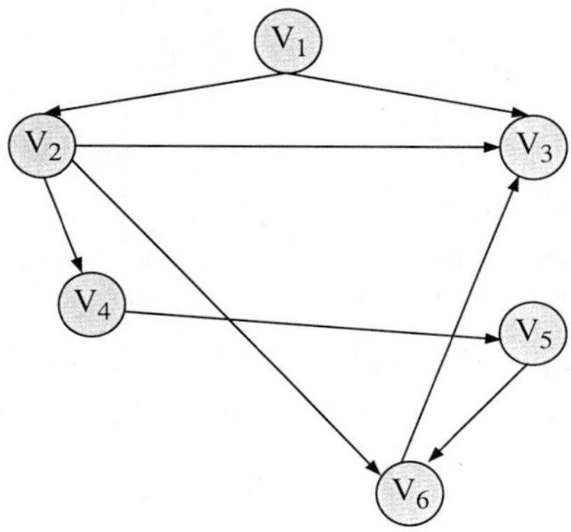

Figure 19.10 A directed acyclic graph.

Solution: Applying TopSort() method to the DAG in Figure 19.10, we have the sequence of the topological ordering as in Figure 19.11. Here numbered, dashed arcs represent sequence of deleting a vertex with no incoming edges with deletion of its outgoing edges. Every time the deleted vertex is a part of the partial topological result.

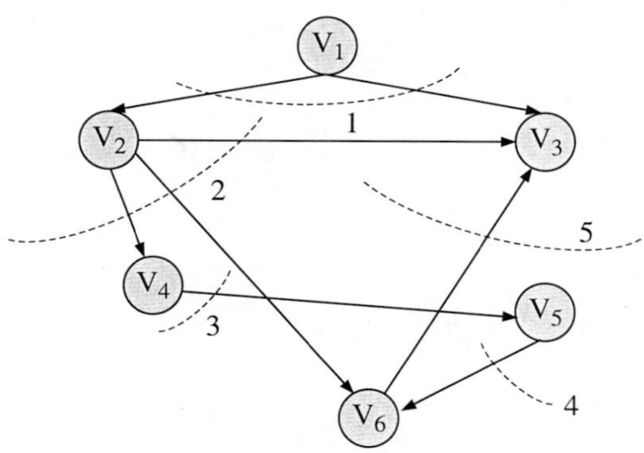

Figure 19.11 Removal of vertices with no incoming edges sequentially.

Thus, the topological order of the vertices is $\{V_1, V_2, V_4, V_5, V_6, V_3\}$.

19.2.1.2 Second Topological Sorting Algorithm

The bottleneck in the previous is step 1. A minor modification however will allow us to do this step in probably constant time. Assume the adjacency list representation of the graph. Consider two adjacency lists are created for each vertex of the input DAG G. First list is for the incoming edges and second list is for the outgoing edges. This representation can easily be done in linear time. In this approach we will maintain a queue Q of vertices with no incoming edges. This will help in reducing the cost of finding a vertex with no incoming edges to constant time. One extra overhead is that of maintaining the queue. The algorithm is stated below.

Algorithm: TopSort2(G)

Where $G = (V, E)$, a DAG

Q = queue of vertices with no incoming edges.

T = a set containing topological sorting of vertices, initially it is empty.

Step 1: Traverse G and insert each v *into* Q if it has no incoming edge.

Step 2: Pick a vertex u off the queue Q and add u as the next vertex in the total order T.

Step 3: Delete u *from* Q and all outgoing edges (u, v).

Step 4: For each such v found in step 2, if it has no incoming edges, put v on the queue Q.

Step 5: Go to step 2.

Step 6: End.

Complexity Analysis: Step 1 takes time $O(|V|)$. Step 2 takes a constant time but it iterates at most $O(|V|)$ times. In step 3 we are removing outgoing edges, it may at most be say m. This operation takes time $O(m)$ over all iterations, since each edge can be deleted at most once. Thus, the overall time complexity of the TopSort2() is $O(m + n)$, where $n = |V|$.

Example 3. Suppose, a job consists of 9 {1 ... 9} tasks with the following precedence rules:

(*a*) The job must start with any of the tasks 7, 5, 4.

(*b*) Task 1 must follow task 7.

(*c*) Tasks 3 and 6 must follow both tasks 7 and 5.

(*d*) Tasks 8 must follow 6 and 4.

(*e*) Task 2 must follow task 4.

(*f*) Task 9 must follow task 2.

Make a directed graph and then a find the topological ordering of the tasks using TopSort2() algorithm.

Answer: Directed graph of the above schedule could be as in Figure 19.12.

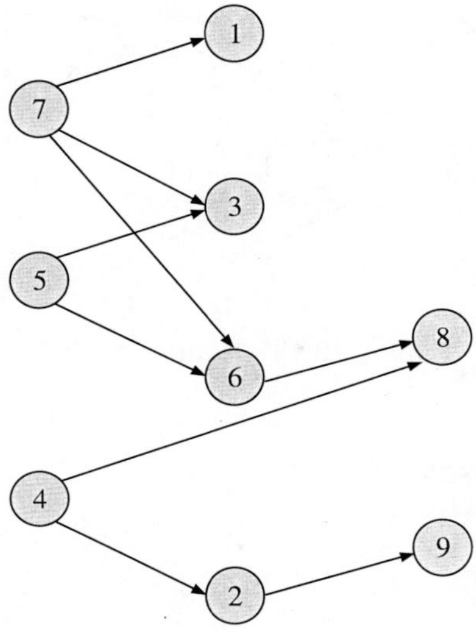

Figure 19.12 Tasks shown as a directed graph.

The key point is that the job must have to start from one of the tasks 7, 5, 4. So, the queue Q = {7, 5, 4}, and T = { }.

Delete 7 from Q and it to T. Add 1 to Q. So, Q ={5, 4, 1}and T = {7}. The graph would look like Figure 19.13.

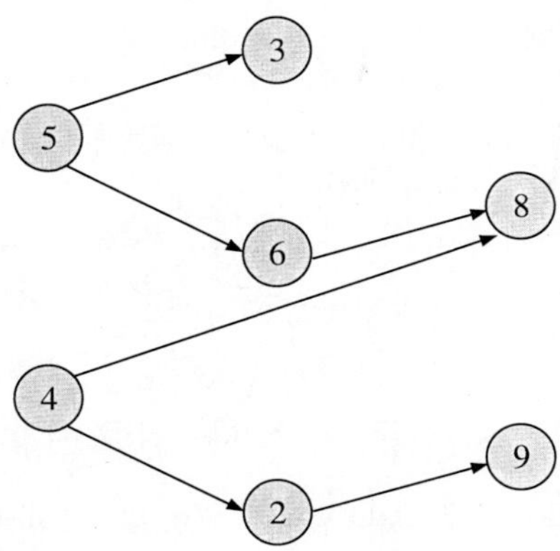

Figure 19.13 Reduced DAG.

Delete 5 from Q and it to T. Add 3, 6 to Q. So, Q = {4, 1, 3, 6} and T = {7, 5}. The graph would look like Figure 19.14.

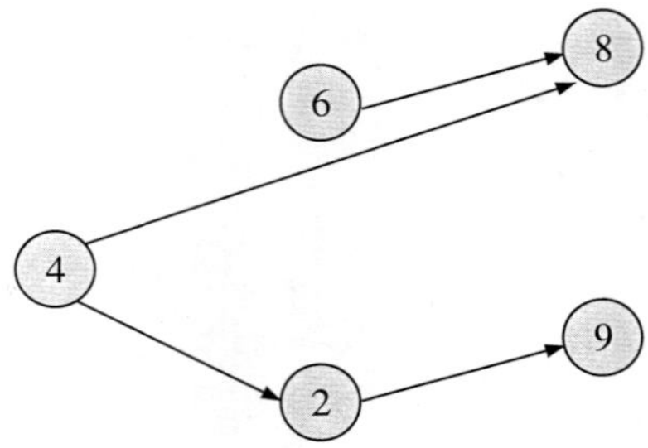

Figure 19.14 Reduced DAG.

Delete 4 from Q and it to T. Add 2 to Q. So, $Q = \{1, 3, 6, 2\}$and $T = \{7, 5, 4\}$. The graph would look like Figure 19.15.

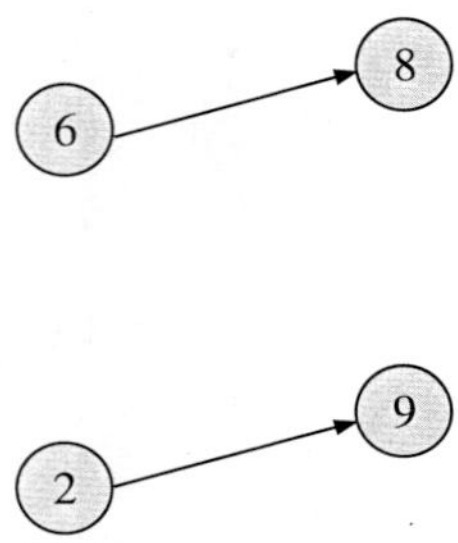

Figure 19.15 Reduced DAG.

Delete 1, 3 from Q and them to T, no adjacent vertex with no incoming edges possible for them. So, $Q = \{6, 2\}$and $T = \{7, 5, 4, 1, 3\}$.

Delete 6 from Q and it to T. Add 8 to Q. So, $Q = \{2, 8\}$and $T = \{7, 5, 4, 1, 3, 6\}$. The graph would look like Figure 19.16.

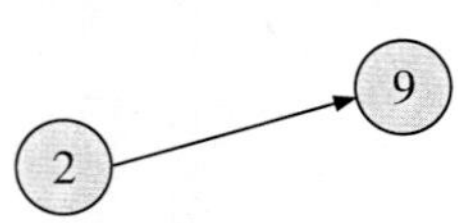

Figure 19.16 Reduced DAG.

Delete 2 from Q and it to T. Add 9 to Q. So, $Q = \{8, 9\}$and $T = \{7, 5, 4, 1, 3, 6, 2\}$.

Delete 8, 9 from Q and them to T. So, $Q = \{\}$ and $T = \{7, 5, 4, 1, 3, 6, 2, 8, 9\}$.

Thus the topological ordering is $T = \{7, 5, 4, 1, 3, 6, 2, 8, 9\}$.

19.2.1.3 Third Topological Sorting Algorithm

In this section we present another simple algorithm for finding topological ordering of vertices of a directed acyclic graph using *depth first search algorithm*. One of the advantages of exploiting DFS algorithm is that if there exists a *back edge* in the DFS spanning tree of the input graph then there does not exist any topological ordering because the graph no longer is acyclic.

Algorithm: TopSortDfs(G)

Where $G = (V, E)$, a DAG

Step 1: Traverse G using DFS algorithm. The step generates DFS spanning tree.

Step 2: Do numbering on the vertices of the DFS spanning tree using Postorder tree traversal algorithm.

Step 3: Insert nodes (vertex of G) of the DFS spanning tree in ascending order of their number (sometimes called finish time) always as first node (front node) of a linked list.

Step 4: Return the linked list of vertices.

Step 5: End.

Complexity Analysis: Time complexity of the algorithm depends mainly on the DFS algorithm. Postorder traversal needs $O(|V|)$ time to do numbering the vertices. Inserting a node at the front of linked list requires $O(1)$ time, in total $O(|V|)$ time. Step 4 does also require $O(|V|)$ time. Therefore, We can perform a topological sort in time $O(|V| + |E|)$.

Example 4. Suppose, a directed acyclic graph is given in Figure 19.17. Find topological ordering of the vertices using DFS traversal starting from vertex 1.

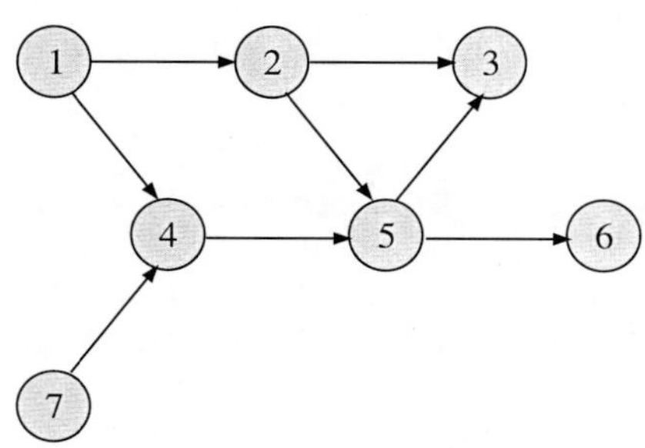

Figure 19.17 A Directed Acyclic Graph.

Solution: The DFS spanning tree of the input graph start at vertex 1would look like Figure 19.18.

After numbering using Postorder traversal on DFS tree in Figure 19.18 would look like Figure 19.21.

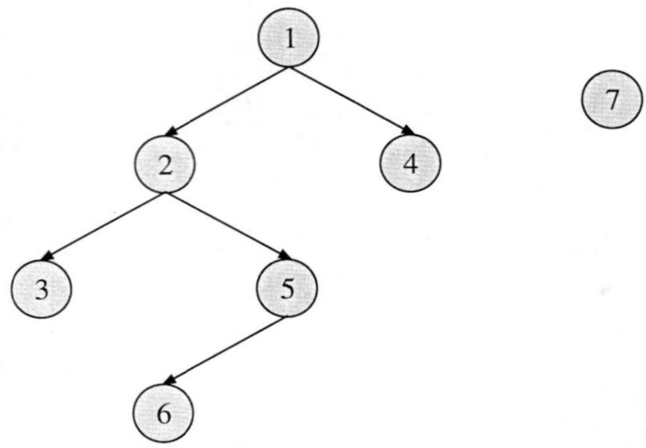

Figure 19.18 DFS tree of the input DAG.

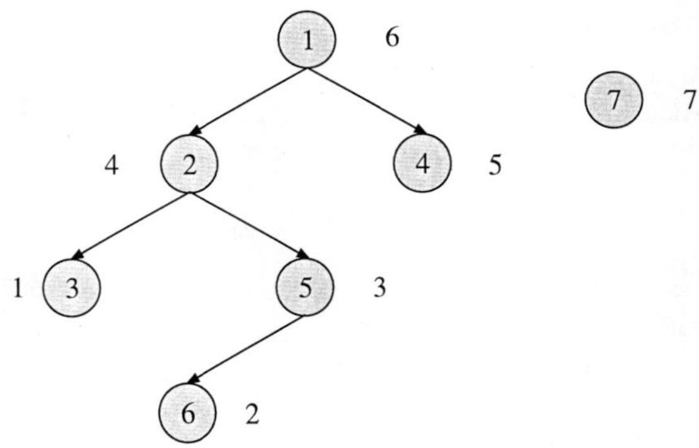

Figure 19.19 DFS tree of after numbering.

Insert vertices in ascending order of their number at front of a linked list. We have the topological order of vertices {7, 1, 4, 2, 5, 6, 3}. This is shown in Figure 19.20 with numbering value.

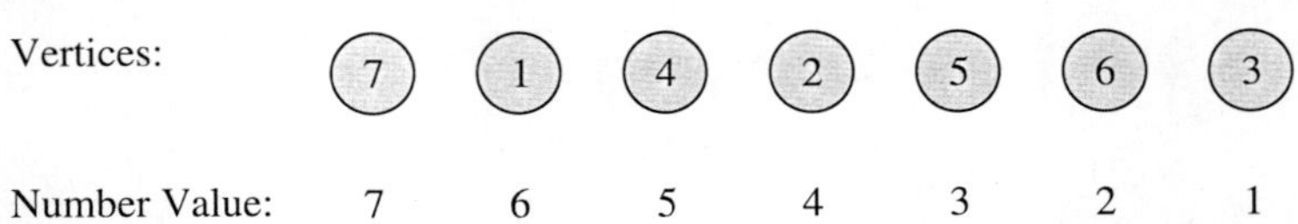

Figure 19.20 Topological Ordering of vertices.

Some books follow the representation of topological sorting in original graph representation. For the readers' purposes, the above sorting can represented as Figure 19.21. The way is to add links between vertices as in input graph.

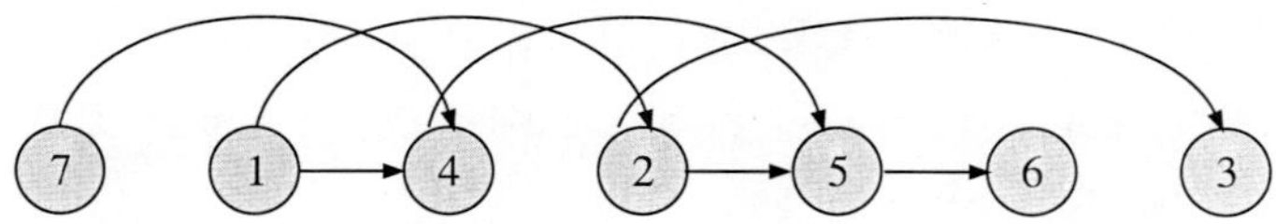

Figure 19.21 Topological Ordering of vertices.

19.2.2 Applications of Topological Sorting

Topological sorting is often useful in scheduling jobs in their proper sequence. We can use it to order things given constraints, such as a set of left-right constraints on the positions of objects. Topological Sorting on Directed Acyclic Graphs have many important applications areas. Some of these include:

(*a*) *Planning* - A set of jobs are needed to be completed with a proper planning. This can be done using topological ordering in some extent.

(*b*) *Scheduling* - Given a number of tasks with a number of constraints between the tasks. Suppose, task A must be completed before task B. These tasks can start scheduling with the constraints that form a directed acyclic graph. A topological sort of the graph gives an order in which the tasks can be scheduled while still satisfying the constraints.

(*c*) *Cycle Detection* - The algorithm can be modified to detect cycles in digraph.

(*d*) A topological sort is useful in something like a course prerequisite system, where a teacher might know what classes he needs to take before he can take others.

(*e*) Consider a routine of a course instructor who is generally getting ready for a dinner out

He must wear the following:

– jacket, shirt, briefs, socks, tie, Blackberry, etc.

There are certain constraints:

– the pants really should go on after the briefs,

– the Blackberry must be clipped on the belt,

– socks are put on before shoes.

This daily routine and similar jobs can easily be solved using topological sorting.

(*f*) Consider a situation of scheduling of jobs. This application would be used for performing these tasks on *m* processors. The topological sort can specifically be used minimizing the total run time of the collection of tasks.

(*g*) Dressing schedule from CLR.

(*h*) Identifying errors in DNA fragment assembly etc.

19.3 SOME COMMON QUESTIONS

Question 1. What happens if we run topological sort on a cyclic graph?

Answer: Two cases may arise.

(*a*) There will be either no vertex with 0 prerequisites to begin with, or at some point in the iteration.

(*b*) If there are vertices left undeleted when we run a topological sort on a graph, the graph contains a cycle.

Question 2. What is meant by lowest depth first number?

Answer: For every vertex V in depth first spanning tree, we compute the L(u) that is reachable from V by taking zero or more edges and then possible one back edge.

Question 3. What is tree edge and cross edge?

Answer: In the breath first search forest, whenever a new unvisited vertex is reached for the first time, it is attached as a child to the vertex from which it is being reached. Such an edge is called tree edge. If an edge is leading to a previously visited vertex other than its immediate predecessor, that is noted as a cross edge.

Exercises

1. Make a comparison study on all the algorithms proposed to find topological ordering of the vertices of a DAG.
2. Prove that: A digraph is acyclic if and only if any DFS forest of G yields no back edges.
3. Order the following items as shown in Figure 19.22 of a food chain.

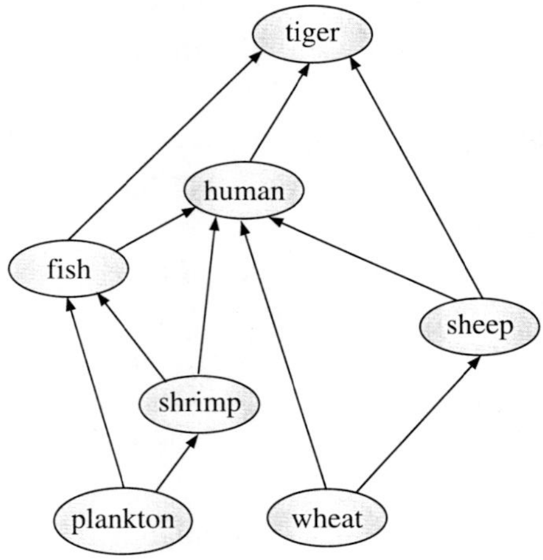

Figure 19.22 Food Chain.

4. What kind of DAG has a unique topological sort?
5. Prove that the topological sorting problem has a solution for a digraph if and only if it is a dag.
6. For a digraph with n vertices, what is the largest number of distinct solutions the topological sorting problem can have?
7. What is the time efficiency of the DFS-based algorithm for topological sorting?
8. List the order in which the nodes of the directed graph are visited by topological order traversal that starts from vertex A.

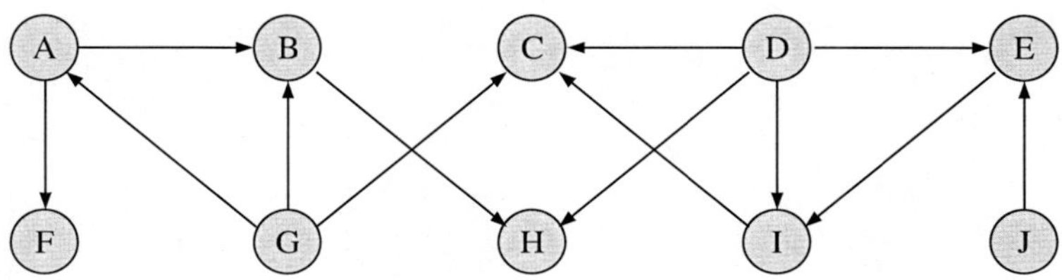

Figure 19.23 A directed acyclic graph.

9. Consider a daily routine of an employee of an organization. The routine is a busy schedule for him that includes from wake up of a bed to get out of his home to his office. The schedule is shown in Figure 19.24. Give a suitable sequence of jobs in the routine so that he can in time reach the office.

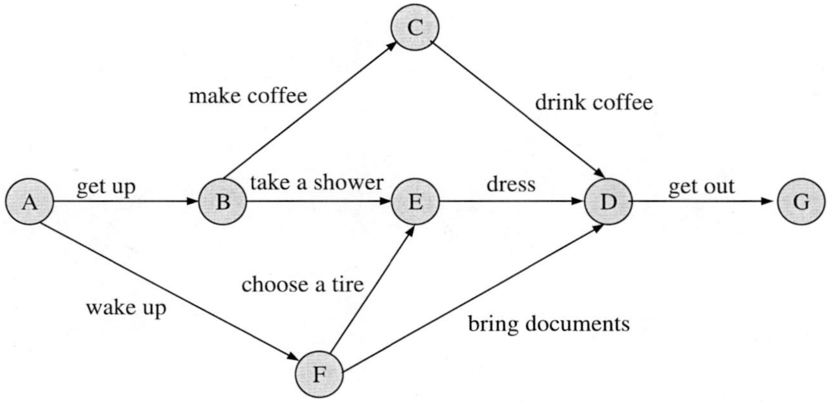

Figure 19.24 Daily routine of an employee.

10. Consider a DAG of dependencies for putting on goalie equipment shown in Figure 19.25.

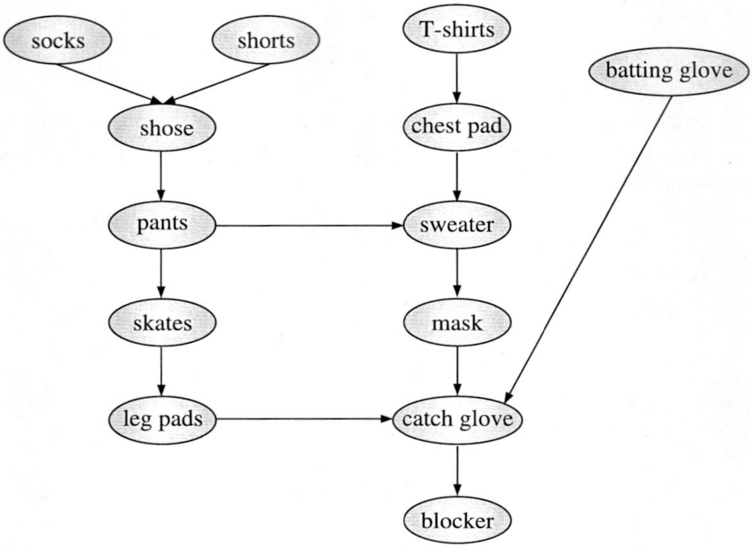

Figure 19.25 Goalie equipment for players.

Find a suitable sequence of the equipments so that a player can easily wear them.

11. Starting at vertex 1, what will be the topological ordering of the DAG given in Figure 19.26.

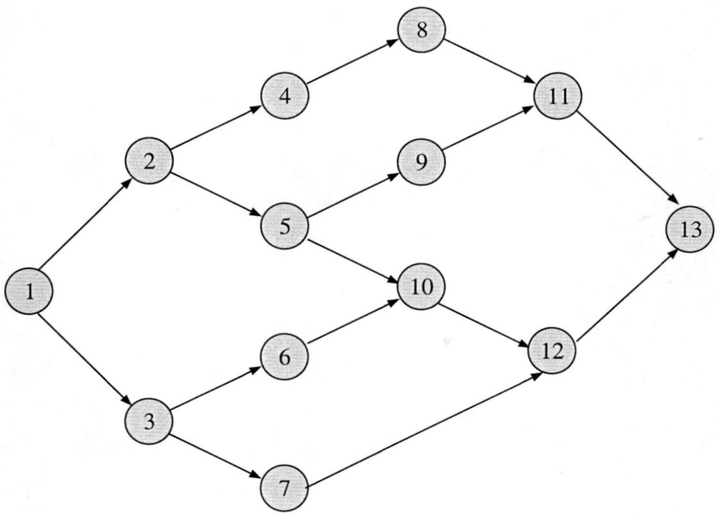

Figure 19.26 A DAG.

University Questions

1. What is the process of topological sorting? Explain how time complexity of topological sorting algorithm, depends on the data structure used to represent the graph. ***[BPUT'05]***
2. Show the ordering of vertices produced by Topological Sort when it runs on the following directed acyclic graph: ***[IETE ALCCS'11]***

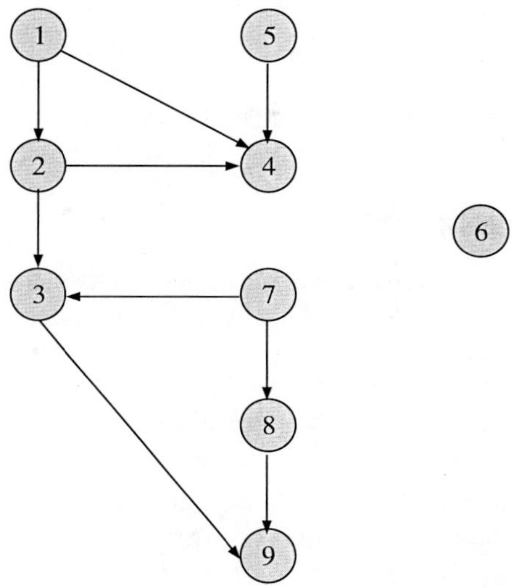

3. Discuss topological sort algorithm. ***[IETE ALCCS'11]***

❑❑❑

20 Chapter Dijkstra's Algorithm

Computer Science is no more about computers than astronomy is about telescopes.

— ***Edsger Wybe Dijkstra***

20.1 INTRODUCTION

Suppose a traveller wishes to visit one place in a city. The place is unknown to him. He has given a road map of the state in which the place belongs. How can he determine a shortest route from his place to the desired place? One possible way is to enumerate all the routes from his place to desired place, add up the distances on each route, and select the shortest. It is easy to see, however, that even if we disallow routes that contain cycles, there are many of possibilities, most of which are simply not worth considering.

Consider that the weight of an edge in a graph represents the cost of travelling that edge. For example, a graph represents a network of airports, and the weights could represent: distance, cost or time. Suppose, A and B are two vertices in a graph. Such a graph could be used to answer any of the following:

(*a*) What is the fastest way to get from A to B?

(*b*) Which route from A to B is the least expensive?

(*c*) What is the shortest possible distance from A to B?

Each of these questions indicates the instance of the same problem — *The Shortest Path Problem*! However, if all the edges in a graph have non-negative weights, then it is perhaps possible to find the shortest path from any two vertices.

20.2 SHORTEST PATH PROBLEM

The *shortest path* problem states a way to find the path in a weighted graph connecting two given vertices u and v with the property that the sum of the weights of all the edges is minimized over all such paths. If P is a shortest path from u to v, and if t is on the path P, the segment from u to t is a shortest path between u and t. It is shown by Figure 20.1.

Mathematically, we can express a shortest-paths problem as: we are given a weighted, directed graph $G = (V, E, W)$. Here V, E, and W the set of vertices, set of edges, and the weight function respectively. The weight function W is defined as $W : E \rightarrow R$ mapping edges to real-valued weights. This edge weight is the path length of the edge. The weight of path or path length $W(P)$ for $P = \{V_1, V_2, \ldots V_n\}$ is the sum of the weights of its constituent edges and is given by:

$$W(P) = \sum_{i=1}^{n}(V_i, V_{i+1})$$

Figure 20.1 Shortest path between *u* and *v* with *t* as intermediate vertex.

The shortest-path weight or length from u to v is thus any path between them which has minimum path weight. A variation of the *Shortest Path Problem* is *Single Source Shortest Path Problem.* Consider the weighted digraph G. We are also given a starting node $s \in V$. The *Single Source Shortest Path Problem* also known as *One-To-All Shortest Path Problem* is the problem of determining the shortest path from node s to all the other nodes in the G. This G may be considered as a network.

In chapter 5 we have discussed more details regarding shortest path problems and how to compute them using Dynamic Programming approach. Here we will discuss *Single Source Shortest Path Problem* using another algorithm design technique – Greedy Approach. We focus on it with the well famous Dijkstra's algorithm.

20.3 ALGORITHMS FOR SHORTEST PATH PROBLEM

The shortest path problem can be solved in various ways. Some important algorithms solving the problem are listed below.

(*a*) **Dijkstra's algorithm** - Solves only the problems with non-negative costs.

(*b*) **Bellman-Ford algorithm** - Applicable to problems with arbitrary costs

(*c*) **Floyd-Warshall algorithm** - Applicable to problems with arbitrary costs, solves a more general all-to-all (all-pairs) shortest path problem. However, Floyd-Warshall and Bellman-Ford algorithm solve the problems on graphs that do not have a cycle with negative cost.

(*d*) **Breadth First Search** - If the weights of edges are all 1. It is to find the path containing the minimum number of edges which connects u and v.

20.4 DIJKSTRA'S ALGORITHM

The general method to solve the single-source shortest-path problem is known as Dijkstra's algorithm. Dijkstra's algorithm was conceived by Dutch computer scientist Edsger Dijkstra in 1959. Dijkstra's algorithm is a graph search algorithm that solves the single-source shortest path problem for a graph with non-negative edge path costs, producing a shortest path tree. This algorithm is often used in routing and as a subroutine in other graph algorithms. Dijkstra's algorithm is sometimes called the single-source shortest path algorithm. It is also known as the single source shortest path problem. It computes length of the shortest path from the source to each of the remaining vertices in the graph. The algorithm is based on Greedy approach. Greedy algorithms generally solve a problem in stages by doing local to be the best thing at each stage.

20.4.1 General Description

Before going into details of the pseudo-code of the algorithm it is important to know how the algorithm works. Suppose we want to find a shortest path from a given node s to other nodes in a network (directed- weighted graph and the edge weights are non-negative). Node s is called a source vertex or a starting node or an initial node. Dijkstra's algorithm starts by assigning some initial values (cost or weight) for the distances from node s and to every other node in the network. It operates in steps, and at each step the algorithm improves the distance values. At each step, the shortest distance from node s to another node is determined.

20.4.2 Formal Description

The general description for Dijkstra algorithm to solve single source shortest path problem can be formally described in steps.

1. Assign a tentative distance value (generally unknown value or infinity) to every node except starting vertex s which is set to zero.
2. Mark all nodes unvisited. Set the initial node as current. Create a set of the unvisited nodes called the *unvisited set U*. This set U consists of all the nodes except s.
3. Mark s visited.
4. From the current node to its unvisited neighbours calculate their *tentative* distances. If the new distance from current to its neighbour is less than the previously stored distance, then overwrite that distance.
5. Select a neighbour with minimum distance, mark this node as visited and remove it from the U. A visited node will never be checked again. The distance recorded now is final and minimal.
6. If the U is empty, then stop. The algorithm has finished.
7. Set the unvisited node marked with the smallest tentative distance as the next "current node" and go back to step 4.

20.4.3 Dijkstra's Pseudocode

Dijkstra's algorithm finds the shortest paths to a vertex to rest of the vertices in a graph. First, it finds the shortest path from the source to a vertex nearest to it, then to a second nearest, and so on. The algorithm works on both directed and undirected graphs. However, all edges must have non-negative weights. Now we can give an algorithm (pseudocode) of Dijkstra's algorithm.

Algorithm: Dijkstra(G)

Where $G = (V, E, W)$, a weighted connected graph.

W = weight matrix. Each entry is the edge(i, j) cost.

S = set of visited vertices from source vertex to destination vertex. Initially it is empty.

s = source vertex.

Dist[v] = distance of a vertex from s.

Step 1: Initialize $S = \{s\}$

Step 2: Initialize Dist[s] = 0 // distance to source vertex is zero

Step 3: Repeat for all $v \in V - \{s\}$

- Dist[v] = ∞ // set distances of other vertices to infinity

Step 4: Repeat while $S \neq V$

(*a*) Find a vertex $w \in V - S$ such that Dist[w] is a minimum distance

(*b*) $S = S \cup \{w\}$ // add u to list of visited vertice

(*c*) Repeat for all $v \in V - S$

- Dist[v] = min{Dist[v], Dist[w] + $W(w, v)$}

Step 5: Return Dist

Step 6: End.

Complexity Analysis: The simplest implementation is to store vertices in an array or linked list. This will produce a running time of $O(|V|^2 + |E|) \approx O(|V|^2)$. If the graph is sparse then an array or linked list representation would not be an efficient way because sparse graphs are with very few edges and many nodes. It can be implemented more efficiently storing the graph in an adjacency list using a binary heap or priority queue. In that case, a running time of the Dijkstra algorithm is $O((|E| + |V|) \log |V|) \approx O(|E| \log |V|)$.

Example 1. Run Dijkstra algorithm on the graph shown in Figure 20.2 to find shortest path from the vertex v_1 to other vertices.

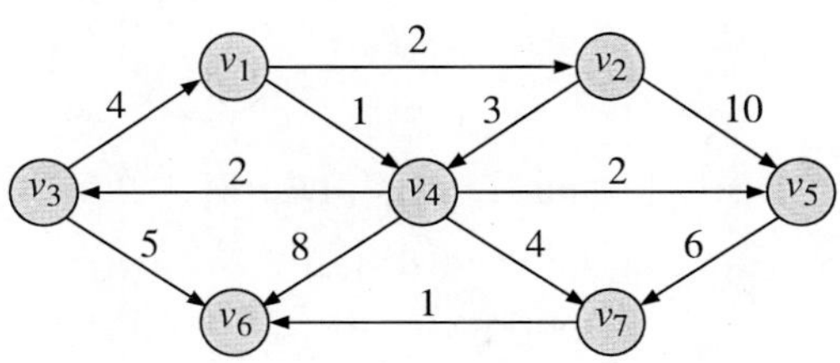

Figure 20.2 A weighted digraph.

Solution: Starting vertex $s = v_1$. That is $S = \{v_1\}$. Dist[s] = Dist[v_1] = 0.

Assume the distance from v_1 to other vertices is the initial arrangement of the Dijkstra's algorithm. No neighbour. So,

Iteration	S	w	Dist[w]	Dist[v_2]	Dist[v_3]	Dist[v_4]	Dist[v_5]	Dist[v_6]	Dist[v_7]
	$\{v_1\}$	—	—	∞	∞	∞	∞	∞	∞

Iteration 1: $V = V - S = \{v_2 \ldots v_7\}$

Now current vertex $w = v_1$. Find distances from v_1 to its adjacent vertices. Only v_2 and v_4 are adjacent.

$$\begin{aligned}
\text{Dist}[v_2] &= \min\{\text{Dist}[v_2], \text{Dist}[w] + W(w, v_2)\}\\
&= \min\{\infty, \text{Dist}[v_1] + W(v_1, v_2)\}\\
&= \min\{\infty, \infty + 2\} \quad \text{// } W(v_1, v_2) \text{ is the cost/weight/ distance of the edge}(v_1, v_2)\\
&= 2
\end{aligned}$$

$$\begin{aligned}
\text{Dist}[v_4] &= \min\{\text{Dist}[v_4], \text{Dist}[w] + W(w, v_4)\}\\
&= \min\{\text{Dist}[v_4], \text{Dist}[v_1] + W(v_1, v_4)\}\\
&= \min\{\{\infty, \infty + 1\}\\
&= 1
\end{aligned}$$

Distances to other vertices remain same. Put these values into the first row the table given below. Now Dist[v_4] has minimum value. So, $w = v_4$. Include v_4 in S and put $w = v_4$ and Dist[v_4] = 1 in the second row of the table.

Iteration	S	w	Dist[w]	Dist[v_2]	Dist[v_3]	Dist[v_4]	Dist[v_5]	Dist[v_6]	Dist[v_7]
	$\{v_1\}$	—	—	2	∞	**1**	∞	∞	∞
1	$\{v_1, v_4\}$	v_4	1						

Iteration 2: $V = V - S = \{v_2, v_3, v_5, v_6, v_7\}$

From $w = v_4$, adjacent vertices are v_3, v_6, v_7, v_5.

$$\begin{aligned}
\text{Dist}[v_3] &= \min\{\text{Dist}[v_3], \text{Dist}[w] + W(w, v_3)\}\\
&= \min\{\infty, \text{Dist}[v_4] + W(v_4, v_3)\}\\
&= \min\{\infty, 1 + 2\}\\
&= 3
\end{aligned}$$

$$\begin{aligned}
\text{Dist}[v_6] &= \min\{\text{Dist}[v_6], \text{Dist}[w] + W(w, v_6)\}\\
&= \min\{\infty, \text{Dist}[v_4] + W(v_4, v_6)\}\\
&= \min\{\infty, 1 + 8\}\\
&= 9
\end{aligned}$$

$$\begin{aligned}
\text{Dist}[v_7] &= \min\{\text{Dist}[v_7], \text{Dist}[w] + W(w, v_7)\}\\
&= \min\{\infty, \text{Dist}[v_4] + W(v_4, v_7)\}\\
&= \min\{\infty, 1 + 4\}\\
&= 5
\end{aligned}$$

$$\begin{aligned}
\text{Dist}[v_5] &= \min\{\text{Dist}[v_5], \text{Dist}[w] + W(w, v_5)\}\\
&= \min\{\infty, \text{Dist}[v_4] + W(v_4, v_5)\}\\
&= \min\{\infty, 1 + 2\}\\
&= 3
\end{aligned}$$

Distances to other vertices remain same. Put these values into the table as iteration 1. $w = v_3$. Include v_3 in S and put $w = v_3$ and Dist$[v_3] = 3$. From the table it is seen that Dist$[v_2]$ has smallest entry we have next $w = v_2$. Include v_2 in S, make entry for w, and Dist$[w]$.

Iteration	S	w	Dist$[w]$	Dist$[v_2]$	Dist$[v_3]$	Dist$[v_4]$	Dist$[v_5]$	Dist$[v_6]$	Dist$[v_7]$
	$\{v_1\}$	—	—	2	∞	**1**	∞	∞	∞
1	$\{v_1, v_4\}$	v_4	1	**2**	3	*1*	3	9	5
2	$\{v_1, v_4, v_2\}$	v_2	2						

Iteration 3: $V = V - S = \{ v_3, v_5, v_6, v_7\}$

From $w = v_2$, adjacent vertices are v_4, v_5. Already v_4 is included. So

$$\begin{aligned} \text{Dist}[v_5] &= \min\{\text{Dist}[v_5], \text{Dist}[w] + W(w, v_5)\} \\ &= \min\{3, \text{Dist}[v_2] + W(v_2, v_5)\} \\ &= \min\{3, 2 + 10\} \\ &= 3 \end{aligned}$$

Similarly make entry in the table. Next w may be v_3 or v_5. Consider v_3. Include v_3 in S.

Iteration	S	w	Dist$[w]$	Dist$[v_2]$	Dist$[v_3]$	Dist$[v_4]$	Dist$[v_5]$	Dist$[v_6]$	Dist$[v_7]$
	$\{v_1\}$	-	-	2	∞	**1**	∞	∞	∞
1	$\{v_1, v_4\}$	v_4	1	**2**	3	*1*	3	9	5
2	$\{v_1, v_4, v_2\}$	v_2	2	*2*	**3**	*1*	3	9	5
3	$\{v_1, v_4, v_2, v_3\}$	v_3	3						

Iteration 4: $V = V - S = \{v_5, v_6, v_7\}$

From $w = v_3$, adjacent vertices are v_1, v_6. Now v_1 is source. So

$$\begin{aligned} \text{Dist}[v_6] &= \min\{\text{Dist}[v_6], \text{Dist}[w] + W(w, v_6)\} \\ &= \min\{9, \text{Dist}[v3] + W(v_3, v_6)\} \\ &= \min\{9, 3 + 5\} \\ &= 8 \end{aligned}$$

Similarly make entry in the table. Next $w = v_5$. Include v_5 in S.

Iteration	S	w	Dist$[w]$	Dist$[v_2]$	Dist$[v_3]$	Dist$[v_4]$	Dist$[v_5]$	Dist$[v_6]$	Dist$[v_7]$
	$\{v_1\}$	—	—	2	∞	**1**	∞	∞	∞
1	$\{v_1, v_4\}$	v_4	1	**2**	3	*1*	3	9	5
2	$\{v_1, v_4, v_2\}$	v_2	2	**2**	3	*1*	3	9	5
3	$\{v_1, v_4, v_2, v_3\}$	v_3	3	*2*	3	*1*	**3**	8	5

4	$\{v_1, v_4, v_2, v_3, v_5\}$	v_5	3						

Iteration 5: $V = V - S = \{v_6, v_7\}$

From $w = v_5$, adjacent vertex is v_7.

$$\begin{aligned} \text{Dist}[v_5] &= \min\{\text{Dist}[v_7], \text{Dist}[w] + W(w, v_7)\} \\ &= \min\{5, \text{Dist}[v_5] + W(v_5, v_7)\} \\ &= \min\{5, 3 + 6\} \\ &= 5 \end{aligned}$$

Similarly make entry in the table. Next $w = v_7$. Include v_7 in S.

Iteration	S	w	Dist[w]	Dist[v_2]	Dist[v_3]	Dist[v_4]	Dist[v_5]	Dist[v_6]	Dist[v_7]
	$\{v_1\}$	-	-	2	∞	**1**	∞	∞	∞
1	$\{v_1, v_4\}$	v_4	1	**2**	3	*1*	3	9	5
2	$\{v_1, v_4, v_2\}$	v_2	2	**2**	3	*1*	3	9	5
3	$\{v_1, v_4, v_2, v_3\}$	v_3	3	*2*	3	*1*	**3**	8	5
4	$\{v_1, v_4, v_2, v_3, v_5\}$	v_5	3	*2*	3	*1*	**3**	8	**5**
5	$\{v_1, v_4, v_2, v_3, v_5, v_7\}$	v_7	5						

Iteration 6: $V = V - S = \{v_6\}$

From $w = v_7$, adjacent vertex is v_6.

$$\begin{aligned} \text{Dist}[v_6] &= \min\{\text{Dist}[v_6], \text{Dist}[w] + W(w, v_6)\} \\ &= \min\{8, \text{Dist}[v_7] + W(v_7, v_6)\} \\ &= \min\{8, 5 + 1\} \\ &= 6 \end{aligned}$$

Similarly make entry in the table. Now $w = v_6$. Include v_6 in S.

Iteration	S	w	Dist[w]	Dist[v_2]	Dist[v_3]	Dist[v_4]	Dist[v_5]	Dist[v_6]	Dist[v_7]
	$\{v_1\}$	—	—	2	∞	**1**	∞	∞	∞
1	$\{v_1, v_4\}$	v_4	1	**2**	3	*1*	3	9	5
2	$\{v_1, v_4, v_2\}$	v_2	2	**2**	3	*1*	*3*	9	5

3	$\{v_1, v_4, v_2, v_3\}$	v_3	3	2	3	1	3	8	5
4	$\{v_1, v_4, v_2, v_3, v_5\}$	v_5	3	2	3	1	3	8	5
5	$\{v_1, v_4, v_2, v_3, v_5, v_7, \}$	v_7	5	2	3	1	3	6	5
6	$\{v_1, v_4, v_2, v_3, v_5, v_7, v_6\}$	V_6	6						

Now V is empty. All the vertex have been computed. So the distance matrix table is

Iteration	S	w	Dist[w]	Dist[v_2]	Dist[v_3]	Dist[v_4]	Dist[v_5]	Dist[v_6]	Dist[v_7]
	$\{v_1\}$	—	—	2	∞	1	∞	∞	∞
1	$\{v_1, v_4\}$	v_4	1	2	3	1	3	9	5
2	$\{v_1, v_4, v_2\}$	v_2	2	2	3	1	3	9	5
3	$\{v_1, v_4, v_2, v_3\}$	v_3	3	2	3	1	3	8	5
4	$\{v_1, v_4, v_2, v_3, v_5\}$	v_5	3	2	3	1	3	8	5
5	$\{v_1, v_4, v_2, v_3, v_5, v_7, \}$	v_7	5	2	3	1	3	6	5
6	$\{v_1, v_4, v_2, v_3, v_5, v_7, v_6\}$	V_6	6	2	3	1	3	6	5

20.4.4 Applications of Dijkstra's Algorithm

Many problems can be cast as shortest path problems, making Dijkstra's algorithm a powerful and general tool. The functionality of Dijkstra's original algorithm can be extended with a variety of modifications.

(*a*) Dijkstra's algorithm is applied to automatically find directions between physical locations, such as driving directions on websites like Mapquest or Google Maps.

(*b*) In telecommunication applications, Dijkstra's algorithm has been used for solving the min-delay path.

(*c*) In networking routing of packets is based on Dijkstra's algorithm. It is usually the working principle behind link-state routing protocols, OSPF and IS-IS being the most common ones.

(*d*) It is also used for solving a variety of shortest path problems arising in plant and facility layout, robotics, transportation, and VLSI design etc.

20.5 SOME COMMON QUESTIONS

Question 1. Why does Dijkstra's Algorithm work?

Answer: To understand how it works, we need two mathematical results.

(*a*) Triangle inequality – If $\delta(u, v)$ is the shortest path length between u and v, then $\delta(u, v) \leq \delta(u, x) + \delta(x, v)$ for some vertex x.

(*b*) The subpath of any shortest path is itself a shortest path.

Question 2. Why does Dijkstra's Algorithm use?

Answer: Dijkstra's algorithm calculates the shortest path source vertex to every other vertex of graph. However, it is about as computationally expensive to calculate the shortest path from vertex u to every vertex using Dijkstra's as it is to calculate the shortest path to some particular vertex v. Therefore, anytime we want to know the optimal path to some other vertex from a determined origin, we can use Dijkstra's algorithm.

Exercises

1. Consider the undirected weighted graph in Figure 20.3. Find the shortest path from vertex a to other vertices.

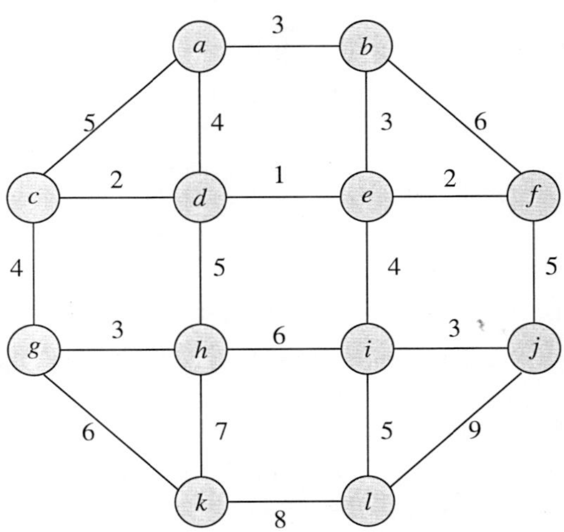

Figure 20.3 Undirected connected weighted graph.

2. Give a simple example of a directed graph with negative-weight edges for which Dijkstra's algorithm produces incorrect answers.
3. Find the shortest path from node 1 to all vertices of the graph given below. Show all the intermediate steps. The numbers on the edges are the weights.

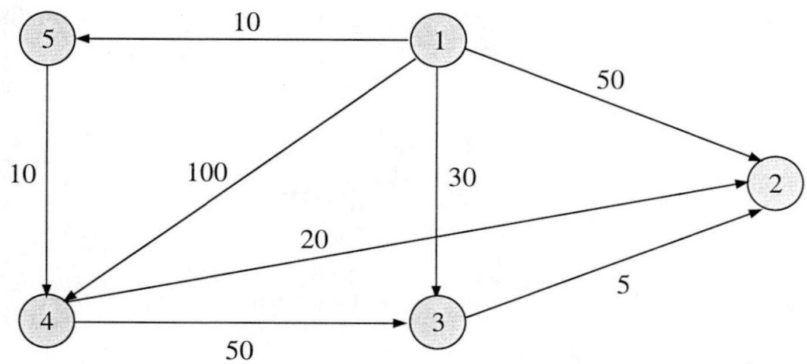

Figure 20.4 Weighted digraph.

4. Consider an example to explain Dijkstra's algorithm.
   (*a*) Does Dijkstra's algorithm work for graphs with negative weights?
   (*b*) Will it be applicable to both undirected and directed graphs?
5. Find shortest path form vertex to vertex using Dijkstra algorithm in the following graph.

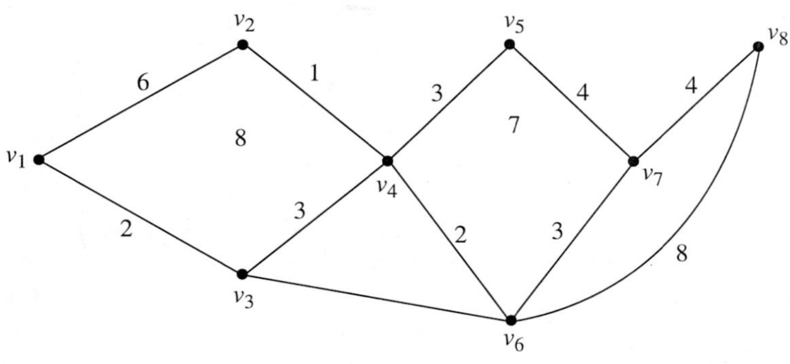

Figure 20.5 Weighted undirected graph.

6. Suppose that we are given a weighted, directed graph $G = (V, E)$ in which edges that leave the source vertex s may have negative weights, all other edge weights are non-negative, and there are no negative-weight cycles. Argue that Dijkstra's algorithm correctly finds shortest paths from s in this graph.
7. Calculate the efficiency of Dijkstra's algorithm in two different representations.

University Questions

1. Prove that the time complexity of Dijkstra algorithm is $3(n - 2)(n - 1)/2 = O(n^2)$. ***[WBUT'08]***
2. Find out the shortest path from vertex 1 to all remaining vertices for the following graph using Dijkstra algorithm. ***[WBUT'08]***

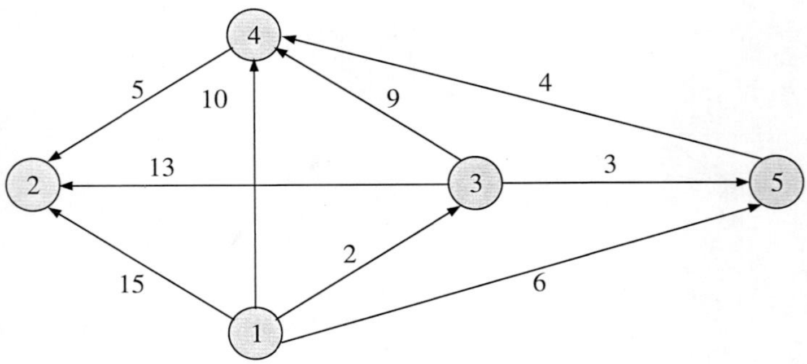

3. Write down lengths for the edges of the following graph, so that Dijkstra's algorithm would not find the correct shortest path from s to t. ***[DOEACC'08]***

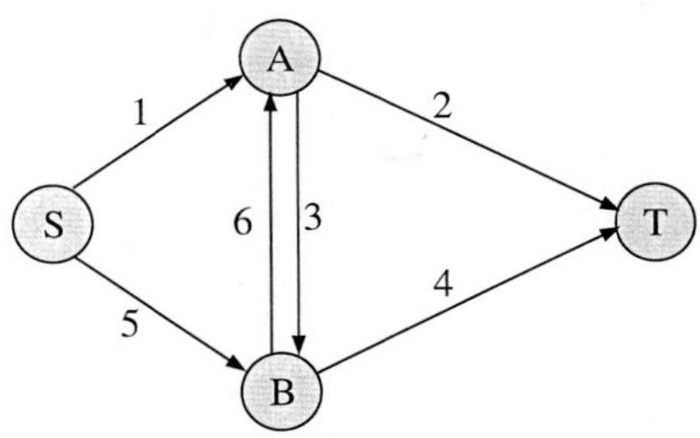

4. Which of the shortest path algorithms would be most appropriate for finding paths in the graph of the previous question? ***[DOEACC'08]***
5. Argue that all edge weights of a graph are positive, then any subset of edges that connects all vertices and has minimum total weight must be a tree. Give an example to show that the same conclusion does not follow if we allow some weights to be non-positive. ***[DOEACC'08]***
6. Write Greedy algorithm to generate shortest path. ***[JNTU'08]***
7. Write a simpler version of Dijkstra's algorithm that finds only the distances from a given vertex to all other vertices of a graph represented by its weight matrix. ***[JNTU'09]***
8. Describe the Dijkstra's algorithm for finding shortest path with help of suitable example. ***[UPTU'10, IETE ALCCS'11]***

9. Use Dijkstra's algorithm to find the single source shortest paths for the following graph taking vertex '*s*' as the source. *[MDU'08]*

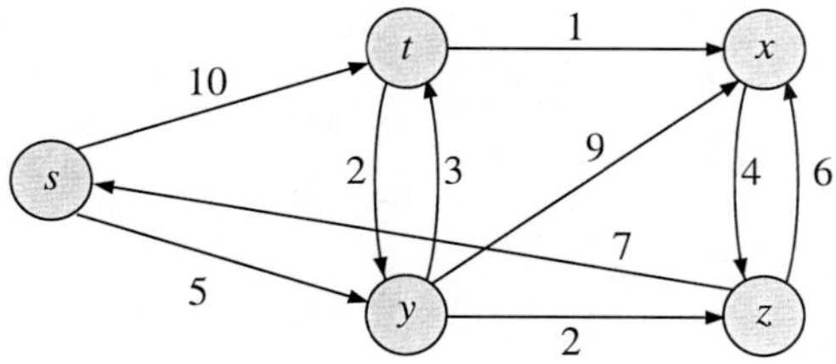

10. Suggest a greedy algorithm to find shortest paths from the designated vertex start to all of the other vertices in connected weighted, *n* vertex graph. *[BPUT'08]*
11. Discuss Dijkstra's algorithm to solve the single source shortest paths on a weighted, directed graph with an example. *[DOEACC'06, IETE ALCCS'11]*
12. How Dijkstra's algorithm is based on greedy approach? What are the assumptions made in it? Give an example where Dijkstra's algorithm fails? Explain the reason. *[DOEAC'06]*
13. Write down the Dijkstra algorithm for finding out the shortest path. *[WBUT'08, PTU'09]*
14. Using Dijkstra's algorithm, find the shortest path between node A and all the remaining nodes in the following graph. *[IETE ALCCS'11]*

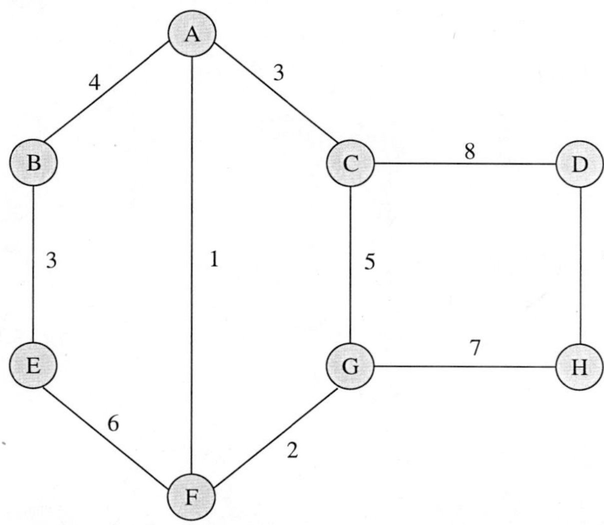

15. What are single source shortest paths? Write down Dijkstra's algorithm for it. *[UPTU'05]*
16. Comments on whether Dijkstra's shortest path algorithm works correctly if weights are negative. *[UPTU'06]*

❑❑❑

GATE QUESTIONS
(1987-2012)

GATE Questions

Design and Analysis of Algorithms

YEAR – 1987

Question 1. Let P be a quick sort program to sort number in ascending order. Let t_1 and t_2 be the time taken by the program for the inputs [1 2 3 4 5] and [5 4 3 2 1] respectively. Which of the following holds?

(a) $t_1 = t_2$ (b) $t_1 > t_2$

(c) $t_1 < t_2$ (d) $t_1 = t_2 + 5 \log 5$

Question 2. Show that the number of odd degree vertex in a finite graph is even.

Question 3. Specify an adjacency-lists representation of the undirected graph.

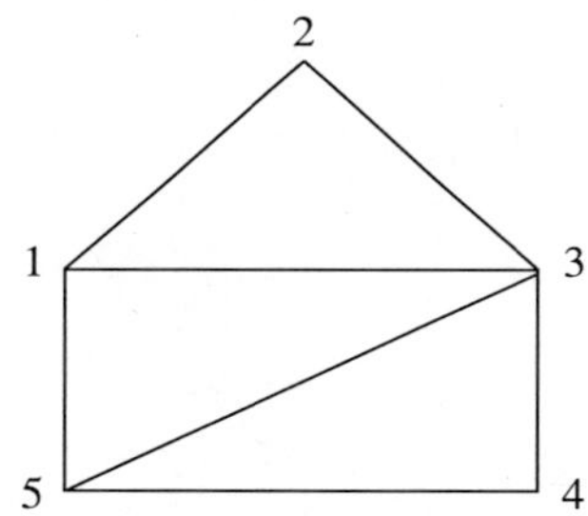

Question 4. Solve the recurrence equations given below.

$$T(n) = T(n - 1) + n$$
$$T(1) = 1$$

YEAR – 1988

Question 1. Quick sort is .... efficient than heapsort in the worst case.

Question 2. Write the adjacency matrix representation of the graph given in figure below.

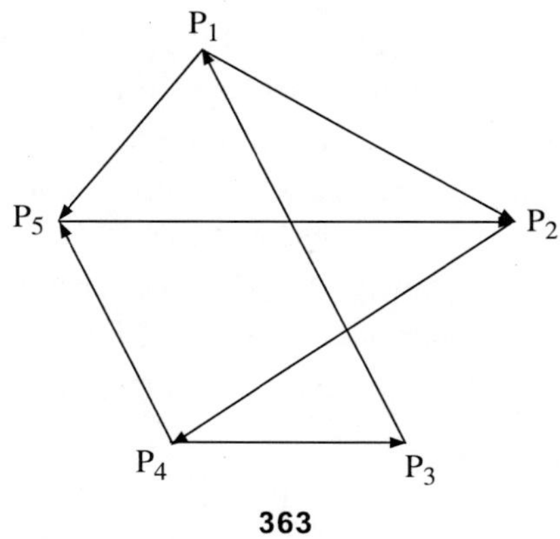

Question 3. Solve the recurrence equations.

$$T(n) = T(n/2) + 1$$
$$T(1) = 1$$

YEAR – 1989

Question 1. Match the pair in the following questions:

(*a*) $O(\log n)$	(*p*) Heapsort
(*b*) $O(n)$	(*q*) Depth-first-search
(*c*) $O(n \log n)$	(*r*) Binary search
(*d*) $O(n^2)$	(*s*) Selection of *k*th smallest element in a set of *n* elements

Question 2. Depth-first search cannot be used to find connected component of a graph. Justify.

Question 3. In the graph shown below, the depth-first spanning tree edges are marked with 'T'. Identify the forward, backward and cross edges.

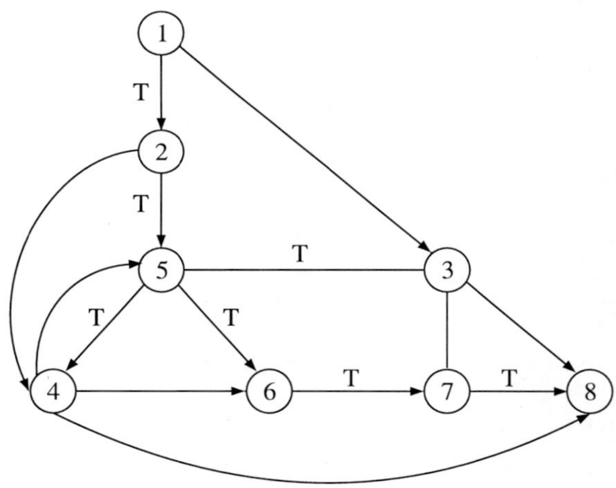

Question 4. An input file has 10 records with keys given as: 25, 7, 34, 2, 70, 9, 61, 16, 49, and 19. This is to be sorted in non-decreasing order.

(*i*) Sort the input file using QUICKSORT by correctly positioning the first element of the file/subfile. Show the subfiles obtained at all intermediate steps. Use square brackets to demarcate subfiles.

(*ii*) Sort the input file using 2-way-MERGESORT showing all major intermediate steps. Use square brackets to demarcate subfiles.

Question 5. Will recursion work correctly in a language with static allocation of all variables? Explain.

Question 6. Find a solution to the following recurrence equation:

$$T(n) = \sqrt{n} + T(n/2)$$
$$T(1) = 1$$

YEAR – 1990

Question 1. Match the pair in the following question.

(*a*) Strassen's matrix multiplication algorithm	(*p*) Greedy algorithm
(*b*) Kruskal's minimum spanning tree algorithm	(*q*) Dynamic programming
(*c*) Bi-connected components algorithm	(*r*) Divide-and-Conquer
(*d*) Floud's shortest path algorithm	(*s*) Depth-first search

Question 2. The complexity of comparison based sorting algorithm is:

(*a*) $\theta(n \log n)$ (*b*) $\theta(n)$

(*c*) $\theta(n^2)$ (*d*) $\theta(n\sqrt{n})$

Question 3. Consider the height balanced tree T_1, with values stored at only the leaf nodes shown in the given figure below.

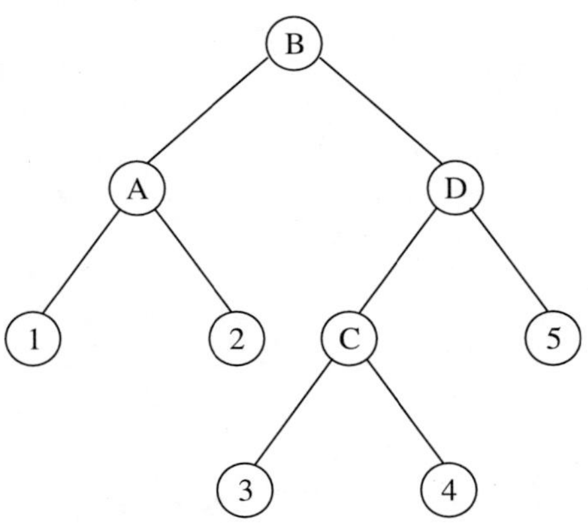

Figure T_1

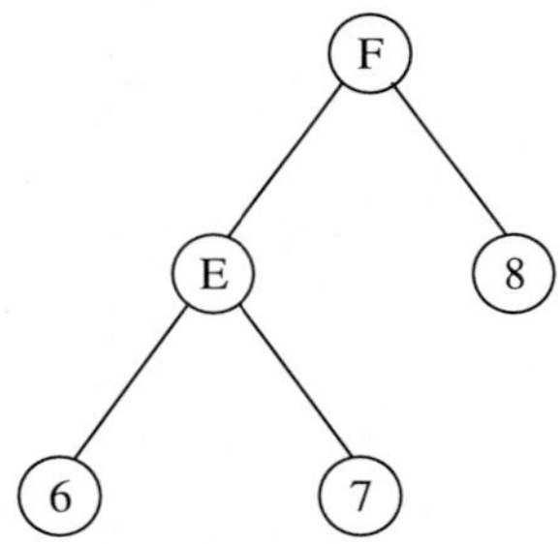

Figure T_2

(*i*) Show how to merge to the tree T_1 elements from tree T_2 shown in the given figure using the node D of tree T_1.

(*ii*) What is the time complexity of a merge operation on balanced trees T_1 and T_2 where T_1 and T_2 are of height h_1 and h_2 respectively, assuming rotation scheme are given. Give reasons.

Question 4. The following algorithm works on a undirected graph G.

```
Program Explore (G)
   procedure Vinih (n)
      begin
         if Adj (TU) is not empty
            {comment : Adj (u) is the list of edges incident
            to u)
         then
            begin
               select an edge front Adj [n} ;
               Let an edge e = (u, v) ;
               remove e from Adj (u) and Adj (v) ;
               Visit (v) ;
            end
         else
            mark u as a finished vertex and remote u from
            LIST;
            {Comment : LIST is tire set of vertices in the
            graph]
      end;
   begin
      while LIST in not empty
         do
            begin
               Let v ∈ LIST;
               Visit (v);
            end
   end
```

Note: Initially Adj (*u*) is the list of all edges incident to *u* and LIST is the set of all vertices in the graph. They are globally accessible.

What kinds of sub-graphs are obtained when this algorithm traverses the graphs G_1 and G_2 shown in Figure 1 and Figure 2, respectively?

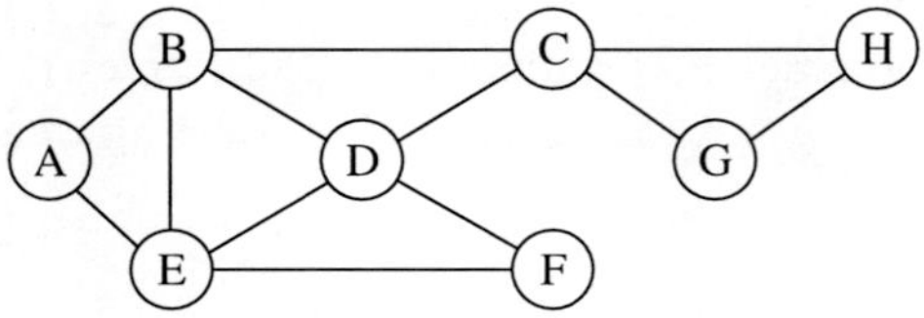

Figure 1

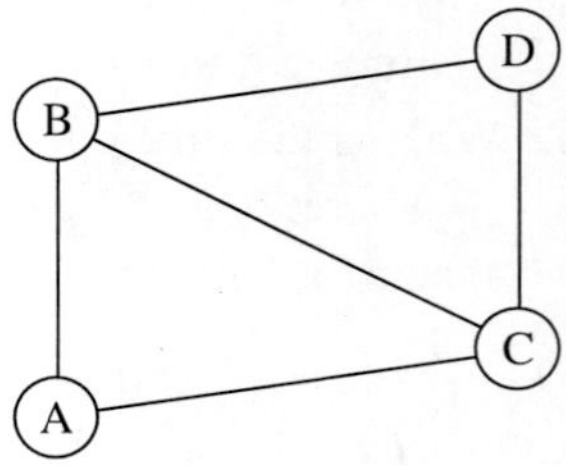

Figure 2

(*i*) What is the commonly known traversal of graphs that can be obtained from the subgraphs generated by *Program Explore*?

(*ii*) Show that the time complexity of the procedure is $O(v + e)$ for a graph with v vertices and e edges, given that each vertex can be accessed and removed from LIST in const time. Also show that all the edges of the graph are traversed.

Question 5: The average case complexity of quick sort algorithm for sorting n elements is

(*a*) $(n \log \log n)$ (*b*) (n^2)

(*c*) $(n \log n)$ (*d*) $(n \log^2 n)$

$(x[l], x[l + 1] \ldots x[m]. m > l$ and

$(y[i], y[i + 1] \ldots y[j], j > i$ where

$x[l].\text{key} \le x[l + 1].\text{key} \le \ldots \le x[m].\text{key}$

and

$y[i].\text{key} \le y[i + 1].\text{key} \le \ldots \le y[j].\text{key}$ and it obtains in a sorted list

$(z[1]. z[2] \ldots z[k]$ where $k = m + j - l - i + 2$.

Complete the procedure by replacing each '?' by the appropriate operator from the procedure set {>, <, =>, <=, <>, or, and, not}

```
procedure MERGE (var x, y, z: a file: 1, m, i, j: integer); var
k, s, t, u: integer:
begin
  s: = 1;
  k: = 1;
  u: = I;
  while[(s?m) ? (j?u)] do
     begin
        if x[s].key?y[u].key
           then
              begin
                 z[k]: = x[g];
                 s: = s+1
              end
           else
              begin
                 z[k]: = y[u];
                 u: = u+1
              end
        k: = k+1;
     end
```

```
if s ? m
   then
         for t: = u to j do
         z[k + t - u] : = y[t]
   else
         for t: = s to m do
         z[k + t - g]: = x[t]
end;
```

YEAR – 1991

Question 1. The minimum number of comparisons required to sort 5 elements is ...

Question 2. The weighted external path length of the binary tree in the given figure is................

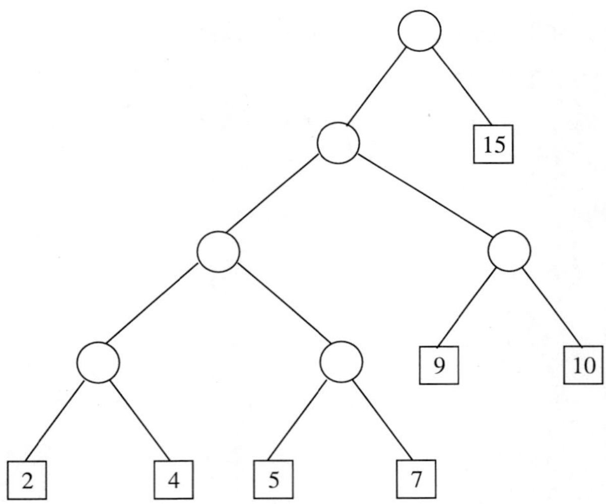

Question 3. Consider the following recursive definition of fib:

fib (n) : = if $n = 0$ then 1
else if $n = 1$ then 1
else fib ($n - 1$) + fib ($n - 2$)

Find the number of times fib is called (including the first call) for an evaluation of fib(7).

Question 4. The maximum number of possible edges in an undirected graph with n vertices and k components is ...

Question 5. Choose the correct alternatives (more than one may be correct) and write the corresponding letters only: Kruskal's algorithm for finding a minimum spanning tree of a weighted graph G with n vertices and m edges has the time-complexity of:

(*a*) (n^2) (*b*) $(m\ n)$

(*c*) $(m + n)$ (*d*) $(m \log n)$ (*e*) (m^2)

Question 6. Show that all vertices in an undirected finite graph cannot have distinct degrees, if the graph has at least two vertices.

Question 7. The worst-case time complexity of mergesort for sorting n elements is

(*a*) (n^2) (*b*) $(n \log n)$

(*c*) $(n^2 \log n)$ (*d*) $(\log n)$

YEAR – 1992

Question 1. Complexity of Kruskal's algorithm for finding the minimum spanning tree of a undirected graph containing n vertices and m edges, if the edges are sorted is _______.

Question 2. Maximum number of edges in a planner graph with n vertices is _______.

Question 3. A non-planer graph with minimum number of vertices has

(*a*) 9 edges, 6 vertices (*b*) 6 edges, 4 vertices

(*c*) 10 edges, 5 vertices (*d*) 9 edges, 5 vertices

Question 4. Following algorithm(s) can be used to sort n integers in the range [1 ... n] in $O(n)$ time.

(*a*) Heapsort (*b*) Quicksort

(*c*) Mergesort (*d*) Radixsort

Question 5. How many edges are there in a forest with p components having n vertices in all?

Question 6: Assume that the last element of the set is used as partition element in Quicksort. If n distinct elements from the set [1 ... n] are to be sorted, give an input for which Quick sort takes maximum time.

Question 7: Consider the function $F(n)$ for which the pseudocode is given below:

```
Function F (n)
begin
F1 ← 1
If (n = 1) then F ← 3
   else For I = 1 to n do
      begin
         C ← 0
```

```
            For j = 1 to F (n - 1) do
                begin C ← C + 1 end
            F1 = F1 * C
        end
        F = F1
    end
[n is positive integer greater than zero].
```

(*a*) Derive a recurrence relation for $F(n)$. (4)

(*b*) Solve the recurrence relation for a closed form solution of $F(n)$.

YEAR – 1993

Question 1. Consider a simple connected graph G with n vertices and n edges ($n > 2$). Then which of the following statements are true?

(*a*) G has no cycles

(*b*) The graph obtained by removing any two edges from G is not considered connected

(*c*) G has at least one cycle

(*d*) The graph obtained by removing any two edges from G is not considered connected

(*e*) None of the above.

Question 2. The number of distinct simple graphs with up to three nodes is

(*a*) 15 (*b*) 10 (*c*) 7 (*d*) 9

Question 3. The recurrence relation that arises in relation to the complexity of binary search is

(*a*) $T(n) = T(n/2) + k$, k is a constant

(*b*) $T(n) = 2T(n/2) + k$, k is a constant

(*c*) $T(n) = T(n/2) + \log n$

(*d*) $T(n) = T(n/2) + n$.

Question 4. Which of the following algorithm design techniques is used in the quicksort algorithm?

(*a*) Dynamic programming

(*b*) Backtracking

(*c*) Divide-and-conquer

(*d*) Greedy method.

Question 5: Consider the following two functions:

$$g1(n) = \begin{cases} n^3 & \text{for } 0 \le n \le 10{,}000 \\ n & \text{for } n \ge 10{,}000 \end{cases}$$

$$g1(n) = \begin{cases} n & \text{for } 0 \le n \le 100 \\ n^3 & \text{for } n > 100 \end{cases}$$

Which of the following is true?

(*a*) $g1(n)$ is $O(g2(n))$

(*b*) $g1(n)$ is $O(n^3)$

(*c*) $g2(n)$ is $O(g1(n))$

(*d*) $g2(n)$ is $O(n)$

YEAR – 1994

Question 1. What are the number of edges in a regular graph of degree d and n vertices?

Question 2. An independent set in a graph is a subset of vertices such that no two vertices in the subset are connected by an edge. An incomplete scheme for greedy algorithm to find a maximum independent set in a tree is given below:

```
V: = Set of all vertices in a tree:
I: = ϕ do
begin
    select a vertex u • V such that
  _______________:
    V: = V-{u}:
    If u is such that
  ______________ then I: = I U{u}
end:
Output (I):
```

(*i*) Complete the algorithm by specifying the property of vertex u in each case.

(*ii*) What is the time complexity of the algorithm?

Question 3. Mergesort uses:

(*a*) Divide-and-conquer strategy

(*b*) Backtracking approach

(*c*) Heuristic search

(*d*) Greedy approach

Question 4. For merging two sorted lists of size m and n into a sorted list of size $m + n$, we require comparisons of

(*a*) $O(m)$ (*b*) $O(n)$

(*c*) $O(m + n)$ (*d*) $O(\log m + \log n)$

Question 5: The minimum number of edges in a connected cyclic graph on n vertices is

(*a*) $n - 1$ (*b*) n

(*c*) $n + 1$ (*d*) none of the above

YEAR – 1995

Question 1. How many minimum spanning trees does the following graph have? Draw them (weights are assigned to the edges).

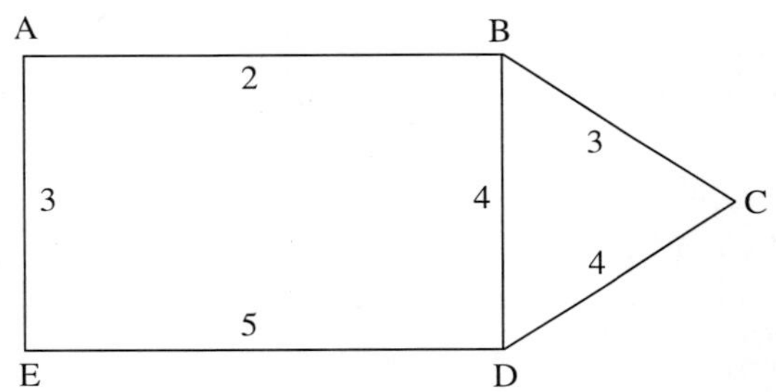

Question 2. Prove that in finite graph, the number of vertices of odd degree is always even.

Question 3. Which of the following is false?

(*a*) $100\, n \log n = O\left(\frac{n \log l}{100}\right)$ (*b*) $\sqrt{\log n} = O(\log \log n)$

(*c*) if $0 < x < y$ then $n^x = O(n^y)$ (*d*) $2^n \neq O(n^k)$

Question 4. The minimum number of interchanges needed to convert the array: 89, 19, 40, 17, 12, 10, 2, 5, 7, 11, 6, 9, 70 into a heap with the maximum element at the root node is

(*a*) 0 (*b*) 1

(*c*) 2 (*d*) 3

Question 5. The recurrence relation

$$T(1) = 2$$

$$T(n) = 3T(n/4) + n$$

has the solution $T(n)$ equal to

(*a*) $O(n)$ (*b*) $O(\log n)$

(*c*) $O(n^{3/4})$ (*d*) none of the above

Question 6. The average number of key comparisons done in successful sequential search in a list of length n is

(*a*) $\log n$ (*b*) $(n - 1)/2$

(*c*) $n/2$ (*d*) $(n + 1)/2$

Question 7. Quicksort is run on two inputs shown below to sort in ascending order

(*i*) 1, 2, 3, ... n

(*ii*) $n, n - 1, n - 2, ..., 1$

Let C_1 and C_2 be the number of comparisons made for inputs (*i*) and (*ii*) respectively. Then,

(*a*) $C_1 < C_2$ (*b*) $C_1 > C_2$

(*c*) $C_1 = C_2$ (*d*) We cannot say anything for arbitrary n

YEAR – 1996

Question 1. A complete undirected, weighted graph G is given on the vertex set $\{0, 1, ..., n - 1\}$ for any fixed 'n'. Draw the minimum spanning tree of G if

(*a*) The weight of the edge (u, v) is $|u - v|$

(*b*) The weight of the edge (u, v) is $(u + v)$

Question 2: Let G be the directed, weighted graph shown below.

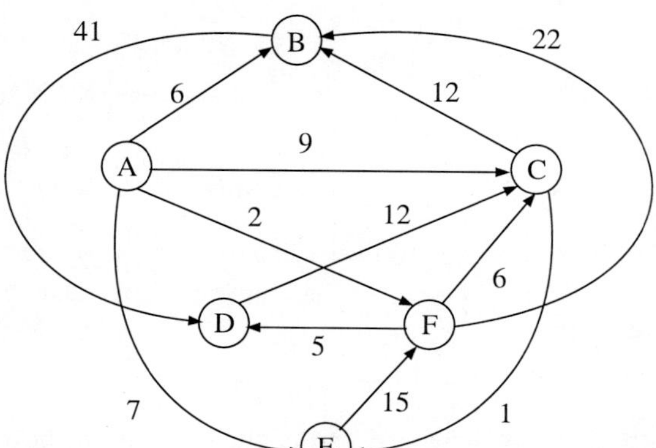

We are interested in the shortest paths from A

(*a*) Output the sequence of vertices identified by the Dijkstra's algorithm for single source shortest path when the algorithm is started at node A.

(*b*) Write down the sequence of vertices in the shortest path from A to E.

(*c*) What is the cost of shortest path from A to E?

YEAR – 1997

Question 1. The correct matching for the following pairs is

(*a*) All pairs shortest paths	1. Greedy
(*b*) Quicksort	2. Depth first search
(*c*) Minimum weight spanning tree	3. Dynamic programming
(*d*) Connected component	4. Divide-and-conquer

(*a*) $a-2, b-4, c-1, d-3$ (*b*) $a-3, b-4, c-1, d-2$

(*c*) $a-3, b-4, c-2, d-1$ (*d*) $a-4, b-1, c-2, d-3$

Question 2. Heap allocation is required for languages

(*a*) that support recursion

(*b*) that support dynamic data structures

(*c*) that use dynamic scope rules

(*d*) none of the above

Question 3. Let $T(n)$ be the function defined by

$$T_1 = 1, T(n) = 2T\left(\left\lfloor \frac{n}{2} \right\rfloor\right) + \sqrt{n} \text{ for } n \geq 2.$$

Which of the following statements is true?

(*a*) $T(n) = O(\sqrt{n})$ (*c*) $T(n) = O(n)$

(*c*) $T(n) = O(\log n)$ (*d*) None of the above

Question 4. Let G be a graph with 100 vertices numbered 1 to 100. Two vertices i and j are adjacent iff $|i-j| = 8$ or $|i-j| = 12$. The number of connected components in G is

(*a*) 8 (*b*) 4

(*c*) 12 (*d*) 25

Question 5. Consider a graph whose vertices are points in the plane with integer co-ordinates (x, y) such that $1 \leq x \leq n$ and $1 \leq y \leq n$, where $n \geq 2$ is an integer. Two vertices (x_1, y_1) and (x_2, y_2) are adjacent iff $|x_1 - x_2| \leq 1$ or $|y_1 - y_2| \leq 1$. The weight of an edge $\{(x_1, y_1), (x_2, y_2)\}$ is

$$\sqrt{(x_1 - x_2)^2 + (y_1 - y_2)^2}$$

(*a*) What is the weight of a minimum weight spanning tree in this graph? Write only the answer without any explanations.

(*b*) What is the weight of a maximum weight spanning tree in this graph? Write only the answer without any explanations.

Question 6. Consider the following function.

Function *F*(*n*, *m*: integer): integer;

begin

if(*n* <= 0) or (*m* <= 0) then

F: = 1

else

F: = *F*(*n* – 1, *m*) + *F*(*n*, *m* – 1);

End

Use the recurrence relation to answer the following questions.

$$\binom{n}{k} = \binom{n-1}{k} + \binom{n-1}{k-1}$$

Assume *n*, *m* are positive integers. Write only the answer without any explanation.

(*a*) What is the value of *F*(*n*, 2)?

(*b*) What is the value of *F*(*n*, *m*)?

(*c*) How many recursive calls are made to the function *F*, including the original call, when evaluating *F*(*n*, *m*)?

YEAR – 1998

Question 1. Which of the following algorithm design techniques is used in finding all pairs of shortest distances in a graph?

(*a*) Dynamic programming
(*b*) Backtracking
(*c*) Greedy
(*d*) Divide-and-Conquer

Question 2. Give the correct matching for the following pairs.

(A) $O(\log n)$	(P) Selection
(B) $O(n)$	(Q) Insertion sort
(C) $O(n \log n)$	(R) Binary search
(D) $O(n_2)$	(S) Mergesort

(*a*) A – R, B – P, C – Q, D – S
(*b*) A – R, B – P, C – S, D – Q
(*c*) A – P, B – R, C – S, D – Q
(*d*) A – P, B – S, C – R, D – S

Question 3: What value would be the following function return for the input $x = 95$?

```
Function fun(x: integer):integer;
  begin
      if x > 100 then
            fun:= x – 10
      else
            fun: = fun(fun (x + 11))
end;
```

(*a*) 89 (*b*) 90

(*c*) 91 (*d*) 92

Question 4. Solve the recurrence relation

$x_n = 2x_n - 1, n > 1$

$x_1 = 2$

Question 5. A sorting technique is called *stable* if

(*a*) It takes $O(n \log n)$ time

(*b*) It maintains the relative order of occurrence of non-distinct elements

(*c*) It uses divide and conquer paradigm

(*d*) It takes $O(n)$ space

Question 6: If one uses straight two-ways mergesort algorithm to sort the following elements in ascending order 20, 47, 15, 8, 9, 4, 40, 30, 12, 17. Then the order of these elements after the second pass of the algorithm is:

(*a*) 8, 9, 15, 20, 47, 4, 12, 17, 30, 40 (*b*) 8, 15, 20, 47, 4, 9, 30, 40, 12, 17

(*c*) 15, 20, 47, 4, 8, 9, 12, 30, 40, 17 (*d*) 4, 8, 9, 15, 20, 47, 12, 17, 30, 40

Question 7: The number of articulation points of the following graph is:

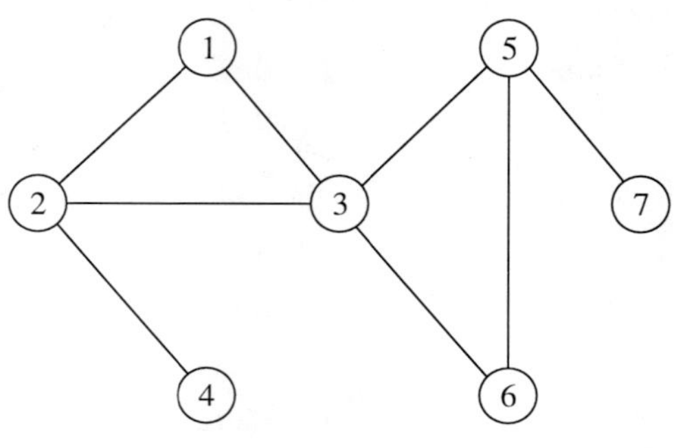

(*a*) 0 (*b*) 1

(*c*) 2 (*d*) 3

Question 8. If n is a power of 2, then the minimum number of multiplications needed to compute an is:

(*a*) $\log n$ (*b*) $\sqrt{n}$

(*c*) $n - 1$ (*d*) n

YEAR – 1999

Question 1. If $T_1 = O(1)$, give the correct matching for the following pairs:

(M) $T_n = T_{n-1} + n$ (U) $T_n = O(n)$

(N) $T_n = T_{n/2} + n$ (V) $T_n = O(n \log n)$

(O) $T_n = T_{n/2} + n \log n$ (W) $T_n = O(n^2)$

(P) $T_n = T_{n-1} + \log n$ (X) $T_n = O(\log^2 n)$

(*a*) M – W, N – V, O – U, P – X (*b*) M – W, N – U, O – V, P – V

(*c*) M – V, N – W, O – X, P – U (*d*) M – W, N – U, O – V, P – X

Question 2. Let A be a $n \times n$ matrix such that the elements in each row and each column are arranged in ascending order. Draw a decision tree which finds 1st, 2nd and 3rd smallest elements in minimum number of comparisons.

Question 3: Consider the following algorithm. Assume, procedure *A* and procedure *B* take $O(1)$ and $O(1/n)$ unit of time, respectively. Derive the time complexity of the algorithm in O-notation.

```
algorithm what(n)
    begin
        if n = 1 then
            call A
        else
            begin
              what(n – 1)
              call B(n)
            end
    end.
```

Question 4. Write a constant time algorithm to insert a node with data *D* just before the node with address *p* of a singly linked list.

Question 5. Draw the min-heap that results from insertion of the following elements in order into an initially empty min-heap: 7, 6, 5, 4, 2, 3, 1. Show the result after the deletion of the root of this heap.

YEAR – 2000

Question 1. The most appropriate matching for the following pairs

(X): Depth-first search	1: heap
(Y): Breadth-first search	2: queue
(Z): Sorting	3: stack

(*a*) X – 1, Y – 2, Z – 3
(*b*) X – 3, Y – 1, Z – 2
(*c*) X – 3, Y – 2, Z – 1
(*d*) X – 2, Y – 3, Z – 1

Question 2. Let *s* be a sorted array of *n* integers. Let $t(n)$ denote the time taken for the most efficient algorithm to determine if there are two elements with sum less than 1000 in *s*. Which of the following statements is true?

(*a*) $T(n)$ is $O(1)$
(*b*) $n \le t(n) \le n \log_2 n$
(*c*) $n \log_2 n \le t(n) < \binom{n}{2}$
(*d*) $t(n) = \binom{n}{2}$

Question 3. Consider the following functions

$$f(n) = 3n^{\sqrt{n}}$$

$$g(n) = 2^{\sqrt{n}\log_2 n}$$

$$h(n) = n!$$

Which of the following is always true?

(*a*) $h(n)$ is $O(f(n))$
(*b*) $h(n)$ is $O(g(n))$
(*c*) $g(n)$ is not $O(f(n))$
(*d*) $f(n)$ is $O(g(n))$

Question 4: Let *G* be an undirected graph with distinct edge weights. Let e_{max} be the edge with maximum weight and the edge with minimum weight. Which of the following is false?

(*a*) Every minimum spanning tree of *G* must contain e_{min}
(*b*) If e_{max} is an minimum spanning tree, then its removal must disconnect *G*
(*c*) No minimum spanning tree contains e_{max}
(*d*) *G* has a unique minimum spanning tree.

Question 5: Let *G* be an undirected graph. Consider a depth first traversal *G*, and let *T* be the resulting depth first search tree. Let *u* be a vertex in *G* and *v* be the first new (unvisited) vertex visited after visiting *u* in the traversal. Which of the following is always true?

(*a*) $\{u, v\}$ must be an edge in G, and u is descendent of v in T

(*b*) $\{u, v\}$ must be an edge in G, and v is descendent of u in T

(*c*) If $\{u, v\}$ is not an edge in G, then u is leaf in T

(*d*) If $\{u, v\}$ is not an edge in G, then u and v must have the same parent in T.

Question 6: A recursive program to compute Fibonacci numbers is shown below. Assume you are also given an array $f[0 \ldots M]$ with all elements initialized to 0.

```
fib (n) {
if (n > M) error () ;
if (n= =0) return 1;
if (n= =1) return 1;
if ( [          ]
        return [          ]
t = fib(n - 1) + fib (n - 2) ;
[          ]
return t;
}
```

(*a*) Fill in the boxes with expressions/statements to make **fib()** store and reuse computed Fibonacci values. Write the box number and the corresponding contents in your answer book.

(*b*) What is the time complexity of the resulting program when computing **fib(n)**?

YEAR – 2001

Question 1. Randomized quick sort is an extension of quicksort where the pivot is chosen randomly. What is the worst-case complexity of sorting n numbers using randomized quicksort?

(*a*) $O(n)$ (*b*) $O(n \log n)$

(*c*) $O(n^2)$ (*d*) $O(n!)$

Question 2. Consider an array representation of an n element binary heap where the elements are stored from index 1 to n of the array. For the element stored at index i of the array ($i \leq n$), the index of the parent is

(*a*) $i - 1$ (*b*) $\lfloor i/2 \rfloor$

(*c*) $\lceil i/2 \rceil$ (*d*) $(i + 1)/2$

Question 3. Let $f(n) = n^2 \log n$ and $g(n) = n(\log)^{10}$ be without positive functions of n. Which of the following statements is correct?

(*a*) $f(n) = O(g(n))$ and $g(n) \neq O(f(n))$
(*b*) $g(n) = O(f(n))$ and $f(n) \neq O(g(n))$
(*c*) $f(n) \neq O(g(n))$ and $g(n) \neq O(f(n))$
(*d*) $f(n) = O(g(n))$ and $g(n) = O(f(n))$

Question 4. Consider an undirected unweighted graph G. Let a breadth-first traversal of G be done starting from a node r. Let $d(r, u)$ and $d(r, v)$ be the lengths of the shortest paths from r to u and v respectively in G. If u is visited before v during the breadth-first traversal, which of the following is correct?

(*a*) $d(r, u) < d(r, v)$
(*b*) $d(r, u) > d(r, v)$
(*c*) $d(r, u) \leq d(r, v)$
(*d*) None of the above

Question 5. How many undirected graphs (not necessarily connected) can be constructed out of a given set And = $\{v_1, v_2, ..., v_n\}$ of n vertices?

(*a*) $n(n - 1)/2$
(*b*) 2^n
(*c*) $n!$
(*d*) $2^{n(n-1)/2}$

Question 6. Consider a weighted undirected graph with vertex set And = $\{n_1, n_2, n_3, n_4, n_5, n_6\}$ and edge set $E = \{(n_1, n_2, 2), (n_1, n_3, 8), (n_1, n_6, 3), (n_2, n_4, 4), (n_2, n_5, 12), (n_3, n_4, 7), (n_4, n_5, 9), (n_4, n_6, 4)\}$. The third value in each tuple represents the weight of the edge specified in the tuple.

(*a*) Lists the edges of a minimum spanning tree of the graph.
(*b*) How many distinct minimum spanning tree does this graph have?
(*c*) Is the minimum among the edge weights of a minimum spanning tree unique over all possible minimum spanning trees of a graph?
(*d*) Is the maximum among the edge weights of a minimum spanning tree unique over all possible minimum spanning trees of a graph?

YEAR – 2002

Question 1. The solution of the recurrence equation $T(2^k) = 3T(2^{k-1}) + 1$, $T(1) = 1$ is

(*a*) 2^k
(*b*) $(3^{k+1} - 1)$
(*c*) $3^{\log_2 k}$
(*d*) $2^{\log_3 k}$

Question 2. The minimum number of colours required to colour the vertices of a cycle with n nodes in such a way that no two adjacent nodes have the same colour is:

(*a*) 2
(*b*) 3
(*c*) 4
(*d*) $n - 2[n/2] + 2$

Question 3. In the worst case, the number of comparisons needed to search a singly linked list of length n for a given element is

(*a*) $\log n$
(*b*) $n/2$
(*c*) $\log n - 1$
(*d*) n

Question 4. The maximum number of edges in a n-node undirected graph without self loops is

(*a*) n^2 (*b*) $n(n - 1)/2$

(*c*) $n - 1$ (*d*) $(n + 1)(n)/2$

Question 5. Consider the following algorithm for searching a given number x in an unsorted array $A[1 \ldots n]$ having n distinct values:

1. Choose an i uniformly at random from $1 \ldots n$;
2. If $A[i] = x$ then Stop else Goto 1;

Assuming that x is present in A, what is the expected number of comparisons made by the algorithm before it terminates?

(*a*) n (*b*) $n - 1$

(*c*) $2n$ (*d*) $n/2$

Question 6: The running time of the following algorithm

Procedure $A(n)$

If $n <= 2$ return (1) else return $(A(\lceil n \rceil))$:

is best described by

(*a*) $O(n)$ (*b*) $O(\log n)$

(*c*) $O(\log \log n)$ (*d*) $O(1)$

Question 7: Fill in the blanks in the following template of an algorithm to compute all pairs of shortest path lengths in a directed graph G with $n \times n$ adjacency matrix A. $A[i, j]$ equals 1 if there is an edge in G from i to j, and 0 otherwise.

```
INITIALIZATION: For I = 1 ... n
    {For j = 1 ... n
        {If A[i, j] = 0 then P[i, j] = ____ else P[i, j] = ____;}
    }
ALGORITHM: For i = 1 . . . n
    {For j = 1 ... n
        {For k = 1 ... n
             {P[ _, _ ] = min{ ______________};}
        }
    }
```

(*a*) Copy the complete line containing the blanks in the Initialization step and fill in the blanks.

(*b*) Copy the complete line containing the blanks in the Initialization step and fill in the blanks.

(*c*) Fill in the blank: The running time of the algorithm is $O($ _____ $)$.

YEAR – 2003

Question 1. Let A $T(n)$ be the number of different binary search trees on n distinct elements. Then

$$T(n) = \sum_{k=1}^{n} T(k-1)T(x)$$

Where x is

(*a*) $n - k + 1$ (*b*) $n - k$

(*c*) $n - k - 1$ (*d*) $n - k - 1$

Question 2. Let G be an arbitrary graph with n nodes and k components. If a vertex is removed from G, the number of components in the resultant graph must necessarily lie between

(*a*) k and n (*b*) $k - 1$ and $k + 1$

(*c*) $k - 1$ and $n - 1$ (*d*) $k + 1$ and $n - k$

Question 3. Consider the following three claims

I. $(n + k)^m = Q(n^m)$, where k and m are constants

II. $2^{n+1} = O(2^n)$

III. $2^{2n} = O(2^n)$

Which of these claims are correct?

(*a*) I and II (*b*) I and III

(*c*) II and III (*d*) I, II and III

Question 4. Consider the following graph

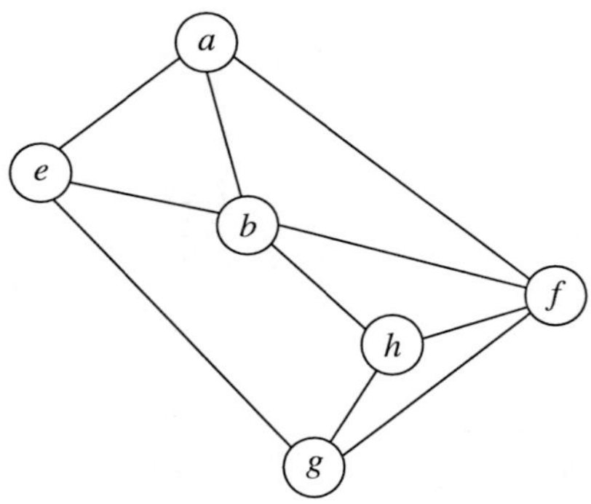

Among the following sequences

I. $a\ b\ e\ g\ h\ f$

II. $a\ b\ f\ e\ h\ g$

III. $a\ b\ f\ h\ g\ e$

IV. $a\ f\ g\ h\ b\ e$

which are the depth-first traversals of the above graph?

(*a*) I, II and IV only
(*b*) I and IV only
(*c*) II, III and IV only
(*d*) I, III and IV only

Question 5: The usual implementation of insertion sort to sort an array uses linear search to identify the position where an element is to be inserted into already sorted part of the array. If, instead, we use binary search to find the position, the worst-case running time will

(*a*) remain $\Theta(n^2)$
(*b*) become $\Theta(n (\log n)^2)$
(*c*) become $\Theta(n \log n)$
(*d*) become $\Theta(n)$

Question 6. In heap with n elements with the smallest element at the root, the 7th smallest element can be found in time

(*a*) $\Theta(n \log n)$
(*b*) $\Theta(n)$
(*c*) $\Theta(\log n)$
(*d*) $\Theta(1)$

Question 7. Consider the following recurrence relation

$$T(1) = 1$$

$$T(n + 1) = T(n) + \lfloor \sqrt{n+1} \rfloor \text{ for all } n \geq 1$$

The value of $T(m^2)$ for $m \geq 1$ is

(*a*) $\frac{m}{6}(21m - 39) + 4$
(*b*) $\frac{m}{6}(4m^2 - 3m - 5)$
(*c*) $\frac{m}{2}(3m^{2.5} - 11m + 20) - 5$
(*d*) $\frac{m}{6}(5m^5 - 34m^2 + 137m - 104) + \frac{5}{6}$

Question 8. How many perfect matchings are there in a complete graph of 6 vertices?

(*a*) 15
(*b*) 24
(*c*) 16
(*d*) 60

Question 9: If all permutations are equally likely, what is the expected number of inversions in a randomly chosen permutation of 1 … n?

(*a*) $n(n - 1)/2$
(*b*) $n(n - 1)/4$
(*c*) $n(n + 1)/4$
(*d*) $2n \log_2 n$

Question 10. What would be the worst-case time complexity of insertion sort algorithm, if the inputs are restricted to permutations of 1 … n with at most n inversions?

(*a*) $\Theta(n_2)$ (*b*) $\Theta(n \log n)$

(*c*) $\Theta(n_{1.5})$ (*d*) $\Theta(n)$

Question 11. A data structure is required for storing a set of integers such that each of the following operations can be done in $O(\log n)$ time, where n is the number of elements in the set.

I. Deletion of the smallest element

II. Insertion of an element if it is not already present

Which of the following data structure can be used for this purpose?

(*a*) A heap can be used but not a balance binary search tree

(*b*) A balance binary search tree can be used but not a heap

(*c*) Both balance binary search tree and heap can be used

(*d*) Neither balance binary search tree nor heap can be used

Question 12. The cube root of a natural number n is defined as the largest natural number such that $n^3 \leq n$. The complexity of computing the cube root of n (n is represented in binary notation) is

(*a*) $O(n)$ but not $O(n^{0.5})$

(*b*) $O(n^{0.5})$ but not $O(\log n)^k)$ for any constant $k > 0$

(*c*) $O(\log n)^k)$ for some constant $k > 0$, but not $O(\log \log n)^m)$ for any constant $m > 0$

(*d*) $O(\log \log n)^k)$ for some constant $k > 0.5$, but not $O(\log \log n)^{0.5})$

Question 13. Let $G = (V, E)$ be an undirected graph with a subgraph $G_1 = (V_1, E_1)$. Weights are assigned to edges of G as follows.

A single-source shortest path algorithm is executed on the weighted graph (V, E, what) with an arbitrary vertex v_1 of V_1 as the source. Which of the following can always be inferred from the path costs computed?

(*a*) The number of edges in the shortest paths from v_1 to all vertices of G

(*b*) G_1 is connected

(*c*) V_1 forms a clique in G

(*d*) G_1 is a tree

Question 14. What is the weight of a minimum spanning tree of the following graph?

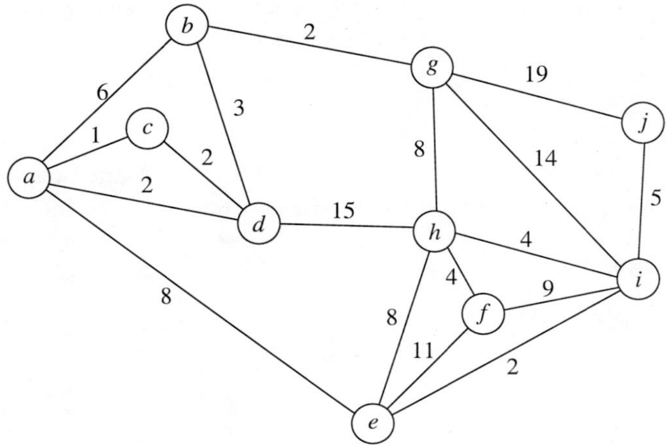

(*a*) 29 (*b*) 31

(*c*) 38 (*d*) 41

Question 15. Let $G = (V, E)$ be a directed graph with n vertices. Path from v_i to v_j in G is a sequence of vertices $(v_i, v_{i+1}, \ldots v_j)$ such that (v_k, v_{k+1}) E for all k in I through $j - 1$. A simple path is the path in which no vertex appears more than once.

Let A be an $n \times n$ array initialized as follows.

$A[j, k] = 1$ if $(j, k) \in$ E

0 otherwise

Consider the following algorithm.

```
for i = 1 to n
    for j = 1 to n
        for k = 1 to n
            A[j, k] = max(A[j, k], A[j, i] + A[i, k]);
```

Which of the following statements is necessarily true for all j and k after termination of the above algorithm?

(*a*) $A[j, k] \le n$

(*b*) If $A[j, j] \ge n - 1$, then G has a Hamiltonian cycle

(*c*) If there exists a path from j to k, $A[j, k]$ contains the longest path length from j to k

(*d*) If there exists a path from j to k, every simple path from j to k contains at most $A[j, k]$ edges.

YEAR – 2004

Question 1. The tightest lower bound on the number of comparisons, in the worst case, for comparison-based sorting is of the order of

(*a*) n (*b*) n^2

(*c*) $n \log n$ (*d*) $n \log^2 n$

Question 2. The elements 32, 15, 20, 30, 12, 25, 16 are inserted one-by-one in the given order into a max heap. The resultant max heap is

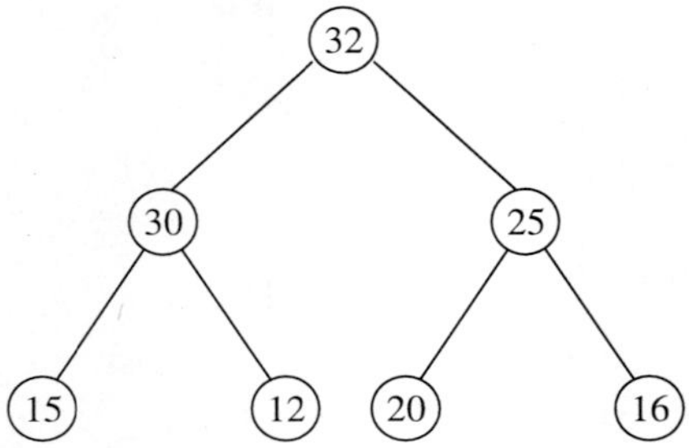

(*a*)

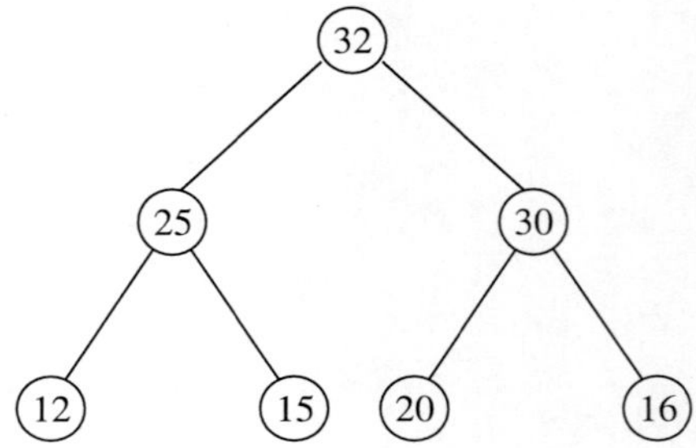

(*b*)

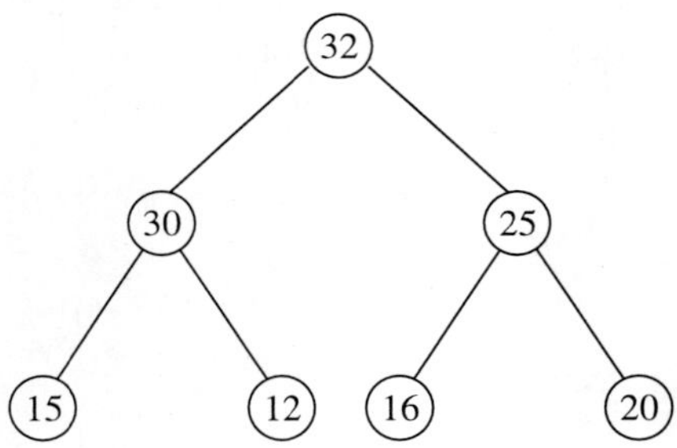

(*c*)

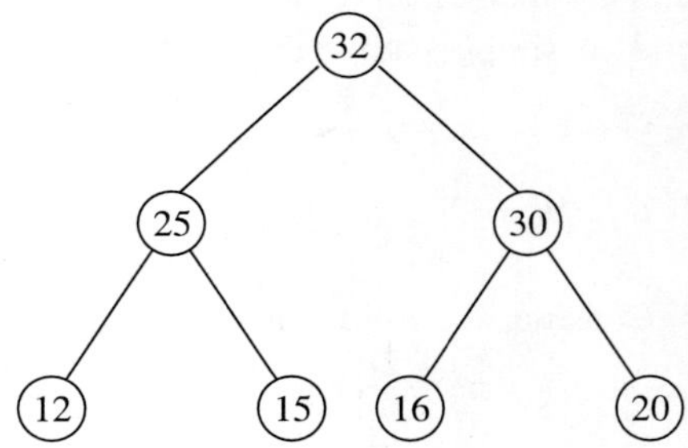

(*d*)

Question 3. What does the following algorithm approximate? (Assume $m > 1, \in > 0$).

```
x = m;
y = 1;
while (x – y > ∈)
{    x = (x + y)/2;
     y = m/x
}
print (x);
```

(*a*) $\log m$ (*b*) m^2

(*c*) $m^{1/2}$ (*d*) $m^{1/3}$

Question 4. Suppose we run Dijkstra's single source shortest path algorithm on the following edgeweighted directed graph with vertex P as the source.

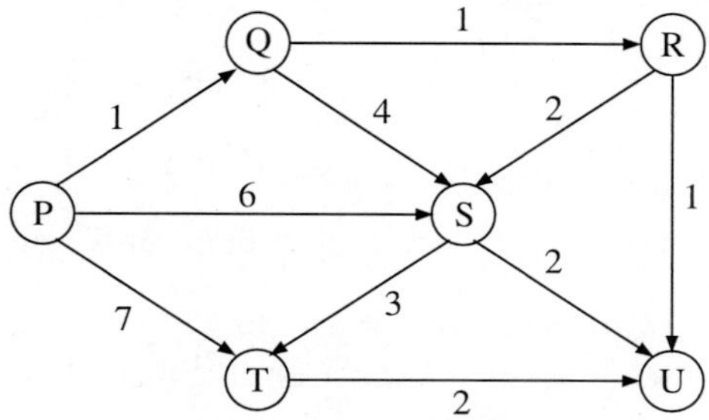

In what order do the nodes get included into the set of vertices for which the shortest path distances are finalized?

(*a*) P, Q, R, S, T, U (*b*) P, Q, R, U, S, T

(*c*) P, Q, R, U, T, S (*d*) P, Q, T, R, U, S

Question 5. The minimum number of colours required to colour the following graph, such that no two adjacent vertices are assigned the same colour, is

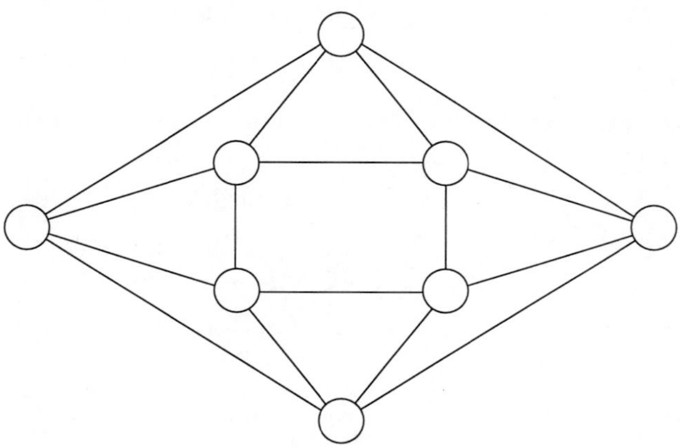

(*a*) 2 (*b*) 3

(*c*) 4 (*d*) 5

Question 6. How many graphs on n-labelled vertices exist which have at least $(n^2 - 3n)/2)$ edges?

(*a*) ${}^{(n^2-n)/2}C_{(n^2-3n)/2}$ (*b*) $\sum_{k=0}^{(n^2-3n)/2} {}^{(n^2-n)}C_k$

(*c*) ${}^{(n^2-n)}C_n$ (*d*) $\sum_{k=0}^{2} {}^{(n^2-n)}C_k$

Question 7. Let $G_1 = (V, E_1)$ and $G_2 = (V, E_2)$ be connected graphs on the same vertex set V with more than two vertices. If $G_1 \cap G_2 = (V, E_1) \cap E_2)$ is not a connected graph, then the graph $G_1 \cup G_2 = (V, E_1 \cup E_2)$

(*a*) cannot have a cut vertex

(*b*) must have a cycle

(*c*) must have a cut-edge (bridge)

(*d*) has chromatic number strictly greater than those of G_1 and G_2

Question 8. Let $A[1 \ldots n]$ be an array storing a bit (1 or 0) at each location, and $f(m)$ is a function whose time complexity is $\Theta(m)$. Consider the following program fragment written in a C like language:

```
counter = 0;
for (i = 1; i <= n; i ++)
    {  if (A[i] == 1) counter++;
       else { f (counter); counter = 0;}
    }
```

The complexity of this program fragment is

(*a*) $\Omega(n^2)$ (*b*) $\Omega(n \log n)$ and $O(n^2)$

(*c*) $\Theta(n)$ (*d*) $O(n)$

Question 9. The time complexity of the following *C* function is (assume $n > 0$):

```
int recursive (int n) {
    if (n = = 1)
       return (1);
    else {f(counter); counter =0;}
       return (recursive (n - 1) + recursive (n - 1) ) ;
    }
```

(*a*) $O(n)$ (*b*) $O(n \log n)$

(*c*) $O(n^2)$ (*d*) $O(2^n)$

Question 10. The recurrence equation

$$T(1) = 1$$

$$T(n) = 2T(n - 1) + n, n \geq 2$$

evaluates to

(*a*) $2^{n+1} - n - 2$ (*b*) $2^n - n$

(*c*) $2^{n+1} - 2n - 2$ (*d*) $2^n + n$

Question 11. A program takes as input a balance binary search tree with n leaf nodes and computes the value of a function $g(x)$ for each node x. If the cost of computing $g(x)$ is minimum {no. of leaf-nodes in left-subtree of x, no. of leaf-nodes in right-subtree of x}, then the worst-time case complexity of the program is

(*a*) $\Theta(n)$ (*b*) $\Theta(n \log n)$

(*c*) $\Theta(n^2)$ (*d*) $\Theta(n^2 \log n)$

Question 12: What is the number of edges in an acyclic undirected graph with n vertices?

(*a*) $n - 1$ (*b*) n

(*c*) $n + 1$ (*d*) $2n - 2$

Question 13. Let P be a singly linked list. Let Q be the pointer to an intermediate node x in the list. What is the worst-case time complexity of the best-known algorithm to delete the node x from the list?

(*a*) $O(n)$ (*b*) $O(\log^2 n)$

(*c*) $O(\log n)$ (*d*) $O(1)$

Question 14. What is the number of vertices in an undirected connected graph with 27 edges, 6 vertices of degree 2, 3 vertices of degree 4 and remaining of degree 3?

(*a*) 10 (*b*) 11

(*c*) 18 (*d*) 19

Question 15. An array of integers of size n can be converted into a heap by adjusting the heaps rooted at each internal node of the complete binary tree starting at the node $\lfloor (n-1) \rfloor$, and doing this adjustment up to the root node (root node is at index 0) in the order $\lfloor (n-1) \rfloor$, $\lfloor (n-3) \rfloor$, ..., 0. The time required to construct a heap in this manner is

(*a*) $O(\log n)$ (*b*) $O(n)$

(*c*) $O(n \log \log n)$ (*d*) $O(n \log n)$

Question 16. Let $f(n)$, $g(n)$ and $h(n)$ be functions defined for positive integers such that $f(n) = O(g(n))$, $g(n) \neq O(f(n))$, $g(n) = O(h(n))$ and $h(n) = O(g(n))$. Which one of the following statements is FALSE?

(*a*) $f(n) = O(g(n) + h(n))$ (*b*) $f(n) = O(h(n))$

(*c*) $h(n) \neq O(f(n))$ (*d*) $f(n)\ h(n) \neq O(g(n)h(n)$

Question 17. Consider the undirected graph below.

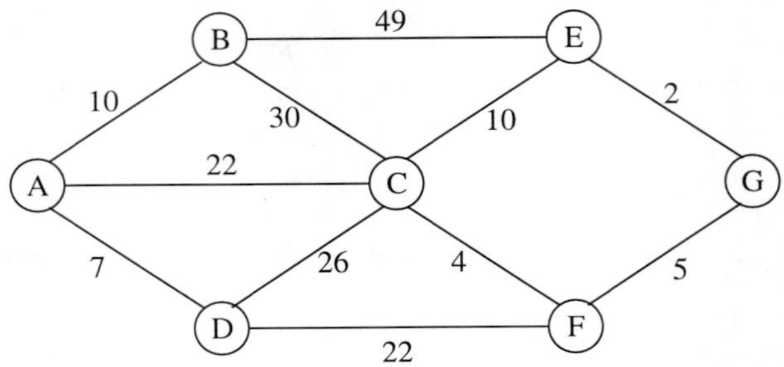

Using Prim's algorithm to construct a minimum spanning tree starting with node A, which one of the following sequences of edges represents a possible order in which the edges would be added to construct the minimum spanning tree?

(*a*) (E, G), (C, F), (F, G), (A, D), (A, B), (A, C)

(*b*) (A, D), (A, B), (A, C), (C, F), (G, E), (F, G)

(*c*) (A, B), (A, D), (D, F), (F, G), (G, E), (F, C)

(*d*) (A, D), (A, B), (D, F), (F, C), (F, G), (G, E)

Question 18: Consider a list of recursive algorithms and a list of recurrence relations as shown below. Each recurrence relation corresponds to exactly one algorithm and is used to derive the time complexity of the algorithm.

P. Binary search	I. $T(n) = T(n - k) + T(k) + c_n$
Q. Mergesort	II. $T(n) = 2T(n - 1) + 1$
R. Quicksort	III. $T(n) = T(n/2) + c_n$
S. Tower of Hanoi	IV. $T(n) = T(n/2) + 1$

(*a*) P – II, Q – III, R – IV, S – I (*b*) P – IV, Q – III, R – I, S – II

(*c*) P – III, Q – II, R – IV, S – I (*d*) P – IV, Q – II, R – I, S – III

YEAR – 2005

Question 1. An undirected graph G has n nodes. Its adjacency matrix is given by an $n \times n$ square matrix whose (*i*) diagonal elements are 0s and (*ii*) non-diagonal elements are 1s. Which one of the following is TRUE?

(*a*) Graph G has no minimum spanning tree (MST)

(*b*) Graph G has a unique MST of cost $n - 1$

(*c*) Graph G has multiple distinct MSTs, each of cost $n - 1$

(*d*) Graph G has multiple minimum spanning trees of different costs.

Question 2. The time complexity of computing the transitive closure of a binary relation on a set of n elements is known to be:

(*a*) $O(n)$ (*b*) $O(n \log n)$

(*c*) $O(n^{3/2})$ (*d*) $O(n^3)$

Question 3. Let G be a simple connected graph with 13 vertices and 19 edges. Then, the number of faces in the planar embedding of the graph is:

(*a*) 6 (*b*) 8

(*c*) 9 (*d*) 13

Question 4: A Priority-Queue is implemented as a Max-Heap. Initially, it has 5 elements. The level-order traversal of the heap is given below:

10, 8, 5, 3, 2

Two new elements '1' and '7' are inserted in the heap in that order. The level-order traversal of the heap after the insertion of the elements is:

(*a*) 10, 8, 7, 5, 3, 2, 1 (*b*) 10, 8, 7, 2, 3, 1, 5

(*c*) 10, 8, 7, 1, 2, 3, 5 (*d*) 10, 8, 7, 3, 2, 1, 5

Question 5. Suppose $T(n) = 2T(n/2) + n$, $T(0) = T(1) = 1$. Which of the following is FALSE?

(*a*) $T(n) = O(n^2)$ (*b*) $T(n) = \Theta(n \log n)$

(*c*) $T(n) = \Omega(n^2)$ (*d*) $T(n) = O(n \log n)$

Question 6. Let $G(V, E)$ be an undirected graph with positive edge weights. Dijkstra's single source shortest path algorithm can be implemented using heap data structure with time complexity:

(*a*) $O(|V|^2)$ (*b*) $O(|E| + |V| \log |V|)$

(*c*) $O(|V| \log |V|)$ (*d*) $O((|E| + |V|) \log |V|)$

Question 7. Suppose there are $\lfloor \log n \rfloor$ sorted lists of $\left\lfloor \frac{n}{\log n} \right\rfloor$ elements each. The time complexity of producing a sorted list of all these elements is:

(**Hint:** Use a heap data structure)

(*a*) $O(n \log \log n)$ (*b*) $\Theta(n \log n)$

(*c*) $\Omega(n \log n)$ (*d*) $\Omega(n^3/2)$

Question 8: Consider the following *C* function:

```
double foo (int n) {
    int i;
    double sum;

    if (n = = 0) return 1.0;
    else {
        sum =0.0;
        for(i = 0; I < n; i++)
            sum += foo (i) ;
        return sum;
    }
}
```

The space complexity of the above function is:

(*a*) $O(1)$ (*b*) $O(n)$

(*c*) $O(n!)$ (*d*) $O(n^n)$

Question 9. Suppose we modify the function foo() in previous question and store the values of foo (*i*), $0 \leq i < n$, as and when they are computed. With this modification, the time complexity for function foo() is significantly reduced. The space complexity of the modified function would be:

(*a*) $O(1)$ (*b*) $O(n)$

(*c*) $O(n^2)$ (*d*) $O(n!)$

Question 10. Let *s* and *t* be two vertices in an undirected graph *G* = (*V*, *E*) having distinct positive edge weights. Let [*X*, *Y*] be partition of *V* such that *s OE X* and *t OE Y*. Consider the edge *e* having the minimum weight amongst all those edges that have one vertex in *X* and one vertex in *Y*.

(*a*) The edge *e* must definitely belong to:

(A) The minimum weighted spanning tree of *G*

(B) The weighted shortest path from *s* to *t*

(C) Each path from *s* to *t*

(D) The weighted longest path from *s* to *t*

(*b*) Let the weight of an edge *e* denote the congestion on that edge. The congestion on a path is defined to be the maximum of the congestions on the edges of the path. We wish to find the path from *s* to *t* having minimum congestion. Which of the following paths is always such a path of minimum congestion?

(A) A path from *s* to *t* in the minimum weighted spanning tree

(B) A weighted shortest path from *s* to *t*

(C) An Euler walk from s to t

(D) A Hamiltonian path from s to t

Question 11. A function f defined on stacks of integers satisfies the following properties.

$f(f) = 0$ and

$f(\text{push}(S, i)) = \max(f(S), 0) + i$ for all stacks S and integers i.

If a stack S contains the integers 2, –3, 2, –1, 2 in order from bottom to top, what is $f(S)$?

(*a*) 6 (*b*) 4

(*c*) 3 (*d*) 2

Question 12. In depth-first traversal of a graph G with n vertices, k edges are marked as tree edges. The number of connected components in G is

(*a*) k (*b*) $k + 1$

(*c*) $n - k - 1$ (*d*) $n - k$

Question 13. In the following table, the left column contains the names of standard graph algorithms and the right column contains the time complexities of the algorithms. Match each algorithm with its time complexity.

1. Bellman-Ford algorithm	A. $O(m \log 11)$
2. Kruskal's algorithm	B. $O(n^3)$
3. Floyd-Warshall algorithm	C. $O(nm)$
4. Topological sorting	D. $O(n + m)$

(*a*) 1 → C, 2 → A, 3 → B, 4 → D

(*b*) 1 → B, 2 → D, 3 → C, 4 → A

(*c*) 1 → C, 2 → D, 3 → A, 4 → B

(*d*) 1 → B, 2 → A, 3 → C, 4 → D

Question 14: Let $T(n)$ be a function defined by the recurrence

$$T(n) = 2T(n/2) + \sqrt{n} \text{ for } n \geq 2 \text{ and}$$

$$T(1) = 1$$

Which of the following statement is TRUE?

(*a*) $T(n) = \Theta(\log n)$ (*b*) $T(n) = \Theta(\sqrt{n})$

(*c*) $T(n) = \Theta(n)$ (*d*) $T(n) = \Theta(n \log n)$

Question 15. Let G be a weighted undirected graph and e be an edge with maximum weight in G. Suppose there is minimum weight spanning tree in G containing edge e. Which of the following statements is always true?

(*a*) There exists a cutest in G having all edges of maximum weight.

(*b*) There exists a cycle in G having all edges of maximum weight.

(*c*) Edge *e* can be contained in a cycle.

(*d*) All edges in *G* have the same weight.

Question 16. Let *G* be a directed graph whose vertex set is the set of number from 1 to 100. There is an edge from a vertex *i* to a vertex *j* iff either $j = i + 1$ or $j = 3i$. The minimum number of edges in a path in *G* from vertex 1 to 100 is:

(*a*) 4 (*b*) 7

(*c*) 23 (*d*) 99

YEAR – 2006

Question 1. In a binary max heap containing *n* numbers, the smallest element can be found in time

(*a*) $\Theta(n)$ (*b*) $\Theta(\log n)$

(*c*) $\Theta(\log \log n)$ (*d*) $\Theta(1)$

Question 2. Consider a weighted complete graph *G* on the vertex set $\{v_1, v_2, \ldots, v_n\}$ such that the weight of the edge (v_i, v_j) is $2|i - j|$. The weight of a minimum spanning tree of *G* is

(*a*) $n - 1$ (*b*) $2n - 2$

(*c*) $\binom{n}{2}$ (*d*) n^2

Question 3. To implement Dijkstra's shortest path algorithm on undirected graphs so that it runs in linear time, the data structure to be used is

(*a*) Queue (*b*) Stack

(*c*) Heap (*d*) *B*–Tree

Question 4: Consider the following C-program fragment in which *i*, *j*, and *n* are integer variables.

for ($i = n, j = 0; i > 0; i /= 2, j += i$);

Let val(*j*) denote the value stored in the variable *j* after termination of the loop. Which of the following is true?

(*a*) $\text{val}(j) = \Theta(\log n)$ (*b*) $\text{val}(j) = \Theta(\sqrt{n})$

(*c*) $\text{val}(j) = \Theta(n)$ (*d*) $\text{val}(j) = \Theta(n \log n)$

Question 5. Let S be an NP–complete problem and *Q* and *R* be two other problems not known to NP. *Q* is polynomial time reducible to *S* and *S* is polynomial time reducible to *R*. Which one of the following statements is true?

(*a*) *R* is NP-complete (*b*) *R* is NP hard

(*c*) *Q* is NP-complete (*d*) *Q* is NP hard

Question 6. Consider the following graph.

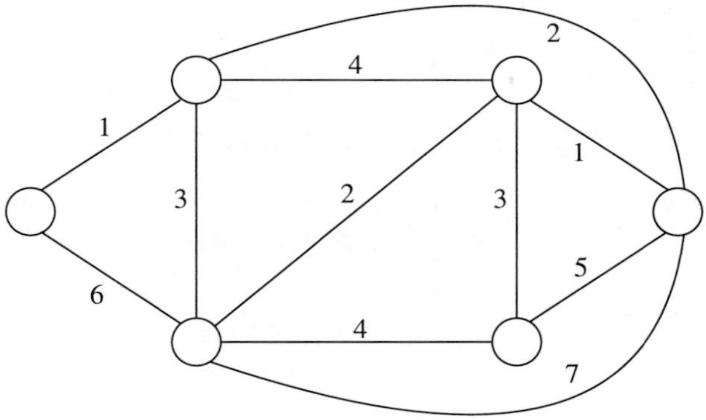

Which one of the following cannot be the sequence of edges added, in that order, to a minimum spanning tree using Kruskal's algorithm?

(*a*) (*a-b*), (*d-f*), (*b-f*), (*d-c*), (*d-e*) (*b*) (*a-b*), (*d-f*), (*d-c*), (*b-f*), (*d-e*)

(*c*) (*d-f*), (*a-b*), (*d-c*), (*b-f*), (*d-e*) (*d*) (*d-f*), (*a-b*), (*b-f*), (*d-e*), (*d-c*)

Question 7. Let *T* be a depth-first search tree in an undirected graph *G*. Vertices *u* and *v* are leaves of this tree *T*. The degrees of both *u* and *v* in *G* are at least 2. Which one of the following statements is true?

(*a*) There must exist a vertex *w* adjacent to both *u* and *v* in *G*

(*b*) There must exist a vertex *w* whose removal disconnects *u* and *v* in *G*

(*c*) There must exist a cycle in *G* containing *u* and *v*

(*d*) There must exist a cycle in *G* containing *u* and *v* and all its neighbours in *G*

Question 8. A set *X* can be represented by an array *x*[*n*] as follows:

$$x[i] = \begin{cases} 1 \text{ if } i \in X \\ 0 \text{ otherwise} \end{cases}$$

Consider the following algorithm in which *x*, *y* and *z* are boolean arrays of size *n*:

```
Algorithm zzz (x[ ], y[ ], z[ ]) {
    int I;
    for (i = 0; i< n; + + i)
        z[i]  = x[i]^~Y[i] v (~ x[i] ^ y[i])
}
```

The set *Z* computed by the algorithm is

(*a*) $(X \cup Y)$ (*b*) $(X \cap Y)$

(*c*) $(X - Y) \cap (Y - X)$ (*d*) $(X - Y) \cup (Y - X)$

Question 9. Consider the following recurrence:

$$T(n) = 2T\left(\left\lceil \sqrt{n} \right\rceil\right) + 1, T(1) = 1$$

Which one of the following is true?

(*a*) $T(n) = \Theta(\log n)$ (*b*) $T(n) = \Theta(\sqrt{n})$

(*c*) $T(n) = \Theta(n)$ (*d*) $T(n) = \Theta(n \log n)$

Question 10. The median of n elements can be found in $O(n)$ time. Which one of the following is correct about the complexity of quicksort, in which median is selected as pivot?

(*a*) $\Theta(n)$ (*b*) $\Theta(n \log n)$

(*c*) $\Theta(n^2)$ (*d*) $\Theta(n^3)$

Question 11. Given two arrays of numbers $a_1, \ldots, a_n$ and $b_1, \ldots, b_n$ where each number is 0 or 1, the fastest algorithm to find the largest span (i, j) such that $a_i + a_{i+1} + \ldots + a_j = b_i + b_{i+1} + \ldots + b_j$ or report that there is no such span,

(*a*) Takes $O(3^n)$ and $\Omega(2^n)$ time if hashing is permitted

(*b*) Takes $O(n^3)$ and $\Omega(n^{2.5})$ time in the key comparison model

(*c*) Takes $\Theta(n)$ time and space

(*d*) Takes $O(\sqrt{2})$ time only if the sum of the $2n$ elements is an even number

Question 12. The $2n$ vertices of a graph G correspond to all subsets of size n, for $n \geq 6$. Two vertices of G are adjacent if and only if the corresponding sets intersect in exactly two elements.

(*a*) The number of vertices of degree zero in G is

(A) 1 (B) n

(C) $n + 1$ (D) 2^n

(*b*) The maximum degree of a vertex in G is

(A) $\binom{n/2}{2} 2^{n/2}$ (B) 2^{n-2}

(C) $2^{n-3} \times 3$ (D) 2^{n-1}

(*c*) The number of connected components in G is

(A) 2 (B) $n + 2$

(C) $2^{n/2}$ (D) $\frac{2^n}{n}$

Question 13. A problem in NP is NP-complete if

(*a*) it can be reduced to the 3-SAT problem in polynomial time

(*b*) 3-SAT problem can be reduced to it in polynomial time

(*c*) it can be reduced to any other problem in NP in polynomial time

(*d*) some problem in NP can be reduced to it in polynomial time

Question 14. If all the edge weights of an undirected graph are positive, then any subset of edges that connects all the vertices and has minimum total weight is a

(*a*) Hamiltonian cycle (*b*) grid

(*c*) hypercube (*d*) tree

Question 15. Consider the undirected graph G defined as follows. The vertices of G are bit strings of length n. We have an edge between vertex u and vertex v if and only if u and v differ in exactly one bit position (in other words, v can be obtained from u by flipping a single bit). The ratio of the chromatic number of G to the diameter of G is

(*a*) $1/2^{n-1}$ (*b*) $1/n$

(*c*) $2/n$ (*d*) $3/n$

Question 16. Which of the following sequences of array elements forms a heap?

(*a*) {23, 17, 14, 6, 13, 10, 1, 12, 7, 5}

(*b*) {23, 17, 14, 6, 13, 10, 1, 5, 7, 12}

(*c*) {23, 17, 14, 7, 13, 10, 1, 5, 6, 12}

(*d*) {23, 17, 14, 7, 13, 10, 1, 12, 5, 7}

Question 17. Suppose that we have numbers between 1 and 100 in a binary search tree and want to search for the number 55. Which of the following sequences CANNOT be the sequence of nodes examined?

(*a*) {10, 75, 64, 43, 60, 57, 55}

(*b*) {90, 12, 68, 34, 62, 45, 55}

(*c*) {9, 85, 47, 68, 43, 57, 55}

(*d*) {79, 14, 72, 56, 16, 53, 55}

Question 18. Which of the following is the correct decomposition of the directed graph given below into its strongly connected components?

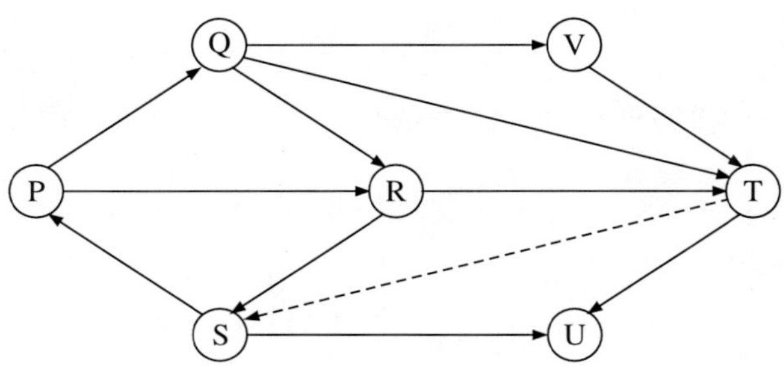

(*a*) {P, Q, R, S}, {T}, {U}, {V} (*B*) {P, Q, R, S, T, V}, {U}

(*c*) {P, Q, S, T, V}, {R}, {U} (*D*) {P, Q, R, S, T, U, V}

Question 19. Consider the depth-first search of an undirected graph with 3 vertices P, Q and R. Let discovery time $d(u)$ represent the time instant when the vertex u is first visited, and finish time $f(u)$ represent the time instant when the vertex u is last visited. Given that

$d(P) = 5$ units $f(P) = 12$ units

$d(Q) = 6$ units $f(Q) = 10$ units

$d(R) = 14$ units $f(R) = 18$ units

Which of the following statements is TRUE about the graph?

(*a*) There is only one connected component

(*b*) There are connected components, and P and R are connected

(*c*) There are connected components, and Q and R are connected

(*d*) There are connected components, and P and Q are connected

Question 20. An array X of n distinct integers is interpreted as a complete binary tree. The index of the first element of the array is 0.

(*i*) The index of the parent element $X[i]$, $i \neq 0$, is

(*a*) $\lfloor i/2 \rfloor$ (*b*) $\lceil (i-1)/2 \rceil$

(*c*) $\lceil i/2 \rceil$ (*d*) $\lceil i/2 \rceil - 1$

(*ii*) If only the root node does not satisfy the heap property, the algorithm to convert the complete binary tree into a heap has the best asymptotic time complexity of

(*a*) $O(n)$ (*b*) $O(\log n)$

(*c*) $O(n \log n)$ (*d*) $O(n \log \log n)$

(iii) If the root node is at level 0, the level of element $X[i]$, $i \neq 0$, is

(*a*) $\log_2 i$ (*b*) $\lceil \log_2(i+1) \rceil$

(*c*) $\lfloor \log_2(i+1) \rfloor$ (*d*) $\lfloor \log_2 i \rfloor$

YEAR – 2007

Question 1. Let G be the non-planar graph with the minimum possible number of edges. Then G has

(*a*) 9 edges and 5 vertices (*b*) 9 edges and 6 vertices

(*c*) 10 edges and 5 vertices (*d*) 10 edges and 6 vertices

Question 2. Consider the DAG with V = {1, 2, 3, 4, 5, 6}, shown below.

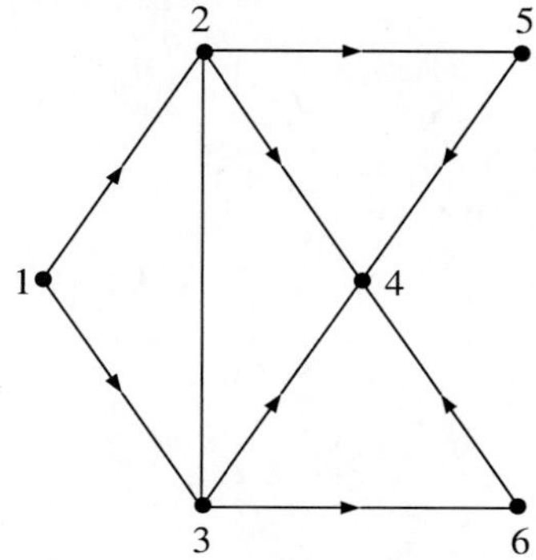

Which one the following is NOT a topological ordering?

(*a*) 1 2 3 4 5 6 (*b*) 1 3 2 4 5 6

(*c*) 1 3 2 4 6 5 (*d*) 3 2 4 1 6 5

Question 3. The height of a binary tree is the maximum number of edges in any root to leaf path. The maximum number of nodes in a binary tree of height h is:

(*a*) 2^h (*b*) $2^{h-1} - 1$

(*c*) $2^{h+1} - 1$ (*d*) 2^{h+1}

Question 4. Which of the following sorting algorithms has the lowest worst-case complexity?

(*a*) Mergesort (*b*) Bubblesort

(*c*) Quicksort (d) Selectionsort

Question 5. Consider the following segment of *C*-code:

```
int j, n;
j = 1;
while     (j <= n)
            j = j * 2;
```

The number of comparisons made in the execution of the loop for any $n > 0$ is:

(*a*) $\lceil \log_2 n \rceil + 1$ (*b*) n

(*c*) $\lceil \log_2 n \rceil$ (*d*) $\lfloor \log_2 n \rfloor + 1$

Question 6. In an unweighted, undirected connect graph, the shortest path from a node *S* to every node is computed most efficiently, in terms of *time complexity,* by

(*a*) Dijkstra's algorithm starting from *S*.

(*b*) Warshall's algorithm.

(*c*) Performing a DFS starting from *S*.

(*d*) Performing a BFS starting from *S*.

Question 7. Consider the following C function:

```
int f(int n)
{
    static int are = 0;
    If (n < = 0) return 1;
    If (n > 3)
    {  r = n;
       return f (n - 2) + 2;
    }
    return f(n - 1) + r;
}
```

What is the value of $f(5)$?

(*a*) 3 (*b*) 7

(*c*) 9 (*d*) 18

Question 8. In the following C function, let $n \geq m$.

```
int gcd(n, m)
{
    If (n % m = = 0) return m;
    n = n % m;
    return gcd(m, n);
}
```

How many recursive calls are made by this function?

(*a*) $\Theta(\log n)$ (*b*) $\Omega(n)$

(*c*) $\Theta(\log \log n)$ (*d*) $\Theta(n)$

Question 9. What is the *time complexity* of the following recursive function:

```
int Dosomething (int n) {
If (n < = 2)
return 1;
else
return Dosomething floor(sqrt (n))) + n);
}
```

(*a*) $\Theta(n)$ (*b*) $\Theta(n \log n)$

(c) $\Theta(\log n)$ (*d*) $\Theta(\log \log n)$

Question 10. Consider the process of inserting an element into a *Max Heap,* where the *Max Heap* is represented by an array. Suppose we perform a binary search on the path from the new leaf to the root to find the position for the newly inserted element, the number of *comparisons* performed is:

(*a*) $\Theta(\log n)$ (*b*) $\Theta(\log \log n)$

(*c*) $\Theta(n)$ (*d*) $\Theta(n \log n)$

Question 11. Let w be the minimum weight among all edge weights in an undirected connected graph. Let e be a specific edge of weight w. Which of the following is FALSE?

(*a*) There is a minimum spanning tree containing e.

(*b*) If e is not in a minimum spanning tree T, then in the cycle formed by adding e to T, all edges have the same weight.

(*c*) Every minimum spanning tree has an edge of weight w.

(*d*) e is present in every minimum spanning tree.

Question 12. Consider the following C code segment:

```
int IsPrime(n)
{
    int i, n;
    for (i = 2; i <= sqrt (n); i++)
    if(n % i = = 0)
    {printf (" Not Prime\n"); return 0;}
     return 1;
}
```

Let $T(n)$ denote the number of times the for loop is executed by the program on input n. Which of the following is TRUE?

(*a*) $T(n) = O(\sqrt{n})$ and $T(n) = \Omega(\sqrt{n})$ (*b*) $T(n) = O(\sqrt{n})$ and $T(n) = \Omega(1)$

(*c*) $T(n) = O(n)$ and $T(n) = \Omega(\sqrt{n})$ (*d*) None of the above.

Question 13. Suppose that a robot is placed on the Cartesian plane. At each step it is allowed to move either one unit up or right, *i.e.,* if it is at (i, j) then it can move to either $(i + 1, j)$ or $(i, j + 1)$.

(*i*) How many distinct paths are there for the robot to reach the point (10, 10) starting from the initial position (0, 0)?

(*a*) $\binom{20}{10}$ (*b*) 2^{20}

(*c*) 2^{10} (*d*) None of the above

(*ii*) Suppose that the robot is not allowed to traverse the line segment from (4, 4) to (5, 4). With this constraint, how many distinct paths are there for the robot to reach (10, 10) starting from (0, 0)?

(*a*) 2^9 (*b*) 2^{19}

(*c*) $\binom{8}{4} \times \binom{11}{5}$ (*d*) $\binom{20}{10} - \binom{8}{4} \times \binom{11}{5}$

Question 14. What is the largest *m* such that every simple connected graph with *n* vertices and *n* edges contains at least *m* different spanning trees?

(*a*) 1 (*b*) 2

(*c*) 3 (*d*) *n*

Question 15. Consider n jobs $J_1, J_2, \ldots, J_n$ such that job J_i has execution time t_i and a non-negative integer weight *w*. The weighted mean completion time of the job is defined to be

$$\frac{\sum_{i=1}^{n} w_i T_i}{\sum_{i=1}^{n} w_i}$$

where T_i is the completion time of job J_i. Assuming that there is only one processor available, in what order must the job be executed in order to minimize the weighted mean completion time of the jobs?

(*a*) Non-decreasing order of t_i (*b*) Non-decreasing order of w_i

(*c*) Non-decreasing order of $w_i t_i$ (*d*) Non-decreasing order of with/t_i

YEAR – 2008

Question 1. The most efficient algorithm for finding the number of connected components in an undirected graph on *n* vertices and *m* edges has time complexity

(*a*) $\Theta(n)$ (*b*) $\Theta(m)$

(*c*) $\Theta(m + n)$ (*d*) $\Theta(mn)$

Question 2. The Breadth First Search algorithm has been implemented using the queue data structure. One possible order of visiting the nodes of the following graph is

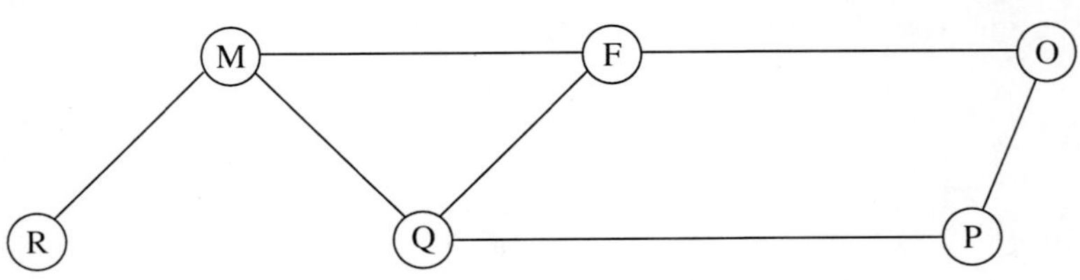

(*a*) MNOPQR (*b*) NQMPOR

(*c*) QMNPRO (*d*) QMNPOR

Question 3. Which of the following statements is true for every planar graph on n vertices?

(*a*) The graph is connected.

(*b*) The graph is Eulerian.

(*c*) The graph has a vertex-cover of size at most $3n/4$.

(*d*) The graph has an independent set of size at least $n/3$.

Question 4. Consider the following functions:

$$f(n) = 2^n$$
$$g(n) = n!$$
$$h(n) = n^{\log n}$$

Which of the following statements about the asymptotic behaviour of $f(n)$, $g(n)$, and $h(n)$ is true?

(*a*) $f(n) = O(g(n));\ g(n) = O(h(n))$ (*b*) $f(n) = \Omega(g(n));\ g(n) = O(h(n))$

(*c*) $f(n) = O(f(n));\ g(n) = O(f(n))$ (*d*) $h(n) = O(f(n));\ g(n) = \Omega(f(n))$

Question 5. The minimum number of comparisons required to determine if an integer appears more than $n/2$ times in a sorted array of n integers is

(*a*) $\Theta(n)$ (*b*) $\Theta(\log n)$

(*c*) $\Theta(\log^2 n)$ (*d*) $\Theta(1)$

Question 6. G is a graph of n vertices and $2n - 2$ edges. The edges of G can be partitioned into two edge-disjoint spanning trees. Which of the following is NOT true for G?

(*a*) For every subset of k vertices, the induced subgraph has at most $2k - 2$ edges.

(*b*) The minimum cut in G has at least two edges.

(*c*) There are two edge-disjoint paths between every pair of vertices.

(*d*) There are two vertex-disjoint paths between every pair of vertices.

Question 7. Consider the Quicksort algorithm. Suppose there is a procedure for finding a pivot element which splits the list into two sub-lists each of which contains at least one-fifth of the elements. Let $T(n)$ be the number of comparisons required to sort n elements. Then

(*a*) $T(n) \le 2T(n/5) + n$

(*b*) $T(n) \le T(n/5) + T(4n/5) + n$

(*c*) $T(n) \le 2T(4n/5) + n$

(d) $T(n) \le 2T(n/2) + n$

Question 8. Dijkstra's single source shortest path algorithm when run from vertex a in the graph, computes the correct shortest path distance to

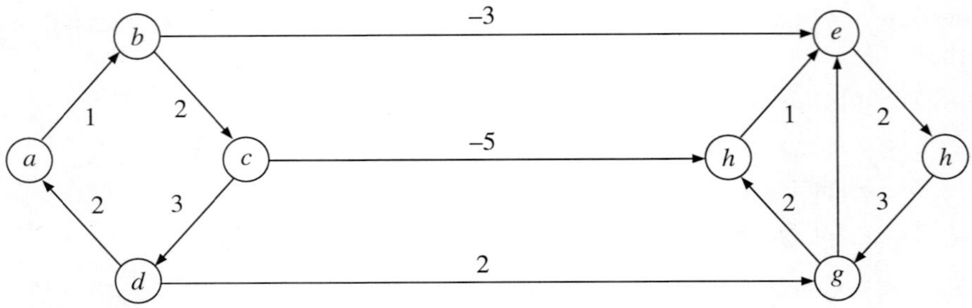

(a) only vertex a (b) only vertices a, e, f, g, h

(c) only vertices a, b, c, d (d) all the vertices

Question 9. You are given the postorder traversal, P, of a binary search tree on the n elements 1, 2, ..., n. You have to determine the unique binary search tree that has P as its postorder traversal. What is the time complexity of the most efficient algorithm for doing this?

(a) Q(log n)

(b) Q(n)

(c) Q(n log n)

(d) none of the above, as the tree cannot be uniquely determined.

Question 10. We have a binary heap on n elements and wish to insert more elements (not necessarily one after another) into this heap. The total time required for this is

(a) $\Theta(\log n)$ (b) $\Theta(n)$

(c) $\Theta(n \log n)$ (d) $\Theta(n^2)$

Question 11. What is the chromatic number of the following graph?

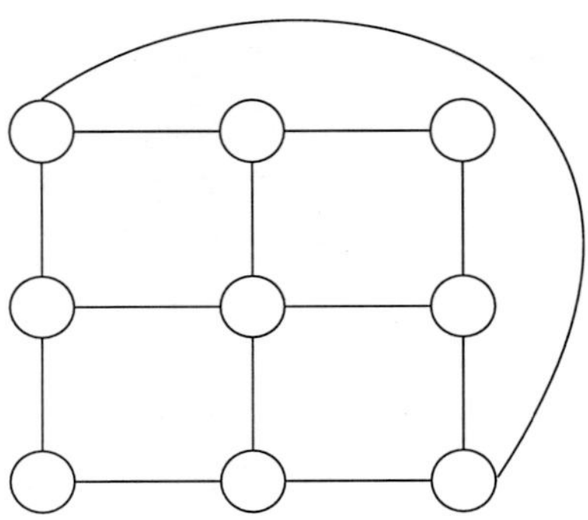

(a) 2 (b) 3

(c) 4 (d) 5

Question 12. G is a simple undirected graph. Some vertices of G are of odd degree. Add a node v to G and make it adjacent to each odd degree vertex of G. The resultant graph is sure to be

(a) regular (b) complete

(c) Hamiltonian (d) Euler

Question 13. Consider the case below.

When $n = 2^{2k}$ for some $k \geq 9$, the recurrence relation

$$T(n) = \sqrt{2T}\ (n/2) + \sqrt{n}, T(1) = 1$$

evaluates to

(a) $\sqrt{n}(\log n + 1)$ (b) $\sqrt{n} \log n$

(c) $\sqrt{n} \log \sqrt{n}$ (d) $n \log \sqrt{n}$

Question 14. For the undirected, weighted graph given below, which of the following sequences of edges represents a correct execution of Prim's algorithm to construct a Minimum Spanning Tree?

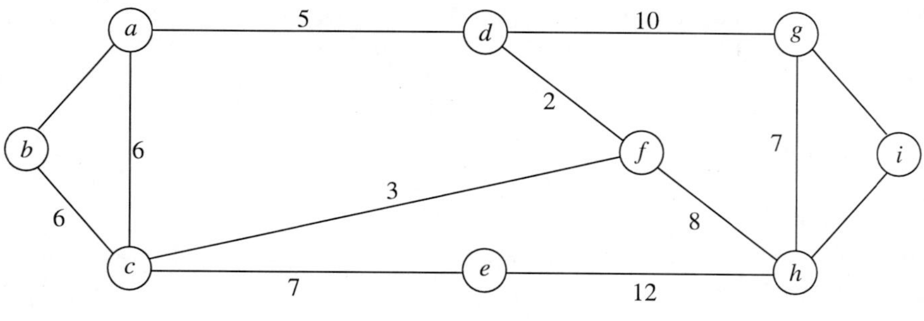

(a) (a, b), (d, f), (f, c), (g, i), (d, a), (g, h), (c, e), (f, h)

(b) (c, e), (c, f), (f, d), (d, a), (a, b), (g, h), (h, f), (g, i)

(c) (d, f), (f, c), (d, a), (a, b), (c, e), (f, h), (g, h), (g, i)

(d) (h, g), (g, i), (h, f), (f, c), (f, d), (d, a), (a, b), (c, e)

Question 15. Consider the following sequences of nodes for the undirected graph given below.

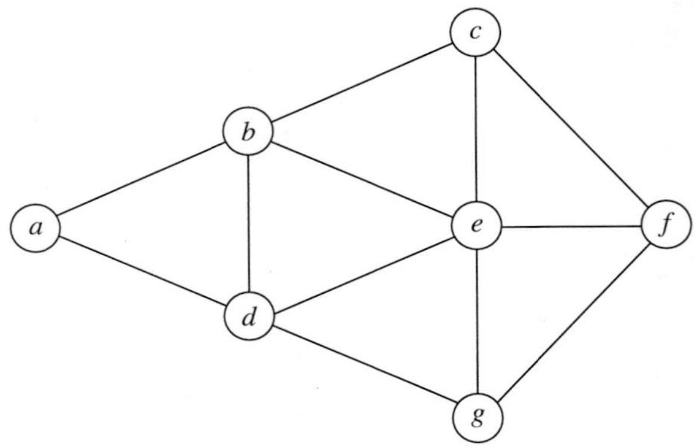

I. *a b e f d g c* II. *a b e f c g d*

III. *a d g e b c f* IV. *a d b c g e f*

A Depth First Search (DFS) is started at node *a*. The nodes are listed in the order they are first visited. Which all of the above is (are) possible output(s)?

(*a*) I and III only (*b*) II and III only

(*c*) II, III and IV only (*d*) I, II and III only

YEAR – 2009

Question 1. What is the number of swaps required to sort *n* elements using selection sort, in the worst case?

(*a*) $\Theta(n)$ (*b*) $\Theta(n \log n)$

(*c*) $\Theta(n^2)$ (*d*) $\Theta(n^2 \log n)$

Question 2. The running time of an algorithm is represented by the following recurrence relation:

$$T(n) = \begin{cases} n & n \leq 3 \\ T(n/3) + cn & \text{otherwise} \end{cases}$$

Which one of the following represents the time complexity of the algorithm?

(*a*) $\Theta(n)$ (*b*) $\Theta(n \log n)$

(*c*) $\Theta(n^2)$ (*d*) $\Theta(n^2 \log n)$

Question 3. Consider the following graph:

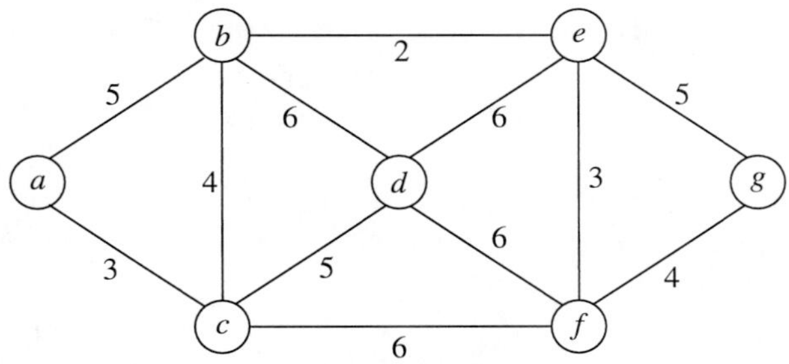

Which one of the following is NOT the sequence of edges added to the minimum spanning tree using Kruskal's algorithm?

(*a*) (*b*, *e*) (*e*, *f*) (*a*, *c*) (*b*, *c*) (*f*, *g*) (*c*, *d*)

(*b*) (*b*, *e*) (*e*, *f*) (*a*, *c*) (*f*, *g*) (*b*, *c*) (*c*, *d*)

(*c*) (*b*, *e*) (*a*, *c*) (*e*, *f*) (*b*, *c*) (*f*, *g*) (*c*, *d*)

(*d*) (*b*, *e*) (*e*, *f*) (*b*, *c*) (*a*, *c*) (*f*, *g*) (*c*, *d*)

Question 4. In quicksort, for sorting *n* elements, the (*n*/4)th smallest element is selected as pivot using an $O(n)$ time algorithm. What is the worst-case time complexity of the quicksort?

(*a*) $\Theta(n)$ (*b*) $\Theta(n \log n)$

(*c*) $\Theta(n^2)$ (*d*) $\Theta(n^2 \log n)$

Question 5. Consider a binary max-heap implemented using an array.

Question 6. Which one of the following array represents a binary max-heap?

(*a*) {25, 12, 16, 13, 10, 8, 14}

(*b*) {25, 14, 13, 16, 10, 8, 12}

(*c*) {25, 14, 16, 13, 10, 8, 12}

(*d*) {25, 14, 12, 13, 10, 8, 16}

Question 7. What is the content of the array after two delete operations on the correct answer to the previous question?

(*a*) {14, 13, 12, 10, 8}

(*b*) {14, 12, 13, 8, 10}

(*c*) {14, 13, 8, 12, 10}

(*d*) {14, 13, 12, 8, 10}

YEAR – 2010

Question 1. Let $G = (V, E)$ be a graph. Define £(G) = where is the number of vertices of degree d in G. If S and T are two different trees with £(S) = £(T), then

(a) $| S | = 2 | T |$ (b) $| S | = | T | - 1$

(c) $| S | = | T |$ (d) $| S | = | T | + 1$

Question 2. The degree sequence of a simple graph is the sequence of the degrees of the nodes in the graph in decreasing order. Which of the following sequences can not be the degree sequence of any graph?

I. 7, 6, 5, 4, 4, 3, 2, 1

II. 6, 6, 6, 6, 3, 3, 2, 2

III. 7, 6, 6, 4, 4, 3, 2, 2

IV. 8, 7, 7, 6, 4, 2, 1, 1

(a) I and II (b) III and IV

(c) IV only (d) II and IV

Question 3. Consider a complete undirected graph with vertex set {0, 1, 2, 3, 4}. Entry W_{ij} in the matrix W below is the weight of the edge $\{i, j\}$.

$$W = \begin{matrix} 0 & 1 & 8 & 1 & 4 \\ 1 & 0 & 12 & 4 & 9 \\ 8 & 12 & 0 & 7 & 3 \\ 1 & 4 & 7 & 0 & 2 \\ 4 & 9 & 3 & 2 & 0 \end{matrix}$$

What is the minimum possible weight of a spanning tree T in this graph such that vertex 0 is a leaf node in the tree T?

(a) 7 (b) 8

(c) 9 (d) 10

Question 4. What is the minimum possible weight of a path P from vertex 1 to vertex 2 in the graph in previous question such that P contains at most 3 edges?

(a) 7 (b) 8

(c) 9 (d) 10

YEAR – 2011

Question 1. K_4 and Q_3 are graphs with the following structures

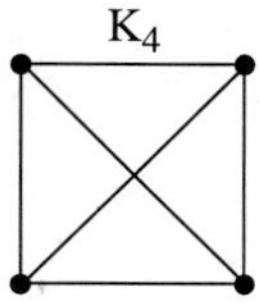

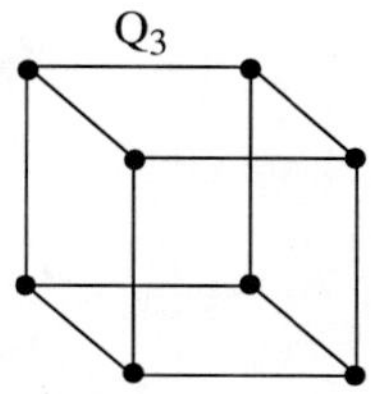

Which one of the following statements is TRUE in relation to these graphs?

(*a*) K_4 is planar while Q_3 is not

(*b*) Both K_4 and Q_3 are planar

(*c*) Q_3 is planar while K_4 is not

(*d*) Neither K_4 not Q_3 is planar

Question 2. A max-heap is a heap where the value of each parent is greater than or equal to the value of its children. Which of the following is a max-heap?

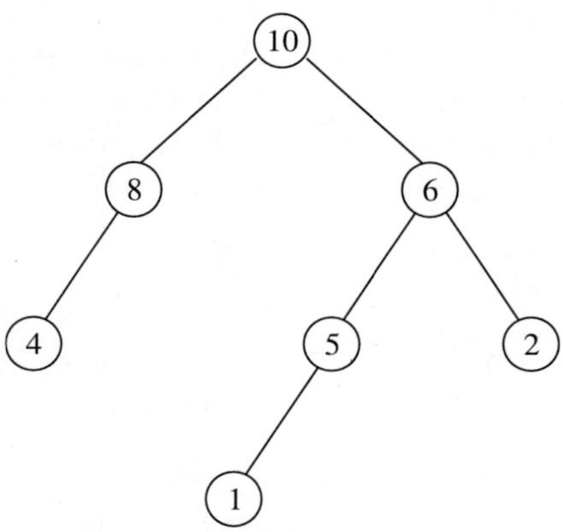

(*a*)

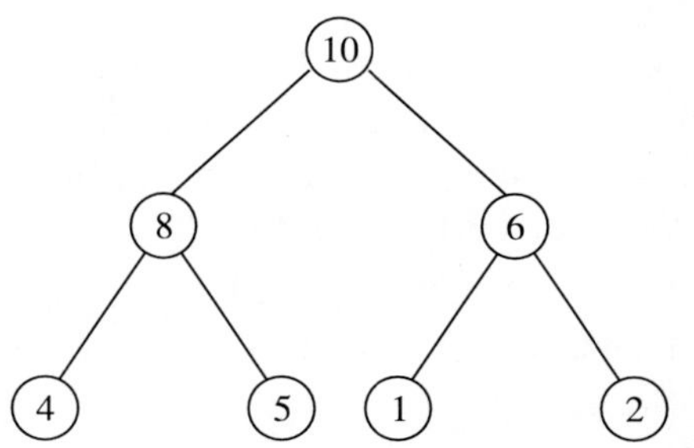

(*b*)

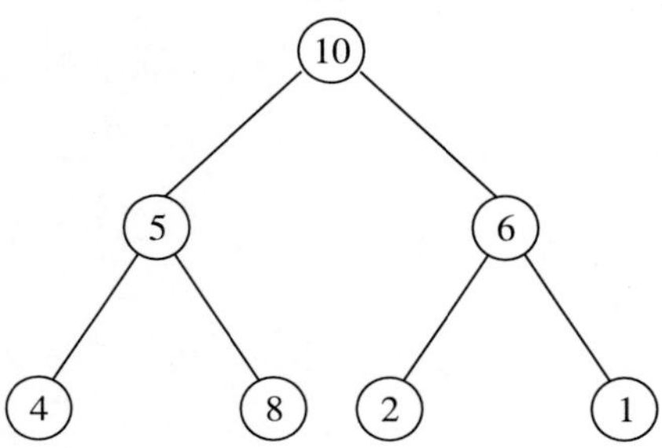

(*c*)

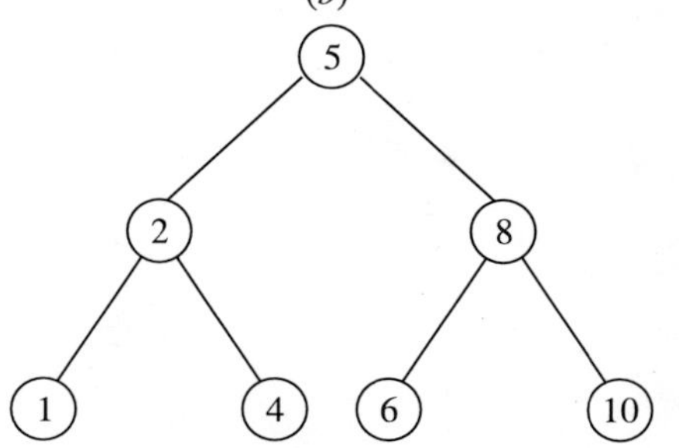

(*d*)

Question 3. An algorithm to find the length of the longest monotonically increasing sequence of numbers in an array $A[0 : n - 1]$ is given below.

Let L_i denote the length of the longest monotonically increasing sequence starting at index i in the array

Initialize $L_{n-1} = 1$

For all i such that $0 \le i \le n - 2$

$$L_i = \begin{cases} 1 + L_{i-1} & \text{if } A[i] < A[i+1] \\ 1 & \text{otherwise} \end{cases}$$

Finally the length of the longest monotonically increasing sequence is Max $\{L_1, L_2, ..., L_{n-1}\}$. Which of the following statement is TRUE?

(*a*) The algorithm uses dynamic programming paradigm

(*b*) The algorithm has a linear complexity and uses branch and bound paradigm

(*c*) The algorithm has a non-linear polynomial complexity and uses branch-and-bound paradigm

(*d*) The algorithm uses divide-and-conquer paradigm.

Question 4. Four matrices M_1, M_2, M_3 and M_4 are dimensions $p \times q$, $q \times r$, $r \times s$ and $s \times t$ respectively can be multiplied in several ways with different number of total scalar multiplications. For example, When multiplied as $((M_1 \times M_2) \times (M_3 \times M_4))$ the total number of scalar multiplications is $pqr+rst+prt$. When multiplied $((M_1 \times M_2) \times M_3) \times M_4)$, the total number of scalar multiplications is $pqr + prs + pst$. If $p = 10$, $q = 100$, $r = 20$, $s = 5$ and $t = 80$, then the minimum number of scalar multiplications needed is

(*a*) 248000 (*b*) 44000

(*c*) 19000 (*d*) 25000

Question 5. An undirected graph $G(V, E)$ contains n $(n > 2)$ nodes named $v_1, v_2, ...$ v_n. Two nodes v_i, v_j are connected if and only if $0 < |i - j| \le 2$. Each edge (v_i, v_j) is assigned a weight $i + j$. A sample graph with $n = 4$ is shown below

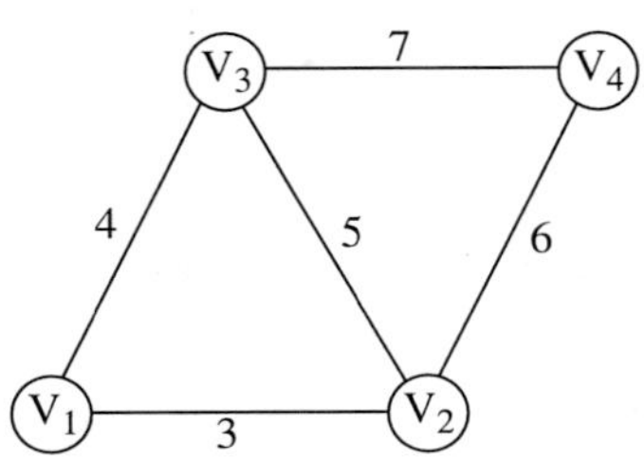

(*i*) What will be the cost of the minimum spanning tree (MST) of such a graph with n nodes?

(a) $\frac{1}{12}(11n^2 - 5n)$ (b) $n^2 - n + 1$

(c) $6n - 11$ (d) $2n + 1$

(ii) The length of the path from v_5 to v_6 in the MST of previous question with $n = 10$ is

(a) 11 (b) 25

(c) 31 (d) 41

Question 6. Consider the following recursive C function that takes two arguments

```
unsigned int foo (unsigned int n, unsigned int r)
{
    if (n>0) return (n%r)+ foo (n / r, r);
    else return 0;
}
```

(i) What is the return value of the function foo when it is called as foo (513, 2)?

(a) 9 (b) 8

(c) 5 (d) 2

(ii) What is the return value of the function foo when it is called as foo (345, 10) ?

(a) 345 (b) 12

(c) 5 (d) 3

YEAR – 2012

Question 1: Let $W(n)$ and $A(n)$ denote respectively, the worst case and average case running time of an algorithm executed on an input of size n. Which of the following is **ALWAYS TRUE**?

(a) $A(n) = \Omega(W(n))$ (b) $A(n) = \Theta(W(n))$

(c) $A(n) = O(W(n))$ (d) $A(n) = o(W(n))$

Question 2: Let G be a simple undirected planar graph on 10 vertices with 15edges. If G is a connected graph, then the number of bounded faces in any embedding of G on the plane is equal to

(a) 3 (b) 4

(c) 5 (d) 6

Question 3: The recurrence relation capturing the optimal execution time of the Towers of Hanoi problem with n discs is

(a) $T(n) = 2T(n - 2) + 2$ (b) $T(n) = 2T(n - 1) + n$

(c) $T(n) = 2T(n/2) + 1$ (d) $T(n) = 2T(n - 1) + 1$

Question 4: The worst case running time to search for an element in a balanced binary search tree with $n2^n$ elements is

(*a*) $\Theta(n \log n)$ (*b*) $\Theta(n2^n)$

(*c*) $\Theta(n)$ (*d*) $\Theta(\log n)$

Question 5: Assuming $P \neq NP$, which of the following is TRUE?

(*a*) NP-Complete = NP (*b*) NP-Complete $\cap$ $P = \Phi$

(*c*) NP-Hard = NP (*d*) P = NP-Complete

Question 6: Consider the directed graph shown in the figure below. There are multiple shortest paths between vertices S and T. Which one will be reported by Dijkstra's shortest path algorithm? Assume that, in any iteration, the shortest path to a vertex v is updated only when a strictly shorter path to v is discovered.

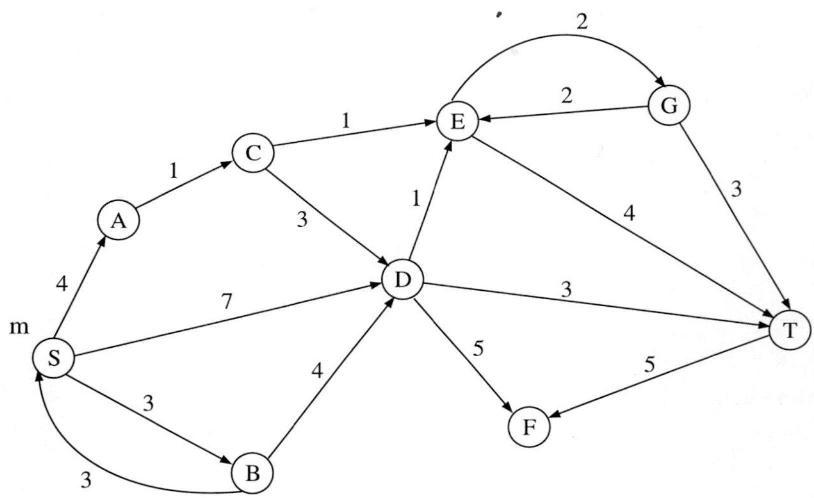

Question 7: A list of n stings, each of length n, is sorted into lexicographic order using the mergesort algorithm. The worst case running time of this computation is

(*a*) $O(n \log n)$ (*b*) $O(n^2 \log n)$

(*c*) $O(n^2 + \log n)$ (*d*) $O(n^2)$

Question 8: Let G be a complete undirected graph on 6 vertices. If vertices of G are labeled, then the number of distinct cycles of length 4 in G is equal to

(*a*) 15 (*b*) 30

(*c*) 90 (*d*) 360

Question 9: Let G be a weighted graph with edge weights greater than one and G' be the graph constructed by squaring the weights of edges in G. Let T and T' be the minimum spanning trees of G and G' respectively, with total weights t and t'. Which of the following statements is TRUE?

(a) $T' = T$ with total weight $t' = t^2$ (b) $T' = T$ with total weight $t' < t^2$

(c) $T \times T$ but total weight $t' = t^2$ (d) None of these

Question 10: Which of the following graph is isomorphic to

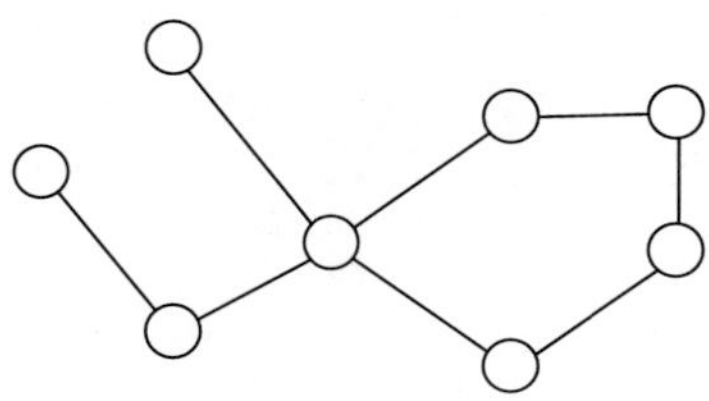

(a)

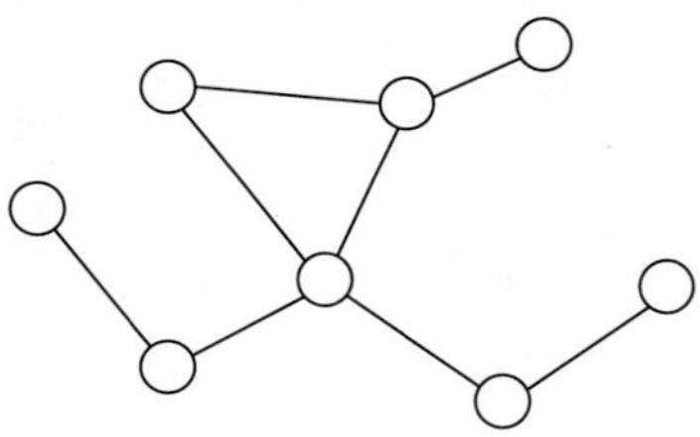

(b)

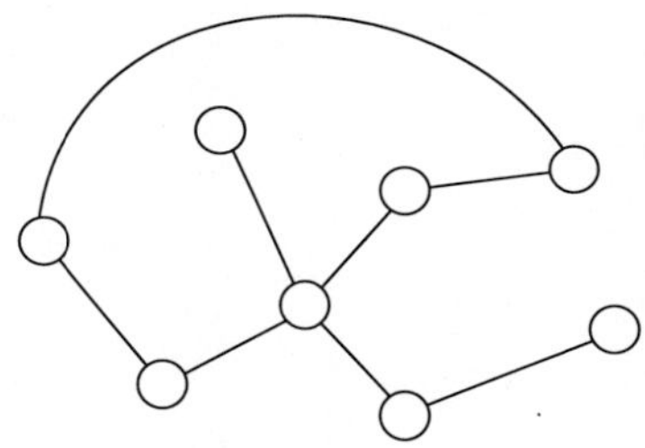

(c)

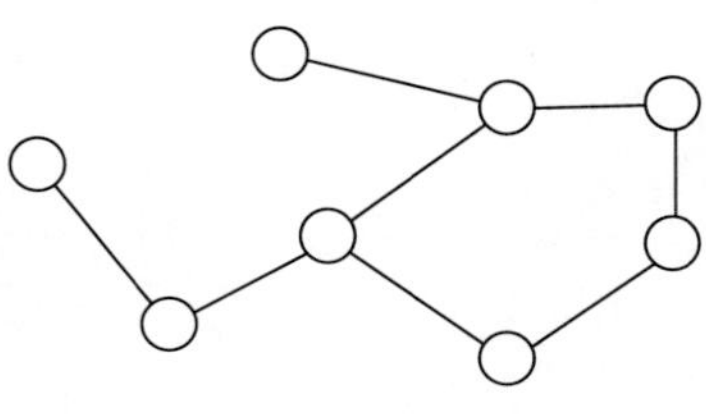

(d)

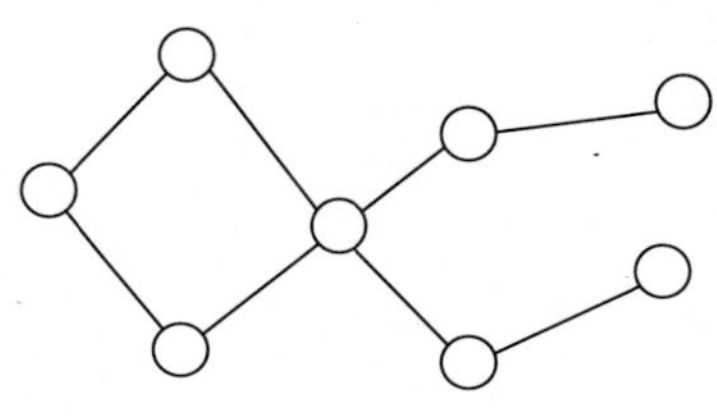

❑❑❑

MODEL QUESTION PAPERS

Model Question Paper–I

Design and Analysis of Algorithms

Time Allotted: 3 hours **Full Marks: 70**

Figure in the margin indicate full marks.
Candidates are required to give their answers in their own words as far as practicable.

GROUP–A
[Multiple Choice Questions]

1. Choose the correct alternatives. Attempt any *Ten*. **$10 \times 1 = 10$**

(*i*) Time complexity of binary searching is order of:
(*a*) log *n* (*b*) *n* (*c*) *n* log *n* (*d*) None of these

(*ii*) Which is(are) the basic property of an algorithm:
(*a*) *i*/*o* (*b*) *o*/*p* (*c*) Finiteness (*d*) All of these

(*iii*) Both space and time complexities depend on:
(*a*) Constant factor (*b*) Variable factor (*c*) Both (*b*) and (*a*) (*d*) None of these

(*iv*) Time complexity of $f(n) = 2^n + 5n^3 + 10 \log n + 7n$ is:
(*a*) $O(2^n)$ (*b*) $O(n^3)$ (*c*) $O(\log n)$ (*d*) $O(n)$

(*v*) The inequality $c_1 g(n) \le f(x) \le c_2 g(n)$ represents big:
(*a*) $\Theta(n)$ (*b*) $O(n)$ (*c*) $\Omega(n)$ (*d*) None of these

(*vi*) Chromatic number of a regular graph is:
(*a*) 1 (*b*) 2 (*c*) 3 (*d*) None of these

(*vii*) A MST is a sub-graph of a graph that contains all the:
(*a*) Vertices (*b*) Edges (*c*) Both (*a*) and (*b*) (*d*) None of these

(*viii*) Fractional Knapsack problem can be solved by:
(*a*) Greedy method (*b*) DP approach (*c*) Both (*a*) and (*b*) (*d*) None of these

(*ix*) The statement $P = NP$ is:
(*a*) Always true (*b*) Sometimes true (*c*) false (*d*) None of these

(*x*) *NP-C* is a subset of
(*a*) *NP-H* (*b*) *NP* (*c*) *P* (*d*) None of these

(*xi*) Which one is asymptotically lower?
(*a*) $O(\log n)$ (*b*) $O(n^2)$ (*c*) $O(n \log n)$ (*d*) None of these

(*xii*) Prim's –Jarnik algorithm is used to construct
(*a*) MST (*b*) BFS search tree (*c*) Both (*a*) and (*b*) (*d*) None of these

GROUP–B
[Short Answer Type Questions]

Answer any *Three* from the followings: **3 × 5 = 15**

2. Apply heap sorting on the following list: < 3, 5, 4, 7, 6, 8, 9, 13, 20, 15, 12, 8 >. 5

3. Find out the DFS traversal result of the following graph. 5

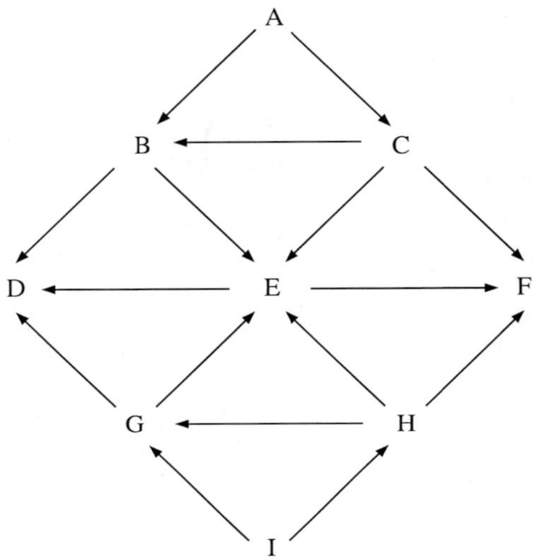

4. Find the optimal solution for the following Knapsack problem with W = 30 kg. 5

Weights	5	10	6	3	7	1	2	9
Values	5	10	6	3	7	1	2	9

5. Draw the Comparison Tree to sort elements in ascending order.

6. Solve the following recurrence using recursion tree method.

$$T(n) = T(n - a) + T(a) + cn, \text{ where } a \geq 1 \text{ and } c > 0 \text{ are constants.}$$

GROUP–C
[Long Answer Type Questions]

Answer any *Three* from the followings: **3 × 15 = 45**

7. (*a*) Define Tail Recursion.
(*b*) Write an algorithm for factorial procedure in tail recursive form.
(*c*) Show a drawback of master theorem.
(*d*) Analyse time complexity of quicksort algorithm. 2 + 4 + 2 + 7

8. (*a*) Derive asymptotic time complexity of tower of Hanoi problem.
(*b*) Write an algorithm to find the second largest element in an array of *n* numbers. Your algorithm must follow divide-and-conquer principle.

(*c*) Derive the time complexity of your algorithm used in question 8(*b*).

(*d*) Show that the space complexity of Quicksort Algorithm varies between $\Theta(\log n)$ and $\Theta(n)$. 3 + 5 + 4 + 3

9. (*a*) Write two primary criteria of dynamic programming.

(*b*) Consider the following graph. Starting from vertex 1, find the optimal tour so that a person may come back to source of journey visiting all the vertices.

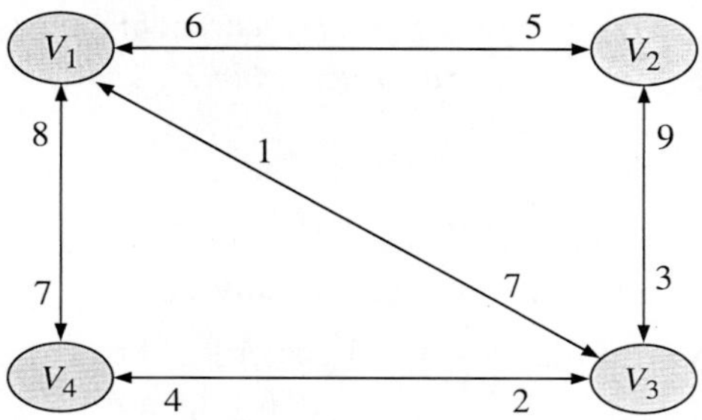

(*c*) Give different types of shortest path problems. 4 + 8 + 3

10. (*a*) What are greedy criteria for Knapsack problem?

(*b*) How does greedy method differ from dynamic programming?

(*c*) Find feasible solution for the following list of jobs. Also find the total profit earned.

Assume that each job needs one unit time in a single machine.

Job No.	I	II	III	IV	V
Profit	50	30	10	20	15
Deadline	2	1	3	2	4

5 + 3 + 7

11. (*a*) Find chromaticity of the following graph.

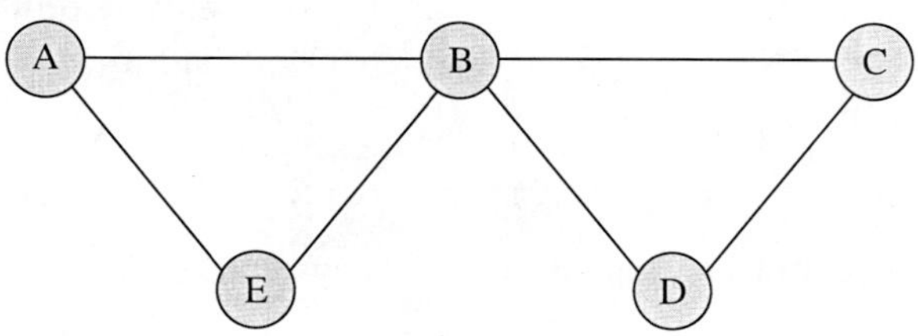

(*b*) Are *B* & *B* and Backtracking similar?

(*c*) Write Union algorithm for disjoint set using tree representation. 8 + 2 + 5

12. Write short notes on (*Any Three*) 3 × 5

(*a*) Pseudocode

(*b*) DFT

(*c*) CDP

(*d*) BFS

(*e*) Reducibility

Model Question Paper–II

Design and Analysis of Algorithms

Time Allotted: 3 hours **Full Marks: 70**

Figure in the margin indicate full marks.
Candidates are required to give their answers in their own words as far as practicable.

GROUP–A
[Multiple Choice Questions]

1. Choose the correct alternatives. Attempt any *Ten*. **10 × 1 =10**

(*i*) Time complexity of binary searching is order of:
(*a*) log *n* (*b*) *n* (*c*) *n* log *n* (*d*) None of these

(*ii*) Time complexity of $f(n) = 2^n + 5n^3 + 10 \log n + 7n$ is:
(*a*) $O(2^n)$ (*b*) $O(n^3)$ (*c*) $O(\log n)$ (*d*) $O(n)$

(*iii*) Chromatic number of a regular graph is:
(*a*) 1 (*b*) 2 (*c*) 3Group-A (*d*) None of these

(*iv*) The statement *P* = NP is:
(*a*) Always true (*b*) Sometimes true (*c*) False (*d*) None of these

(*v*) Which one is asymptotically lower?
(*a*) $O(\log n)$ (*b*) $O(n^2)$ (*c*) $O(n \log n)$ (*d*) None of these

(*vi*) C-Sat problem is:
(*a*) NP-H (*b*) NP-C (*c*) NP (*d*) None of these

(*vii*) MCM problem can be solved efficiently using:
(*a*) DP approach (*b*) Strassen's algorithm
(*c*) Greedy (*d*) None of these

(*viii*) Complexity of HCP is:
(*a*) $O(n^2 \log n)$ (*b*) $O(n^2 2^n)$ (*c*) $O(n^n)$ (*d*) None of these

(*ix*) Minimum cost of a spanning tree is always equal when computed with any method.
(*a*) True (*b*) False (*c*) Sometimes (*d*) None of these

(*x*) Running time of binary searching for 10 elements will be
(*a*) 2.6 (*b*) 2.7 (*c*) 2.8 (*d*) 2.9

(*xi*) Which of the followings is more suitable for graph representation
(*a*) Adjacency Matrix (*b*) Adjacency List
(*c*) Both (*a*) and (*b*) (*d*) None of these

GROUP–B
[Short Answer Type Questions]

Answer any *Three* from the followings: **3 × 5 = 15**

2. Analyse Time complexity of mergesort.

3. Compare *B*-Search with *D*-Search

4. Prove that: $f(n) = a_m n^m + a_{m-1} n^{m-1} + \ldots + a_1 n + a_0$ implies $f(n) = O(n^m)$.

5. Write down algorithm for 15-puzzle problem.

6. Define tail recursion with an example.

GROUP–C
[Long Answer Type Questions]

Answer any *Three* from the followings: **3 × 15 = 45**

7. (*a*) Illustrate the operation of Max-Heapify on the array *A* = < 7, 17, 3, 16, 15, 11, 13, 10, 1, 5, 7, 12, 4, 14, 8, 9, 10, 18.

(*b*) Let H be the heap shown in Figure 1. What will be the heap result when

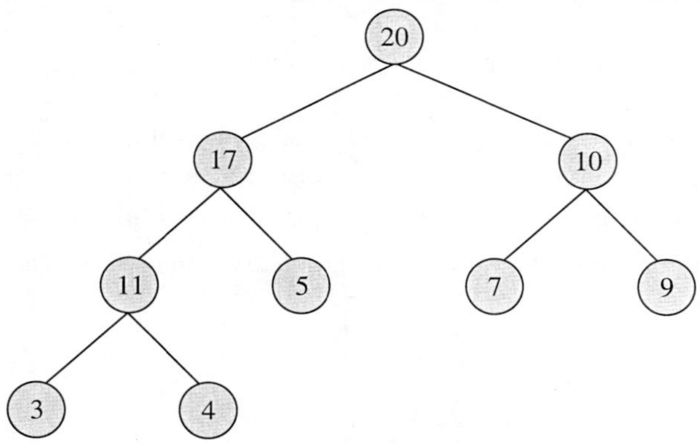

Figure 1

(*i*) Deleting the element with key 17. (*ii*) Inserting an element with key 15.

(*iii*) Inserting an element with key 19. (*iv*) Deleting the element with key 20.

(*c*) Given a set of integers. Give an efficient algorithm that determines whether two of them are equal. 5 + 6 + 4

8. (*a*) Fibonacci series is defined as follows:

$f(0) = 0$ $f(1) = 1$

$f(n) = f(n-1) + f(n-2)$. Analyze its running-time.

(*b*) Suppose $T(n) = 8T(n/2) + n^2$ describes the running time of an algorithm *A*. Another competing algorithm *A′* has running time $T'(n) = aT'(n/4) + n^2$. Find largest integer value of *a* such that *A′* is asymptotically faster than *A*?

(*c*) Solve the following recurrences.

(*i*) $T(n) = \begin{cases} \Theta(1), n = 1 \\ 2T\dfrac{n}{2} + \Theta(n), n > 1 \end{cases}$ (*ii*) $T(n) = \begin{cases} \Theta(1), n = 1 \\ T(n-1) + \Theta(n), n > 1 \end{cases}$

(*iii*) $T(n) = 16T(n/4) + n!$ 5 + 5 + 5

9. (*a*) Find shortest path between all pairs on the weighted directed graph shown in Fig 2.

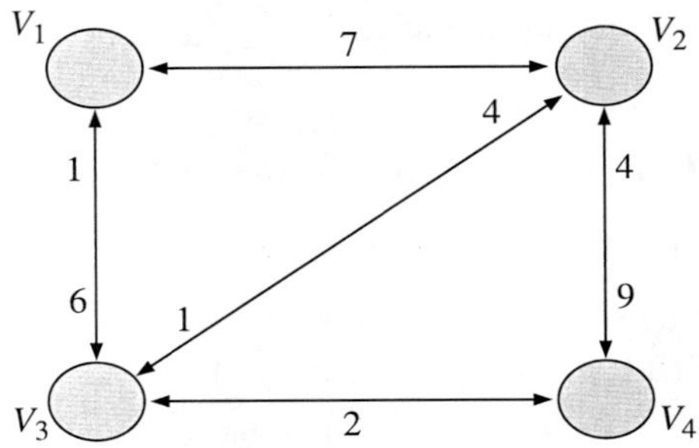

Figure 2

(*b*) Consider the *money change* problem. We have a currency system that has n coins with values $v_1, v_2 \ldots v_n$ where $v_1 = 1$, and we want to pay change of value X in such a way that the total number of coins is minimised.

(*i*) Give a dynamic programming algorithm to solve this problem.

(*ii*) What are the time and space complexities of your algorithm?

(*c*) Consider the table that represents a graph given below. Find a MST if possible.

$$\begin{bmatrix} 0 & 2 & 4 & 6 \\ 2 & 0 & 3 & 1 \\ 5 & 9 & 0 & 4 \\ 8 & \infty & 3 & 0 \end{bmatrix}$$

5 + 5 + 5

10. (*a*) Find Hamiltonian Cycle if exists in the graph given below in Figure 3.

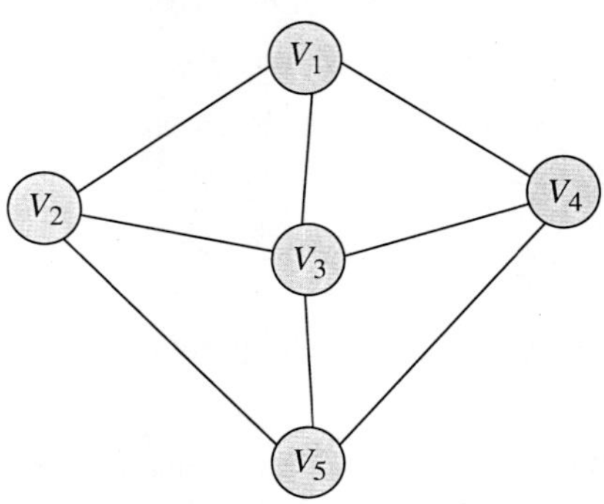

Figure 3

(*b*) Find chromatic number of the following graph in Figure 4.

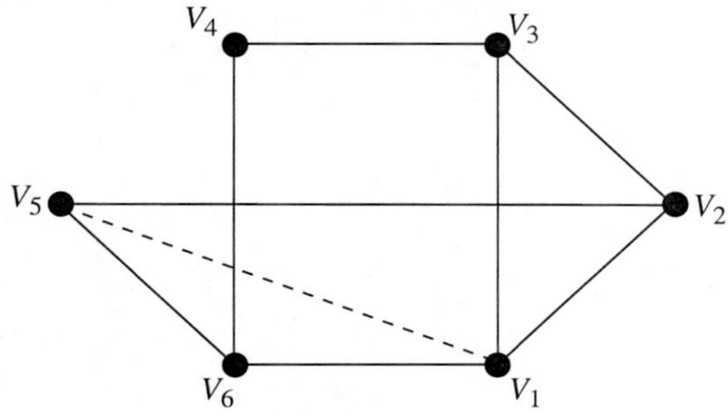

Figure 4

(*c*) Draw an optimum space search tree for 4-queens' problem. 5 + 5 + 5

11. (*a*) Draw the decision tree for the algorithm: Linear search on five elements.

(*b*) What are the ranks of nodes 3, 7 and 5 in Figure 5.

(*c*) Let x be a node in tree T resulting from a sequence of unions and finds using both the heuristics of union by rank and path compression. Prove that rank (x) is an upper bound on the height of x.

(*d*) Run depth-first search on the undirected graph shown in Figure 6. Starting vertex is A. Show the result.

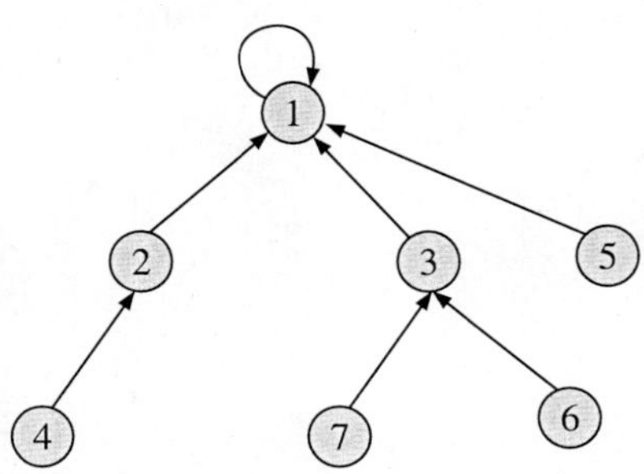

Figure 5

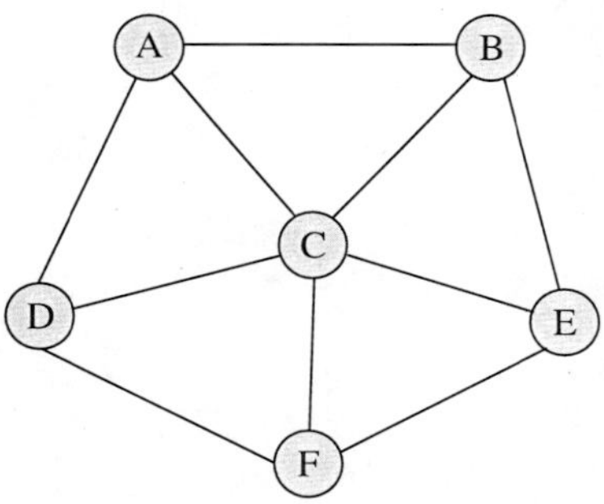

Figure 6

4 + 3 + 4 + 4

12. (*a*) State Horner's rule. Why does it necessary?

(*b*) Show that the Clique problem is *NP*.

(*c*) Write short notes on:

(*i*) Strassen's Algorithm.

(*ii*) Oracles and Adversary Arguments 4 + 3 + 4 + 4

PREVIOUS YEARS EXAMINATION PAPERS

CS/B.Tech(CSE)/SEM-5/CS-503/2009-10
Design and Analysis of Algorithms

Time Allotted : 3 Hours **Full Marks : 70**

The figure in the margin indicate full marks.
Candidates are required to give their answers in their own words as far as practicable.

GROUP–A
(Multiple Choice Type Questions)

1. Choose the correct alternatives for the following: **10 × 1 = 10**

(*i*) Lower bound of any comparison sort is
(*a*) $O(\log n)$ (*b*) $O(n^2)$ (*c*) $O(n \log n)$ (*d*) $O(n^2 \log n)$.

(*ii*) $o(g(n))$ is [Read as small *oh* of $g(n)$ is]
(*a*) Asymptotically loose (*b*) Asymptotically tight
(*c*) Same as big *oh* (*d*) None of these.

(*iii*) Kruskal algorithm is a:
(*a*) Divide-and-conquer algorithm (*b*) Branch-and-bound algorithm
(*c*) Greedy algorithm (*d*) Dynamic programming.

(*iv*) Travelling salesman problem belongs to:
(*a*) *P* class (*b*) *NP* class (*c*) *NP*-Hard (*d*) *NP*-complete class.

(*v*) Time complexity of insertion sort is:
(*a*) Linear (*b*) Quadratic (*c*) Cubic (*d*) Exponential.

(*vi*) Which one of the following function is asymptotically smallest?
(*a*) 2^n (*b*) $n^{\log n}$
(*c*) $n^{\sqrt{n}}$ (*d*) $(100)^{(\log n)^{1/3} + (\log \log n)^{2/3}}$.

(*vii*) Which one of the *following* statements is correct ?
(*a*) If $A \le_p B$ and $B \in P$ then $A \in P$ (*b*) If $A \le_p B$ and $A \notin P$ then $B \notin P$
(*c*) If $A \le_p B$ and $B \le_p C$ then $A \le_p C$ (*d*) All of these.

(*viii*) Consider the following statements
(*I*) JVP hard problem is a subset of *NP* complete problem,
(*II*) An algorithm to multiply two matrices has complexity $O(n^3)$.
Which, of the following alternatives is true,
(*a*) I-True, II-False (*b*) Both true
(*c*) Both False (*d*) I-False, II-True.

(*ix*) Optimal substructure property is exploited by
(*a*) Dynamic programming (*b*) Greedy method
(*c*) Both (*a*) and (*b*) (*d*) None of these.

(*x*) Which of the following approaches is adopted in Divide-and-Conquer algorithms?
(*a*) Top-down (*b*) Bottom-up
(*c*) Both (*a*) and (*b*) (*d*) None of these.

GROUP–B
(Short Answer Type Questions)

2. (*a*) Derive the complexity of mergesort.
(*b*) What is the difference between a 0-1 Knapsack problem and a fractional Knapsack problem? 4 + 1

3. Write an algorithm for eight queens problem.

4. State master's theorem and find the time complexity for the following recurrence:
$$T(n) = 2T(n^{1/2}) + \log n$$

5. (*a*) What are the basic characteristics of dynamic programming?
(*b*) Write an algorithm for matrix-chain multiplication. 2 + 3

6. Apply backtracking technique to solve the 3-colouring problem for the following graph.

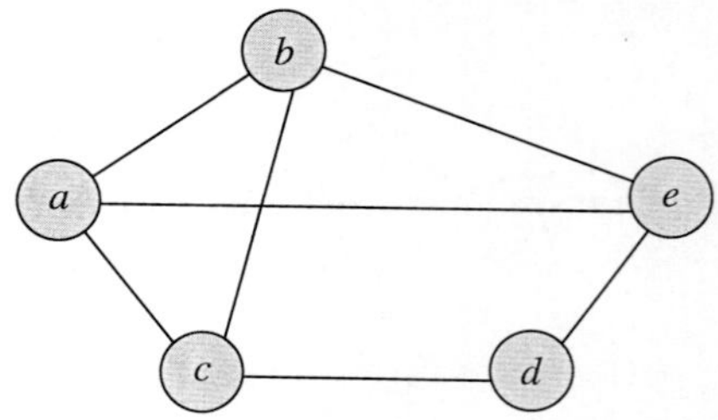

GROUP–C
Long Answer Type Questions

Answer any *three* of the following: 3 × 15 = 45

7. (*a*) Define the classes *P* and *NP*.
(*b*) Discuss what you mean by polynomial reductions.
(*c*) Discuss diagrammatically the relations among *P* class *NP* class, *NP*-hard and *NP*-complete.
(*d*) Describe Clique Decision Problem (CDP).
(*e*) Prove the CDP is *NP*-complete. 2 + 2 + 2 + 2 + 7

8. (*a*) State the general Knapsack problem. Write a greedy algorithm for this problem and derive its time complexity.
(*b*) Given the weight vector (2, 3, 5, 7, 1, 4, 1) and the profit vector { 10, 5, 15, 7, 6, 8, 18, 3) and a Knapsack of capacity 15. find at least three feasible solutions including optimal one for the knapsack problem of seven objects. 10 + 5

9. Write the algorithm of Quicksort. Find the best case, worst case arid average case time complexities of this algorithm. 5 + 10

10. (*a*) Explain how do you attempt to solve 15-puzzle problem using branch and bound strategy. Draw a portion of the state space generated by it.
(*b*) Write an algorithm for finding the minimum spanning , tree of a graph. Discuss its time complexity. 8 + 7

11. (*a*) What do you mean by non-deterministic algorithms?
(*b*) How are graphs represented in computer?
(*c*) Write short notes on any three of the following:
(*i*) Recursion tree (*ii*) Approximation schemes
(*iii*) Turing machines (*iv*) FFT algorithm. 3 + 3 + (3 × 3)

Solutions 2009: CS - 503

GROUP–A

1. The possible correct answers are the following:

(*i*) (*c*)	(*ii*) (*a*)	(*iii*) (*c*)	(*iv*) (*d*)	(*v*) (*b*)
(*vi*) (*b*)	(*vii*) (*d*)	(*viii*) (*d*)	(*ix*) (*a*)	(*x*) (*a*)

GROUP–B

2. (*a*) Refer chapter 4 section 4.7.1 analysis part.
(*b*) Refer chapter 6 section 6.13.

3. Refer chapter 7 section 7.4.3. Consider 8 in place of *N* as input to the algorithm.

4. Refer chapter 3 section 3.8.3.4.

Given, $T(n) = 2T(\sqrt{n}) + \log n$

Say, $m = \log n \Rightarrow n = 2^m$

Now, $\sqrt{n} = 2^{m/2}$

Thus, the recurrence is:

$$T(2^m) = T(2^{m/2}) + m$$

Let, $T(2^m) = S(m) \Rightarrow T(2^{m/2}) = S(m/2)$

Then, the recurrence will be:

$S(m) = 2S(m/2) + m = \Theta(m \log m)$ [using master theorem]

$= \Theta(\log n \log (\log n))$

[putting $m = \log n$]

5. (*a*) Refer chapter 5 section 5.10.
(*b*) Refer chapter 5 section 5.5.4.

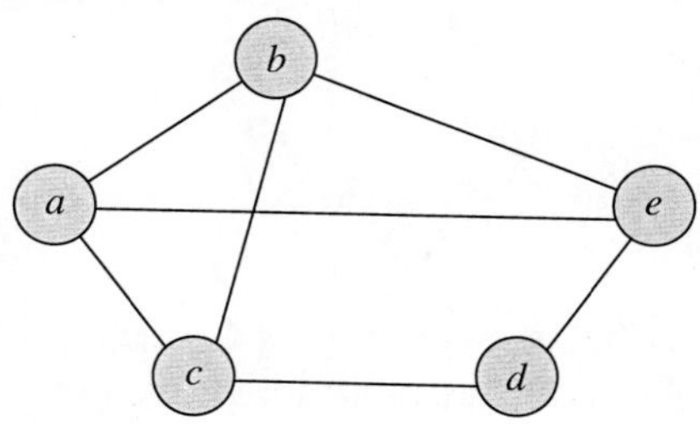

6. Given, the graph

Say, starting vertex is *a*.

Let C = Set of different colours used and
S = Set of vertices having same colour.

Both are initially empty ().

Using backtracking,

Step 1: Colour vertex *a* with colour say, Black.

$C = C \cup \{\text{Black}\} = \{\emptyset\} \cup \{\text{Black}\}$
$= \{\text{Black}\}$

$S = S \cup \{a\} = \{\emptyset\} \cup \{a\} = \{a\}$

Explore vertex *a*.

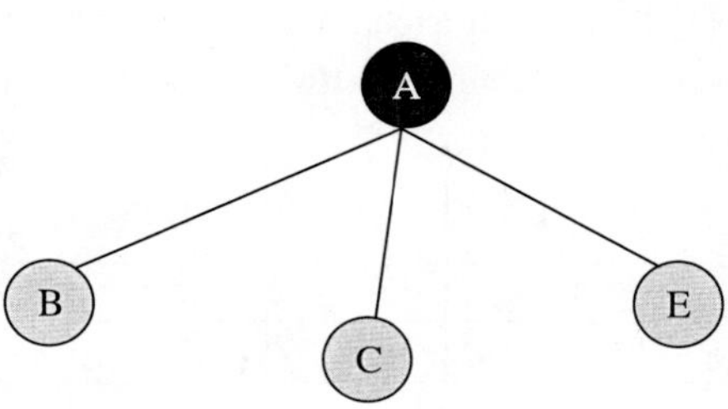

Step 2: Colour vertex b with a new colour say, Green as it is adjacent of vertex a and there is only one colour in C. C = {Black, Green} $S = \{a, b\}$ Explore vertex b.

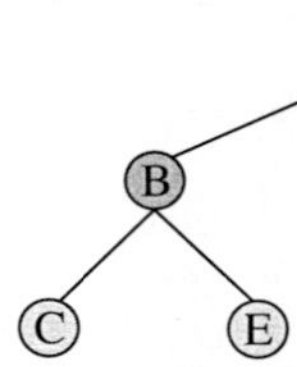

Step 3: Colour vertex c with a new colour say, Red as it is adjacent of vertices a and b.
C = {Black, Green, Red}, $S = \{a, b, c\}$ Explore vertex c.

Step 4: Colour vertex d with colour either Black or Green as it is neither adjacent vertex a nor vertex b. Say colour considered is Green.

C = {Black, Green, Red}, $S = \{a, b, c, d\}$ Explore vertex d. No exploration possible. So backtrack to vertices c and b.

Step 5: Colour vertex e with colour Red as it is neither adjacent vertex c nor vertex d. Green cannot be considered as it is used by vertex b.

C = {Black, Green, Red}, $S = \{a, b, c, d, e\}$. Thus, all the vertices have been coloured. No exploration requires.

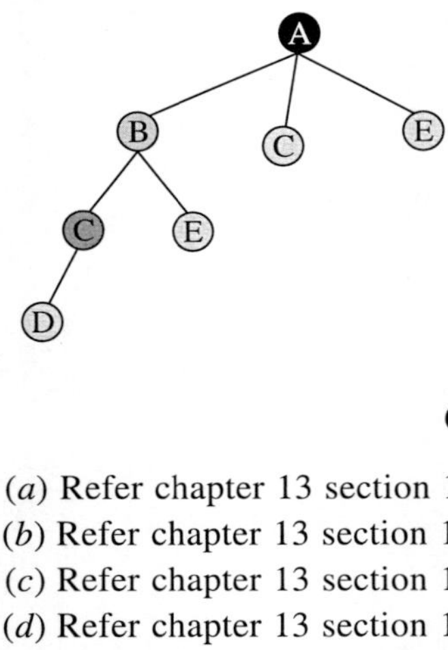

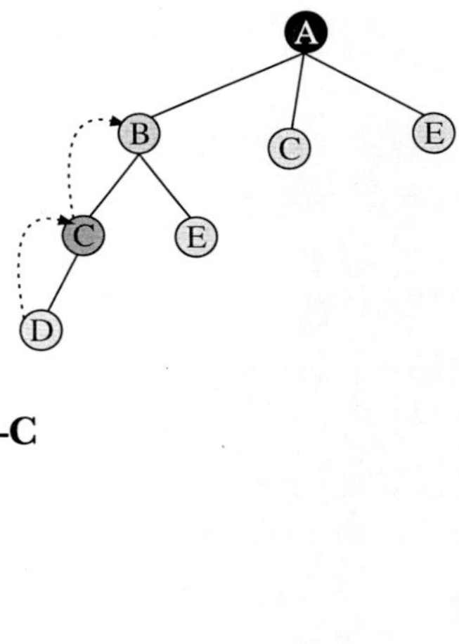

GROUP–C

7. (*a*) Refer chapter 13 section 13.4
   (*b*) Refer chapter 13 section 13.6
   (*c*) Refer chapter 13 section 13.5
   (*d*) Refer chapter 13 section 13.10
   (*e*) Refer chapter 13 section 13.10

8. (*a*) Refer chapter 6 section 6.10
   (*b*) Construct the following table of then given data as.

Index	1	2	3	4	5	6	7
Items	I_1	I_2	I_3	I_4	I_5	I_6	I_7
Weights	2	3	5	7	1	4	1
Profit	10	5	15	7	6	18	3

Knapsack capacity $W = 15$. Say P be the earned profit value, initially 0.

Solution: using greedy criterion I (Arrange items in descending order of their profit values), the table will be as.

Index	1	2	3	4	5	6	7
Items	I_6	I_3	I_1	I_4	I_5	I_6	I_7
Weights	4	5	2	7	1	3	1
Profit	18	15	10	7	6	5	3

Now, include first item (*i.e.*, I_6) into Knapsack.

$$W = W - W[I_6] = 15 - 4 = 11$$
$$P = P + \text{Profit }[I_6] = 0 + 18 = 18$$

Take next item I_3 into Knapsack.

$$W = W - W[I_3] = 11 - 5 = 6$$
$$P = P + \text{Profit }[I_3] = 18 + 15 = 33$$

Take next item I_1 into Knapsack.

$$W = W - W[I_1] = 6 - 2 = 4$$
$$P = P + \text{Profit }[I_1] = 33 + 10 = 43$$

Next item I_4 cannot be put into Knapsack because its weight > remained Knapsack capacity.

First feasible solution: items included are I_6, I_3, and I_1.

Similarly apply greedy criteria II and III for 2nd and 3rd feasible solutions.

9. Refer chapter 4 section 4.6.3

10. (*a*) Refer chapter 8 section 8.5
(*b*) Refer chapter 6 section 6.12

11. (*a*) Refer chapter 13 section 13.3
(*b*) Refer chapter 11 section 11.5
(*c*) For short notes refer
(*i*) chapter 3 section 3.8.3.3
(*ii*) chapter 14 section 14.3
(*iii*) chapter 1 section 1.10.1
(*iv*) chapter 12 section 12.7

Design and Analysis of Algorithms
CS/B.Tech(CSE)/SEM-8/IT-803C/2010

Time Allotted : 3 Hours **Full Marks : 70**

The figure in the margin indicate full marks.
Candidates are required to give their answers in their own words as far as practicable.

GROUP–A
(Multiple Choice Type Questions)

1. Choose the correct alternatives for the following: **10 × 1 = 10**

(*i*) To move 5 disk from BEG to END using AUX peg. The Total no. of Movement required:

(*a*) 32 (*b*) 64 (*c*) 31 (*d*) 63.

(*ii*) O-notation (Big Oh) is Asymptotically.

(*a*) Tight (*b*) Upper (*b*) Lower (*c*) Loose.

(*iii*) Average case the complexity of binary search is:

(*a*) $O(n)$ (*b*) $O(n \log n)$ (*b*) $O(\log n)$ (*d*) $O(n^2)$.

(*iv*)

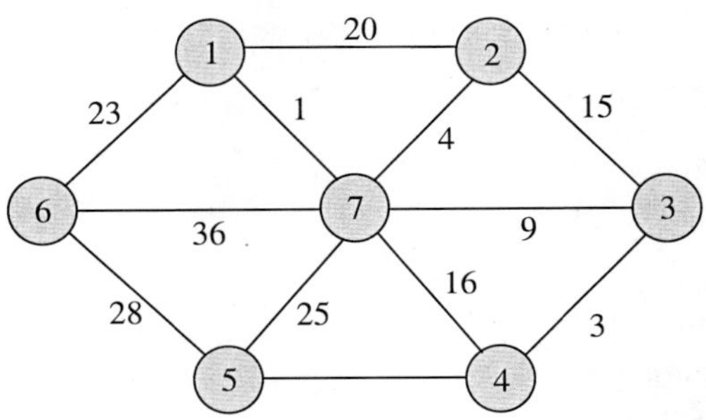

Total cost of the spanning tree for the above graph using Kruskal's algorithm is:

(*a*) 57 (*b*) 55

(*c*) 64 (*c*) 52.

(*v*) Lower bound of time camplexity for any comparison based sorting algorithm is:

(*a*) $O(\log n)$ (*b*) $O(n)$

(*c*) $O(n \log n)$ (*d*) $O(n^2)$.

(*vi*) Space complexity of Mergesort is:

(*a*) $O(\log n)$ (*b*) $O(n)$

(*c*) $O(n \log n)$ (*d*) $O(n^2)$.

(*vii*)

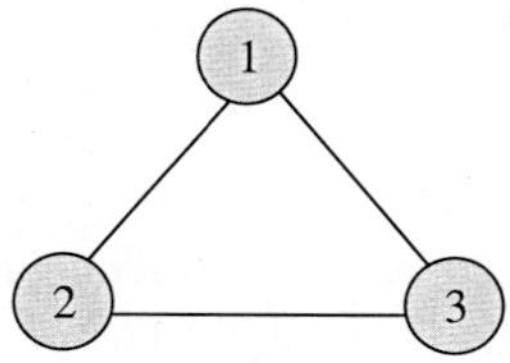

The chromatic number of the above graph is:

(*a*) 1 (*b*) 2

(*c*) 3 (*d*) 4.

(*viii*) ...., is used in BFS.

(*a*) Stack (*b*) Queue

(*c*) Link (*d*) None of these.

(*ix*) Time complexity of the following recurrence $T(n) = 9T(n/3) + n$ is:

(*a*) $O(n \log n)$ (*b*) $\theta(n^2)$

(*c*) $\theta(\log n)$ (*d*) $O(n^3)$.

(*x*) Which edge removal makes the graph disconnected?

(*a*) Pendant vertex (*b*) Bridge

(*c*) Articulation (*d*) Coloured vertex.

GROUP–B
(Short Answer Type Questions)

2. (*a*) Derive the complexity of mergesort.

(*b*) What is the difference between a 0-1 Knapsack problem and a fractional Knapsack problem? 4 + 1

3. Write an algorithm for eight queens problem. 5

4. State Master's theorem and find the time complexity for the following recurrence

$$T(n) = 3T\ (n/4) + n \log\ (n).$$ 5

5. (*a*) What are the basic characteristics of dynamic programming?

(*b*) Write down Bellman and Ford algorithm for single source shortest path.

2 + 3

6. Write an algorithm to find a minimal spanning tree of undirected graph. Estimate the time complexity of your algorithm. 5

GROUP–C
(Long Answer Type Questions)

Answer any *three* of the following **3 × 15 = 45**

7. (*a*) Define the classes *P* and *NP* and *NO* complete.

(*b*) Discuss what you mean by polynomial reductions.

(*c*) Discuss diagrammatically the relations among *F* class, *NP* class, *NP* hard and *NP* complete.

(*d*) Describe circuit satisfiability prove that circuit satisfiability is in *NP*.

3 + 2 + 3 + 7

8. (*a*) What do you mean by dynamic programming ? What is the difference between dynamic programming and greedy method ?

(*b*) Write an algorithm for Quicksort and find the average case time complexity of this algorithm.

(*c*) Find an optimal parenthesisation of a matrix-chain product whose sequence of dimension is (5, 10, 3, 12, 4, 50). (1 + 2) + 5 + 7

9. (*a*) Discuss the procedure for Strassen's matrix multiplication to evaluate the product of *n* matrices. Find the resulting recurrence relation for the same and analyse its time complexity. Justify that this method is an improvement over the conventional matrix multiplication method.

(*b*) What is *UNION-FIND* algorithm? (7 + 1 + 2 + 2) + 3

10. (*a*) How can you find a solution of simultaneous linear equation using LUP Decomposition ?

(*b*) Given weight vector (15, 25, 35, 45, 55) and the profit vector (10, 20, 30, 40, 50) and a Knapsack of capacity 100, find at least three feasible solutions including optimal one for the Knapsack problem of 5 items.

(*c*) Discuss activity selection problem for job sequencing with an example.

3 + 7 + 5

11. Write short notes on any three of the following: 3 × 5

(*a*) RAM model

(*b*) Turing Machine

(*c*) Eight queens problem

(*d*) Approximations algorithm

(*e*) Asymptotic notation.

Solutions 2010: IT-803C

GROUP–A

1. Correct solutions are the followings.

(*i*) (*c*) (*ii*) (*b*) (*iii*) (*c*) (*iv*) (*a*) (*v*) (*c*)
(*vi*) (*b*) (*vii*) (*b*) (*viii*) (*b*) (*ix*) (*b*)

GROUP–B

2. (*a*) Refer chapter 4 section 4.7.1 analysis part.
(*b*) Refer chapter 6 section 6.13

3. Refer chapter 7 section 7.4.3. Consider 8 in place of N as input to the algorithm.

4. Refer chapter 3 section 3.8.3.4

Given, $T(n) = 3T(n/4) + n \log n$

Using master theorem,

$$a = 3,\ b = 4 \Rightarrow E = \log_b a = \log_4 3 = 0.792$$

$$f(n) = n \log n > n^E \Rightarrow f(n) = O(n^{E+e}) \qquad [e \text{ is constant} > 0]$$

$$af(n/b) = 3\left(\frac{n}{4}\log\frac{n}{4}\right) = \frac{3}{4}n(\log n - \log 4) \le \frac{3}{4}n \log n \le cn \log n \le cf(n) \qquad [c \text{ is const.}]$$

Thus, third condition is applicable.

Therefore, $T(n) = \Theta(f(n)) = \Theta(n \log n)$.

5. (*a*) Refer chapter 5 section 5.10. (*b*) Refer chapter 5 section 5.8.1.

6. Refer chapter 6 section 6.12.5.

GROUP–C

7. (*a*) Refer chapter 13 section13.4. (*b*) Refer chapter 13 section13.6.
(*c*) Refer chapter 13 section 13.5. (*d*) Refer chapter 13 section 13.8 and 13.9

8. (*a*) Refer chapter 5 section 5.1 and chapter 6 section 6.8
(*b*) Refer chapter 4 section 4.6.3.
(*c*) Given the matrices dimension sequence < 5, 10, 3, 12, 4, 50 >. Suppose the matrices are as given below:

Matrix	*Dimension*
A	5×10
B	10×3
C	3×12
D	12×4
E	4×50

Now, $P = <5, 10, 3, 12, 4, 50>$.

$m[i][j]$ using $m[i][j] = m[i][k] + m[k+1][j] + Pi\text{-}1PkPj$ for $k = i$ to $j - 1$ is computed as

$m\ [1]\ [2] = P_0P_1P_2 = 5 \times 10 \times 3 = 150 \quad k = 1$

$m\ [2]\ [3] = P_1P_2P_3 = 10 \times 3 \times 12 = 360 \quad k = 2$

$m\ [3]\ [4] = P_2P_3P_4 = 3 \times 12 \times 4 = 144 \quad k = 3$

$m\ [4]\ [5] = P_3P_4P_5 = 12 \times 4 \times 50 = 2400\ k = 4$

$m[1]\ [3] = \min\ \{m[1]\ [1] + m[2]\ [3] + P_0P_1P_3,\ m[1]\ [2] + m[3]\ [3] + P_0P_2P_3\}$

$= \min\ \{0 + 360 + 5 \times 10 \times 12,\ 150 + 0 + 5 \times 3 \times 12\}$

$= 330 \qquad k = 2$

$m[2]\ [4] = \min\ \{m[2]\ [2] + m[3]\ [4] + P_1P_2P_4,\ m[2]\ [3] + m[4]\ [4] + P_1P_3P_4\}$

$= \min\ \{0 + 144 + 10 \times 3 \times 4,\ 360 + 0 + 10 \times 12 \times 4\}$

$= 264 \qquad k = 2$

$m[3]\ [5] = \min\ \{m[3]\ [3] + m[4]\ [5] + P_2P_3P_5,\ m[3]\ [4] + m[5]\ [5] + P_2P_4P_5\}$

$= \min\{0 + 2400 + 3 \times 12 \times 50,\ 144 + 0 + 3 \times 4 \times 50\}$

$= 744 \qquad k = 4$

$m[1]\ [4] = \min\ \{m[1]\ [1] + m[2]\ [4] + P_0P_1P_4,\ m[1]\ [2] + m[3]\ [4] + P_0P_2P_4,\ m[1]\ [3] + m[4]\ [4] + P_0P_3P_4\}$

$= \min\{0 + 264 + 5 \times 10 \times 4,\ 150 + 144 + 5 \times 3 \times 4,\ 330 + 0 + 5 \times 2 \times 4\}$

$= 354 \qquad k = 2$

$m[2]\ [5] = \min\ \{m[2]\ [2] + m[3]\ [5] + P_1P_2P_5,\ m[2]\ [3] + m[4]\ [5] + P_1P_3P_5,\ m[2]\ [4] + m[5]\ [5] + P_1P_4P_5\}$

$= \min\ \{0 + 744 + 10 \times 3 \times 50,\ 360 + 2400 + 10 \times 12 \times 50,\ 264 + 0 + 10 \times 4 \times 50\}$

$= 2244 \qquad k = 2$

$m[1]\ [5] = \min\ \{m[1]\ [1] + m[2]\ [5] + P_0\ P_1P_5,\ m[1]\ [2] + m[3]\ [5] + P_0P_2P_5,\ m[1]\ [3] + m[4]\ [5] + P_0P_3P_5,\ m[1]\ [4] + m[5]\ [5] + P_0P_4P_5\}$

$= \min\ \{0 + 2244 + 5 \times 10 \times 50,\ 150 + 744 + 5 \times 3 \times 50,\ 330 + 2400 + 5 \times 12 \times 50,\ 354 + 0 + 5 \times 4 \times 50\ \}$

$= 1354 \qquad k = 4$

Therefore, the cost of scalar multiplications = 1354

The solution matrix *S* can be formed as below.

i / *j*	2	3	4	5
1	1	2	2	4
2		2	2	2
3			3	4
4				4

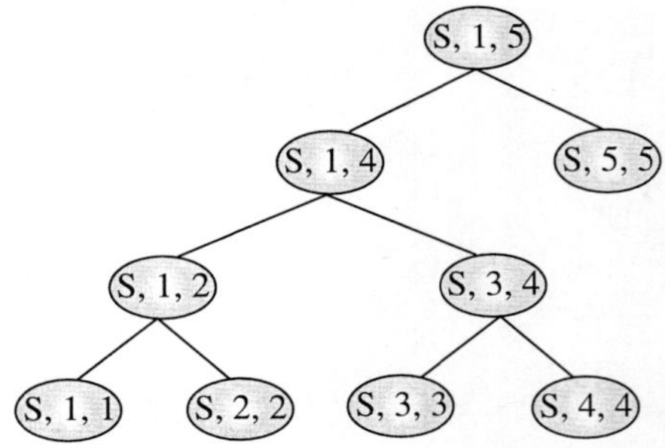

Now applying OS_MCM (*S*, *i*, *j*) to parenthesize ABCDE, *i* = 1 and *j* = 5 the implicit tree for the parenthesization is drawn below.

These leaves are like the sequence of the matrices from left to right. Thus the optimal parenthesisation is ((AB)(CD)E).

9. (*a*) Refer chapter 12 section 12.9

(*b*) Refer chapter 10 sections 10.5.3 and 10.6.1 for union and find algorithms respectively.

10. (*a*) A set of linear equations can be represented in a matrix form. Say A, X, and C are the matrices for co-efficients, variables, and constants respectively. Then $AX = C$. Using matrix inversion method it can be solved. But the method is not suitable one. LUP decomposition is another tool in the place.

The idea in this method is to find three $n \times n$ matrices L, U, and P such that

$$PA = LU$$

Where L = Unit lower triangular matrix. U = Unit upper triangular matrix.
P = Permutation matrix.

These three matrices are called LUP decomposition of A. One advantage of this method is that linear systems are solved easily when the matrices are triangular.

Now, $AX = C$

$\Rightarrow$ $PAX = PC$ [multiplying P both sides]

$\Rightarrow$ $LUX = PC$ [putting PA]

$\Rightarrow$ $LY = PC$ [Let $Y = UX$]

Now, solving Y using "forward substitution" we have the solution for lower triangular system,

$$LY = PC$$

Then we solve upper triangular system $UX = Y$ for X using "back substitution" method. This X vector is our solution to $AX = C$.

(*b*) Given the Knapsack capacity $W = 100$ and

Index	1	2	3	4	5
Item	I_1	I_2	I_3	I_4	I_5
Weight	15	25	35	45	55
Profit	10	20	30	40	50

Now, Profit/Weight ratio for each item can be calculated as

Profit/Weight = < 10/15, 20/25, 30/35, 40/45, 50/55 >
= < 0.67, 0.8, 0.86, 0.89, 0.9 >

Arranging the items in descending order of profit/weight ratio (greedy criteria *III*), the above table will become:

Index	1	2	3	4	5
Item	I_5	I_4	I_3	I_2	I_1
Weight	55	45	35	25	15
Profit	50	40	30	20	10
Profit/Weight	0.9	0.89	0.86	0.8	0.67

Let P be the profit earned, initially 0. Include first item (I_5) from the current sequence into the Knapsack.

$$W = W - W(I_5) = 100 - 55 = 45$$
$$P = P + \text{Profit } (I_5) = 0 + 50 = 50$$

Include 2nd item, $W = W - W(I_4) = 45 - 45 = 0$

$$P = P + \text{Profit } (I_4) = 50 + 40 = 90$$

Thus, feasible solution = $< I_4, I_5 >$ and profit earned = 90

Similarly apply greedy criteria *I* and *II* for other two feasible solutions.

(*c*) An activity-selection problem is the problem of scheduling a resource amongst several competing activities. The problem can be stated as: *Given a set S of n activities with start time S_i and finish time f_i for an ith activity. Find the maximum number of mutually compatible activities.*

Solution of the above statement lays on the assumption that the activities i and j are compatible iff the half-open internal $[s_i, f_i)$ and $[s_j, f_j)$ do not overlap. It means that activities i and j are compatible iff $s_i \geq f_j$ and $s_j \geq f_i$. The problem can be solved by both dynamic programming and greedy. Here we consider only greedy version.

Algorithm: Activity_Selection_Problem_ Greedy (S, n)

Input: S = set of activities n = total number of activities $|S|$

Output: F = feasible solution.

Step 1: Sort the activities by increasing finishing time.

$$f_1 \leq f_2 \leq \ldots \geq f_n$$

Step 2: Initialise

(*i*) $F = \{i\}$ // include first job (*ii*) $j = 1$

Step 3: Repeat for $i = 2$ to n

(*a*) If $s_i \geq f_j$

(*i*) $F = F \cup \{i\}$ // include ith activity (*ii*) Set $j = i$

Step 4: Return F

Step 5: Exit

Analysis: If we are to sort the activities as in step 1 then it can be done in $O(n \log n)$ time using any standard sorting algorithm. The feasible solution F *i.e.,* subset of S is obtained by step 3. Since a few activities form F, this step executes in $\Theta(n)$ time.

Example: Let $S = \{p, q, r, s, t, u, v, w, x, y, z\}$ be the set of activities. Their start and finish times are (1, 4), (3, 5), (0, 6), 5, 7), (3, 8), (5, 9), (6, 10), (8, 11), (8, 12), (2, 13), and (12, 14).

Applying the above algorithm the feasible solution will be $F = \{p, s, w, z\}$.

11. (*a*) Refer chapter 1 section 1.10.2

(*b*) Refer chapter 1 section 1.10.1

(*c*) Refer chapter 7 section 7.4.2

(*d*) Refer chapter 14 sections 14.1, 14.2 and 14.3

(*e*) Refer chapter 2 section 2.1 and give definitions of Big-O, Big-Ω, and Big-Θ with graphical representations and two features of each.

Design and Analysis of Algorithms
CS/B.Tech(CSE)/SEM-5/CS-503/2010-11

Time Allotted : 3 Hours **Full Marks : 70**

The figure in the margin indicate full marks.
Candidates are required to give their answers in their own words as far as practicable.

GROUP–A
(Multiple Choice Type Questions)

1. Choose the correct alternatives for the following: **10 × 1 = 10**

(*i*) Which of the following problems is solved by using Branch-and-Bound method?

(*a*) Knapsack problem (*b*) Hamiltonian Problem
(*c*) Travelling salesman problem (*d*) 15 puzzle problem.

(*ii*) Which of the following algorithm design technique is used in the quicksort algorithm?

(*a*) Dynamic programing (*b*) Backtracking
(*c*) Divide-and-conquer (*d*) Greedy method.

(*iii*) 'Small *o*' g(*n*) is

(*a*) Asymptotically loose (*b*) Asymptotically tight
(*c*) Same as 'Big *O*' (*d*) None of these.

(*iv*) The time factor when determining the efficiency of an algorithm is measured by

(*a*) Counting microseconds
(*b*) Counting the number of key operations
(*c*) Counting the number of statements
(*d*) Counting the kilobyte of algorithm.

(*v*) Which of the following is used to depict the working of algorithm?

(*a*) Flowchart (*b*) Pseudocode
(*c*) Source code (*d*) All of these.

(*vi*) The complexity of Bubble sort is:

(*a*) $O(n)$ (*b*) $O(\log n)$
(*c*) $O(n^2)$ (*d*) $O(n \log n)$.

(*vii*) Which one is true of the following?

(*a*) All NP-hard problems are NP-complete
(*b*) All NP-complete problems are NP-hard
(*c*) Some NP-complete problems are NP-hard
(*d*) None of these.

(*viii*) BFS on a graph $G = (V, E)$ has running time:
(*a*) $O(|V| + |E|)$ (*b*) $O(|V|)$
(*c*) $O(|E|)$ (*d*) None of these.

(*ix*) The fractional Knapsack problem can be solved by using:
(*a*) Greedy method (*b*) Divide-and-Conquer method
(*c*) Dynamic programming (*d*) None of these.

(*x*) Time complexity of Binary Search algorithm on *n* items is:
(*a*) $O(n)$ (*b*) $O(n \log n)$
(*c*) $O(n^2)$ (*d*) $O(n \log n)$.

GROUP–B
(Short Answer Type Questions)

Answer any *three* of the following: **3 × 5= 15**

2. Let $A[1..\ n]$ be a sorted array of *n* distinct integers, Give a divide-and-conquer algorithm that can find an index *i* such that $A[t] = i$ (if one exists) with running time $O(\log n)$.

3. What do you mean by dynamic programming? Write the algorithm of matrix-chain multiplication.

4. Write the FFT algorithm and find the computational complexity of this algorithm.

5. Write an algorithm of eight queens problem and find time complexity of the algorithm. Explain the algorithm using an example,

6. Show that 2SAT is in *P* but 3SAT is NP-complete.

GROUP–C
(Long Answer Type Questions)

Answer any *three* of the following: **3 × 15 = 45**

7. Suppose we have a recurrence relation

$T(n) = aT\left(\frac{n}{b}\right) + f(n)$. Show that the followings:

(*a*) $af\left(\frac{n}{b}\right) = k\, f(n)$ for some constant $k < 1$, then

$$T(n) = O(f(n)).$$

(*b*) If $af\left(\frac{n}{b}\right) = k\, f(n)$ for some constant $k > 1$, then

$$T(n) = O(n \log_b a).$$

(*c*) If $af\left(\frac{n}{b}\right) = k\, f(n)$ for some constant $k = 1$, then

$$T(n) = O(f(n)) \log_b n.$$

8. What is Heap property? Write an algorithm to make a Heap containing elements. Then, show that how can you insert an element into a Heap. Then, write the algorithm of Heapsort and find the running time of this algorithm. Write an algorithm to find the existence of an element into a Heap.

9. Write the algorithm of Quicksort. Then find the best case, worst case and average case time complexities of this algorithm. You also find the kth smallest element of n elements with time complexity less than that of quick sort algorithm.

10. Using any method you like, compute the following subgraphs for the weighted graph below:

(*a*) a depth-first search tree, starting at the top vertex;

(*b*) a breadth-first search tree, starting at the top vertex;

(*c*) the minimum spanning tree.

11. Answer any three of the following:

(*a*) Show that 3-colouring problem is *NP*-complete.

(*b*) Write the travelling salesman problem with an algorithm.

(*c*) What is 0/1 Knapsack problem? Explain it with an example.

(*d*) How can you convert the following recursion algorithm (Algorithm 1) to tall recursion ?

Algorithm 1 Fib (n)	
(*i*)	if $n \leq 1$ then
(*ii*)	Return (n)
(*iii*)	else
(*iv*)	Return Fib (n – 1) + Fib (n – 2)
(*v*)	end if

Solutions 2010: CS-503

GROUP–A

1. Possible correct solutions are the followings:

(*i*) (*d*) (*ii*) (*c*) (*iii*) (*a*) (*iv*) (*b*) (*v*) (*d*)

(*vi*) (*c*) (*vii*) (*b*) (*viii*) (*a*) (*ix*) (*a*)

(*x*) All the answers are wrong. It will be $O(\log n)$.

GROUP–B

2. The algorithm is binary searching algorithm. Refer section 2.6.2.2 or 4.8.2.

3. Refer section 5.1 and section 5.5.4.

4. Refer section 12.7

5. Refer sections 7.4.2 and 7.4.3. Take $N = 8$ as input.

6. Refer section 13.11 Question 3.

GROUP–C

7. Given, $T(n) = aT(n/b) + f(n)$

$$= a[aT(n/b^2) + f(n/b)] + f(n) \qquad \text{// using iteration method}$$

$$= a^2T(n/b^2) + af(n/b) + f(n)$$

$$= a^3T(n/b^3) + a^2f(n/b^2) + af(n/b) + f(n) \ldots$$

$$= a^hT(n/b^h) + \ldots + a^2f(n/b^2) + af(n/b) + f(n)$$

Suppose, base case is $T(1) = \Theta(1)$, then $n/b^h = 1 \Rightarrow h = \log_b n$

Thus, $T(n) = f(n) + af(n/b) + a^2f(n/b^2) + \ldots + a^h\, T(n/b^h)$

$$= f(n) + af(n/b) + a^2f(n/b^2) + \ldots + a^{\log_b n}$$

$$= f(n) + af(n/b) + a^2f(n/b^2) + \ldots + n^{\log_b a}$$

Also given, $af(n/b) = kf(n)$ for some constant k.

Now, if $k < 1$, then $T(n) = f(n) + af(n/b) + a^2f(n/b^2) + \ldots + n^{\log_b a}$

$$\le cf(n) \qquad \text{// } c \text{ is a constant} > 0$$

$\Rightarrow$ $T(n) = O(f(n))$

Now, if $k > 1$, then $T(n) = f(n) + af(n/b) + a^2f(n/b^2) + \ldots + n^{\log_b a}$

$$\le c\, n^{\log_b a} \qquad \text{// } c \text{ is a constant} > 0$$

$\Rightarrow$ $T(n) = O(n^{\log_b a})$

Now, if $k = 1$, then $af(n/b) = kf(n) \Rightarrow a = b = 1$

Then, $T(n) = f(n) + af(n/b) + a^2f(n/b^2) + \ldots + n^{\log_b a}$

$$= hf(n) \qquad \text{// } h \text{ times the } f(n)$$

$$= O(f(n)\log_b n) \qquad \text{// since } h = \log_b n$$

8. (*a*) Refer section 2.6.1.

(*b*) Refer section 2.6.2.3.

(*c*) Refer section 2.6.2.4.

(*d*) Existence of an element from the heap can be done in the following way:

(*i*) Build a Max-Heap in $O(n)$ time.

(*ii*) Set k = the element to be searched.

(*iii*) Compare k with root of the tree. If found then stop otherwise delete this root. Rebuild Max-Heap. Repeat the process until k equals with root.

9. (*a*) Refer section 4.6.3.

(*b*) Finding kth smallest element from a list of n elements can be done by

(*i*) Build a Min-Heap tree in $O(n)$

(*ii*) Delete the root from the above Min-Heap. Rebuild Min-Heap after deletion. Repeat it $k - 1$ times. Root of the current Min-Heap is the kth smallest element. Complexity will be $O(k \log n)$.

10. Consider any weighted undirected graph. Then apply the following algorithms on the graph.

(a) DFS (*b*) BFS (*c*) Kruskal's

11. Answers of the followings.

(*a*) 3-colouring problem is NP-Complete

First: Suppose $G = (V, E)$ be undirected connected graph. Then colouring of G with 3 different colours, can be checked in $O(|E|)$ whether the Colouring is allowed. Thus, 3-COLOR problem is NP.

Second: Show that 3-SAT $\leq_p$ 3-Colour *i.e.,* 3-colouring is NP-Hard.

(*b*) Refer sections 5.6 and 5.6.4

(*c*) Refer section 6.10.4 and its example

(*d*) Refer section 3.4.4.

Design and Analysis of Algorithms
CS/B.Tech(CSE)/SEM-5/CS-503/2011-12

Time Allotted : 3 Hours **Full Marks : 70**

The figure in the margin indicate full marks.
Candidates are required to give their answers in their own words as far as practicable.

GROUP–A
(Multiple Choice Type Questions)

1. Choose the correct alternatives for the following: **$10 \times 1 = 10$**

(*i*) The running time of an algorithm $T(n)$, where n is the input size given by $T(n) = 8T(n/2) + qn$, if $n > 1$, and $T(n) = p$, if $n = 1$, where p and q are constants. The order of this algorithm is

(*a*) $\Theta(n^2)$ (*b*) $\Theta(n^n)$ (*c*) $\Theta(n^3)$ (*d*) $\Theta(n^{\log n})$

(*ii*) Which of the following algorithms solves All-Pairs Shortest Path Problem?

(*a*) Dijkstra's (*b*) Floyd-Warshall's
(*c*) Prim's (*d*) Kruskal's

(*iii*) The minimum number of colours needed in a graph having $n > 3$ vertices and 2 edges is

(*a*) 2 (*b*) 3 (*c*) 4 (*d*) 1

(*iv*) The diagonal of the adjacency matrix of a graph with a self-loop contains only

(*a*) 1 (*b*) 0 (*c*) –1 (*d*) ∞

(*v*) Which of the following design techniques is used in the Quicksort algorithm?

(*a*) Dynamic Programming (*b*) Backtracking
(*c*) Greedy method (*d*) Divide-and-Conquer

(*vi*) The average number of comparisons performed by mergesort algorithm in merging '2' sorted list of lengths '2' is

(*a*) 8/5 (*b*) 11/7 (*c*) 11/6 (*d*) 8/3

(*vii*) Which of the following is useful in traversing a given graph using BFS?

(*a*) Stack (*b*) Linked list (*c*) Array (*d*) Queue

(*viii*) Which of the following can not be performed recursively?

(*a*) Binary search (*b*) Quicksort (*c*) DFS (*d*) None of these

(*ix*) The time complexity of TSP is

(*a*) $O(n^2 2^n)$ (*b*) $\Theta(n^2 2^n)$ (*c*) $\Omega(n^2 2^n)$ (*d*) None of these

(*x*) In which sorting technique, is an element placed in its proper position at each step?

(*a*) Bubblesort (*b*) Quicksort (*c*) Mergesort (*d*) Heapsort

GROUP–B
(Short Answer Type Questions)

Answer any *three* of the following: **3 × 5= 15**

2. Find the best and worst case time complexity for mergesort.

3. Solve the following Knapsack problem with the given conditions: $n = 3$, weight of the Knapsack $M = 20$, Profits $(P_1, P_2, P_3) = (25, 24, 15)$, and weights $(w_1, w_2, w_1) = (18, 15, 10)$.

4. Differentiate between divide-and-conquer and dynamic programming.

5. Solve the following recurrence relation using generating function:

$a_n = 6a_{n-1} - 11a_{n-2} + 6a_{n-3}$ for $n \geq 3$ with initial conditions $a_0 = 1$, $a_1 = -1$, and $a_2 = 1$.

6. Define different asymptotic notations (O, Θ, Ω) with suitable examples.

GROUP–C
(Long Answer Type Questions)

Answer any *three* of the following: **3 × 15 = 45**

7. Answer the following questions w.r.t. divide-and-conquer:

(*a*) Discuss the procedure for Strassen's matrix multiplication to evaluate the product of n matrices. Find the resulting recurrence relation for the same and analyze its time complexity. Is this method an improvement over the conventional matrix multiplication method? If so, why? 7+1+2+2

(*b*) The solution of recursive MAX-MIN problem is based on some assumptions. Briefly state the assumptions and its effect on the algorithm in comparison to the reality. 3

8. Find the optimal parenthesization of the matrix chain product whose sequence of dimensions is < 5, 10, 3, 12, 5, 50, and 6 >. 5

(*a*) Give an algorithm for the above procedure. 3

(*b*) Analyze its complexity. 2

(*c*) What is the union-find algorithm? Explain with an example. 5

9. (*a*) Solve the single source shortest path problem for the following graph considering '1' as source vertex using Dijkstra's algorithm.

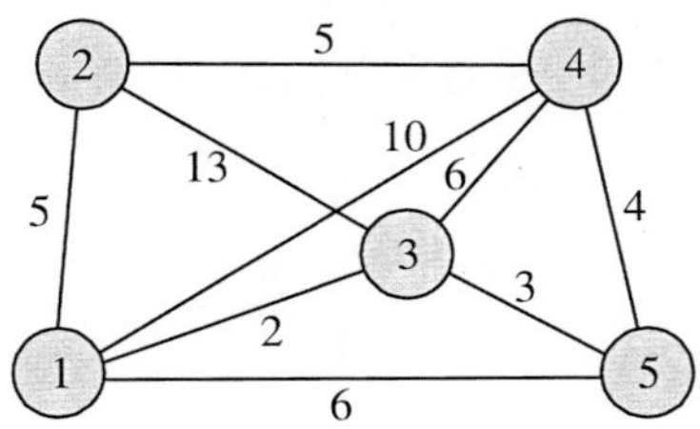

(*b*) Prove that the time complexity of Dijkstra's algorithm is $O(n^3)$.

(*c*) Describe Floyd's algorithm for all pairs shortest path problem. Prove that the time complexity of the algorithm is cubic. 5+3+7

10.(*a*) Describe the Breadth first search algorithm for a given graph and explain its time complexity.

(*b*) Explain the graph colouring problem and write the algorithm.

(*c*) Apply backtracking technique to solve the 3-colouring problem for the following graph. 6+6+3

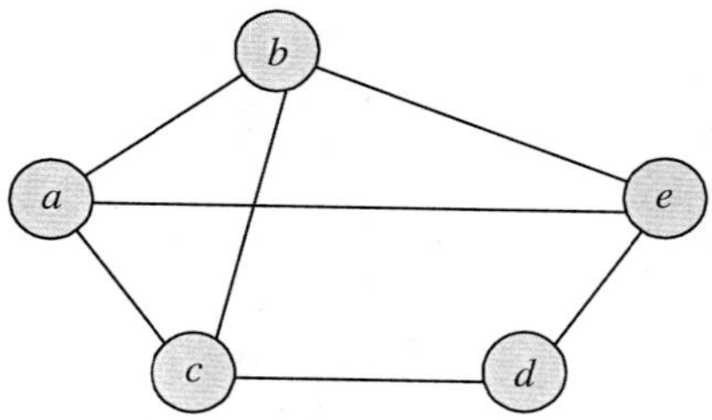

11.(*a*) What is Non-deterministic algorithm? Differentiate between Deterministic and Non-deterministic algorithm.

(b) Write algorithm to sort an array using Deterministic and Non-deterministic techniques. Compare the two techniques and show that the time complexity of Non-deterministic technique is better than Deterministic.

(c) Describe *P* class, NP class NP-Hard class, and NP-complete class. Describe their relationships. 3+(6+2)+4

Solutions 2011: CS-503

GROUP–A

1. Posible correct solutions are the followings:

(*i*) (*c*)	(*ii*) (*b*)	(*iii*) (*a*)	(*iv*) (*a*)	(*v*) (*d*)
(*vi*) (*d*)	(*vii*) (*d*)	(*viii*) (*d*)	(*ix*) (*b*)	(*x*) (*d*)

GROUP–B

2. Refer chapter 4 section 4.7.1.2

3. Refer solution of 2009 (CS-503)- Q. 8(b)

4. Refer chapter 5 section 5.10

5. Characteristic equation of the generating function is

$$x^3 - 6x^2 + 11x - 6 = 0 \quad \text{...(1)}$$

Solving this equation, $x = 1, 2, 3$.

Now the general form of the solution is given by:

$$a_n = C_1 \propto_1^n + C_2 \propto_2^n + C_3 \propto_3^n \quad \text{...(2)}$$

Now, $\propto_1 = 1, \propto_2 = 2, \propto_3 = 3, n \geq 1, a_0 = 1, a_1 = -1$, and $a_2 = 1$

Therefore, $a_n = C_1(1)^n + C_2(2)^n + C_3(3)^n$ ...(3)

Now, at $n = 0$, $C_1 + C_2 + C_3 = a_0 = 1$ ...(4)

$n = 1$, $C_1 + 2C_2 + 3C_3 = a_1 = -1$ ...(5)

$n = 2$, $C_1 + 4C_2 + 9C_3 = a_2 = 1$ ...(6)

Solving equations (4), (5), and (6) we get, $C_1 = 6$, $C_2 = -8$, $C_3 = 3$

Thus, general solution $a_n = 6 - 8.(2)^n + 3.(3)^n$.

$$= 3^{n+1} - 2^{n+3} + 6$$

6. Refer chapter 2 sections 2.2, 2.3, and 4.4.

GROUP–C

7. (*a*) Refer chapter 12 section 12.9

(*b*) Iteratively the problem can be solved in the following way

– Find MAX ($n - 1$ compares)

– Find MIN of the remaining elements ($n - 2$)

– Number of comparisons is $2n - 3$.

Using divide-and-conquer (nontrivial strategy) we can solve the following problem as below.

– Split the list of elements in half.

– Find the MAX and MIN of both halves.

– Compare the 2 MAXes and compare the 2 MINs to get overall MAX and MIN.

– We only consider the number of comparisons and ignore all other operations. This is given by the recurrence below.

$$T(n) = \begin{cases} 0 & \text{if } n = 1 \\ 1 & \text{if } n = 2 \\ T(\lfloor n/2 \rfloor) + T(\lceil n/2 \rceil) + 2 & \text{if } n = 3 \end{cases}$$

$$= \left(3\frac{n}{2} - 2\right), \text{ if } n \text{ is power of 2.}$$

$$= O(n).$$

The accurate number of comparisons is $\left\lceil \frac{3}{2} n \right\rceil$.

8. (*a*) Refer chapter 5 similar example in section 5.5.4
   (*b*) Refer chapter 5 section 5.5.4
   (*c*) Refer chapter 10 sections 10.5 and 10.6
9. (*a*) Refer chapter 20 similar example in section 20.4.3
   (*b*) Refer chapter 20 section 20.4.3
   (*c*) Refer chapter 5 section 5.9.1
10. (*a*) Refer chapter 11 section 11.7.1
    (*b*) Refer chapter 7 section 7.6
    (*c*) Refer solution to 2009(CS – 503) Question 6.
11. (*a*) Refer chapter 13 section 13.3
    (*b*) **Hints** – consider any comparison based sorting algorithm for deterministic algorithm.

Consider the following algorithm for non-deterministic technique.

Algorithm: NDSort(A, B, n)

Where A = input list of n elements.

B = secondary list of size n.

n = number of elements.

Step 1: Repeat for $i = 1$ to n

• $B[i] = 0$

Step 2: Repeat for $i = 1$ to n

(*i*) Select any element j from the set of n elements

(*ii*) If $B[j] \neq 0$

(*a*) Output "Failure".

(*b*) Exit

(*iii*) Initialize, $B[j] = A[i]$

Step 3: Repeat for $i = 1$ to $n - 1$

- If $B[i] > B[i + 1]$

  (*a*) Output "Failure".

  (*b*) Exit

Step 4: Repeat for i = 1 to *n*

- Output B[i] //sorted list

Step 5: End

A general comparison can be made. Comparison based sorting runs in $\Omega(n \log n)$ time whereas the above non-deterministic algorithm runs in $O(n)$ time.

(*c*) Refer chapter 13 section 13.4

❑❑❑

Bibliography

BOOKS

1. Thomas H. Coremen, Charles E. Leiserson, Ronald L. Rivest, Clifford Stein, "Introduction to Algorithms", PHI, 2nd Edn, 2008.
2. Ellis Horowitz, Sartaj Sahani, Sanguthevar Rajasekaran, "Fundamentals of Computer Algorithms", Golgotia, 1998.
3. M.H. Alsuwaiyel , "Algorithms Design Techniques and Analysis", PHEI, 2002.
4. Richard Neapolitan and Kumarss Naimipour, "Foundations of Algorithms Using C++ Pseudocode", Jones and Bartlett Publishers, 3rd Edition, 2004.
5. Dexter C. Kozen, "The Design and Analysis of Algorithms", Springer-Verlag, 1991.
6. Mark Allen Weiss, "Data Structures and Algorithm Analysis in C", Addison Wesley, 1996.
7. Levitin. Anany, "Introduction to the Design & Analysis of Algorithms", Pearson Education, 2nd Edition, 2009.
8. Alan Parker, "Algorithms and Data Structures in C++", CRC Press, 1993.
9. Steven S. Skiena, "The Algorithm Design Manual", Springer, 2nd Edition, 2008.
10. Gilles Brassard and Paul Bratley, "Algorithmics -Theory and Practice", Prentice Hall, 1988.
11. Jeffrey J. McDonnell, "Analysis of Algorithms: An Active Learning Approach", Jones and Bartlett Publishers, 2001.
12. Vijay V. Vazirani, "Approximation Algorithms", Springer, 2001.
13. Peter G´acs, L´aszl´o Lov´asz, "Complexity of Algorithms", Lecture Notes, 1999.
14. Oded Goldreich, "Computational Complexity- A Conceptual Perspective", Cambridge University Press, 2008.
15. Sanjeev Arora and Boaz Barak, "Computational Complexity: A Modern Approach", Cambridge University Press, 2006.
16. Nicola Santoro, "Design and Analysis of Distributed Algorithms", John Wiley & Sons, 2007.

17. David P. Williamson and David B. Shmoys, "The Design of Approximation Algorithms", Cambridge University Press, 2010.
18. Oded Goldreich, "P, NP, and NP-Completeness The Basics of Computational Complexity", Cambridge University Press, 2010.
19. Nitin Upadhayay, "The Design and Analysis of Algorithms", S.K Kataria & Sons, 2nd Edition, 2006.
20. S. Dasgupta, C.H. Papadimitriou, and U.V. Vazirani, "Algorithms", TMH, 5th reprint, 2009.
21. Goodman "Introduction to Design and Analysis of Algorithms", TMH, 2nd Edn, 2008.
22. Aho, Hopcroft &.Ullman, "The Design and Analysis of Algorithms, Narosa.
23. Aho, Hopcroft &.Ullman:"Data Structures and Algorithms", Pearson Education, 2003.
24. Dave P. H, Dave H. B, "Design and Analysis of Algorithms", Pearson Education, 1st edition, 2008.
25. Vijaya Ramachandran, "ALGORITHMS", The University of Texas, Austin, Spring 2006.
26. Herbert S. Wilf, "Algorithms and Complexity", University of Pennsylvania , Internet Edition, Summer, 1994
27. Éva Tardos and Jon Kleinberg, "Algorithm Design", Addison Wesley, 2005.
28. Robert Sedgewick, "Algorithms", Addison-Wesley Publishing Company, August 1984
29. David M. Mount, "Analysis of Computer Algorithms", University of Maryland, Fall 2003.
30. N. Deo, "Graph Theory with Applications to Engineering and Computer Science", PHI, 23rd edition, 2002.
31. N. Guruprasad, "Data Structures using C", Schitech, 2nd Edition.
32. D. Samanta, "Classic Data Structures", PHI, 2nd reprint, 2002.
33. Rogger S. Pressman, "Software Engineering – A Practioner's Approach", MGH, 5th Edition, 2001.
34. Behrouza A. Forouzan, "Data Communications and Networking", MGH, 4th Edition, 2008.
35. Pal B. K, Das K, "Engineering Mathematics", U. N. Dhur & Sons, 4th edition, 2009.
36. Mishra K.L.P, Chandrasekharan N, "Theory of Computer Science", PHI, Third Edition, 2007.

JOURNALS

37. Biswajit Bhowmik, Sreyasi Nag Chowdhury, "Prograph Based Analysis of Single Source Shortest Path Problems with Few Distinct Positive Lengths", Engineering, Technology & Applied Science Research, Vol. 1, No. 4, 2011, Pages: 90 – 97.

38. Biswajit Bhowmik, "Dynamic Programming – Its Principles, Applications, Strengths, and Limitations", International Journal of Engineering Science and Technology, Vol. 2(9), 2010, Pages: 4822 – 4826.
39. Biswajit Bhowmik, "Studies on Dimultigraph and Prograph Based Applications of Graph Theory in Computer Science", International Journal of Computer and Communication Technology, Vol. 1(2, 3, 4), 2010, Pages: 57-61.
40. Biswajit Bhowmik, "Simplified Optimal Parenthesisation Scheme for Matrix Chain Multiplication Problem using Bottom-up Practice in 2-Tree Structure", Journal of Applied Computer Science & Mathematics, Vol. 11, No. 5, 2011, Pages: 9 – 14.
41. William S. Bowie, "Applications of Graph Theory in Computer Systems", International Journal of Parallel Programming, Volume 5, Number 1 / March, 1976, pages 9-31.
42. David B. Wagner, "Dynamic Programming", The Mathematica Journal, Miller Freeman Publications, 1995.
43. J. N. Tsitsiklis and B. Van Roy, "Optimal Stopping of Markov Processes: Hilbert Space Theory, Approximation Algorithms, and an Application to Pricing Financial Derivatives", IEEE Transactions on Automatic Control; Vol. 44, No. 10, October 1999, pp. 1840-1851.
44. Nocedal, J., Wright, S. J.: Numerical Optimization, page 9, Springer, 2006.
45. Ronald A. Howard. , "Dynamic Programming and Markov Processes", John Wiley & Sons, 1960
46. Steven Skiena, Applications of Dynamic Programming, State University of New York Stony Brook, NY 11794–4400
47. Bob Sedgewick, Kevin Wayne, "Algorithms and Data Structures", Fall 2007.
48. Jared Saia, " Lectures Notes on Algorithms", University of New Mexico, 2005.
49. Venkatesh Raman, "Some introductory notes on Design and Analysis of Algorithms" The Institute of Mathematical Sciences, India, 2008.
50. Barath Raghavan, "Asymptotic analysis, algorithm analysis", Berkeley, Summer 2002.
51. Srini Devadas and Eric Lehman, "Recurrences", Mathematics for Computer Science March 17, 2005
52. Miguel A. Lerma, "Recurrence Relations", October, 2003.
53. Tamara Smyth,"Sorting and Searching", CMPT 126, Simon Fraser University, October 16, 2007.
54. David K. Smith, "Dynamic programming: an introduction", PASS Maths, September 1997.
55. John W. Chinneck, "Practical Optimization: a Gentle Introduction", 2006.
56. Khadija Riaz, Malik Sikander Hayat Khiyal, "Finding Hamiltonian Cycle in Polynomial Time", Information Technology Journal 5 (5): 851-859, 2006.
57. Ian Parberry, "Backtracking", Fall 2001.
58. Richard Hayes, "The Sam Loyd 15-Puzzle", June 2001.
59. Richard Karp, "Great Algorithms", Cheng Tien Ee, Spring 2006.

60. Randal E. Bryant," Theory of NPCompleteness", April 11, 2001.
61. Sushanth Sivaram Vallath, Z. Joseph, "Non-deterministic Polynomial Completeness", 2005.
62. Philip N. Klein, Neal E. Young, "Approximation algorithms for NP-hard optimization problems", CRC Press, 1999.
63. Dr. XuanWang, "Sorting in Linear Time", Intelligence Computing Research Center, Autumn 2007.
64. Sandeep Sen, "Lecture Notes for Algorithm Analysis and Design", IIT Delhi, India, March 16, 2008.

WEB RESOURCES

65. www.en.wikipedia.org
66. http://www.cs.umu.se/kurser
67. www.cse.ucsd.edu
68. http://www.cs.umd.edu
69. http://www.gap-system.org
70. http://tclab.kaist.ac.kr
71. http://www.math.ku.edu
72. http://www.itl.nist.gov
73. http://www.cs.auckland.ac.nz
74. http://triton.towson.edu
75. http://www.csse.monash.edu.au
76. http://mitpress.mit.edu
77. http://www.allinterview.com
78. www.eagle.tamut.edu/
79. www.cc.nctu.edu.tw
80. http://www.omninerd.com
81. http://www.brpreiss.com
82. http://www.brpreiss.com
83. http://www.devarticles.com
84. http://cs482.elliottback.com
85. http://www.ecst.csuchico.edu
86. http://www.users.csbsju.edu
87. http://www.cs.cmu.edu
88. http://datastructures.itgo.com
89. http://www.allaboutcircuits.com
90. http://icrc.hitsz.edu.cn
91. http://numbers.computation.free.fr

❑❑❑

Index

K

L

M

N

O

P

Q

R

❑❑❑

Win Prizes!

ATTENTION READERS

We request you for your frank assessment regarding some of the aspects of the book given as under:

Design and Analysis of Algorithms

By

Biswajit R Bhowmik

Please fill up the given space in neat capital letters.

(i) What topics of your syllabus that are important from you examination point of view are not covered in the book ?

....
....
....
....

(ii) What are the chapters wherein the treatment of the subject matter is not systematic or organised or updated ?

....
....
....
....

(iii) Have you come across misprints/mistakes/factual inaccuracies in the book? Please specify the chapters and the page numbers.

....
....
....
....

(iv) What other books on the same subject have you found/heard better that the present book? Please specify in terms of price and quality.

....
....
....
....

(v) Further suggestions and comments for the improvement of the book:

....
....
....
....

(PS: If need be please attach a separate sheet to write your views.)

Other Details :

(i) Who recommended you the book ? (Please tick in the box near the option relevant to you.)

☐ Teachers ☐ Friends ☐ Bookseller

(ii) Name of the recommending teacher, his designation and address.

....

....

....

....

(iii) Name and address of the bookseller you purchased the book form:

....

....

....

....

(iv) Name and address of your institution (Please mention the University or Board, as the case may be)

....

....

....

....

(v) Your name and complete postal address:

....

....

....

....

(vi) Course you are enrolled in:

....

....

....

....

The best assessment will be awarded each month. The award will be in the form of our publications as decided by the Editorial Board.

Please mail the filled up coupon at your earliest to:

S.K. Kataria & Sons®
4885/109, Prakash Mahal, Dr. Subhash Bhargav Lane,
Daryaganj, New Delhi-110002 (India)
Phone: +91-11-43551243, Mobile: +91-9871775858, 7678425209
e-mail: katariabook@gmail.com, katariabooks@yahoo.com
Website: www.skkatariaandsons.com